Microwave Engineering and Technology

Microwave Engineering and Technology

Edited by **Alessandro Torello**

NY RESEARCH
PRESS

New York

Published by NY Research Press,
23 West, 55th Street, Suite 816,
New York, NY 10019, USA
www.nyresearchpress.com

Microwave Engineering and Technology
Edited by Alessandro Torello

International Standard Book Number: 978-1-63238-518-5 (Hardback)

The publisher's policy is to use permanent paper from mills that operate a sustainable forestry policy. Furthermore, the publisher ensures that the text paper and cover boards used have met acceptable environmental accreditation standards.

Printed in the United States of America.

Contents

Preface

Microwaves are electromagnetic radiations with varied wavelengths. These waves are used for point-to-point telecommunications, navigation, radar, heating and power application, radio astronomy and spectroscopy. Microwave engineering aids the manufacturing and design of components, circuits and systems for deploying different technologies. Some of the frequently used technologies in this field are antennas, transmission lines, measurements, remote sensing, etc. This book presents researches and studies performed by experts across the globe. While understanding the long-term perspectives of the topics, the book makes an effort in highlighting their impact as a modern tool for the growth of the discipline. It will help the readers in keeping pace with the rapid changes in this field. This text attempts to assist those with a goal of delving deeper into the subject of microwave engineering and technology. It will benefit researchers, professionals and students alike.

After months of intensive research and writing, this book is the end result of all who devoted their time and efforts in the initiation and progress of this book. It will surely be a source of reference in enhancing the required knowledge of the new developments in the area. During the course of developing this book, certain measures such as accuracy, authenticity and research focused analytical studies were given preference in order to produce a comprehensive book in the area of study.

This book would not have been possible without the efforts of the authors and the publisher. I extend my sincere thanks to them. Secondly, I express my gratitude to my family and well-wishers. And most importantly, I thank my students for constantly expressing their willingness and curiosity in enhancing their knowledge in the field, which encourages me to take up further research projects for the advancement of the area.

Editor

Design of a Novel Ultrawide Stopband Lowpass Filter Using a DMS-DGS Technique for Radar Applications

Ahmed Boutejdar,[1] Ahmed A. Ibrahim,[2] and Edmund P. Burte[3]

[1]*Microwave Engineering Department, University of Magdeburg, 39106 Magdeburg, Germany*
[2]*Electronic and Communication Engineering Department, Minia University, Minia 61519, Egypt*
[3]*Micro and Sensor Department, University of Magdeburg, 39106 Magdeburg, Germany*

Correspondence should be addressed to Ahmed Boutejdar; ahmed.boutejdar@ovgu.de

Academic Editor: Giancarlo Bartolucci

A novel wide stopband (WSB) low pass filter based on combination of defected ground structure (DGS), defected microstrip structure (DMS), and compensated microstrip capacitors is proposed. Their excellent defected characteristics are verified through simulation and measurements. Additionally to a sharp cutoff, the structure exhibits simple design and fabrication, very low insertion loss in the pass band of 0.3 dB and it achieves a wide rejection bandwidth with overall 20 dB attenuation from 1.5 GHz up to 8.3 GHz. The compact low pass structure occupies an area of $(0.40\lambda g \times 0.24\lambda g)$ where $\lambda g = 148$ mm is the waveguide length at the cut-off frequency 1.1 GHz. Comparison between measured and simulated results confirms the validity of the proposed method. Such filter topologies are utilized in many areas of communications systems and microwave technology because of their several benefits such as small losses, wide reject band, and high compactness.

1. Introduction

With the rapid progress in modern communications systems, design goals such as compact size, low cost, good quality factor, and high performance components are highly considered. To achieve these targets, many filtering structures as open-circuited stubs, hi-low impedances, parallel coupled, and end coupled filters have been investigated. Nevertheless these methods keep the satisfactory results unattainable. In order to approach the desired results, a DGS-DMS filter could be an effective solution. Due to their improved performance characteristics, many filter techniques and methodologies have been proposed and successfully realized. Defected ground structures (DGSs) with and without periodic array have been realized by etching a pattern in the backside of the metallic ground plane to obtain the stopband effect [1–13]. DGS often consisted of two large defected areas and a narrow connecting slot channel, which corresponds to its equivalent L-C elements [14]. The DGS with periodic or nonperiodic topology leads to a reject band in some frequency range due to the slow wave effect, as a result of increasing the

effective capacitance and inductance of the transmission line. In general DMS-unit [15–18] is used as a complementary element for the DGS-unit to achieve required filter response. DMS compared to DGS is etched on the microstrip line and exhibits same frequency behavior. Additionally, the design is simpler than the DGS and is more easily integrated with other microwave circuits. Moreover, it has an effective reduced circuit size compared to DGS.

In this paper, a new compact microstrip low pass filter using coupled DMS, DGS resonators, and compensated capacitors is reported. The compensated capacitors are added on the top layer in order to get a sharp transmission domain and to regenerate transmission zeros to obtain a large reject band. Dimensions of the microstrip capacitors were computed according to the desired equivalent circuit, and using Richard's-Kuroda transformation and TX-Line software [19]; afterwards they were optimized by AWR EM simulator [20]. The measured results agree well with simulated results. The DGS-DMS technique (see Figure 1) in this research can be applied in microwave coupler, antennas, and in MRI technology.

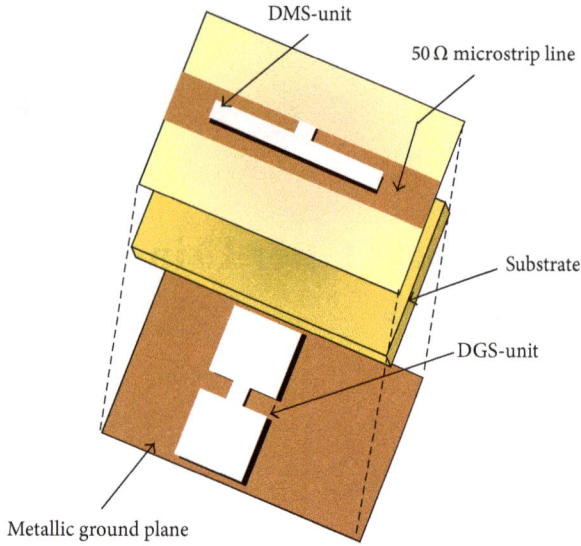

FIGURE 1: Three-dimensional view of the DMS-unit and DGS cell.

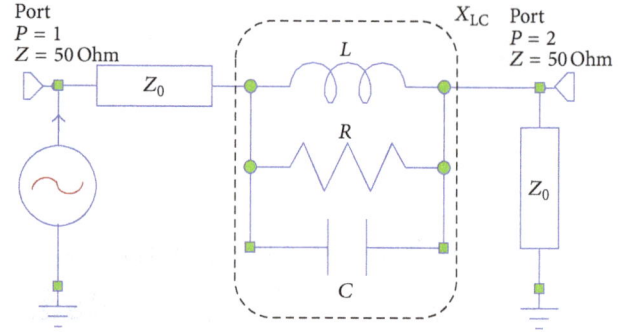

FIGURE 2: Equivalent circuit of the DMS/DGS-unit.

2. Characteristics and Modeling of DMS Resonators

Top layer of Figure 1 shows the proposed DMS cell, which is composed of wide and narrow etched sections in the feed line placed on the top layer. The extremes of this resonator are connected through microstrip line with SMA connectors. The widths of the microstrip lines at port 1 and port 2 are designed to match the characteristic impedance of 50 Ω. The etched surface presents the capacitance, while the arms correspond to the inductance. The DMS cell acts as a band stop element with a resonance frequency of 4.8 GHz and an insertion loss of −0.5 dB as shown in Figure 4. The structure has been designed on RO4003 substrate with a relative dielectric constant $\varepsilon_r = 3.38$ and thicknesses $h = 0.813$ mm and a loss tangent of 0.0027. The equivalent circuit of the DMS cell acts as a parallel LC resonator as shown in Figure 2.

The values R, L, and C of the circuit parameters can be computed using result that is matched to the one-pole Butterworth-type low pass response [7]. Furthermore, radiation effects are more or less neglected. The reactance values of DMS and filter first order can be expressed as

$$X_{\mathrm{LC}} = \left[\omega_0 C \left(\frac{\omega_0}{\omega} - \frac{\omega}{\omega_0} \right) \right]^{-1}. \tag{1}$$

The series inductance (reactance) of one-pole Butterworth low pass filter can be derived as follows:

$$X_L = \omega L = \omega \left(\frac{g_1 Z_0}{\omega_g} \right), \tag{2}$$

where ω_0, ω_g, Z_0, and g_1 are the resonant frequency, cutoff frequency, the scaled characteristic impedance, and prototype value of the Butterworth-type LPF, respectively. By matching the two previous reactance values, the parallel

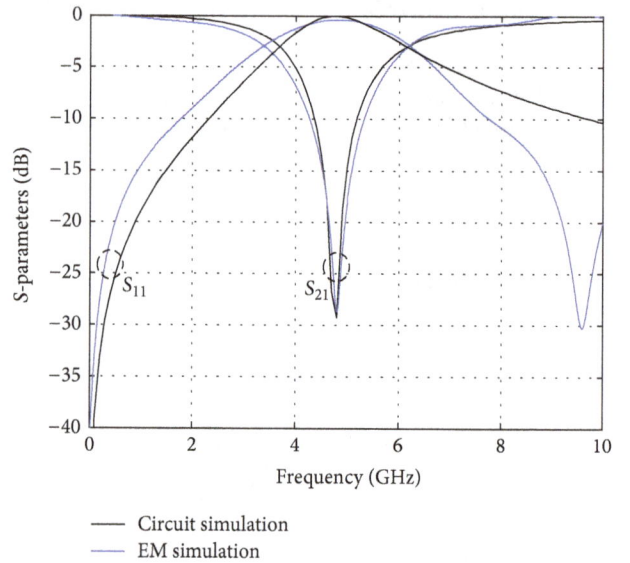

FIGURE 3: S-parameters of the DMS-element and its equivalent circuit.

capacitance and the inductance of the equivalent DMS-circuit can be derived using the following equations:

$$C = \frac{\omega_c}{2Z_0 \left(\omega_0^2 - \omega_c^2 \right)},$$

$$L = \frac{1}{\omega_0^2 C}, \tag{3}$$

$$R = \frac{2Z_0}{\sqrt{1/\left| S_{11} \left(\omega_0 \right) \right|^2 - \left(2Z_0 \left(\omega_0 C - 1/\omega_0 L \right) \right)^2 - 1}}.$$

The computed values of parameters C, L, and R are, respectively, 0.64 pF, 1.73 nH, and 7.42 kΩ. The simulation results of the investigated EM structure and its corresponding circuit are illustrated in Figure 3, which shows identical values of 3 dB cutoff frequency (f_c) and pole frequency (f_p) at 3.37 GHz and 4.83 GHz, respectively. The transmit band shows an insertion loss pass of 0.5 dB. All dimensions of DMS-unit are depicted in Table 1. The proposed DMS resonator is shown in Figure 4.

TABLE 1: Dimensions of the defected microstrip structure- (DMS-) element.

Dimensions of DMS-unit	Values (mm)
h	0.50
p	1.88
g	0.40
k	0.60
l	9.50

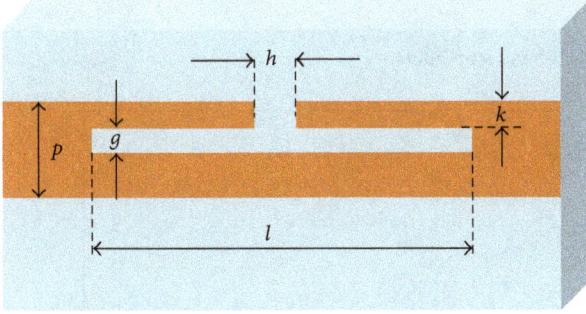

FIGURE 4: Layout of the DMS-element.

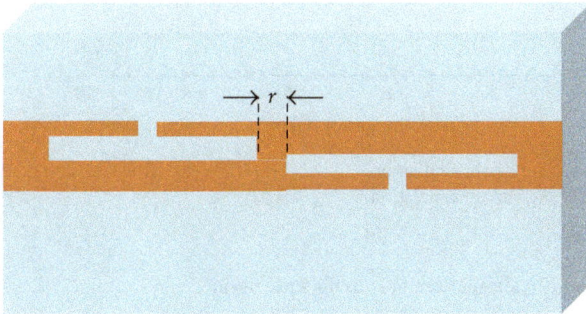

FIGURE 5: Layout of the cascaded DMS-band stop filter.

3. Design of Band Stop Filter Using Cascaded DMS

A new BSF was designed using two cascaded DMS resonators, which are positioned one to the other by 180 degrees and are directly connected with the ports through 50 Ω microstrip lines. Figure 5 shows the 3D view of the proposed BSF. The geometry of each DMS-unit is equal to the dimensions indicated in Table 1, while the microstrip distance (r) between two DMS resonators is 0.5 mm. The 50 Ω feed line has a line width of w. The band stop structure is simulated and optimized by using Microwave Studio CST [21], Microwave Office AWR. The dimensions are calculated using filter theory, TX-Line software, and EM simulator. Figure 6 shows the designed equivalent circuit of the BSF employing circuit simulator AWR. The extracted circuit parameters are computed based on the EM simulations and empirical method and defined as follows: $L = 6.2$ nH, $C = 0.77$ pF, $R = 0.51$ kΩ, $C_p = 0.96$ pH, and $C_0 = 3.32$ pF.

The simulated results depicted in Figures 7(a) and 7(b) prove that the proposed filter has a 3 dB cutoff frequency at

2.7 GHz and a suppression level of 20 dB from 4.5 to 5.5 GHz; the insertion loss in the pass band is about 0.65 dB. Good agreement is verified between the EM simulations and the circuit simulations.

4. Band Stop to Low Pass Using Compensated Capacitors

In order to demonstrate the effectiveness of the compensated microstrip capacitor in transforming a structure with band stop to low pass behaviors, the added parallel microstrip capacitors to the previous structure (Figure 5) are designed and optimized as shown in Figure 8. A new DMS low pass filter is composed of three compensated parallel microstrip capacitors, which are separated through two identical DMS resonators. All components are cascaded on the top layer and directly connected with the SMAs through the two 50 Ω feed lines of width 1.88 mm as shown in Figure 8. The filter has been designed and simulated in order to improve the reject band and to minimize the pass band losses. The DMS low pass structure is designed for cutoff frequency at 1.55 GHz and is simulated on the Rogers RO4003 substrate with the dielectric constant of 3.38 and thickness of 0.813 mm. The total size of the filter is 59×35 mm^2. Simulations have been performed using the full-wave EM Microwave Studio CST and Microwave Office AWR.

Figures 9 and 10 represent the equivalent circuit and the S-parameters of the DMS low pass filter. According to the simulation response, we can conclude that the equivalent circuit has the characteristics of quasielliptic function, because the frequency response of the elliptic function filters is known with its generated transmission zeros in pass band and thus its high sharpness in transition response as shown in Figure 10.

The values of R, L, and C are obtained as 4 kΩ, 1.3 nH, and 0.6 pF, respectively, after using an optimization technique, while the values of three parallel open-circuit capacitors are calculated using TX-Line software or exactly calculated using the following equation:

$$C = \frac{1}{Z_{0C}\omega_c}\sin\left(\beta_C l_C\right) + \frac{2}{Z_{0L}\omega_c}\tan\left(\beta_L l_L\right), \quad (4)$$

where $Z_{0C}, \beta_C, l_C, \beta_L$, and l_L are the characteristic impedance, the phase constant and the physical length of the compensated capacitor (low-impedance line), and the phase constant and the physical width of the series reactance (high-impedance line), respectively. Both series reactance values of the low-impedance line are negligible; thus the length of stub capacitors is approximated as the following:

$$l_C = \frac{\lambda g_C}{2\pi}\sin^{-1}\left(\omega_c Z_{0L} C\right),$$

$$W_C \approx f\left(Z_{0C}, \theta, t, h, \varepsilon_r\right), \quad (5)$$

where W_C, λg_C, θ, t, h, and ε_r are the width of the open-circuit stub capacitance, the guided wavelengths, the electrical length, the thickness of metal, thickness of the substrate, and the relative dielectric constant, respectively. As shown in Figure 10, two reflection zeros at 1 GHz and 1.5 GHz are

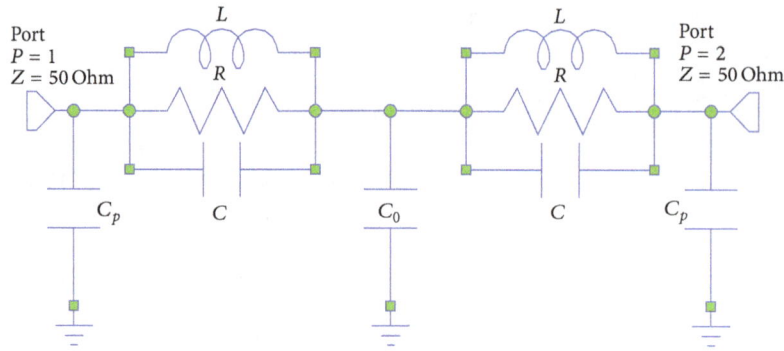

FIGURE 6: Equivalent circuit of the DMS-band stop filter.

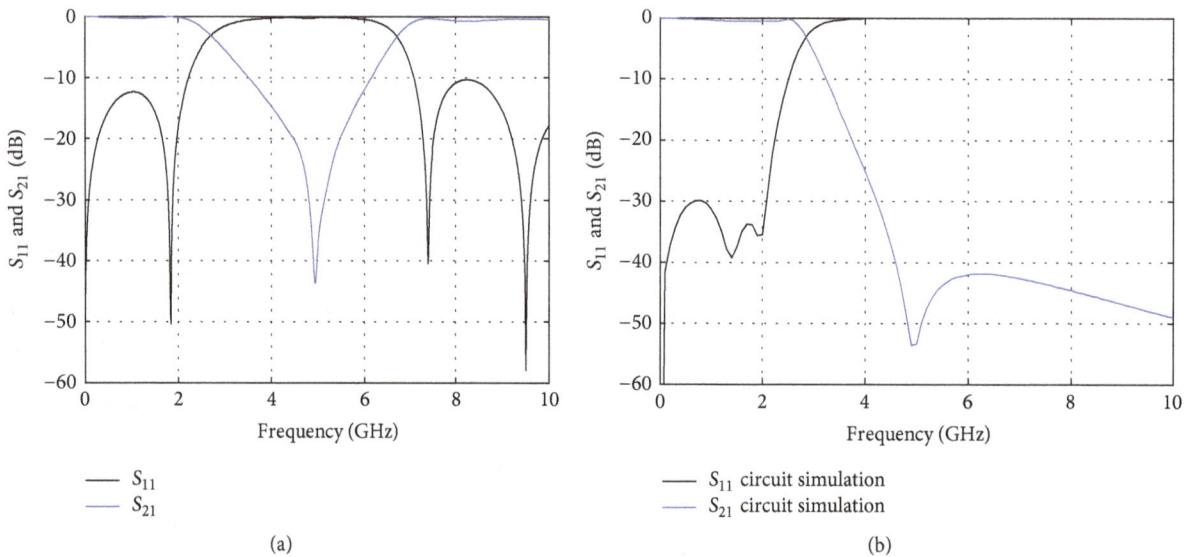

(a)

(b)

FIGURE 7: Simulation results of the DMS-band stop filter. (a) EM simulation, (b) circuit simulation.

TABLE 2: Dimensions of the DMS low pass filter structure.

Dimensions of DMS-unit	Values (mm)
a	9.05
b	10
c	3
e	2
i	20

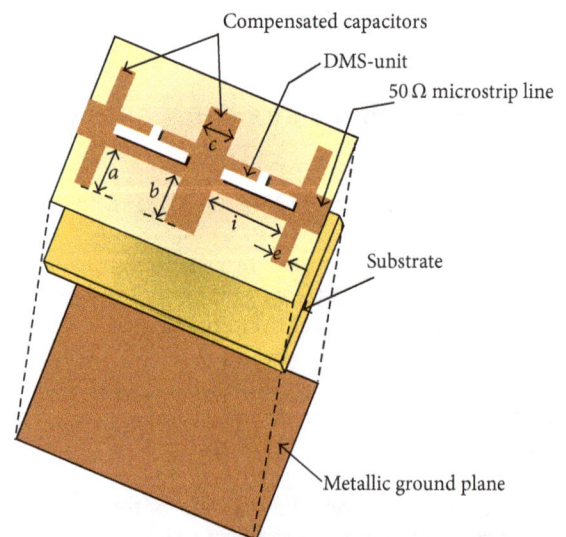

FIGURE 8: 3D view of the proposed DMS low pass filter.

generated; thus the insertion loss is less than 0.2 dB from DC up to 1.55 GHz. The return loss in the pass band is less than −12 dB. The stopband rejection is higher than −20 dB from 2.15 GHz up to 4.25 GHz. As illustrated in Figure 10, an undesired peak appeared around the frequency of 4.4 GHz. In order to suppress this undesired harmonic, another technique based on DGS will be used. The dimensions of the proposed structure are depicted in Table 2.

5. Improvement of the Low Pass Filter

In order to suppress the undesired peak at 4.4 GHz of the DMS low pass filter, a pair of DGS-units has been used. The idea is to choose DGS resonators having resonance frequency around the unwanted frequency 4.4 GHz, thus to realize structure with a wide reject band. As shown in Figure 11, a multilayer structure is used to improve the performance of

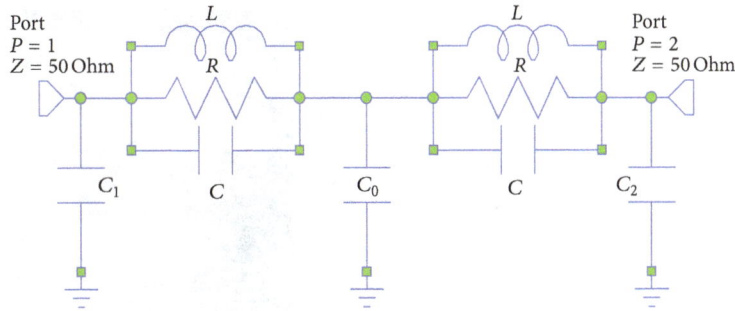

FIGURE 9: Equivalent circuit of DMS low pass filter.

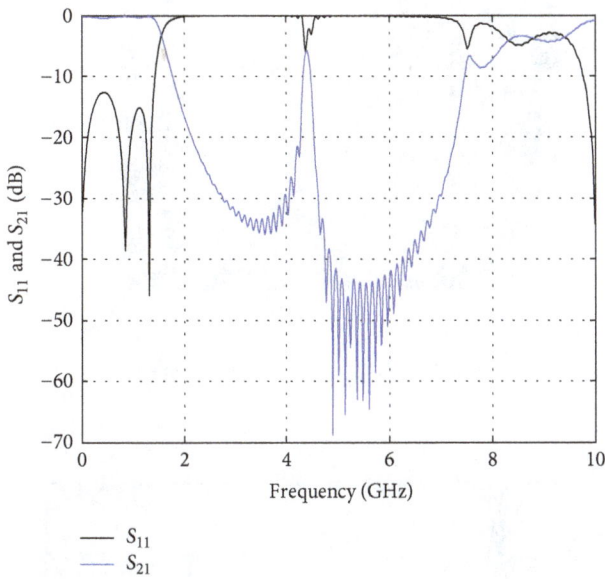

FIGURE 10: Comparison of simulated results of DMS low pass filter.

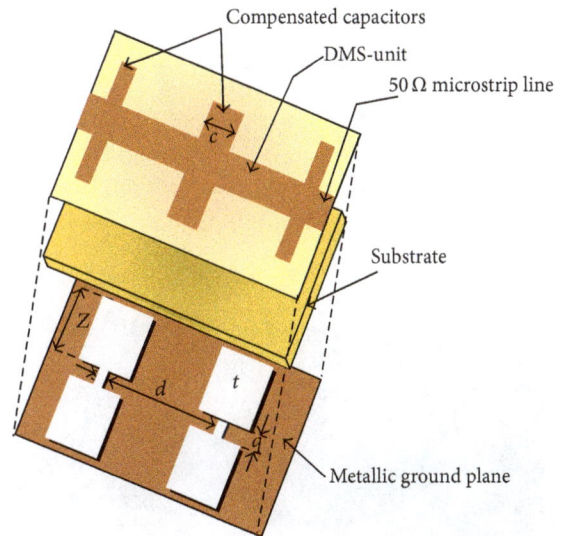

FIGURE 11: 3D view of the proposed DMS-DGS low pass filter.

the previous low pass filter. The proposed element is similar to the DMS-unit with the difference that the new structure consists of additional two DGS-units, which are located between the microstrip capacitors as shown in Figure 11.

The dimensions of the two DGS shapes, which are etched in the ground, have been defined as follows: $s = 0.6$ mm, $q = 6$ mm, $t = 10$ mm, and $z = 10$ mm. The coupled distance ($d = 26$ mm) between the cascaded DGS resonators is obtained based on the empirical method. This proposed geometrical idea is based on using several stacked layers and it was able to improve the performance of the filter. The proposed DGS-DMS low pass filter has been simulated on a Rogers RO4003 substrate with a relative dielectric constant ε_r of 3.38 and a thickness h of 0.813 mm. As depicted in Figure 12, the proposed LPF behaves well in the pass band and the stopband. The filter has a -3 dB cutoff frequency at 1.1 GHz, an insertion loss of 0.1 dB, and a return loss less than -20 dB in the whole pass band.

In addition, an ultrawide suppression level approximately equal to -20 dB in the frequency stopband ranging from 1.5 GHz to more than 8.5 GHz is achieved. Simulations were performed using Microwave Office AWR and CST Microwave Studio simulators.

6. Field Distribution along the Filter Structure

The investigation of this EM field distribution has an objective of showing the frequency behaviour of this proposed filter and to prove the validity of the intuitive equivalent circuit. Figure 13(a) shows the field distribution in the pass band region at the frequency of 0.5 GHz. The magnetic field is concentrated along the DMS resonators and on the 50 Ω lines, while a negligible electric field appears between both poles of this DMS structure. The transmission power between both ports is magnetic. The arm of the DMS represents an inductor. Figure 13(b) shows a filter with stopband behaviour at a resonant frequency of 4 GHz. The electric and magnetic fields show same distribution densities. The electric field is concentrated between extremities of the first slot, which represents the capacity. Based on this EM field investigation, the parallel LC circuit can be an approach model for the DMS-unit.

7. Fabrication and Measurement

Figure 14 shows photographs of the fabricated LPF filter. The simulations of the proposed DMS-DGS-UWRB-LPF are carried out using CST Microwave Studio and AWR Microwave

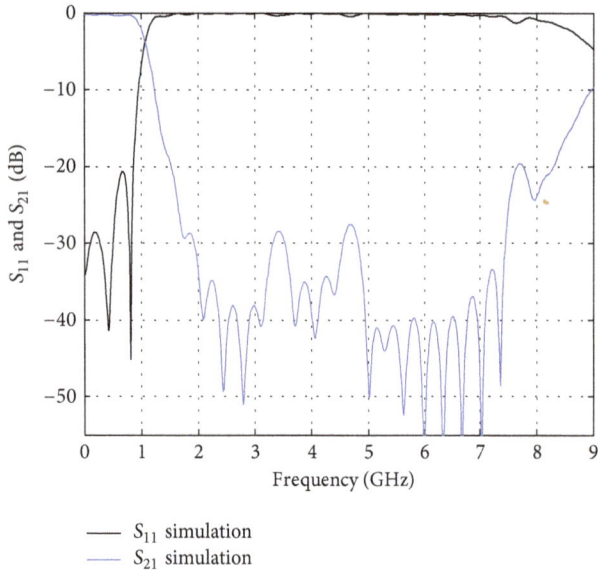

FIGURE 12: Simulation results of the proposed DMS-DGS low pass filter.

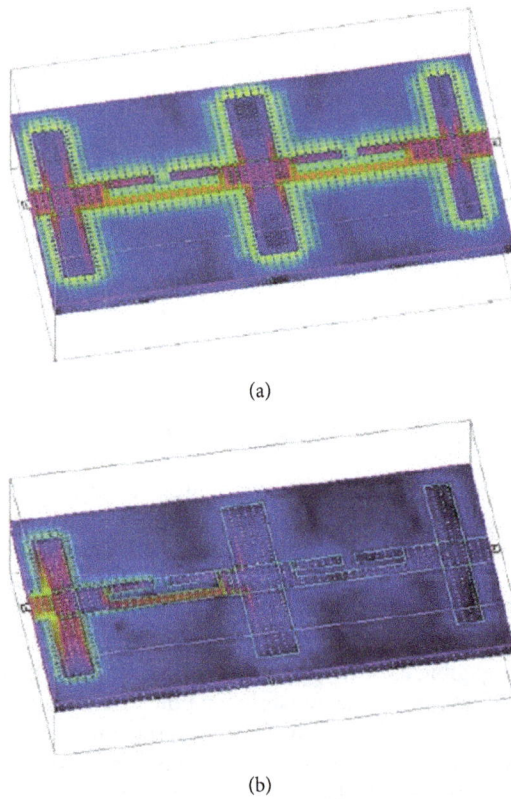

(a)

(b)

FIGURE 13: Field distribution: (a) magnetic field at f = 0.5 GHz, (b) magnetic field at f = 4 GHz.

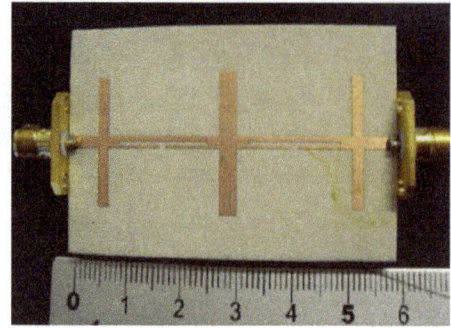

(a)

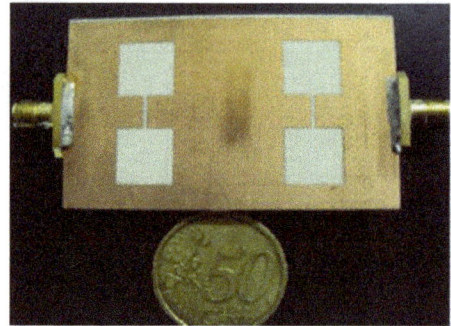

(b)

FIGURE 14: Photograph of the fabricated DMS-DGS LPF, (a) top layer, (b) bottom layer.

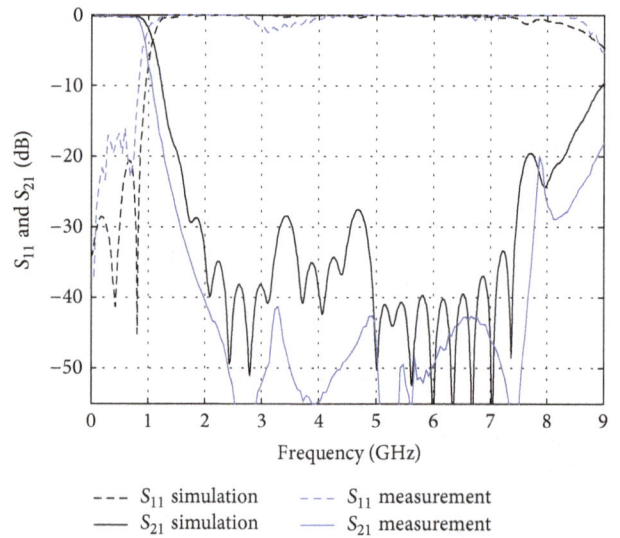

FIGURE 15: Comparison of simulation and measurement results of proposed DMS-DGS LPF.

Office. The simulation results show that the designed filter has a high sharpness factor, small losses in the pass band, and a wide reject band as shown in Figure 15. In order to verify the validity of the proposed DMS-DGS combination idea, the filter has been fabricated and measured using an HP8722D network analyzer. The LPF has been fabricated on a substrate with a relative dielectric constant ε_r of 3.38 and a thickness

h of 0.813 mm. The comparison between measured and simulated results is depicted in Figure 15. In the pass band, the measured insertion and return losses are less than −0.3 dB and −17 dB, respectively. The stopband rejection is higher than −20 dB from 1.3 GHz up to 8.9 GHz The compact low pass structure occupies an area of ($0.40\lambda g \times 0.24\lambda g$), where λg = 148 mm is the guided wavelength at cutoff frequency.

A very good agreement between simulations and measurements has been obtained. Some discrepancy between them can be interpreted as unexpected fabrication tolerances. The observed deviation between simulation and experimental results, which also means the loss of transmission line, has been caused by mismatching effects and the manufacturing tolerance errors.

8. Conclusion

In this work, a novel DMS-DGS wide stopband low pass filter has been introduced and investigated. The filter structure with strong suppression of undesired harmonic responses has been presented, which is based on stopband behaviors of the DMS and DGS cells. It is demonstrated that low insertion loss (0.3 dB), deep return loss (greater than 17 dB), and a wide rejection bandwidth with overall 20 dB attenuation from 1.3 GHz up to 8.9 GHz and bandwidth with overall 40 dB attenuation from 1.9 GHz up to 7.7 GHz have been achieved in this type of filter. It has been shown that the simulated results achieved by full-wave EM were in excellent agreement with the measured ones. The newly proposed DMS-DGS-LPF and the related design method are compatible with monolithic microwave integrated circuit (MMIC) or multilayer technology and can be used in a wide range of microwave and millimeter wave applications.

Conflict of Interests

The authors declare that there is no conflict of interests regarding the publication of this paper.

Acknowledgments

The authors thank the German Research Foundation (DFG) for financial support. The authors thank Ms. Eng. Sonja Boutejdar for her assistance and help and Mr. Harald Dempewolf, the Lab Manager of the Institute for Electronics, Signal Processing and Communication (IIKT) at the University of Magdeburg, Germany, for his support.

References

[1] H. Liu, X. Sun, and Z. Li, "Novel two-dimensional (2-D) defected ground array for planar circuits," *Active and Passive Electronic Components*, vol. 27, no. 3, pp. 161–167, 2004.

[2] A. K. Arya, M. V. Kartikeyan, and A. Patnaik, "Defected ground structure in the perspective of microstrip antennas: a review," *Frequenz*, vol. 64, no. 5-6, pp. 79–84, 2010.

[3] A. Boutejdar, "Design of broad-stop band low pass filter using a novel quasi-Yagi-DGS-resonators and metal box-technique," *Microwave and Optical Technology Letters*, vol. 56, no. 3, pp. 523–528, 2014.

[4] F. Karshenas, A. R. Mallahzadeh, and J. Rashed-Mohassel, "Size reduction and harmonic suppression of parallel coupled-line bandpass filters using defected ground structure," *Applied Computational Electromagnetics Society Journal*, vol. 25, no. 2, pp. 149–155, 2010.

[5] A. Boutejdar, A. Omar, and E. Burte, "High-performance wide stop band low-pass filter using a vertically coupled DGS-DMS-resonators and interdigital capacitor," *Microwave and Optical Technology Letters*, vol. 56, no. 1, pp. 87–91, 2014.

[6] M. K. Mandal and S. Sanyal, "A novel defected ground structure for planar circuits," *IEEE Microwave and Wireless Components Letters*, vol. 16, no. 2, pp. 93–95, 2006.

[7] A. Boutejdar, "New method to transform band-pass to low-pass filter using multilayer- and U-slotted ground structure-technique," *Microwave and Optical Technology Letters*, vol. 53, no. 10, pp. 2427–2433, 2011.

[8] H. Liu, B. Ren, X. Xiao, Z. Zhang, S. Li, and S. Peng, "Harmonic-rejection compact bandpass filter using defected ground structure for GPS application," *Active and Passive Electronic Components*, vol. 2014, Article ID 436964, 4 pages, 2014.

[9] M. Challal, A. Boutejdar, M. Dehmas, A. Azrar, and A. Omar, "Compact microstrip low-pass filter design with ultra-wide reject band using a novel quarter-circle dgs shape," *Applied Computational Electromagnetics Society Journal*, vol. 27, no. 10, pp. 808–815, 2012.

[10] A. Boutejdar, A. Batmanov, M. H. Awida, E. P. Burte, and A. Omar, "Design of a new bandpass filter with sharp transition band using multilayer-technique and U-defected ground structure," *IET Microwaves, Antennas and Propagation*, vol. 4, no. 9, pp. 1415–1420, 2010.

[11] G. E. Al-Omair, S. F. Mahmoud, and A. S. Al-Zayed, "Lowpass and bandpass filter designs based on DGS with complementary split ring resonators," *Applied Computational Electromagnetics Society Journal*, vol. 26, no. 11, pp. 907–914, 2011.

[12] Z. Pan and J. Wang, "Design of the UWB bandpass filter by coupled microstrip lines with U-shaped defected ground structure," in *Proceedings of the International Conference on Microwave and Millimeter Wave Technology (ICMMT '08)*, vol. 1, pp. 329–332, April 2008.

[13] A. Boutejdar, A. B. A. Omar, and E. Burte, "LPF builds on quasi-yagi DGS," *Microwaves & RF*, vol. 52, no. 9, pp. 72–77, 2013.

[14] M. Al Sharkawy, A. Boutejdar, F. Alhefnawi, and O. Luxor, "Improvement of compactness of lowpass/bandpass filter using a new electromagnetic coupled crescent defected ground structure resonators," *Applied Computational Electromagnetics Society Journal*, vol. 25, no. 7, pp. 570–577, 2010.

[15] H.-H. Xie, Y.-C. Jiao, B. Wang, and F.-S. Zhang, "DMS structures stop band pass filter harmonics," *Microwaves and RF*, vol. 50, pp. 72–76, 2011.

[16] J.-K. Xiao and W.-J. Zhu, "New defected *microstrip* structure bandstop filter," in *Proceedings of the Electromagnetics Research Symposium*, pp. 1471–1474, Suzhou, China, September 2011.

[17] J.-K. Xiao and W.-J. Zhu, "Non-uniform defected microstrip structure lowpass filter," *Microwave Journal*, vol. 14, 2012.

[18] J.-K. Xiao, X. Xi'an, and W.-J. Zhu, "New bandstop filter using simple defected microstrip structure," *Microwave Journal*, vol. 11, 2011.

[19] AWR Corporation, *TX-LINE: Transmission Line Calculator*, AWR Corporation, San Jose, Claif, USA, 2015.

[20] Microwave Office AWR, http://web.awrcorp.com/.

[21] Microwave Studio Software, version 12, CST Corporation.

2

A New Method of Designing Circularly Symmetric Shaped Dual Reflector Antennas Using Distorted Conics

Mohammad Asif Zaman and Md. Abdul Matin

Department of Electrical and Electronic Engineering, Bangladesh University of Engineering and Technology, Dhaka 1000, Bangladesh

Correspondence should be addressed to Mohammad Asif Zaman; asifzaman13@gmail.com

Academic Editor: Ramon Gonzalo

A new method of designing circularly symmetric shaped dual reflector antennas using distorted conics is presented. The surface of the shaped subreflector is expressed using a new set of equations employing differential geometry. The proposed equations require only a small number of parameters to accurately describe practical shaped subreflector surfaces. A geometrical optics (GO) based method is used to synthesize the shaped main reflector surface corresponding to the shaped subreflector. Using the proposed method, a shaped Cassegrain dual reflector system is designed. The field scattered from the subreflector is calculated using uniform geometrical theory of diffraction (UTD). Finally, a numerical example is provided showing how a shaped subreflector produces more uniform illumination over the main reflector aperture compared to an unshaped subreflector.

1. Introduction

Reflector antennas are widely used in radars, radio astronomy, satellite communication and tracking, remote sensing, deep space communication, microwave and millimetre wave communications, and so forth [1–3]. The rapid developments in these fields have created demands for development of sophisticated reflector antenna configurations. There is also a corresponding demand for analytical, numerical, and experimental methods of design and analysis techniques of such antennas.

The configuration of the reflectors depends heavily on the application. The dual reflector antennas are preferred in many applications because they allow convenient positioning of the feed antenna near the vertex of the main reflector and positioning of other bulky types of equipment behind the main reflector [3]. Also, the feed waveguide length is reduced [4]. They also have some significant electromagnetic advantage over single reflector systems [5]. Although many dual reflector configurations exist, the circularly symmetric dual reflector antennas remain one of the most popular choices for numerous applications [1].

One of the most common circularly symmetric dual reflector antennas is the Cassegrain antenna. The Cassegrain antenna is composed of a hyperboloidal subreflector and a paraboloidal main reflector. A feed antenna (usually a horn antenna) illuminates the subreflector which in turn illuminates the main reflector. The main reflector produces the radiated electric field that propagates into space. The radiation performance of the dual reflector antennas depends on the radiation characteristics of the feed and the geometrical shapes of the main reflector and the subreflector. Modern wireless communication and RADAR applications enforce stringent requirements on the far-field characteristics of the antenna. For example, satellite communications impose limitations on maximum beamwidth and maximum sidelobe levels of the antenna to avoid interference with adjacent satellites [2]. The traditional Cassegrain antennas have fixed geometries and offer limited flexibilities to antenna designers. As a result, the maximum performance that can be extracted from these antennas is limited by geometrical constraints.

For high performance applications, the traditional hyperboloid/paraboloidal geometry must be changed. Reflector shaping is the method of changing the shape of the reflecting surfaces to improve the performance of the antenna. Shaped reflector antennas outperform conventional unshaped reflector antennas. Reflector shaping allows the designers additional flexibility. The antenna designers have independent control over relative position of the reflectors, diameter of the reflectors, and the curvature of the reflectors when shaped reflectors are used instead of conventional reflectors. This makes reflector shaping an essential tool for designing high performance reflector antennas.

Many methods of designing shaped reflectors are present in literature. One of the first major articles related to reflector shaping was published by Galindo in 1964 [6]. The method is based on geometrical optics (GO). Galindo's method required solution of multiple nonlinear differential equations, which sometimes may be computationally demanding. A modified version of this method was presented by Lee [7]. Lee divided the reflector surfaces into small sections and assumed the sections to be locally planar. This assumption converted the differential equations to algebraic equations, which are much easier to solve. However, the reflector surface must be divided into a large number of sections to increase the accuracy of this method.

Another popular method for designing shaped reflectors involves expanding the shaped surfaces using a set of orthogonal basis functions [8, 9]. Rahmat-Samii has published multiple research papers on this area [8–10]. The expansion coefficients determine the shape of the surface. A small number of terms of the expansion set are sufficient to accurately describe a shaped surface. So, a few expansion coefficients must be determined to define the surface. The differential equation based methods determine the coordinates of the points on the reflector surfaces, whereas the surface expansion based method only determines the expansion coefficients. Due to the decrease in number of unknowns, the surface expansion method is computationally less demanding. The surface expansion method can be incorporated with geometrical theory of diffraction (GTD) or its uniform version, uniform theory of diffraction (UTD), to produce an accurate design algorithm [10]. These design procedures are known as diffraction synthesis [8–10]. This method has been successfully used in many applications. Recently, a few new efficient methods for designing circularly symmetric shaped dual reflector design have been developed. One of the first significant works on this method was reported by Kim and Lee in 2009 [11]. This method divides the shaped reflector surfaces into electrically small sections. Each section is assumed to be a conventional unshaped dual reflector system. Since well-established methods for analyzing conventional dual reflector system exist, the radiation characteristics of each section can easily be evaluated. The shaped surface is defined by combining all the local conventional surfaces. The method requires solutions of several nonlinear algebraic equations. So it is computationally convenient. Another method based on the same principle was proposed by Moreira and Bergmann in 2011 [12]. This method also divides the shaped surface into small local sections. The local sections are represented by unshaped conics. Each conic section is optimized to produce a desired aperture distribution, which is formulated by GO method. As these methods have recently appeared in literature, most

of the advantages and drawbacks of the method have not been investigated. Reduction in computational complexity is an obvious advantage. The proposed work concentrated on presenting an alternative method rather than improving the existing methods.

A design method that requires lesser number of parameters to define the shaped reflector surfaces without decreasing accuracy of the obtained results is a challenging goal for reflector antenna designers. In this paper, the surface of the shaped reflector is defined using a novel equation. The shaped subreflector surfaces are assumed to be distorted forms of unshaped surfaces. As most shaped subreflectors resemble their unshaped counterparts [11–13], the assumption is logical. The shaped surfaces can therefore be represented by modified versions of the equations that represent the conventional unshaped surfaces. This method of visualizing the shaped surface as perturbed/distorted form of unshaped surface has not been reported in literature yet. As the general form of the surface is generated from the conventional conics, only a small number of parameters need be used to represent the shaped surface. Once the subreflector surface is defined, the main reflector surface is defined using GO method and equal optical path length criterion.

The paper is organized as follows. Section 2 describes the geometry of the shaped subreflector. Surface equations and differential geometric analysis are presented in this section. The synthesis method of the main reflector is discussed in Section 3. Section 4 covers the numerical results. Concluding remarks are made in Section 5.

2. Geometry of the Shaped Subreflector

2.1. Surface Equations. In a conventional Cassegrain geometry, the subreflector is hyperboloidal. In shaped-Cassegrain geometry, the subreflector is shaped to provide a desired illumination over the aperture of the main reflector. However, the geometrical features of this shaped subreflector are very similar to the geometrical features of the unshaped hyperboloid. Due to these similarities, the shaped surfaces can be considered a distorted form of the unshaped surfaces. So, to define the geometry of the shaped subreflector, the conventional Cassegrain geometry must be described first.

The geometry of a Cassegrain dual reflector antenna is shown in Figure 1. The parameters describing the geometry of the Cassegrain system are

d_p: diameter of the main reflector = 10 m,

f_p: focal length of the main reflector = 5 m,

d_s: diameter of the subreflector = 1.25 m,

$2c$: distance between the foci = 4 m,

$e = c/a$: subreflector eccentricity = 1.4261,

Δ_p: depth of the paraboloid = 1.25 m,

l_p: distance from feed to paraboloid vertex = 1 m,

ψ_o, ϕ_o: opening half angle of the main reflector and subreflector = 53.13° and 10.037°, respectively.

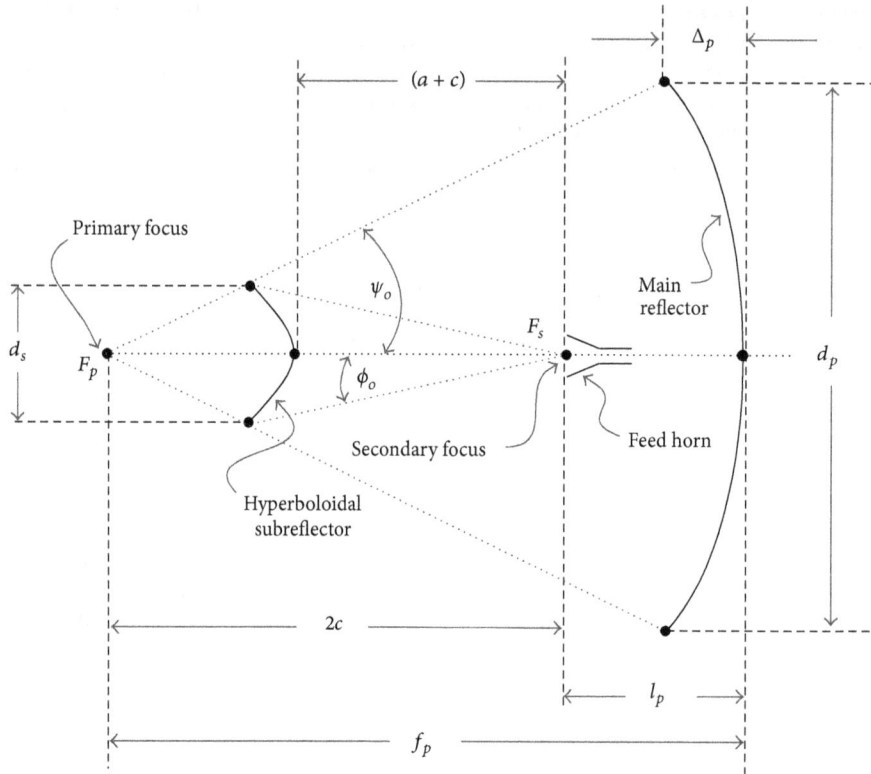

FIGURE 1: Geometry of a Cassegrain antenna.

Some of these parameter values are selected from standard values. The other parameters are found using geometrical relations found in literature [3, 14].

It is assumed that the feed antenna is located at the origin and direction of feed radiation is towards the negative z axis. So the hyperboloid must be located at the negative side of the z axis. One of the focuses of the hyperboloid must be at origin so that it coincides with the feed. The equation of such a hyperboloidal surface symmetric around the z axis is given by the following equation [15]:

$$\frac{(z+c)^2}{a^2} - \frac{x^2+y^2}{b^2} = 1, \tag{1}$$

where

$$b^2 = c^2 - a^2. \tag{2}$$

The parameters a and c are related to the position of the vertex and focus of the hyperboloid as shown in Figure 1 [15]. The parameter ρ_s is defined as the radius of surface point projected on the xy plane. It is related to x and y by

$$\rho_s^2 = x^2 + y^2. \tag{3}$$

Equation (1) can be modified as

$$\frac{(z+c)^2}{a^2} - \frac{\rho_s^2}{b^2} = 1, \tag{4}$$

$$z = -c - \frac{a}{b}\sqrt{b^2 + \rho_s^2}. \tag{5}$$

In (5), the negative square root is taken because the defined coordinate system places the hyperboloid in the negative z axis.

For a circularly symmetric shaped subreflector, (4) and, correspondingly, (5) need be modified. A distortion function, $\delta(\cdot)$, is introduced in the equations to get the shaped surface:

$$\frac{(z+c)^2}{a^2} - \frac{\rho_s^2}{b^2}\delta(\rho_s) = 1, \tag{6}$$

$$z = -c - \frac{a}{b}\sqrt{b^2 + \rho_s^2\delta(\rho_s)}. \tag{7}$$

The distortion function, $\delta(\cdot)$, must be a function of ρ_s to maintain circular symmetry. The shape of the surface depends on the expression of $\delta(\cdot)$. Through literature review, it is found that shaped hyperboloidal subreflectors are usually different from unshaped hyperboloids near the edges [13]. The shaped surface curves towards the vertex at the edges. An exponential function with arguments containing even powers of ρ_s can give the desired shaped. The following generalized expression of the distortion function is developed:

$$\delta(\rho_s) = \exp\left(\sum_{n=1}^{N}\tau_n\rho_s^{2n\zeta_n}\right). \tag{8}$$

The function contains $2N$ number of parameters denoted by τ_n and ζ_n. Here, τ_n denotes the nth amplitude distortion parameter and ζ_n denotes the nth exponent distortion parameter. The parameters together are termed distortion parameters.

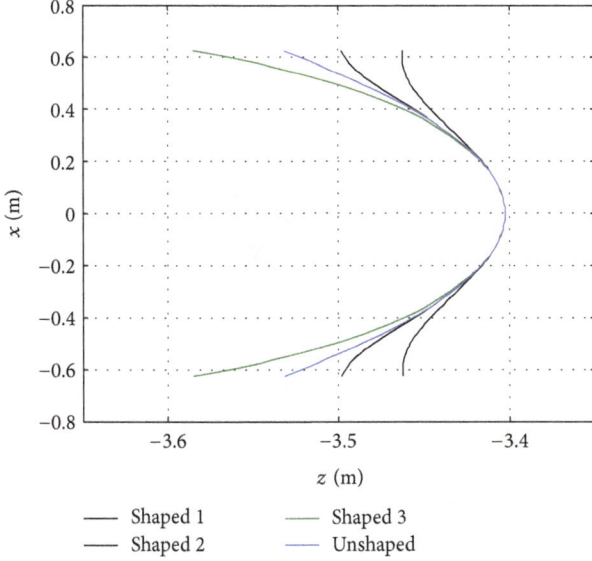

FIGURE 2: Multiple shaped subreflectors generated by using different sets of distortion parameter values.

For high values of N, the control over the curvature of the surface is more precise. But it increases the number of parameters required to define the surface. It is found that, for practical subreflector surfaces, $N = 2$ is sufficient and only 4 parameters are required to define the surface.

Two-dimensional and three-dimensional representations of shaped surfaces obtained using (7) are shown in Figures 2 and 3. The figures are generated using arbitrary values of the distortion parameters. However, it is necessary to establish that practical shaped subreflector surfaces can be accurately represented by these equations. Practical data of a shaped hyperboloidal and subreflector surfaces are taken from literature [13]. The dimensions are normalized with respect to subreflector diameter in [13]. By adjusting the distortion parameters in (8) and using them in (7), a surface closely resembling the shaped subreflector surface of [13] is generated. The results are shown in Figure 4.

It is clear that the unshaped surface is very different from the shaped surface. With $\tau_1 = -3.3363$, $\zeta_1 = 0.5706$, $\tau_2 = 2.1697$, and $\zeta_2 = 0.1951$, the sum of squares of the error between the defined shaped hyperboloidal surface and the practical data of [13] is only 3.2×10^{-5}. So the derived equations represent the practical shaped surface very closely.

2.2. Differential Geometric Analysis. The design procedures of reflector antennas are interrelated with the numerical techniques that are used to analyse those antennas. GO and UTD are two common numerical techniques employed for these purposes. To apply these methods, it is necessary that the reflecting surfaces be expressed in differential geometric form. Also, the normal vector at each point of the surface must also be defined to compute scattered field using GO or UTD method [16, 17].

The shaped hyperboloidal surface defined by (7) can be represented using parameters ρ_s and ϕ_s in differential geometrical form as [16, 18]

$$\mathbf{r} = \rho_s \cos\phi_s \hat{\mathbf{x}} + \rho_s \sin\phi_s \hat{\mathbf{y}} - \left[c + \frac{a}{b}\sqrt{b^2 + \rho_s^2 \delta(\rho_s)} \right] \hat{\mathbf{z}}. \quad (9)$$

Here, (ρ_s, ϕ_s) are the polar coordinates of the projection of a surface point on the xy plane. The unit normal vector, $\hat{\mathbf{n}}_s$, is defined by the following equation [19]:

$$\hat{\mathbf{n}}_s = \frac{(\partial \mathbf{r}/\partial \rho_s) \times (\partial \mathbf{r}/\partial \phi_s)}{|(\partial \mathbf{r}/\partial \rho_s) \times (\partial \mathbf{r}/\partial \phi_s)|}. \quad (10)$$

The differential geometrical parameters along with the coordinate system are shown in Figure 5. The incident ray vector ($\hat{\mathbf{s}}_i$) and the reflected ray vector ($\hat{\mathbf{s}}_r$) are also shown in Figure 5. As the unit normal vector is required for later calculations, (10) needs to be evaluated. The partial derivatives can be calculated from (9) using (8):

$$\frac{\partial \mathbf{r}}{\partial \rho_s} = \cos\phi_s \hat{\mathbf{x}} + \sin\phi_s \hat{\mathbf{y}}$$

$$- \left[\frac{a}{2b} \frac{\rho_s \delta(\rho_s) \left\{ \sum_{n=1}^{N} \tau_n \rho^{2n\zeta_n} + 2 \right\}}{\sqrt{b^2 + \rho_s^2 \delta(\rho_s)}} \right] \hat{\mathbf{z}}, \quad (11)$$

$$\frac{\partial \mathbf{r}}{\partial \phi_s} = -\rho_s \sin\phi_s \hat{\mathbf{x}} + \rho_s \cos\phi_s \hat{\mathbf{y}}.$$

Substituting these values in (10),

$$\hat{\mathbf{n}}_s = \frac{\Lambda(\rho_s)\cos\phi_s}{\sqrt{\Lambda^2(\rho_s) + \Delta_H^2(\rho_s)}}\hat{\mathbf{x}} + \frac{\Lambda(\rho_s)\sin\phi_s}{\sqrt{\Lambda^2(\rho_s) + \Delta_H^2(\rho_s)}}\hat{\mathbf{y}}$$

$$+ \frac{\Delta_H(\rho_s)}{\sqrt{\Lambda^2(\rho_s) + \Delta_H^2(\rho_s)}}\hat{\mathbf{z}}, \quad (12)$$

where

$$\Delta_H(\rho_s) = 2b\sqrt{b^2 + \rho_s^2 \delta(\rho_s)},$$

$$\Lambda(\rho_s) = a\delta(\rho_s)\varepsilon(\rho_s), \quad (13)$$

$$\varepsilon(\rho_s) = \rho_s \left\{ \sum_{n=1}^{N} \tau_n \rho^{2n\zeta_n} + 2 \right\}.$$

For a given set of distortion parameter values, the shaped subreflector surface and the normal vectors can be evaluated using these equations.

3. Synthesizing the Main Reflector

In a dual reflector system, the shape of the main reflector is completely dependent on the shape of the subreflector. Once the shape of the subreflector is defined, the main reflector must be shaped so that the path lengths of the rays are

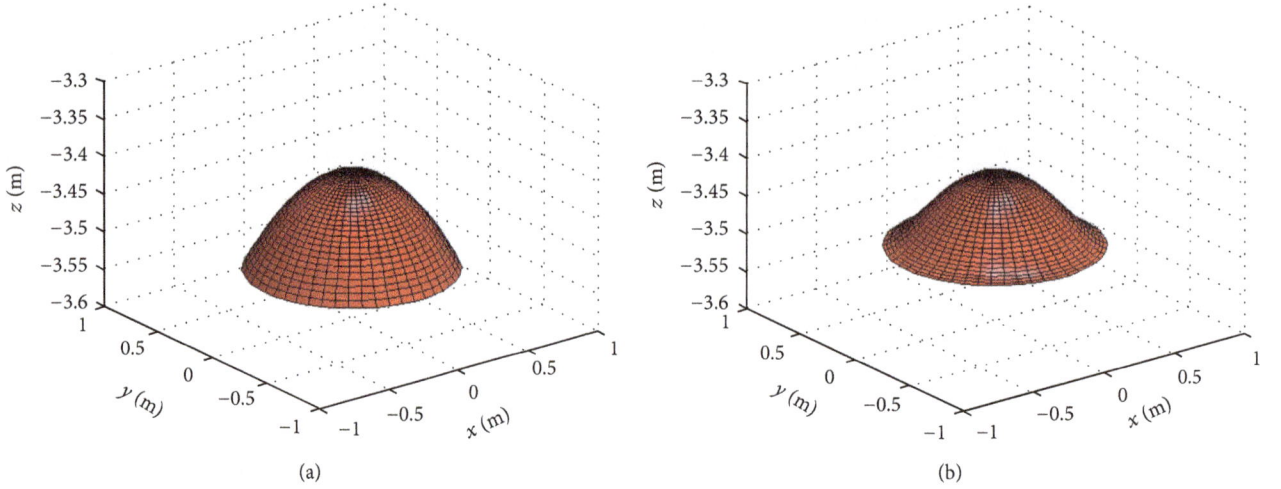

(a) (b)

FIGURE 3: Three-dimensional representation of (a) conventional hyperboloid and (b) shaped hyperboloid.

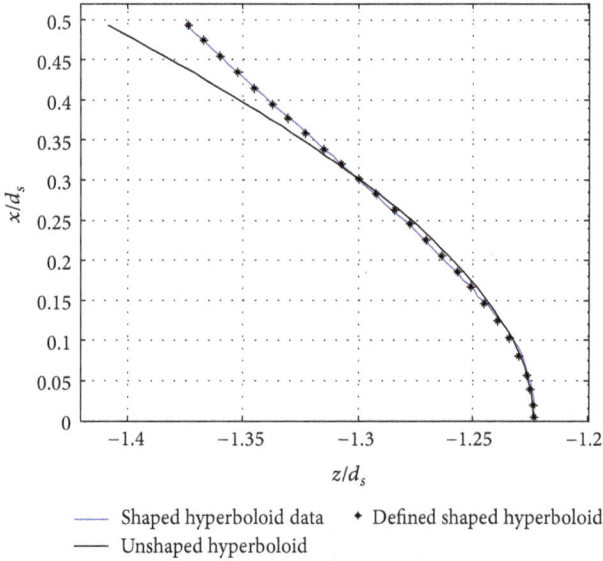

—— Shaped hyperboloid data • Defined shaped hyperboloid
—— Unshaped hyperboloid

FIGURE 4: Representation of practical hyperboloidal shaped subreflector using the proposed equations.

constant at an observation plane perpendicular to the main reflector axis. Using GO methods to calculate the incident and reflected ray vectors, the required position of a point on the main reflector surface for a given point on the subreflector surface can be formulated.

The geometry of dual reflector system with a shaped hyperboloidal subreflector is shown in Figure 6. An observation plane is defined parallel to the xy plane. In the two-dimensional diagram of Figure 6, the observation plane is shown as line parallel to x axis going through the point $(0, 0, z_{ref})$. For a ray emitted from the feed, the sum of the distances d_1, d_2, and d_3 must be constant (k_1), irrespective of the position of reflection on the subreflector surface. So

$$d_1 + d_2 + d_3 = k_1. \tag{14}$$

The distance k_1 can be calculated by considering an axial ray from the feed (along the z axis). From Figure 6, for the axial ray, the distance from feed to subreflector is $(a + c)$; then the distance from the subreflector to main reflector is $(a + c + l_p)$, and finally the distance from main reflector to observation point is $(l_p - z_{ref})$, when considering z_{ref} to be a negative quantity. So k_1 is calculated as

$$k_1 = (a + c) + (a + c + l_p) + (l_p - z_{ref})$$
$$= 2(a + c + l_p) - z_{ref}. \tag{15}$$

The parameter l_p can be related to the geometrical parameters of the unshaped subreflector, c, and the unshaped main reflector f_p as seen in Figure 6:

$$l_p = f_p - 2c. \tag{16}$$

Using (16) in (15),

$$k_1 = 2(a - c + f_p) - z_{ref}. \tag{17}$$

The distance d_1 is related to the coordinates of the subreflector surface point (x_s, y_s, z_s) as

$$d_1 = \sqrt{x_s^2 + y_s^2 + z_s^2}. \tag{18}$$

The reflected ray from the main reflector is parallel to the z axis. The distance d_3 can easily be related to the coordinates of the main reflector surface point (x_m, y_m, z_m) as

$$d_3 = z_m - z_{ref}. \tag{19}$$

To calculate distance d_2, it is necessary to calculate the reflected ray vector. The unit vector in the direction of the reflected ray is found from GO method and is given by [16]

$$\hat{s}_r = \hat{s}_i - 2(\hat{n}_s \cdot \hat{s}_i)\hat{n}_s, \tag{20}$$

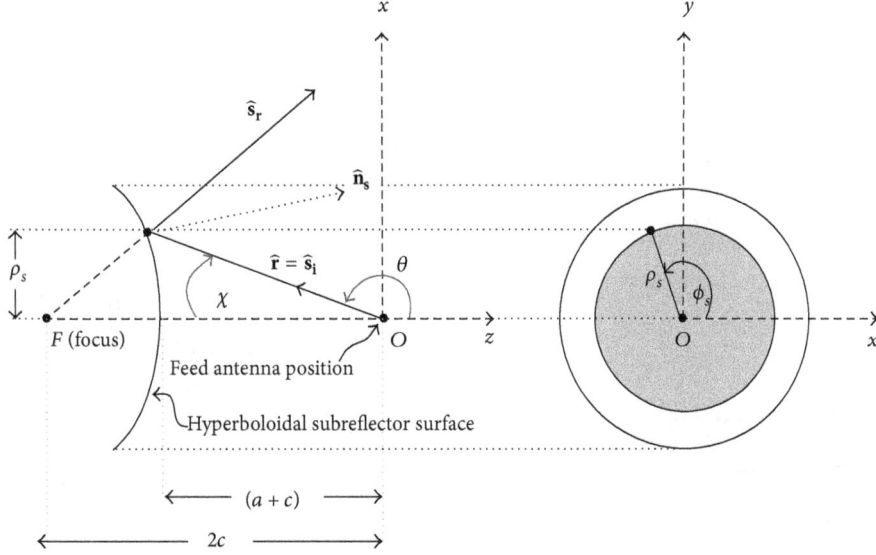

FIGURE 5: Differential geometric representation of the shaped hyperboloidal subreflector.

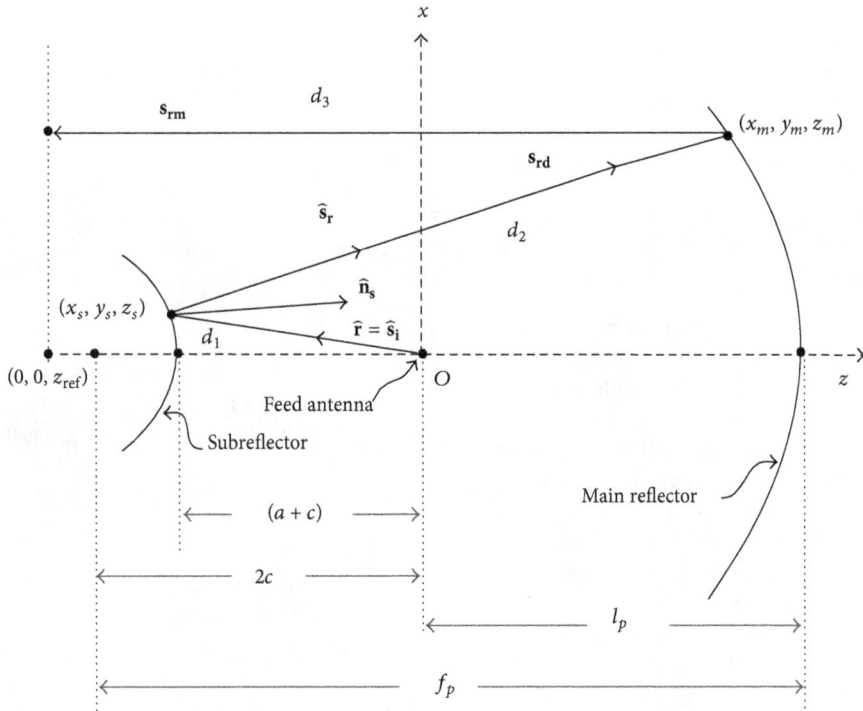

FIGURE 6: Synthesis of main reflector surface for a given shaped subreflector.

where

$$\hat{\mathbf{s}}_{\mathbf{i}} = \hat{\mathbf{r}} = \frac{\mathbf{r}}{|\mathbf{r}|}. \tag{21}$$

The vector $\mathbf{r}$ is defined by (9) and $\hat{\mathbf{n}}_{\mathbf{s}}$ is defined by (12). Once the unit vector along the reflected ray is formulated, the reflected ray vector, $\mathbf{s}_{\mathbf{rd}}$, defined from the reflection point on the subreflector surface can be defined as

$$\mathbf{s}_{\mathbf{rd}} = \hat{\mathbf{s}}_{\mathbf{r}} d_2 + x_s \hat{\mathbf{x}} + y_s \hat{\mathbf{y}} + z_s \hat{\mathbf{z}}. \tag{22}$$

In (22), the subreflector surface point vector appears as additive term. This term is used to shift the vector from origin to the subreflector surface point (x_s, y_s, z_s).

Equation (22) can be reorganized as

$$\mathbf{s}_{\mathbf{rd}} = \{(\hat{\mathbf{s}}_{\mathbf{r}} \cdot \hat{\mathbf{x}}) d_2 + x_s\} \hat{\mathbf{x}} + \{(\hat{\mathbf{s}}_{\mathbf{r}} \cdot \hat{\mathbf{y}}) d_2 + y_s\} \hat{\mathbf{y}} \\ + \{(\hat{\mathbf{s}}_{\mathbf{r}} \cdot \hat{\mathbf{z}}) d_2 + z_s\} \hat{\mathbf{z}}. \tag{23}$$

The components of this vector indicate the coordinates of the main reflector [18, 19]. So

$$x_m = (\hat{\mathbf{s}}_\mathbf{r} \cdot \hat{\mathbf{x}})\, d_2 + x_s,$$
$$y_m = (\hat{\mathbf{s}}_\mathbf{r} \cdot \hat{\mathbf{y}})\, d_2 + y_s, \qquad (24)$$
$$z_m = (\hat{\mathbf{s}}_\mathbf{r} \cdot \hat{\mathbf{z}})\, d_2 + z_s.$$

Substituting the value of z_m from (24) to (19),

$$d_3 = (\hat{\mathbf{s}}_\mathbf{r} \cdot \hat{\mathbf{z}})\, d_2 + z_s - z_{\text{ref}}. \qquad (25)$$

Now, using (18) and (25) in (14) and solving for d_2,

$$d_2 = \frac{2\left(a - c + f_p\right) - \sqrt{x_s^2 + y_s^2 + z_s^2} - z_s}{1 + (\hat{\mathbf{s}}_\mathbf{r} \cdot \hat{\mathbf{z}})}. \qquad (26)$$

Equation (26) can be used to calculate d_2 for a given subreflector surface point and it can be replaced in (24) to calculate corresponding main reflector surface point. Thus, the derived equations can synthesize the main reflector surface for an arbitrary subreflector surface.

To verify the method, the main reflector for an unshaped hyperboloid is synthesized. It is found that the synthesized main reflector is an exact paraboloid, which is expected. Thus, the method is verified.

4. Numerical Example

For numerical analysis, the feed is assumed to be a conical corrugated horn antenna. The radiated field from the horn is calculated using standard equations [1]. The fields scattered from the subreflector are calculated using UTD method. The observation points are taken on the main reflector surface. First, the scattered field from an unshaped hyperboloid is calculated. The results are shown in Figure 7. The rapid fall of the reflected field indicates the reflection shadow boundary (RSB) of the subreflector [16]. It can be observed from Figure 7 that the scattered field tapers gradually as the observation angle increases. This creates a tapered illumination of the main reflector. Using the proposed method, a shaped hyperboloidal subreflector is defined which produces a more uniform field distribution. An optimization algorithm can be used to find the optimum set of distortion parameter values that will create a subreflector surface that will create a desired illumination. The scattered field for a shaped subreflector defined by the parameters $\tau_1 = -3.3928$, $\tau_2 = 2.5015$, $\zeta_1 = 1.7212$, and $\zeta_2 = 0.7184$ is shown in Figure 8. The values of the distortion parameters are calculated using differential evolution (DE) optimization algorithm [20–22]. The fitness function in the optimization process is defined in terms of uniformity of the scattered field. As a result, the algorithm converges to a set of distortion parameter values that result in a uniform aperture distribution. It can be observed that the scattered field is more uniform when the subreflector is shaped.

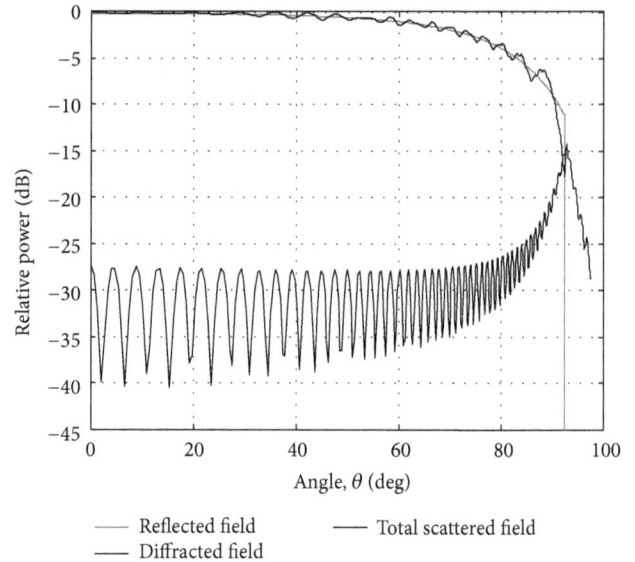

Figure 7: Scattered field from an unshaped hyperboloidal subreflector.

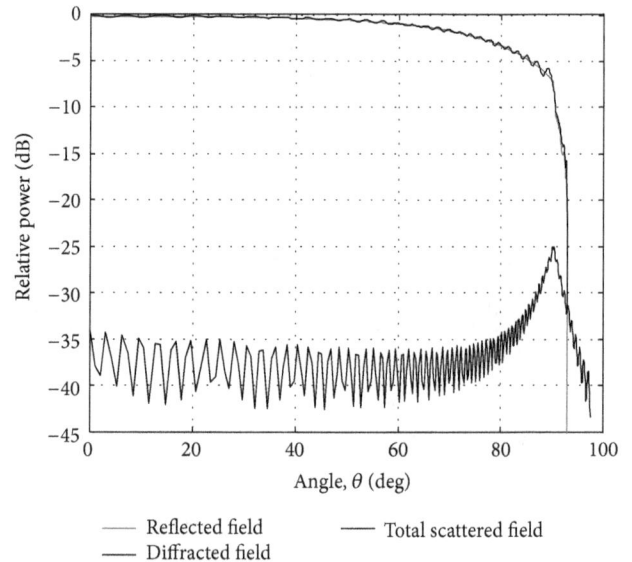

Figure 8: Scattered field from a shaped hyperboloidal subreflector.

5. Conclusion

A novel method of expressing the shaped reflector surfaces has been presented in this paper. The method expresses the shaped subreflector surface as distorted form of the unshaped conventional conic surface. This allows us to mathematically express the shaped subreflector surface with only a few parameters. The shaped main reflector surface is synthesized employing GO method and using the derived equation of the shaped subreflector. A numerical example is provided to show how a shaped reflector system defined by the proposed method can create a uniform field distribution over the main reflector aperture. As only a few parameters completely describe the shaped reflector surfaces, the computational burden of design and optimization process of shaped reflectors can be reduced using the proposed method.

Conflict of Interests

The authors declare that there is no conflict of interests regarding the publication of this paper.

References

[1] J. L. Volakis, Ed., *Antenna Engineering Handbook*, McGraw-Hill, 4th edition, 2007.

[2] C. A. Balanis, Ed., *Modern Antenna Handbook*, John Wiley & Sons, New York, NY, USA, 2008.

[3] J. W. M. Baars, *The Paraboloidal Reflector Antenna in Radio Astronomy and Communications: Theory and Practice*, Springer, 2007.

[4] W. L. Stutzman and G. A. Thiele, *Antenna Theory and Design*, John-Wiley & Sons, 1981.

[5] S. Silver, Ed., *Microwave Antenna Theory and Design*, McGraw-Hill, New York, NY, USA, 1st edition, 1949.

[6] V. Galindo, "Design of dual-reflector antennas with arbitrary phase and amplitude distributions," *IEEE Transactions on Antennas and Propagation*, vol. 12, pp. 403–408, 1964.

[7] C. S. Lee, "A simple method of dual-reflector geometrical optics synthesis," *Microwave and Optical Technology Letters*, vol. 1, no. 10, pp. 367–371, 1988.

[8] Y. Rahmat-Samii and J. Mumford, "Reflector diffraction synthesis using global coefficients optimization techniques," in *Proceedings of the Antennas and Propagation Society International Symposium*, vol. 3, pp. 1166–1169, IEEE, San Jose, Calif, USA, June 1989.

[9] D.-W. Duan and Y. Rahmat-Samii, "Generalized diffraction synthesis technique for high performance reflector antennas," *IEEE Transactions on Antennas and Propagation*, vol. 43, no. 1, pp. 27–40, 1995.

[10] Y. Rahmat-Samii and V. Galindo-Israel, "Shaped reflector antenna analysis using the jacob-bessel series," *IEEE Transactions on Antennas and Propagation*, vol. 28, no. 4, pp. 425–435, 1980.

[11] Y. Kim and T.-H. Lee, "Shaped circularly symmetric dual reflector antennas by combining local conventional dual reflector systems," *IEEE Transactions on Antennas and Propagation*, vol. 57, no. 1, pp. 47–56, 2009.

[12] F. J. S. Moreira and J. R. Bergmann, "Shaping axis-symmetric dual-reflector antennas by combining conic sections," *IEEE Transactions on Antennas and Propagation*, vol. 59, no. 3, pp. 1042–1046, 2011.

[13] M. S. Narasimhan, P. Ramanujam, and K. Raghavan, "GTD analysis of the radiation patterns of a shaped subreflector," *IEEE Transactions on Antennas and Propagation*, vol. 29, no. 5, pp. 792–795, 1981.

[14] C. Granet, "Designing axially symmetric cassegrain or gregorian dual-reflector antennas from combinations of prescribed geometric parameters," *IEEE Antennas and Propagation Magazine*, vol. 40, no. 2, pp. 76–81, 1998.

[15] A. D. Polyanin and A. V. Manzhirov, *Handbook of Mathematics for Engineers and Scientists*, Chapman & Hall/CRC, Taylor & Francis, 2007.

[16] D. A. McNamara, C. W. I. Pistorius, and J. A. G. Malherbe, *Introduction to the Uniform Geometrical Theory of Diffraction*, Artech House, London, UK, 1990.

[17] K. Lim, H. Ryu, and J. Choi, "UTD analysis of a shaped subreflector in a dual offset-reflector antenna system," *IEEE Transactions on Antennas and Propagation*, vol. 46, no. 10, pp. 1555–1559, 1998.

[18] B. O'Niell, *Elementary Differential Geometry*, Academic Press, Elsevier, 2nd edition, 2006.

[19] D. V. Widder, *Advanced Calculus*, Prentice-Hall, New Delhi, India, 2nd edition, 2004.

[20] R. Storn and K. Price, "Differential evolution—a simple and efficient heuristic for global optimization over continuous spaces," *Journal of Global Optimization*, vol. 11, no. 4, pp. 341–359, 1997.

[21] K. V. Price, R. M. Storn, and J. A. Lampinen, *Differential Evolution: A Practical Approach to Global Optimization*, Springer, 2005.

[22] A. Qing and C. K. Lee, *Differential Evolution in Electromagnetics*, Springer, Berlin, Germany, 2010.

3

Design and Investigation of Disk Patch Antenna with Quad C-Slots for Multiband Operations

J. A. Ansari, Sapna Verma, and Ashish Singh

Department of Electronics and Communication, University of Allahabad, Allahabad, Uttar Pradesh 211002, India

Correspondence should be addressed to Sapna Verma; sverma.ece@gmail.com

Academic Editor: Xianming Qing

An investigation into the design and fabrication of multiband disk patch antenna with symmetrically quad C-slots is presented in this paper. The proposed antenna shows multiband resonance frequencies which highly depend on substrate thickness, dielectric constant, and radius of the disk patch. By incorporating two pairs of C-slots in optimum geometry on the radiating patch, the proposed antenna operates between 2 and 12 GHz at different frequency bands centered at 2.27, 7.505, 9.34, 10.33, and 11.61 GHz. The other antenna parameters are studied like gain, antenna efficiency, and radiation pattern. The proposed antenna may find applications in S-, C-, and X-band. The results are carried out with the aid of HFSS and MOM-based IE3D simulator. The measured and simulated results are in good agreement with each other.

1. Introduction

There is increasing demand for antennas having compact size and multiband operation. Since each communication protocol may operate in a distinctive frequency band, instead of using several antennas, it is highly desirable to have one broadband or multiband antenna to meet the needs of multiple systems. Microstrip disk patch antennas are suitable for many wireless communication applications and frequency bands such as S-band (2–4 GHz), C-band (4–8 GHz), and X-band (8–12 GHz). In addition, compact size with multiband operation is a demand factor for several applications such as mobile communication. The demand of multiband antennas is fulfilled by integrating the diode switches, cutting slots of different geometries, multiple narrow slits in rectangular patch antenna, slot antenna with edge-fed, compact fork shaped antenna, and compact slot on the radiating patch [1–4]. There are some other techniques reported in literature to obtain multiband behaviour of the antennas such as employing the two microstrip line feeds placed in orthogonal directions [5] and circular ring antenna with a defected ground plane excited by Y-shape-like microstrip feed [6]. Recently, it has been shown that multiband resonance can be achieved by using edge feeding in two circular slots each

having T-shaped patch which highly depends on the diameter [7] and cutting two pairs of orthogonal narrow slits on stacked circular disk [8]. However, some of these antenna structures involve complex calculation, sophisticated design structures, and large size as compared to proposed antenna.

In this paper, a quad C-slots disk patch antenna optimized for simplicity in design is proposed for multiband operation. The numerical analysis and geometry refinement of the proposed antenna structure are performed by using method of moment (MOM) based IE3D and finite element method (FEM) based HFSS simulator [9, 10]. A measured result shows that the proposed antenna is able to operate in multiband. The details of antenna design and simulated as well as measured results are carefully examined and discussed in the following sections.

2. Antenna Structure

Figure 1 shows the top view and side view of the proposed antenna.

The disk patch antenna is printed on a 1.6 mm thick inexpensive FR4 dielectric substrate of relative permittivity of 4.4 and loss tangent of 0.02. The radius of the probe fed patch antenna is $R = 15.0$ mm. This antenna provides

(a)

(b)

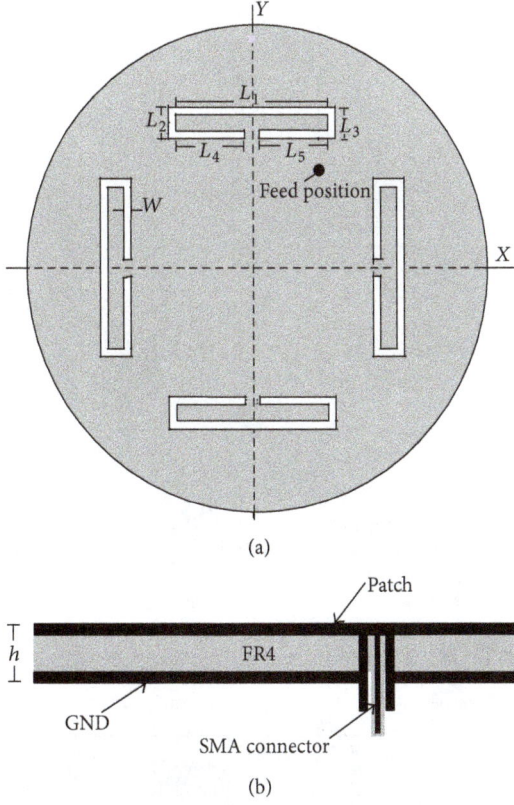

FIGURE 1: Geometry of the disk patch antenna with quad symmetrical C-slots. (a) Top view of the proposed antenna. (b) Side view of the proposed antenna.

FIGURE 2: Fabricated photo of the proposed antenna.

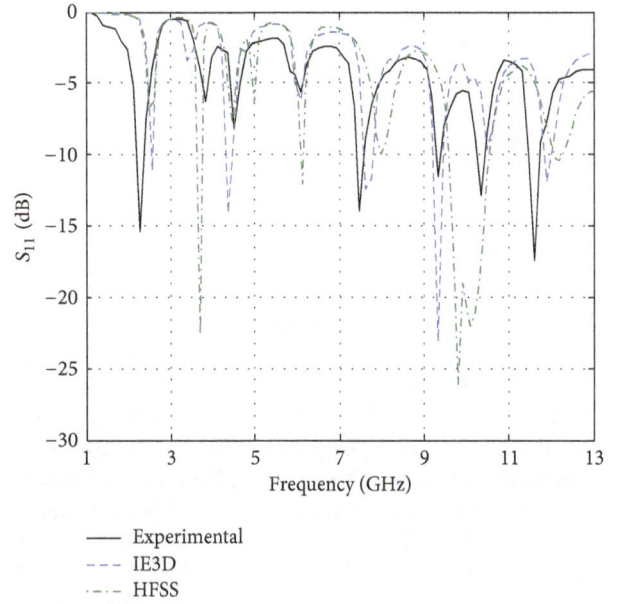

FIGURE 3: Simulated and measured S_{11} of proposed antenna.

the multiband operation due to the quad symmetrical C-slots etched on the edge of the radiating patch. The optimal parameters of the C- and inverted C-slots are kept equal. The $L_1 = 11.0$ mm, $L_4 = L_5 = 5.0$ mm, and $L_2 = L_3 = 2.0$ mm are horizontal and vertical slots of the proposed antenna, respectively. The $W = 0.5$ mm is the width of the horizontal and vertical slots of C- and inverted C-slots.

Theoretically the effect of embedded C-slots on disk patch can be approximately obtained by the impedance of a short dipole via well-known Babinet's principle [11, 12] given as

$$Z_{\text{slot}} \cdot Z_{\text{dipole}} = \frac{\eta_0^2}{4},$$

Z_{dipole}

$$\approx f_1\left(\beta l\right) - j\left(120\left(\ln\frac{2L}{W} - 1\right)\cot\left(\beta l\right) - f_2\left(\beta l\right)\right),$$

$f_1\left(\beta l\right)$

$$= -0.4787 + 7.3246\left(\beta l\right) + 0.3963\left(\beta l\right)^2 + 15.613\left(\beta l\right)^3,$$

$f_2\left(\beta l\right)$

$$= -0.4456 + 17.0082\left(\beta l\right) - 8.6793\left(\beta l\right)^2 + 9.6031\left(\beta l\right)^3,$$

$$(1)$$

where $\eta_0 = 120\pi$, L = length of the dipole, and W = width or diameter of the dipole.

3. Results and Discussion

A photograph of the fabricated quad C-slots loaded disk patch antenna is shown in Figure 2. The Agilent E5071C Network Analyzer is used to measure the measured S_{11} of the proposed antenna and the simulation performed by IE3D and HFSS simulator. The measured and simulated S_{11} parameters of the proposed antenna are compared as shown in Figure 3, which exhibit good agreement and meet the -10 dB requirement in most resonance frequencies that are 2.27, 7.505, 9.34, 10.33, and 11.61 GHz. An important feature of the proposed antenna is capable of impedance matching at five resonance frequencies by cutting one by one quad C-slots without changing the feeding point and any other dimensions on the disk patch antenna.

It is observed that the proposed antenna has multifrequency characteristics. This happens due to the slots inserted on the radiating patch. In the first case, first C-slot is cut on the radiating patch which gives two resonating frequencies. In the second case, second C-slot loaded to the patch which produces third resonating frequency. Thereafter in the next two steps, three and four C-slots are inserted and the four and five resonance frequency bands were observed. Thus, the slots play vital role in obtaining multiband.

The multiband nature of the proposed structure is found to be confirmed by the fabricated results with some errors within the acceptable limits.

Apparently, the measured return loss below −10 dB band-widths ranges from 2.17 to 2.41 GHz, 7.38 to 7.64 GHz, 9.21 to 9.485 GHz, 10.19 to 10.45 GHz, and 11.46 to 11.75 GHz with the relative bandwidth of 10.45%, 3.46%, 2.94%, 2.51%, and 2.498%, respectively, which show the approximate agreement with the simulated results. The differences may be due to the effect of the SMA connector and mismatching tolerance.

The first resonance frequency band operates under the S-band, second resonance frequency band operates at C-band, and third, fourth, and fifth resonance frequency bands operate at X-band.

4. Parametric Study and Discussion

Figure 4 shows the variation of S_{11} versus frequency for different values of radius (R) of the quad C-slots loaded disk patch antenna. When other parameters of the antenna W = 0.5 mm, L_1 = 11.0 mm, L_2 = 2.0 mm, L_3 = 2.0 mm, L_4 = 5.0 mm, and L_5 = 5.0 mm are fixed, it is observed that the resonance frequencies are shifting towards the lower frequency side with increasing the value of radius and corresponding bandwidths are 5.865% (3.31–3.51 GHz), 2.73% (4.34–4.46 GHz), 3.0% (7.53–7.76 GHz), 2.72% (9.05–9.3 GHz), and 2.06% (11.53–11.77 GHz). On the other hand, the resonance frequencies are shifting towards the upper frequency side with decreasing the value of the radius of the proposed antenna and corresponding bandwidths are 3.11% (3.48–3.59 GHz), 4.09% (4.31–4.49 GHz), 3.06% (7.72–7.96 GHz), 2.63% (9.38–9.63 GHz), and 1.48% (12.04–12.22 GHz). It is also observed that when the resonance frequencies are shifting to the lower and upper side, there are some changes in impedance bandwidths.

The variation of S_{11} versus frequency for different values of the substrate thickness is shown in Figure 5, when other parameters R = 15.0 mm, W = 0.5 mm, L_1 = 11.0 mm, L_2 = 2.0 mm, L_3 = 2.0 mm, L_4 = 5.0 mm, and L_5 = 5.0 mm are fixed of the antenna.

It is observed that when the value of substrate thickness increases, the resonances are shifting to the upper frequency side and corresponding bandwidths are 9.52% (2.5–2.75 GHz), 5.95% (4.4–4.67 GHz), 3.26% (7.84–8.1 GHz), 3.02% (9.45–9.74 GHz), and 1.879% (12.12–12.36 GHz). Similarly, when the value of substrate thickness decreases, the resonance is shifting to the lower frequency side and corresponding bandwidths are 4.37% (4.02–4.2 GHz), 3.73% (7.35–7.63 GHz), 3.05% (9.04–9.32 GHz), 1.064% (10.28–10.39 GHz), and 2.475% (11.57–11.86 GHz).

Figure 6 shows the variation of S_{11} versus frequency for different values of relative permittivity (ε_r), when other

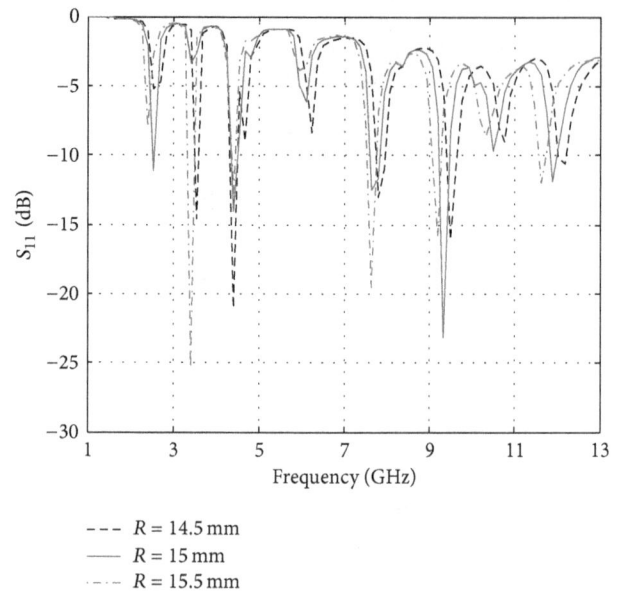

FIGURE 4: Simulated variation of S_{11} versus frequency for different values of radius of the disk (R).

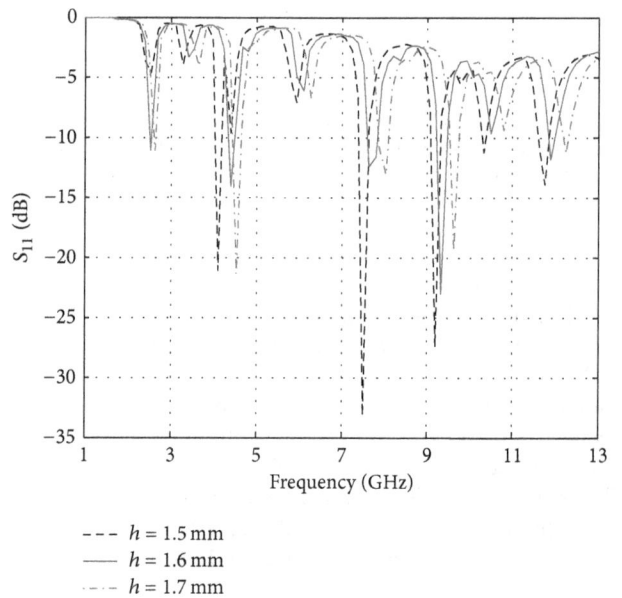

--- R = 14.5 mm
——— R = 15 mm
–·– R = 15.5 mm

FIGURE 5: Simulated variation of S_{11} versus frequency for different values of substrate thickness (h).

--- h = 1.5 mm
——— h = 1.6 mm
–·– h = 1.7 mm

parameters R = 15.0 mm, h = 1.6 mm, W = 0.5 mm, L_1 = 11.0 mm, L_2 = 2.0 mm, L_3 = 2.0 mm, L_4 = 5.0 mm, and L_5 = 5.0 mm are fixed of the antenna. Figure shows the multiband behaviour when dielectric constant is high, that is, 4.4, which gives the five operating bands with bandwidths that are 9.07% (2.42–2.65 GHz), 5.675% (4.28–4.53 GHz), 6.60% (7.32–7.82 GHz), 2.99% (9.2–9.48 GHz), and 2.09% (11.78–12.03 GHz). The rest of dielectric values 2.3 and 1.07 give the single and dual band operation, respectively.

Figure 7 shows the frequency versus gain plot for proposed antenna. From this figure, the measured and simulated

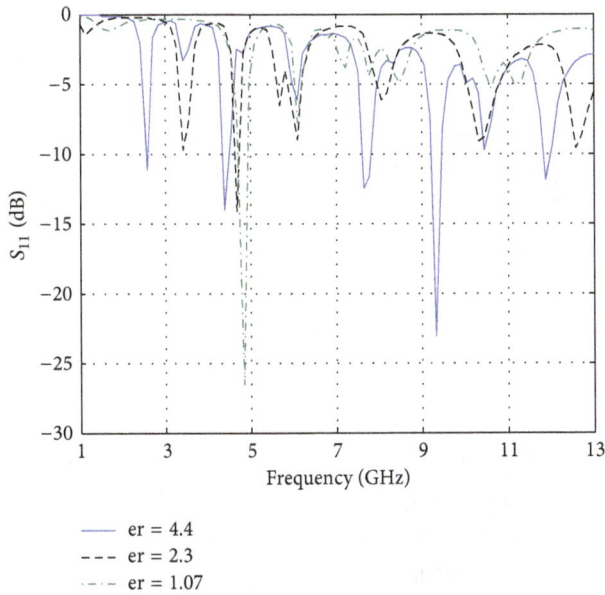

FIGURE 6: Simulated variation of S_{11} versus frequency for different values of substrate material (er).

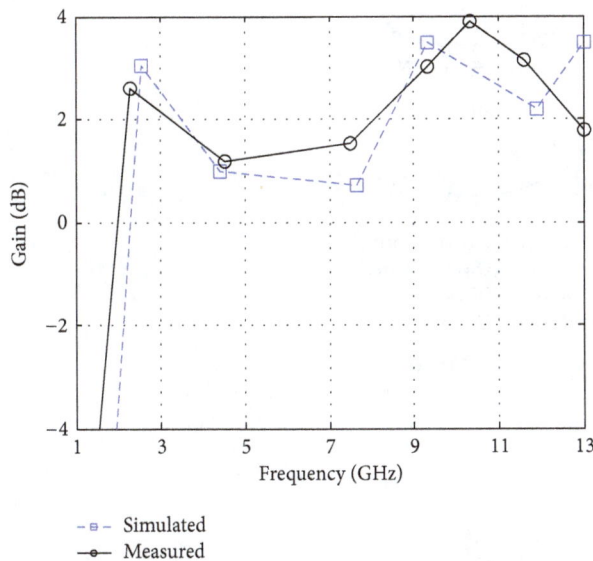

FIGURE 7: Simulated and measured gain at the various frequencies.

gain values at all of the frequencies are shown in figure. The antenna gains are 2.598, 1.518, 3.015, 3.877, and 3.121 dBi at the five resonant frequencies that are 2.27, 7.505, 9.34, 10.33, and 11.61 GHz, respectively. The simulated gains are in good agreement with measured results. Most of the gain variations observed here are less than 1 dB and these slight differences of antenna gains can be attributed to the effects of conductor and dielectric loss.

The comparative plot of simulated and measured antenna efficiency for the proposed antenna is shown Figure 8. From this figure, the simulated and measured results of the antenna efficiency are in good agreement with each other. It is

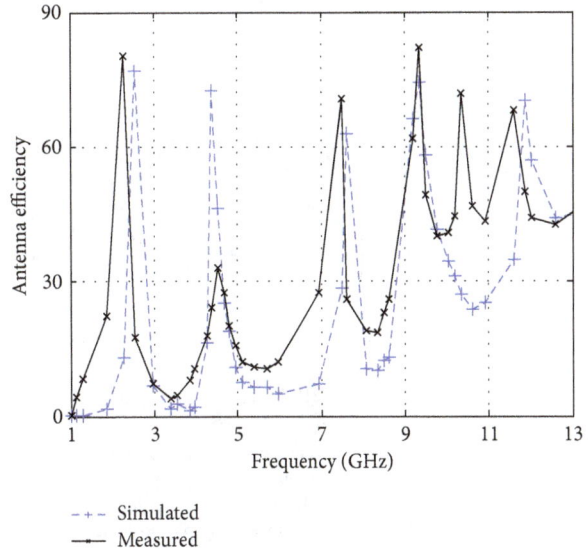

FIGURE 8: Simulated and measured antenna efficiency at the various frequencies.

observed that efficiencies of proposed C-slot loaded disk patch antenna are 80.2%, 70.67%, 81.91%, 71.92%, and 68.15% at five resonance frequencies that are 2.27, 7.505, 9.34, 10.33, and 11.61 GHz, respectively. The radiation pattern of the proposed multiband microstrip antenna is also investigated.

Figure 9 shows the measured and simulated radiation pattern of the antenna at five resonance frequencies 2.27, 7.505, 9.34, 10.33, and 11.61 GHz. It is observed that measured and simulated results are in close agreement. Results are plotted for E_θ, $\varphi = 0°$ in x-z plane, and E_θ, $\varphi = 90°$ in x-y plane. Higher order modes are considered because higher order harmonics does not decay much as it can be observed from Figures 9(c)–9(e), which is for 9.34, 1033, and 11.61 GHz, respectively.

5. Conclusion

The characteristics of a multiband patch antenna with symmetrically quad C-slots have been proposed and verified with simulation and measurement. From the analysis, it is clear that the performance of proposed antenna depends on substrate thickness, radius of the disk patch, and different dielectric material. In proposed antenna, the C- and inverted C-slots play an important role in controlling resonant frequencies and bandwidth of the antenna and achieving the multiple resonant operations. The proposed antenna can provide sufficient bandwidths along with the moderate gain. This antenna is capable of satisfying the requirements of S-, C-, and X-frequency band.

Conflict of Interests

The authors declare that there is no conflict of interests regarding the publication of this paper.

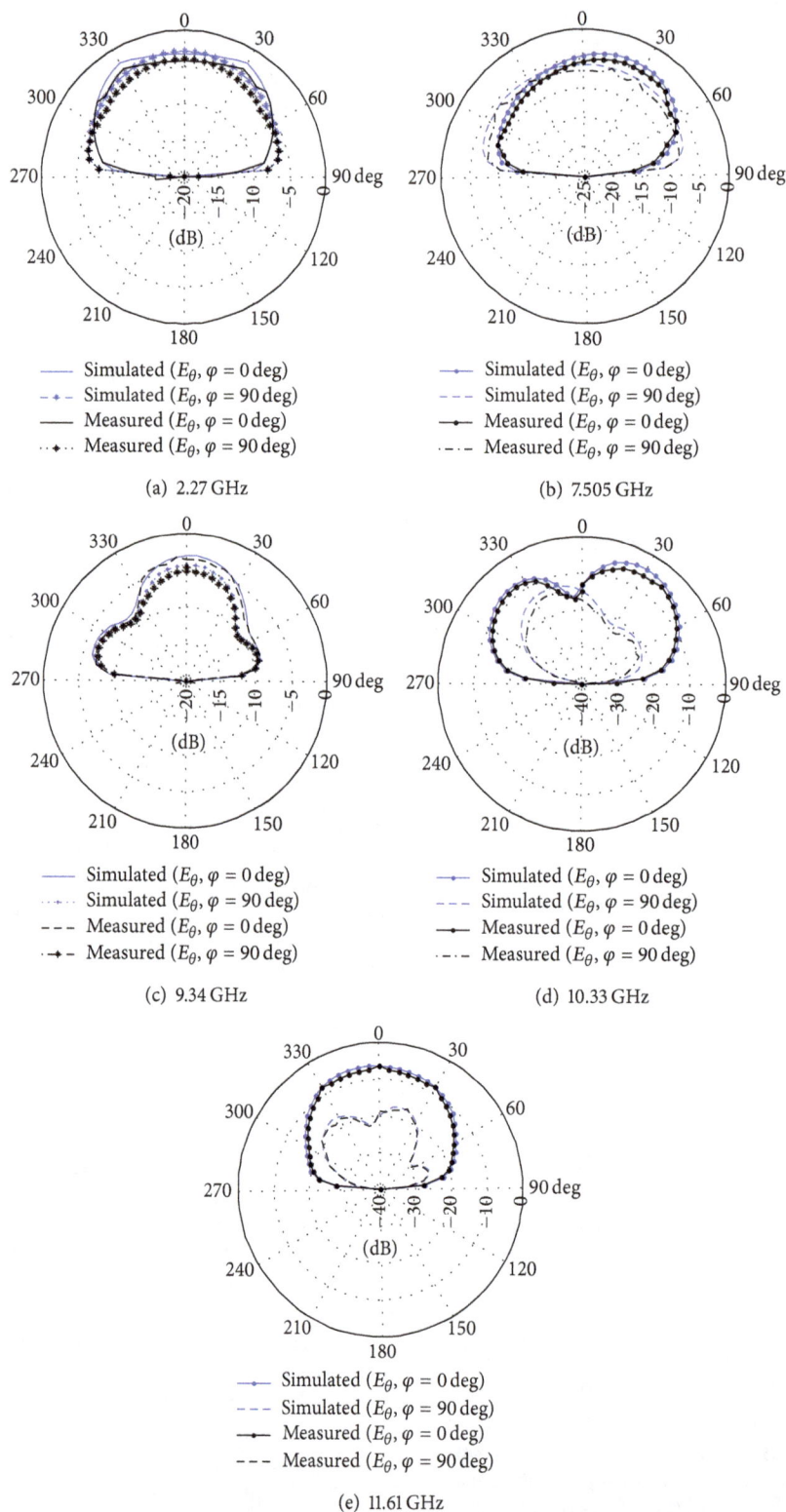

(a) 2.27 GHz

(b) 7.505 GHz

(c) 9.34 GHz

(d) 10.33 GHz

(e) 11.61 GHz

FIGURE 9: Simulated and measured radiation pattern at the various frequencies.

Acknowledgment

Sapna Verma is also grateful to University Grant Commission (UGC), India, for providing financial assistance (SRF).

References

[1] D. N. Elsheakh, H. A. Elsadek, E. A. Abdallah, M. F. Iskander, and H. Elhenawi, "Reconfigurable single and multiband inset feed microstrip patch antenna for wireless communication devices," *Progress in Electromagnetics Research C*, vol. 12, pp. 191–201, 2010.

[2] S. Verma, J. A. Ansari, and M. K. Verma, "A novel compact multi-band microstrip antenna with multiple narrow slits," *Microwave and Optical Technology Letters*, vol. 55, no. 6, pp. 1196–1198, 2013.

[3] L. Xu, Z. Y. Xin, and J. He, "A compact triple-band fork-shaped antenna for WLAN/WiMAX applications," *Progress in Electromagnetics Research Letters*, vol. 40, pp. 61–69, 2013.

[4] L. Dang, Z. Y. Lei, Y. J. Xie, G. L. Ning, and J. Fan, "A compact microstrip slot triple-band antenna for WLAN/WiMAX applications," *IEEE Antennas and Wireless Propagation Letters*, vol. 9, pp. 1178–1181, 2010.

[5] J. P. Thakur, J.-S. Park, B.-J. Jang, and H.-G. Cho, "Small size quad band microstrip antenna," *Microwave and Optical Technology Letters*, vol. 49, no. 5, pp. 997–1001, 2007.

[6] J. Pei, A.-G. Wang, S. Gao, and W. Leng, "Miniaturized triple-band antenna with a defected ground plane for WLAN/WiMAX applications," *IEEE Antennas and Wireless Propagation Letters*, vol. 10, pp. 298–301, 2011.

[7] X. Sun, G. Zeng, H.-C. Yang, Y. Li, X.-J. Liao, and L. Wang, "Design of an edge-fed quad-band slot antenna for GPS/ WIMAX/WLAN applications," *Progress in Electromagnetics Research Letters*, vol. 28, pp. 111–120, 2012.

[8] S. K. Gupta, A. Sharma, B. K. Kanaujia, S. Rudra, R. R. Mishra, and G. P. Pandey, "Orthogonal slit cut stacked circular patch microstrip antenna for multiband operations," *Microwave and Optical Technology Letters*, vol. 55, no. 4, pp. 873–882, 2013.

[9] Zeland Software, *IE3D Simulation Software, Version 14.05*, Zeland Software, 2008.

[10] "HFSS simulator version 12," Ansoft Corporation, Pittsburg, Pa, USA.

[11] Y. I. Huang and K. Boyle, *Antenna from Theory to Practice*, John Wiley & Sons, 2008.

[12] M. Kominami, D. M. Pozar, and D. H. Schaubert, "Dipole and slot elements and array on semi-infinite substrate," *IEEE Transactions on Antennas and Propagation*, vol. 33, no. 6, pp. 600–607, 1985.

An Effective Math Model for Eliminating Interior Resonance Problems of EM Scattering

Zhang Yun-feng,[1,2] Zhou Zhong-shan,[1,2] Su Zhi-guo,[2] Wang Rong-zhu,[2] and Chen Ze-huang[2]

[1]*Collaborative Innovation Center on Forecast and Evaluation of Meteorological Disasters, Nanjing University of Information Science & Technology, Nanjing 210044, China*
[2]*Key Laboratory for Aerosol-Cloud-Precipitation of China Meteorological Administration, Nanjing University of Information Science, No. 219, Ningliu Road, Nanjing 210044, China*

Correspondence should be addressed to Zhang Yun-feng; zhangyunfeng@nuist.edu.cn

Academic Editor: Giancarlo Bartolucci

It is well-known that if an E-field integral equation or an H-field integral equation is applied alone in analysis of EM scattering from a conducting body, the solution to the equation will be either nonunique or unstable at the vicinity of a certain interior frequency. An effective math model is presented here, providing an easy way to deal with this situation. At the interior resonant frequencies, the surface current density is divided into two parts: an induced surface current caused by the incident field and a resonance surface current associated with the interior resonance mode. In this paper, the presented model, based on electric field integral equation and orthogonal modal theory, is used here to filter out resonant mode; therefore, unique and stable solution will be obtained. The proposed method possesses the merits of clarity in concept and simplicity in computation. A good agreement is achieved between the calculated results and those obtained by other methods in both 2D and 3D EM scattering.

1. Introduction

Electric field integral equation (EFIE) and magnetic field integral equation (MFIE) have been widely employed to analyze electromagnetic scattering of conducting bodies [1–3]. However, interior resonance phenomena exist in solving electromagnetic scattering problems with surface integral equations. When working frequencies of conductors are near (or exactly at) to the frequencies associated with interior resonances, the single equation will become highly ill-conditioned (or singular) which makes the solution unstable or nonunique. Also, the interior resonance behavior has significant influence on the late time stability associated with time domain EFIE and MFIE [4, 5].

Several ways of dealing with this numerical problem have been proposed. Nowadays, the popular combined field integral equation (CFIE) technique to overcome this problem is a proper combination of the electric field integral equation and the magnetic field equation [6, 7]. The CFIE technique requires the calculation of both E and H impedance matrices and it is not suitable for aperture problems. The combined source integral equation (CSIE) [8, 9] technique makes up for aperture structures. Also a technique has been proposed by Mittra and Klein [10], involving application of the generalized boundary condition [11], and consists of additional points in the interior of the conductor and forces the field to be zero at those points. The problem with this technique is the fact that the chosen interior points must be carefully selected so as not to lie on nodal lines, which is not too practical for large bodies of simple shape, for which a slight change in frequency can take us from one resonance to the next, of different modal distribution. There are also some iterative methods reported in the works of Sarkar and Ergül [12, 13] to deal with this situationand they are used to compute the minimum norm solution (which produces the correct scattered fields but not the true tangential fields) and to calculate the LQSR. Another work related to the use of extended integral equations has been presented by Mautz and Harrington [8], which involves application of the boundary element method, with observation points lying on an internal closed surface near the boundary of the scatterer. Unfortunately, the resulting matrix equations are also ill-conditioned since the internal contour can resonate by itself. The authors' proposed scheme of allowing the internal contour to vary with the wave number seems impractical and in reality does not solve the problem except in some isolated cases of electrically small simple shapes whose resonances are known. More recently, Canning [14] illustrated a matrix algebra technique known as the singular value decomposition (SVD) has been proposed for moment method calculations involving perfect conductors. Such a technique diagonalizes the matrix equation, isolating the resonant contribution, which is then omitted in the calculation. We also refer to the interesting work of Yaghjian, who originally presented his augmented electric or magnetic field equation [15], and more recently Tobin et al. [16] pointed out their drawbacks for an arbitrarily shaped, multiwavelength body and, most importantly, introduced a modification, the so-called dual-surface integral equation, which is applicable to the perfectly conducting bodies and supposedly eliminates all the spurious solutions. Finally, the modal orthogonal characteristics [17, 18] were applied to solve the same problem in two-dimensional EM scattering.

Here, we present an effective method to solve the scattering problem of conductor bodies at or in the neighborhood of the resonant frequencies in both 2D and 3D EM scattering. At the resonant frequencies, Inagaki modes, firstly applied to analyze the antenna array [19], are employed here to be validated away from resonance with a unique and stable solution to the equation. It can both stabilize the numerical calculation and yield to reliable results of the surface current density and exterior field for conductors at interior resonances.

2. Theory

The principle is elaborated as follows. The solution of anill-conditioned system of the equation will consist of (1) the correct physical solution to the problem and (2) the resonant solution, which is in conjunction with Green's Theorem which produces the nonzero complementary resonant modes. When the orthogonal modes are used to solve the electric field integral equation, the solutions will be divided into the induced modes and the resonant modes corresponding to eigenvalues. We can easily obtain the current density and exterior field from the induced modes.

2.1. Why Is the Solution Unstable? Consider a PEC scatterer excited by an incident wave and the scatterer defined by the surface S, in an impressed electric field $\overline{E}^s$ (see Figure 1).

According to the boundary condition, we get

$$\hat{n} \times \left(\overline{E}^i + \overline{E}^s \right) = 0, \quad \text{on } S. \tag{1}$$

An operator equation for the current J on S is

$$T\left(\overline{J}\right) = \hat{n} \times \overline{E}^i, \quad \text{on } S. \tag{2}$$

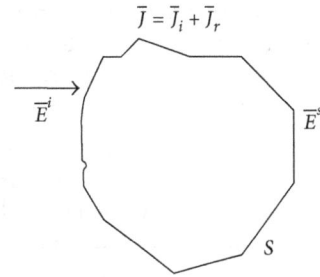

FIGURE 1: PEC scatterer impressed by electromagnetic field.

The operator T is defined by [2]

$$T\left(\overline{J}\right) = j\omega\mu\hat{n} \times \int_S \left[\overline{J}\left(\overline{r}'\right) + \frac{1}{k^2}\nabla' \cdot \overline{J}\left(\overline{r}'\right)\nabla \right] G\left(\overline{r},\overline{r}'\right) ds', \tag{3}$$

where $\overline{r}$ denotes a field point, $\overline{r}'$ denotes a source point, and ω, μ, and k denote angular frequency, permeability, and wave number, respectively, of free space; $G(\overline{r},\overline{r}')$ is the free space Green's function,

$$G\left(\overline{r},\overline{r}'\right) = \frac{\exp\left(-jk\left|\overline{r}-\overline{r}'\right|\right)}{4\pi\left|\overline{r}-\overline{r}'\right|}. \tag{4}$$

Equation (2) is discretized into a set of simultaneous linear algebraic equations. This set of equations can be compactly rewritten in a matrix form as

$$[S][J] = [Y], \tag{5}$$

where $[S]$ is the general impedance matrix, $[Y]$ denotes the general voltage vector, and $[J]$ stands for the coefficient of the current density to be determined. Usually (noninterior resonant condition) the current density of the conductor can be obtained from (5) with a unique and stable solution, but, at the frequencies associated with interior resonances, the situation will be on the contrary.

As a result of interior resonance, the current density $\overline{J}$ on S at discrete resonant frequencies will be composed of two parts: an induced surface caused by the incident field and a resonance surface current associated with the interior resonance mode:

$$\overline{J} = \overline{J}_i + \overline{J}_r, \tag{6}$$

where $\overline{J}_r$ is the resonant current density which is determined by the conducting body alone and it is independent of the incident wave; $\overline{J}_i$ is the induced current density which is jointly determined by the incident field and the conducting body itself.

If the incident $\overline{E}^i$ is removed, the electric field $\overline{E}^s$ produced by the resonant current $\overline{J}_r$ on S satisfies the following homogeneous E-field integral equation:

$$T\left(\overline{J}_r\right) = 0, \quad \text{on } S. \tag{7}$$

In other words, the solution to the inhomogeneous EFIE is not unique since the non-zero- solution to its corresponding homogeneous equation exits at some discrete frequencies. Namely, in addition to the induced current $\bar{J}_i$ determined jointly by the incident field $\bar{E}^i$ and the conducting body, there exists also the resonant current $\bar{J}_r$ at some discrete frequencies determined alone by the conducting body itself. It follows from the nonuniqueness of the solution to EFIE $T(\bar{J}) = \hat{n} \times \bar{E}^i$ that the solution to its corresponding moment matrix equation (5) will also not be unique at the interior resonant condition. The nonuniqueness of the moment matrix equation infers that the moment matrix $[S]$ is singular.

Similar to the behavior of a conducting cavity, the resonant current density $\bar{J}_r$ on a conducting body is also not able to produce scattered field $\bar{E}^s$ in space external to S. Therefore, the scattered field and the radar cross section (RCS) external to S determined by the electric field integral equation should be unique theoretically. Although resonant current theoretically does not change the exterior scattered field, it does make the method of moments analysis unstable and unreliable since the homogeneous operator equation $T(\bar{J}_r) = 0$ has non-zero-solution $\bar{J}_r$.

2.2. How to Stabilize the Solution? For the sake of simplicity, only the electric field integral equation is involved here. To solve the EFIE by method of moments, first of all, we should choose a set of expansion functions and a set of weighting functions. Here, we choose basis functions aiming at the diagonalization of the moment matrix which was proposed by Inagaki and Garbacz [19], in which the expansion functions $\{\bar{u}_n\}$ are chosen to be the eigenfunctions of a composite Hermitian operator T^*T, where T^* is the adjoint operator of T,

$$T^*T\,\bar{u}_n = \lambda_n \bar{u}_n, \tag{8}$$

the weighting functions to be the response of them

$$\bar{v}_n = T\,\bar{u}_n. \tag{9}$$

The orthogonal property of Inagaki modes [13] leads to the satisfaction of the moment matrix diagonalization condition; that is,

$$s_{mn} = \langle \bar{v}_m, T\,\bar{u}_n \rangle = \lambda_n \delta_{mn}; \tag{10}$$

namely,

$$[S] = \begin{bmatrix} \lambda_1 & 0 & \cdots & 0 \\ 0 & \lambda_2 & \cdots & 0 \\ \cdots & \cdots & \cdots & \cdots \\ 0 & 0 & \cdots & \lambda_N \end{bmatrix}. \tag{11}$$

So the modal solution to the operator equation becomes

$$\bar{J} = \bar{U}^T [S]^{-1} \bar{Y}. \tag{12}$$

Here $\bar{U}$ is the expansion function vector,

$$\bar{U} = [\bar{u}_1, \bar{u}_2, \ldots, \bar{u}_N], \tag{13}$$

where $\bar{Y}$ is a column excitation vector given by

$$y_n = \langle T\,\bar{u}_n, \hat{n} \times \bar{E}^i \rangle. \tag{14}$$

Then method of moments is employed here for computation of the modes. To do so, functions $f_1, f_2, \ldots, f_N$ are taken as both expansion functions and weighting functions for a moment method analysis to the operator eigenvalue equation, giving

$$u_n = \sum_{j=1}^{N} b_j^{(n)} f_j, \quad n = 1, 2, \ldots, N,$$

$$\sum_{j=1}^{N} b_j^{(n)} T^* T f_j = \lambda_n \sum_{j=1}^{N} b_j^{(n)} f_j, \tag{15}$$

$$\sum_{j=1}^{N} b_j^{(n)} \langle f_i, T^* T f_j \rangle = \lambda_n \sum_{j=1}^{N} b_j^{(n)} \langle f_i, f_j \rangle,$$

whose matrix form is

$$[Q]\,\vec{B}_n = \lambda_n [P]\,\vec{B}_n, \quad n = 1, 2, \ldots, N. \tag{16}$$

In this matrix eigenvalue equation, both $[P]$ and $[Q]$ are $N \times N$ matrices with their elements, respectively, as

$$p_{ij} = \langle f_i, f_j \rangle,$$

$$q_{ij} = \langle f_i, T^* T f_j \rangle = \langle T f_i, T f_j \rangle. \tag{17}$$

And $\vec{B}_n$ is a column vector composed of the unknown coefficients $\{b_j^{(n)}\}$ as

$$\vec{B}_n = \left[b_1^{(n)}, b_2^{(n)}, \ldots, b_N^{(n)} \right]^T. \tag{18}$$

The matrix $[Q]$ is simply the moment method matrix $[S]$ only if a least-squares method of moments is applied to the general operator equation in which $f_1, f_2, \ldots, f_N$ are taken to be the expansion functions and their responses $Tf_1, Tf_2, \ldots, Tf_N$ the weighting functions.

At the interior resonances, the surface current is composed of the resonant current and the induced current. The resonant current is the solution to the corresponding homogeneous E-field integral equation. It follows from the nonuniqueness of the solution to E-field integral equation that the solution to its corresponding moment matrix equation will also not be unique at the interior resonance condition. The nonuniqueness of the moment matrix equation infers that the moment matrix $[S]$ is singular.

Actually, the moment matrix will not be exactly singular, but rather, it will be highly ill-conditioned since the round-off and the truncated error exist during the computation of the matrix elements. According to the orthogonal property of Inagaki modes [13], the surface current of conductors is composed of serials of modal currents corresponding to the eigenvalue $\{\lambda_n\}$,

$$\bar{J} = \sum_{n=1}^{N} \bar{J}_n = \sum_{n=1}^{N} \frac{\langle T\,\bar{u}_n, \hat{n} \times \bar{E}^i \rangle \,\bar{u}_n}{\lambda_n}. \tag{19}$$

At the interior resonances, when eigenvalue λ_m is much smaller than the other eigenvalue, the corresponding modal current is resonant current and the other is the induced current. These modal currents are mutually orthogonal, so to eliminate the interior resonant mode in the Inagaki mode solution, what we need to do is just to replace the number $1/\lambda_m$ in $[S]^{-1}$ by zero. The incident current $\bar{J}_i$ can be acquired from the following equation:

$$\bar{J}_i = \sum_{n=1}^{N} \frac{\left\langle T\bar{u}_n, \hat{n} \times \bar{E}^i \right\rangle \bar{u}_n}{\lambda_n} - \frac{\left\langle T\bar{u}_m, \hat{n} \times \bar{E}^i \right\rangle \bar{u}_m}{\lambda_m}. \quad (20)$$

Theoretically, the scattered field and the radar cross section external to S determined by the electric field integral equation should be unique. However, due to the ill-condition of the moment matrix, the exterior fields from conductors will be unstable and unreliable. After filtering out the resonant current, the exterior field, obtained from the induced current, is stable and reliable.

In fact, the resonance current does not really exist on the surface of the conductors. The interior resonance problem is just caused by the deficiency of the selected mathematical model. After getting rid of the virtual resonance current, we can obtain the stable and reliable current density and exterior field of conductors at interior resonances.

3. Numerical Results

Here we confine our attention to not only two-dimensional structures but also three-dimensional structures. The presented method is applied here to get the surface current density and exterior field of conductors when they are at (or near) the interior resonant frequencies.

3.1. An Infinitely Long Circular Cylinder. As the first case for testing the approach described above, scattering from a circular conducting cylinder was examined here. It was found that when the TM plane wave normally illuminates an infinite circular cylinder, the interior resonance takesplace if $ka = 3.8214$, where k is the wave number and a is the radius, which is the numerical resonant frequency point of E_{11} mode of the same surface circular cylinder cavity.

From Figures 2 and 3, we can see that the magnitude of induced current is much smaller than that of resonant current, so the interior resonant current has completely masked the true current responsible for the scattered field. The true surface current, computed by the presented method, agrees well with the exact analytical solution. The bistatic RCS of the circular cylinder computed from the induced current is compared with the exact solution. It is clear that there are excellent agreements between the calculated results and the analytic results, as it is shown in Figure 4.

3.2. An Infinitely Long Square Cylinder. Also, an infinitely long square conducting cylinder was considered here with a TM wave incident along the axis of the cylinder. The first and the second resonant frequency of the square cylinder are, respectively, at $d/\lambda = \sqrt{2}/2$ and at $d/\lambda = \sqrt{5}/2$ theoretically.

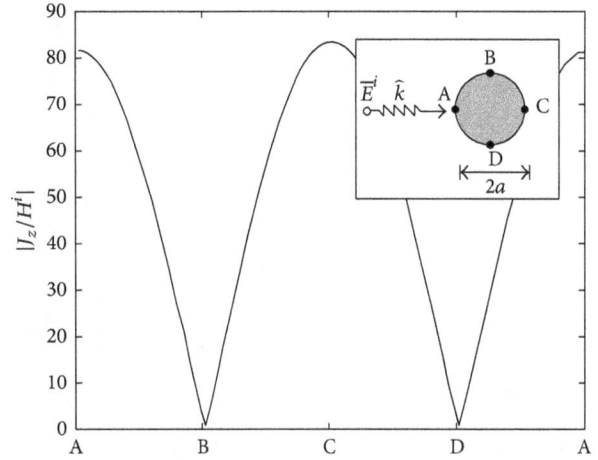

FIGURE 2: Resonant current on a circular cylinder, $ka = 3.8214$.

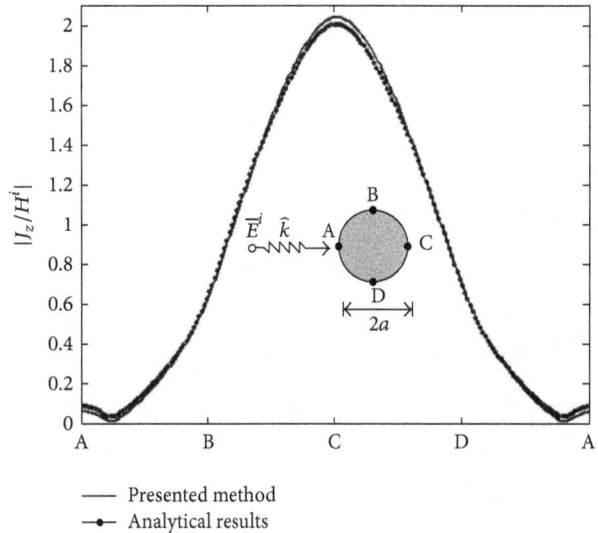

—— Presented method
—•— Analytical results

FIGURE 3: Induced current on a circular cylinder, $ka = 3.8214$.

We can get the first numerical resonant point at $d/\lambda = 0.70775$, using the presented method, as shown in Figure 5. Here, only the first resonant frequency is considered. At the first resonant frequency, the resonant current and induced current are shown, respectively, in Figures 6 and 7. From Figure 7, we can see that the induced current is compared well with that obtained by CFIE technique.

In Figure 8, the bistatic RCS of the square cylinder obtained through three different methods is depicted. Due to the ill-conditioned equation, the RCS of square cylinder computed by single EFIE (dash line) obviates the true result. After we filtered out the resonant modal current, the RCS (dot-dash line) produced by the induced current coincides with that obtained by CFIE method (solid line).

3.3. Two Conducting Spheres. In the end, two conducting spheres, respectively, at the $ka = 2.768$ and $ka = 4.518$, are analyzed by the presented method, where a denotes the radius

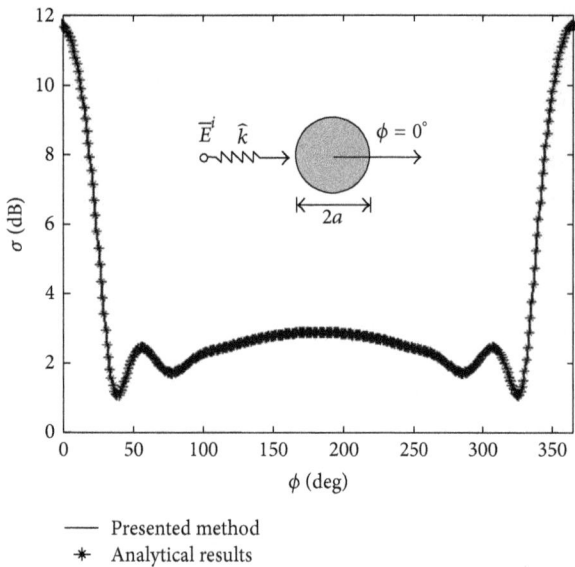

FIGURE 4: Bistatic RCS of a circular cylinder, $ka = 3.8214$.

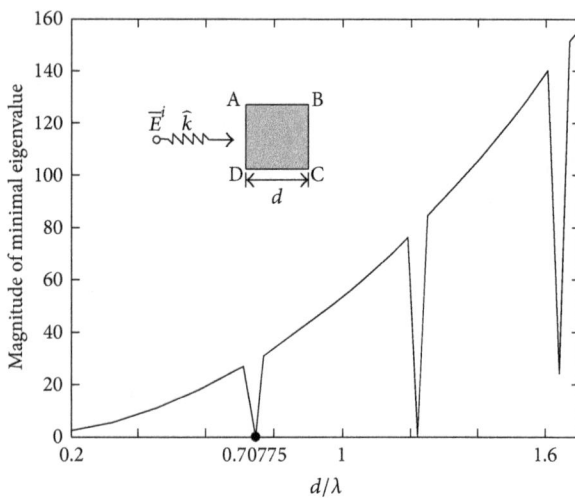

FIGURE 5: Minimal eigenvalue of square cylinder.

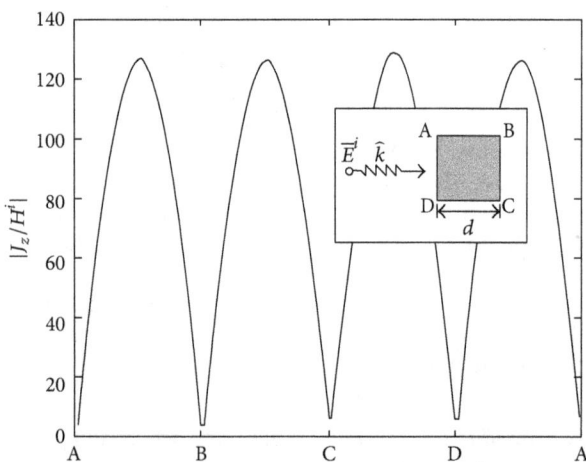

FIGURE 6: Resonant current on a square cylinder, $d/\lambda = 0.70775$.

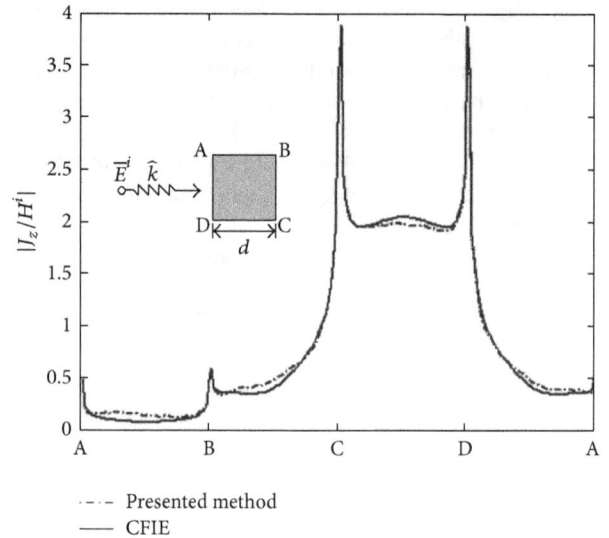

FIGURE 7: Induced current on a square cylinder, $d/\lambda = 0.70775$.

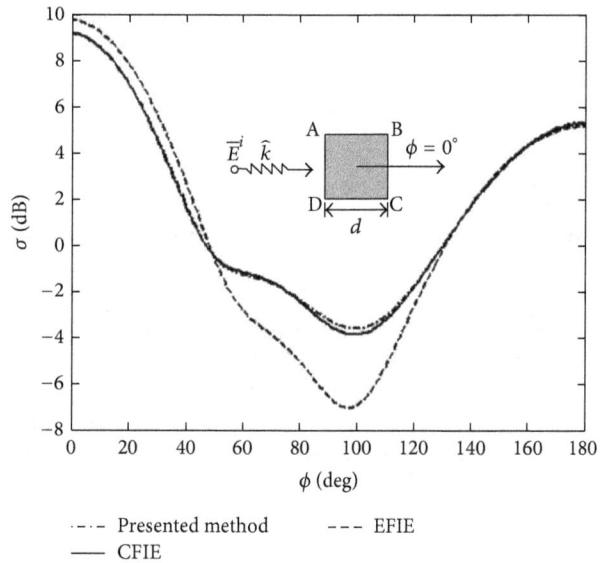

FIGURE 8: Bistatic RCS on square cylinder, $d/\lambda = 0.70775$ obtained by presented method (labeled IM), CFIE method, and EFIE.

of the sphere. At the interior resonant frequency, the bistatic RCS, obtained by the single EFIE method (dash line), is away from the right value. After being corrected by the presented method (dot-dash line), the bistatic RCS coincides with that calculated by CFIE technique (solid line), as shown in Figures 9 and 10.

4. Conclusion

A new scheme for eliminating interior resonance problems associated with surface integral equation is presented. The orthogonal property of Inagaki modes is used here to isolate the resonant mode, which is then omitted in the computation to obtain the right property of the conductors at the interior

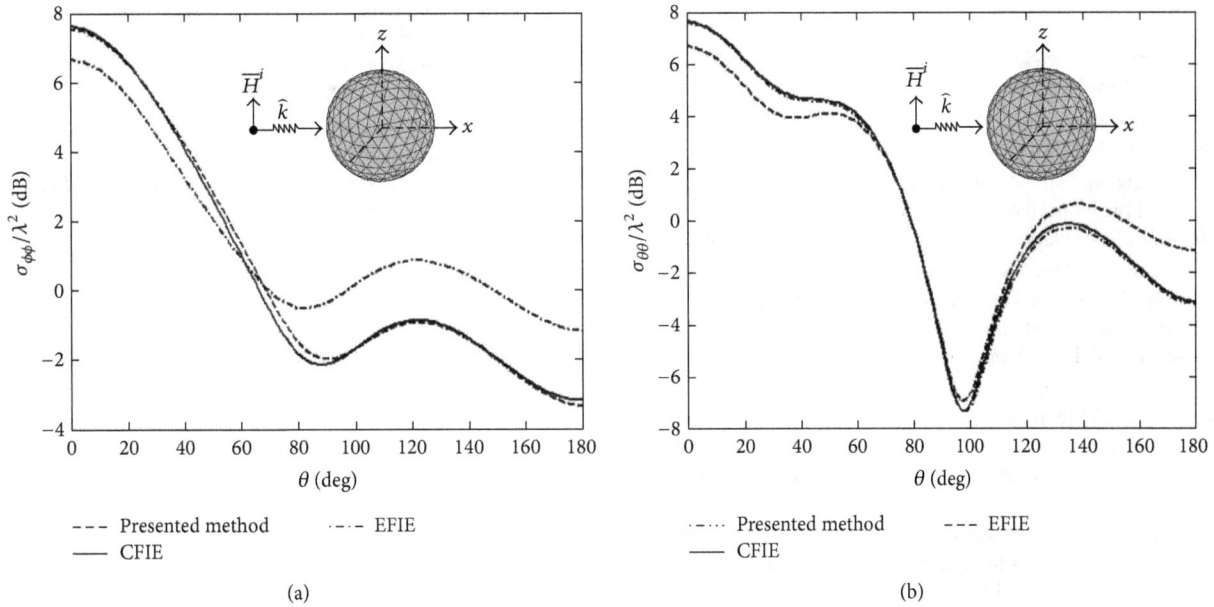

FIGURE 9: Bistatic RCS of a resonant sphere, $ka = 2.768$ (a) $\phi\phi$-polarization, (b) $\theta\theta$-polarization obtained by presented method (labeled IM), CFIE method, and EFIE.

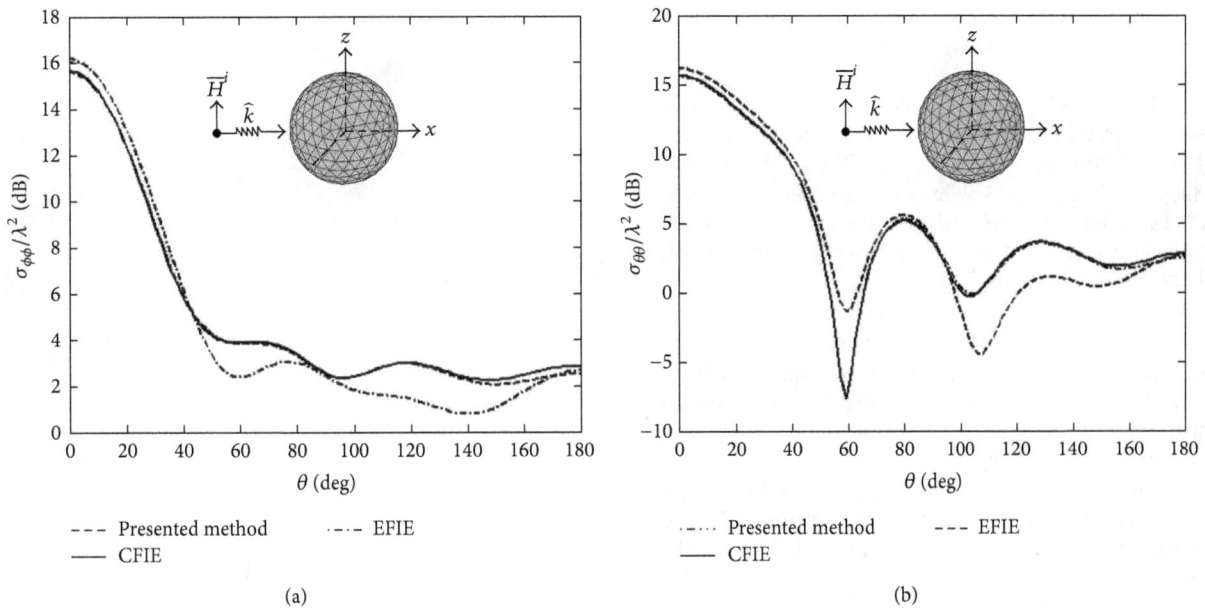

FIGURE 10: Bistatic RCS of a resonant sphere, $ka = 4.518$ (a) $\phi\phi$-polarization, (b) $\theta\theta$-polarization obtained by presented method (labeled IM), CFIE method, and EFIE.

resonances. This simple technique has been proven to be effective in attenuating the resonant modes and getting the unique and stable solution to EFIE when conductors are at (or near) the frequencies associated with interior resonances. Excellent numerical results, away from resonance problems, have been obtained for some shapes not only in two dimensions but also in three dimensions. Compared with other techniques, the advantages of this approach are only involved EFIE equation and easy to get the right results, but,

due to the determination of eigenvalues and eigenvectors, there is a bit time-consuming for property computation at or in the interior resonance.

Conflict of Interests

The authors declare that there is no conflict of interests regarding the publication of this paper.

References

[1] R. F. Harrington, *Field Computation by Moment Methods*, Macmillan, New York, NY, USA, 1968.

[2] R. F. Harrington, *Time Harmonic Electromagnetic Field*, McGraw-Hill, New York, NY, USA, 1968.

[3] W. C. Chew and J. M. Song, "Gedanken experiments to understand the internal resonance problems of electromagnetic scattering," *Electromagnetics*, vol. 27, no. 8, pp. 457–471, 2007.

[4] H. Lianrong, G. Tian, J. Fang, and G. Xiao, "The behavior of MFIE and EFIE at interior resonances and its impact in MOT late time stability," *American Journal of Electromagnetics and Applications*, vol. 1, no. 2, pp. 30–37, 2013.

[5] F. P. Andriulli, K. Cools, F. Olyslager, and E. Michielssen, "Time domain Calderón Identities and their application to the integral equation analysis of scattering by PEC objects part II: stability," *IEEE Transactions on Antennas and Propagation*, vol. 57, no. 8, pp. 2365–2375, 2009.

[6] J. R. Mautz and R. F. Harrington, "H-field, E-field and combined-field solutions for conducting bodies of revolution," *AEU*, vol. 32, pp. 157–164, 1978.

[7] F. P. Andriulli and E. Michielssen, "A regularized combined field integral equation for scattering from 2-D perfect electrically conducting objects," *IEEE Transactions on Antennas and Propagation*, vol. 55, no. 9, pp. 2522–2529, 2007.

[8] J. R. Mautz and R. F. Harrington, "A combined-source solution for radiation and scattering from a perfectly conducting body," *IEEE Transactions on Antennas and Propagation*, vol. 27, no. 4, pp. 445–454, 1979.

[9] K. F. A. Hussein, "Effect of internal resonance on the radar cross section and shield effectiveness of open spherical enclosures," *Progress in Electromagnetics Research*, vol. 70, pp. 225–246, 2007.

[10] R. Mittra and C. A. Klein, "Stability and convergence of moment method solutions," in *Numerical and Asymptotic Techniques in Electromagnetics*, vol. 3 of *Topics in Applied Physics*, pp. 129–163, Springer, Berlin, Germany, 1975.

[11] P. C. Waterman, "Matrix formulation of electromagnetic scattering," *Proceedings of the IEEE*, vol. 53, no. 8, pp. 805–812, 1965.

[12] T. K. Sarkar and S. M. Rao, "A simple technique for solving E-Field integral equations for conducting bodies at internal resonances," *IEEE Transactions on Antennas and Propagation*, vol. 30, no. 6, pp. 1250–1254, 1982.

[13] Ö. Ergül and L. Gürel, "Efficient solution of the electric-field integral equation using the iterative LSQR algorithm," *IEEE Antennas and Wireless Propagation Letters*, vol. 7, pp. 36–39, 2008.

[14] F. X. Canning, "Singular value decomposition of integral equations of EM and applications to the cavity resonance problem," *IEEE Transactions on Antennas and Propagation*, vol. 37, no. 9, pp. 1156–1163, 1989.

[15] A. D. Yaghjian, "Augmented electric and magnetic field equations," *Radio Science*, vol. 16, no. 6, pp. 987–1001, 1981.

[16] A. R. Tobin, A. D. Yaghjian, and M. M. Bell, "Surface integral equations for multiwavelength, arbitrarily shaped, perfectly conducting bodies," in *Proceedings of the URSI National Radio Science Meeting Digest*, pp. 12–15, Boulder, Colo, USA, January 1987.

[17] W. Cao and J. Chen, "The application of bi-orthogonal mode analysis approach to in electromagnetic scattering of conducting bodies at interior resonances," *Chinese Journal of Radio Science*, vol. 10, pp. 16–22, 1995.

[18] Y.-F. Zhang, C.-G. Jiang, and W. Cao, "Inagaki mode approach to electromagnetic scattering of conducting bodies at interior resonances," *Journal of Electronics and Information Technology*, vol. 28, no. 9, pp. 1735–1739, 2006.

[19] N. Inagaki and R. J. Garbacz, "Eigenfunctions of composite Hermitian operator with application to discrete and continuous radiating systems," *IEEE Transactions on Antennas and Propagation*, vol. 30, no. 4, pp. 571–575, 1982.

5

Novel Notched UWB Filter Using Stepped Impedance Stub Loaded Microstrip Resonator and Spurlines

Ramkumar Uikey,[1] **Ramanand Sagar Sangam,**[1]
Kakumanu Prasadu,[2] **and Rakhesh Singh Kshetrimayum**[1]

[1]*Department of Electronics & Electrical Engineering, IIT Guwahati, Assam 781039, India*
[2]*Ford Motor Pvt. Ltd., Dr. MGR Road, Perungundi, Chennai 600096, India*

Correspondence should be addressed to Rakhesh Singh Kshetrimayum; krs@iitg.ernet.in

Academic Editor: Giancarlo Bartolucci

This paper presents a novel ultrawideband (UWB) bandpass filter using stepped impedance stub loaded microstrip resonator (SISLMR). The proposed resonator is so formed to allow its four resonant frequencies in the UWB passband, which extends from 3.1 GHz to 10.6 GHz. Moreover, two spurline sections are employed to create a sharp notched-band filter for suppressing the signals of 5 GHz WLAN devices. Experimental results of the fabricated filters are in good agreement with the HFSS simulations and validate the design.

1. Introduction

Since the release of UWB spectrum by the Federal Communications Commission for unlicensed commercial applications in early 2002 [1], compact size UWB bandpass filters with good in-band transmission, sharp selectivity, and flat group delays are highly demanded to realize UWB radio systems. A number of methods have been reported in the literature [2–8] to design the UWB bandpass filters. An early method reported in [2] is based on cascaded low-pass-high-pass filter sections. A hybrid microstrip/CPW structure with back-to-back transition configuration is used in [3] to achieve the UWB bandpass response. In [4], a stepped impedance four-mode resonator is developed on the method of network analysis and optimization in Z-domain. Today, a major class of available UWB filters are based on the multiple-mode resonators (MMR) and are quite popular due to their easy design methods and simple structures. The concept of MMR based UWB filter was initially presented in [5]. In this work, a triple-mode MMR is integrated with dual-pole overenhanced parallel coupled lines to realize the UWB filter. This work is extended in [6] to get a more feasible filter with relaxed fabrication tolerance. In [7, 8], open ended stub loaded resonator based UWB filters are designed to widen the upper stopband.

In recent years, researchers are more attracted towards filters with notch bands embedded in the UWB passband. These notch bands are required to suppress the strong narrowband emissions in the WLAN and WiMAX bands which coincide with the 3.1 GHz to 10.6 GHz UWB spectrum. Many efforts have been put forward by researchers in [9–12], to design notched-band UWB filters. A Meander line slot is developed in [9] to reject the undesired IEEE 802.11a signals. Symmetrical pairs of defected ground structure with embedded open stubs are employed to create the WLAN notch [10]. In [11], five short circuited stubs are incorporated in the design to exhibit highly selective filtering characteristics. In our early work [12], we designed a notched-band bandpass filter using complementary single split ring resonators in the ground plane. All above mentioned filters possess good notch band filtering but they are either based on defected ground structures or suffer from fabrication difficulties due to via holes. Hence, emphasis is given to planar structures, which are free from via holes and defected ground structures.

In this paper, we are proposing a novel quad-mode stepped impedance stub loaded microstrip resonator

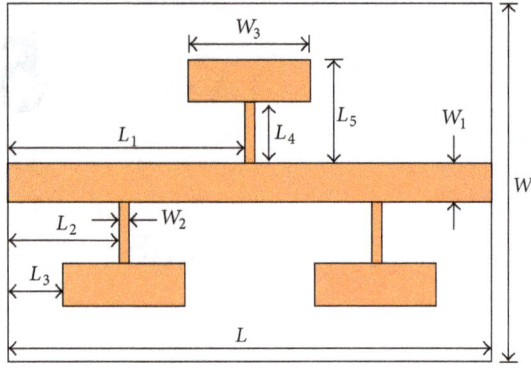

FIGURE 1: Basic structure of the proposed stepped impedance stub loaded UWB bandpass filter.

TABLE 1: Design parameter values for Figure 1.

$L = 38$ mm	$W = 32$ mm	FR4-epoxy substrate
$L_1 = 18.75$ mm	$W_1 = 3.1$ mm	$\epsilon_r = 4.4$
$L_2 = 7.75$ mm	$W_2 = 0.5$ mm	$h = 1.6$ mm
$L_3 = 3.5$ mm	$W_3 = 9.5$ mm	$t = 0.035$ mm
$L_4 = 6.5$ mm		$\tan\delta = 0.002$
$L_5 = 9.5$ mm		

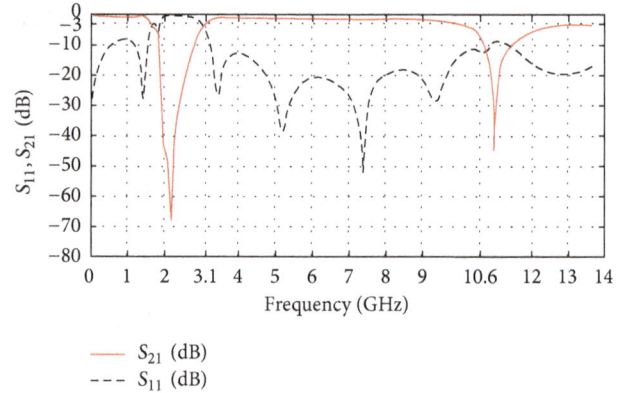

FIGURE 2: Simulated frequency response of the designed UWB bandpass filter.

(SISLMR) to design the UWB bandpass filters. It is constructed by loading a uniform $50\,\Omega$ transmission line with three symmetrical stepped impedance stubs, that is, one at the center and two at the symmetrical side locations. Then two symmetrical spurline sections are developed around the central stepped impedance stub to create a sharp notch band for suppressing the WLAN radio systems operating in the 5 GHz frequency bands. Finite element based Anosoft HFSS software is used for deriving the filter's electrical performance; later, two filter prototypes, one with notch band and one without notch band, are fabricated and measured for experimental verification of the predicted results.

2. Initial UWB Filter Design

Figure 1 depicts the schematic of the proposed resonator. It consists of a conventional transmission line resonator ($50\,\Omega$) of width W_1 and length $2L_1 + W_2$ in the horizontal plane and three vertically loaded stepped impedance stubs, that is, one stepped impedance stub of width W_2, W_3, length L_4, L_5 at the center and the other two stubs of the same dimensions at the symmetrical sides, located at a distance of about $\lambda_g/4$ from the central stub. In the initial design, we load a $50\,\Omega$ transmission line ($\approx 1.5\lambda_g$) with single stepped impedance at the center. When impedance ratio for this stepped impedance stub is set close to 0.2, we observe some UWB filtering characteristics with two transmission zeros near 2 and 11 GHz, respectively, and a pole near 6.5 GHz. Later, this initial resonator structure is modified by introducing two more stepped impedance stubs to have a higher degree of freedom in the design. A parametric analysis is performed using FEM based HFSS software and resonator parameters are optimized to achieve the UWB bandpass response. Optimized electrical parameters for the proposed SISLMR are given in Table 1 and HFSS simulated results are shown in Figure 2. Simulation results depict that the proposed resonator has four resonant frequencies in the desired passband, located at 3.43, 5.20, 7.40, and 9.38 GHz, respectively. Transmission zeros in the lower and upper stopband are located at 2.17 and 10.97 GHz, respectively. Designed filter demonstrates good UWB filtering with $|S_{21}| \geq$

-3 dB and $|S_{11}| \leq -10$ dB and flat group delay in the desired UWB spectrum.

2.1. Approximate Theoretical Analysis. This section describes an approximate theoretical analysis of the proposed resonator. It considers the case of lossless transmission lines and ignores the effects of step discontinuities, frequency dispersion, and edge capacitances at the open stubs.

Figure 3 shows the transmission line model of the proposed resonator. The overall ABCD matrix, $[R]$, in Figure 3 can be obtained by multiplying the ABCD matrices of the terminal lines, connecting lines between stepped impedance stubs and stepped impedance stubs in sequence; that is,

$$[R] = [A][B][C][B][C][B][A], \qquad (1)$$

where

$$[A] = \begin{bmatrix} \cos\left(\theta_1'\right) & \dfrac{j\sin\left(\theta_1'\right)}{Y_1} \\ jY_1 \sin\left(\theta_1'\right) & \cos\left(\theta_1'\right) \end{bmatrix},$$

$$[B] = \begin{bmatrix} 1 & 0 \\ jY_2 \dfrac{Y_2 \tan\left(\theta_3\right) + Y_1 \tan\left(\theta_2\right)}{Y_1 - Y_2 \tan\left(\theta_2\right)\tan\left(\theta_3\right)} & 1 \end{bmatrix}, \qquad (2)$$

$$[C] = \begin{bmatrix} \cos\left(\theta_1\right) & \dfrac{j\sin\left(\theta_1\right)}{Y_1} \\ jY_1 \sin\left(\theta_1\right) & \cos\left(\theta_1\right) \end{bmatrix}.$$

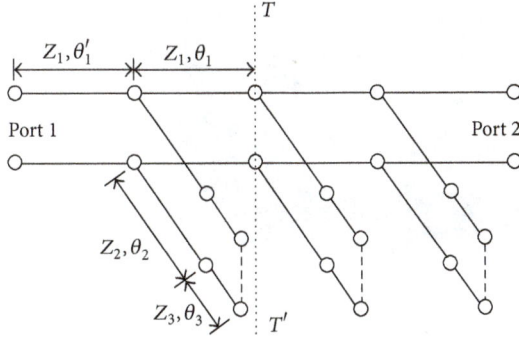

FIGURE 3: Transmission line model of the proposed UWB bandpass filter.

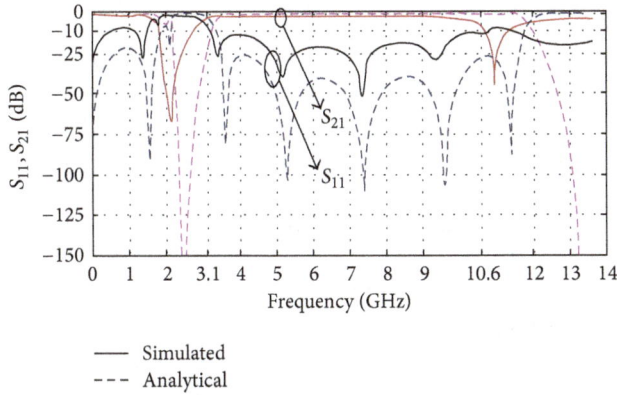

— Simulated
- - - Analytical

FIGURE 4: Comparison of simulated and analytical results.

Using (1) and matched load condition at port 2, $|S_{21}|$ (dB) and $|S_{11}|$ (dB) for the resonator in Figure 1 are deduced as

$$|S_{21}| = 20$$
$$\cdot \log\left(\frac{2\sqrt{Z_{01}/Z_{02}}}{R_{11} + R_{12}/Z_{02} + R_{21}Z_{01} + R_{22}(Z_{01}/Z_{02})}\right) \text{dB},$$

$$|S_{11}| = 20$$
$$\cdot \log\left(\frac{R_{11} + R_{12}/Z_{02} - R_{21}Z_{01} - R_{22}(Z_{01}/Z_{02})}{R_{11} + R_{12}/Z_{02} + R_{21}Z_{01} + R_{22}(Z_{01}/Z_{02})}\right) \text{dB}.$$

(3)

Z_{01} and Z_{02} in the above equations are source and load impedances, respectively. In Figure 4, analytical results of (3) are plotted in MATLAB and compared with the HFSS simulated results. This plot shows good match between the two results and further supports the validity of the proposed SISLMR for designing the UWB bandpass filters.

3. Realization of Notched-Band UWB Filter

UWB filter designed in the previous section is modified by introducing two symmetrical spurlines to create a sharp notch function in the UWB passband. Figure 5 shows the circuit of modified UWB filter. In the modified filter, all the parameters of the initial UWB filter are kept unchanged while dimensions of the two spurlines are set as length (a), gap (b),

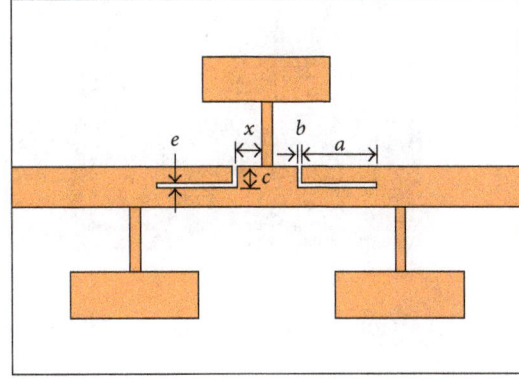

FIGURE 5: UWB filter with spurlines for realizing WLAN notch band.

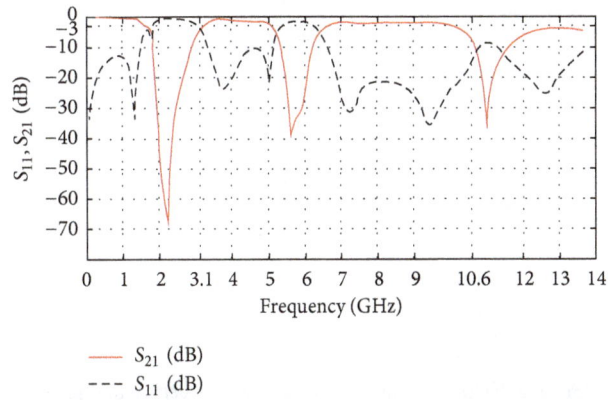

— S_{21} (dB)
- - - S_{11} (dB)

FIGURE 6: Simulated S-parameter of the notched-band UWB bandpass filter.

TABLE 2: Design parameter values for the spurlines.

$a = 7$ mm	$b = 0.3$ mm
$c = 1$ mm	$e = 0.3$ mm
$x = 3.1$ mm	

and height (c), respectively. These are etched symmetrically at a distance of x, around the central stepped impedance stub. Length (a) and gap (b) of spurlines play a key role in adjusting the stopband center frequency [13], and therefore these are properly optimized using HFSS software so that the resulting notched-band filter can completely suppress the interference from WLAN devices, operating in the 5 GHz band. Table 2 lists the optimized parameters of the proposed spurline sections.

Figure 6 shows the variation of HFSS simulated S-parameters with frequency for the notched-band UWB filter. This filter exhibits a 10 dB stopband from 5.15 to 6.6 GHz with minimum $|S_{21}|$ of −39 dB at 5.6 GHz and completely eliminates the 5 GHz WLAN bands. Furthermore, it has two 10 dB passbands in the desired UWB spectrum: the first passband covers a frequency range extending from 3.2 to 5.15 GHz while the second passband covers 6.6 to 10.8 GHz band.

FIGURE 7: Fabricated filters. (a) Initial UWB bandpass filter and (b) notched-band UWB filter.

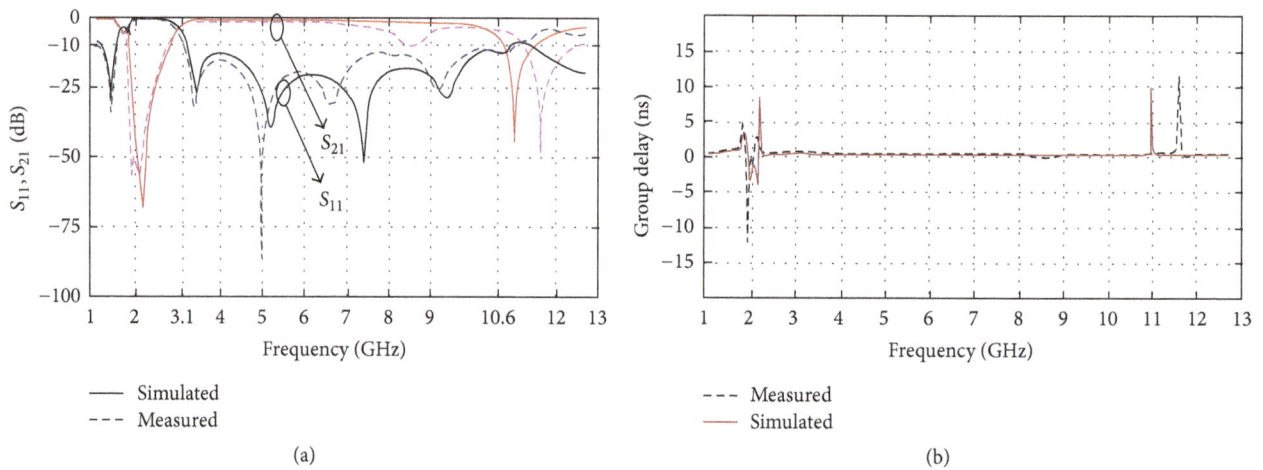

FIGURE 8: Comparison of simulated and experimental results for the initial UWB filter: (a) S-parameters and (b) group delay.

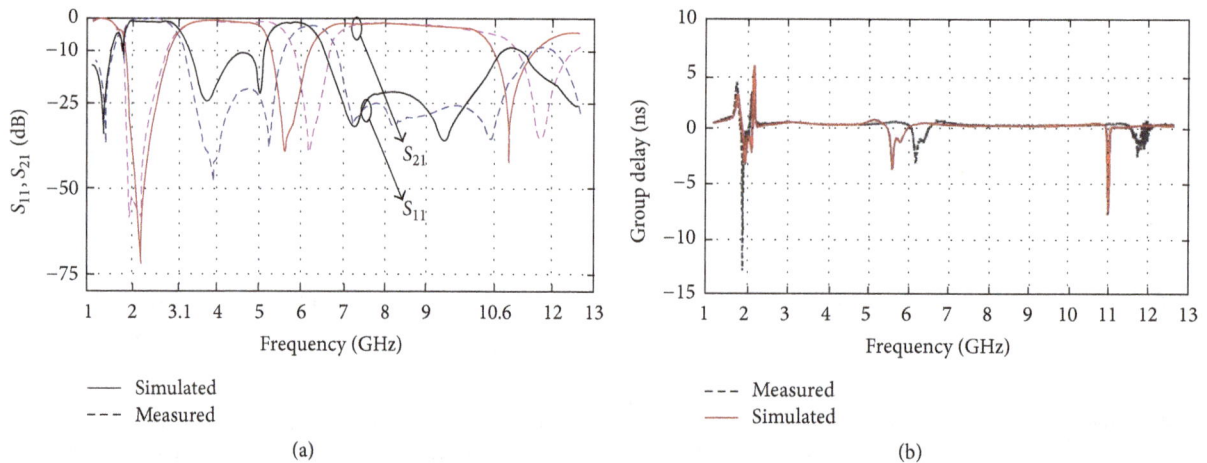

FIGURE 9: Comparison of simulated and experimental results for the notched-band UWB filter: (a) S-parameters and (b) group delay.

4. Comparison with Experimental Results

After deriving the optimized electrical parameters for the proposed UWB filters, the two filter prototypes are fabricated and measured using VNA for experimental verification. Figure 7 shows the photograph of fabricated filters. Simulated and measured S-parameters and group delay for the initially designed UWB filter and notch band UWB filter are plotted together in Figures 8 and 9, respectively, for quantitative comparison. These graphs show a good match between the two results. For the initial UWB filter, measured 10 dB passband extends from 3.1 to 11 GHz, while for the notched-band filter measured stopband extends from 5.5 to 6.8 GHz and the two passbands cover 3.15–5.5 GHz and 6.8–11.4 GHz bands, respectively. For both filters, maximum group delay variation is better than 0.7 ns. The two fabricated filters are compact in size and do not incorporate any defected ground structures or via hole connections. Some minor discrepancies are observed in the measured results which may be caused due to unexpected tolerances in fabrication and substrate parameters similar to what has been reported in [14].

5. Conclusion

In this paper, we proposed a novel stepped impedance stub loaded microstrip resonator (SISLMR) for designing the UWB bandpass filters. Initial UWB filter is designed by loading three double section stepped impedance stubs to the main $50\,\Omega$ microstrip line, and an approximate theoretical analysis is also presented. Later on, the required WLAN notch band is designed using the spurline method. After optimization of filter parameters in HFSS software, two compact size UWB filters were fabricated and measured for experimental verification. Measured and simulated results are in good agreement with each other and validate the design.

Conflict of Interests

The authors declare that there is no conflict of interests regarding the publication of this paper.

References

[1] Federal Communications Commission, "Revision of part 15 of the commission's rules regarding ultra-wideband transmission systems," Tech. Rep. ET-Docket 98-153, FCC02-48, Federal Communications Commission (FCC), Washington, DC, USA, 2002.

[2] C.-L. Hsu, F.-C. Hsu, and J.-T. Kuo, "Microstrip bandpass filters for Ultra-Wideband (UWB) wireless communications," in *Proceedings of the IEEE MTT-S International Microwave Symposium Digest*, pp. 679–682, IEEE, Long Beach, Calif, USA, June 2005.

[3] H. Wang, L. Zhu, and W. Menzel, "Ultra-wideband bandpass filter with hybrid microstrip/CPW structure," *IEEE Microwave and Wireless Components Letters*, vol. 15, no. 12, pp. 844–846, 2005.

[4] P. Cai, Z. Ma, X. Guan, Y. Kobayashi, T. Anada, and G. Hagiwara, "A novel compact ultra-wideband bandpass filter using a microstrip stepped-impedance four-modes resonator," in *Proceedings of the IEEE MTT-S International Microwave Symposium (IMS '07)*, pp. 751–754, IEEE, Honolulu, Hawaii, USA, June 2007.

[5] L. Zhu, S. Sun, and W. Menzel, "Ultra-Wideband (UWB) band-pass filters using multiple-mode resonator," *IEEE Microwave and Wireless Components Letters*, vol. 15, no. 11, pp. 796–798, 2005.

[6] L. Zhu and H. Wang, "Ultra-wideband bandpass filter on aperture-backed microstrip line," *Electronics Letters*, vol. 41, no. 18, pp. 1015–1016, 2005.

[7] R. Li and L. Zhu, "Compact UWB bandpass filter using stub-loaded multiple-mode resonator," *IEEE Microwave and Wireless Components Letters*, vol. 17, no. 1, pp. 40–42, 2007.

[8] Q.-X. Chu and S.-T. Li, "Compact UWB bandpass filter with improved upper-stopband performance," *Electronics Letters*, vol. 44, no. 12, pp. 742–743, 2008.

[9] G.-M. Yang, R. Jin, C. Vittoria, V. G. Harris, and N. X. Sun, "Small Ultra-wideband (UWB) bandpass filter with notched band," *IEEE Microwave and Wireless Components Letters*, vol. 18, no. 3, pp. 176–178, 2008.

[10] W. Zong, X. Zhu, C. You, and J. Wang, "Design and implement of compact UWB bandpass filter with a frequency notch by consisting of coupled Microstrip line structure, DGS and EOS," in *Proceedings of the International Conference on Advanced Technologies for Communications*, pp. 179–182, Haiphong, Vietnam, October 2009.

[11] H. Shaman and J.-S. Hong, "Ultra-wideband (UWB) bandpass filter with embedded band notch structures," *IEEE Microwave and Wireless Components Letters*, vol. 17, no. 3, pp. 193–195, 2007.

[12] S. S. Karthikeyan and R. S. Kshetrimayum, "Notched UWB bandpass filter using complementary single split ring resonator," *IEICE Electronics Express*, vol. 7, no. 17, pp. 1290–1295, 2010.

[13] R. N. Bates, "Design of microstrip spur-line band-stop filters," *IEE Journal on Microwaves, Optics and Acoustics*, vol. 1, no. 6, pp. 209–214, 1977.

[14] P. Cai, Z. Ma, X. Guan et al., "A compact UWB bandpass filter using two-section open-circuited stubs to realize transmission zeros," in *Proceedings of the Asia-Pacific Microwave Conference (APMC '05)*, vol. 5, Suzhou, China, December 2005.

6

Characterization of $Ni_xZn_{1-x}Fe_2O_4$ and Permittivity of Solid Material of NiO, ZnO, Fe_2O_3, and $Ni_xZn_{1-x}Fe_2O_4$ at Microwave Frequency Using Open Ended Coaxial Probe

Fahmiruddin Esa,[1,2] Zulkifly Abbas,[1] Fadzidah Mohd Idris,[3] and Mansor Hashim[3]

[1]*Department of Physics, Faculty of Science, Universiti Putra Malaysia, 43400 Serdang, Malaysia*
[2]*Faculty of Sciences, Technology and Human Development, Universiti Tun Hussein Onn Malaysia, 86400 Batu Pahat, Malaysia*
[3]*Advanced Materials and Nanotechnology Laboratory, Institute of Advanced Technology, Universiti Putra Malaysia, 43400 Serdang, Malaysia*

Correspondence should be addressed to Fahmiruddin Esa; fahmir@uthm.edu.my

Academic Editor: Samir Trabelsi

This paper describes a detailed study on the application of an open ended coaxial probe technique to determine the permittivity of $Ni_xZn_{1-x}Fe_2O_4$ in the frequency range between 1 GHz and 10 GHz. The x compositions of the spinel ferrite were 0.1, 0.3, 0.5, 0.7, and 0.9. The $Ni_xZn_{1-x}Fe_2O_4$ samples were prepared by 10-hour sintering at 900°C with 4°C/min increment from room temperature. Particles showed phase purity and crystallinity in powder X-ray diffraction (XRD) analysis. Surface morphology measurement of scanning electron microscopy (SEM) was conducted on the plane surfaces of the molded samples which gave information about grain morphology, boundaries, and porosity. The tabulated grain size for all samples was in the range of 62 nm–175 nm. The complex permittivity of Ni-Zn ferrite samples was determined using the Agilent Dielectric Probe Kit 85070B. The probe assumed the samples were nonmagnetic homogeneous materials. The permittivity values also provide insights into the effect of the fractional composition of x on the bulk permittivity values $Ni_xZn_{1-x}Fe_2O_4$. Vector Network Analyzer 8720B (VNA) was connected via coaxial cable to the Agilent Dielectric Probe Kit 85070B.

1. Introduction

Electromagnetic (EM) waves at microwave frequencies have many applications in various fields such as wireless telecommunication system, radar, local area network, electronic devices, mobile phones, laptops, and medical equipment [1, 2]. The effect of growth in various applications has led to electromagnetic interference (EMI) problems that have to be suppressed to acceptable limits. EMI reducing materials (absorbers) may be dielectric or magnetic [3] and the design depends on the frequency range, the desired quantity of shielding, and the physical characteristics of the devices being shielded. Thus it is important to determine their high frequency characteristics for the applications of EM in the high GHz ranges [4, 5]. Ni-Zn ferrite ceramics are the preferred ceramic material for high frequency applications in

order to suppress generation of Eddy current [6]. Although Ni-Zn ferrite ceramics have high electrical resistivity to prevent Eddy current generation, they have moderate magnetic permeability compared to Mn-Zn ferrites. However, the electrical and magnetic properties of these ferrite ceramics are heavily influenced by its microstructural features such as grain size, nature of grain boundaries, nature of porosity, and crystalline structure. The microstructural features of interest could be attained via chemical composition and high temperature processing [7]. However, the detailed electrical properties of Ni-Zn ferrite at different Ni-Zn ratio in a wideband frequency using open ended coaxial probe have not been studied yet. Thus, the aim of this work is to determine electrical properties of Ni-Zn ferrites prepared at different chemical composition based on chemical formula $Ni_xZn_{1-x}Fe_2O_4$ with $0.1 \leq x \leq 0.9$ that sintered at constant temperature. The variations in the microstructures, surface morphology, and alterations in reflection coefficient as well as their electrical properties of the Ni-Zn ferrites are the concern of this study.

2. Basic Principle

2.1. Loss Mechanism by Oscillating Electric Field. Materials can be categorized into two types which are the nonmagnetic materials and the magnetic materials. The core loss mechanisms for nonmagnetic materials are dielectric (dipolar) loss and conduction loss. The conduction and dipolar losses usually occur in metallic, high conductivity materials and dielectric insulators, respectively. The loss mechanisms for magnetic materials are also the conductive loss with addition magnetic loss such as hysteresis, eddy current, and the resonance losses (domain wall and electron spin). Loss condition of the materials is greatly influenced by microwave absorption.

The microwave absorption is caused by external electrical field and related to the material's complex permittivity ε:

$$\varepsilon^* = \varepsilon_o \left(\varepsilon' - j\varepsilon'' \right), \tag{1}$$

where ε_o is the permittivity of free space (ε_o = 8.86 × 10^{-12} F/m) and the real part ε' and the imaginary part ε'' are the relative dielectric constant and the effective relative dielectric loss factor, respectively. The real part of permittivity controls the amount of electrostatic energy stored per unit volume for a given applied field in a material. The imaginary part defines the energy loss caused by the lag in the polarization upon wave propagation when it passes through a material.

The translational motions of free or bound charges and rotating charge complexes are induced by the internal field generated when the microwaves penetrate and propagate through a material. These induced motions are resisted by inertial, elastic, and frictional forces, thus causing energy losses.

2.2. Open Ended Coaxial Probe. For the open ended coaxial probe measurement technique the complex relative permittivity is determined by inverting the expression of $Y(\varepsilon^*)$ where Y is the aperture admittance of the probe [8]:

$$Y = Y_o \frac{1 - \Gamma}{1 + \Gamma}, \tag{2}$$

where Y_o is the characteristic admittance of the coaxial line and Γ is the reflection coefficient at the aperture. The aperture admittance of open ended coaxial probes has several analytical expressions which contains the complex permittivity and can be compared to the measured admittance [9–12]. Some are from the computational points which may contribute to convergence problems because of the presence of multiple integrals, Bessel functions, and sine integrals when numerically solved. The expression for the aperture admittance is

given by [13], found by matching the electromagnetic field around the probe aperture, and can be adopted:

$$Y = \frac{\sqrt{\varepsilon_r^*}\gamma_o}{\sqrt{\varepsilon_{cl}^*}\ln(b/a)} \left\{ \int_0^{\pi/2} \left[J_o \left(\gamma_o a \sqrt{\varepsilon_r^*} \sin\theta \right) \right. \right.$$

$$\left. - J_o \left(\gamma_o b \sqrt{\varepsilon_r^*} \sin\theta \right) \right]^2 \frac{d\theta}{\sin\theta} + \frac{j}{\pi}$$

$$\cdot \int_0^\pi \left[2\mathrm{Si} \left(\gamma_o a \sqrt{\varepsilon_r^*} \sqrt{1 + \frac{b^2}{a^2} - 2\frac{b}{a}\cos\theta} \right) \right. \tag{3}$$

$$- \mathrm{Si} \left(2\gamma_o a \sqrt{\varepsilon_r^*} \sin\frac{\theta}{2} \right)$$

$$\left. \left. - \mathrm{Si} \left(2\gamma_o b \sqrt{\varepsilon_r^*} \sin\frac{\theta}{2} \right) \right] d\theta \right\},$$

where ε_r^* is the complex relative permittivity of the material under test, ε_{cl}^* is the relative permittivity of the coaxial line, a and b are the inner and outer radii of the coaxial line, respectively, γ_o is the absolute value of the propagation constant in free space, and Si and J_o are the sine integral and the Bessel function of zero order, respectively. This integral expression can be evaluated numerically by series expansion as in [10, 11] or numerical integration.

A different procedure for the extraction of material parameters involves minimizing the distance between the calculated aperture admittance (3) and the corresponding measured quantities through fitting algorithms, which may be based on either deterministic or stochastic optimization procedures. The minimization can be performed over the whole frequency range or on a point-by-point basis (i.e., at individual frequency points). Optimization procedure is needed to determine parameters for the point-by-point basis since it consists of modelling the complex relative permittivity and magnetic permeability with a prespecified functional form. Laurent series can be used for complex relative permittivity and magnetic permeability models [14], as well as dispersive laws, such as Havriliak-Negami and its special cases Cole-Cole and Debye to model dielectric relaxation [15], or the Lorentz model for both dielectric and magnetic dispersion [16]. The Havriliak-Negami model is an empirical modification of the single-pole Debye relaxation model:

$$\varepsilon_r^* = \varepsilon_\infty + \frac{\varepsilon_s - \varepsilon_\infty}{\left[1 + (j\omega\tau)^{1-\alpha} \right]^\beta}, \tag{4}$$

where ε_s and ε_∞ are the values of the real part of the complex relative permittivity at low and high frequency, respectively, τ is the relaxation time, and α and β are positive real constants ($0 \le \alpha, \beta \le 1$). From this model, the Cole-Cole equation can be derived setting $\beta = 1$; the Debye equation is obtained with $\alpha = 1$ and $\beta = 1$. This empirical model has the ability to give a better fit to the behaviour of dispersive materials over a wide frequency range.

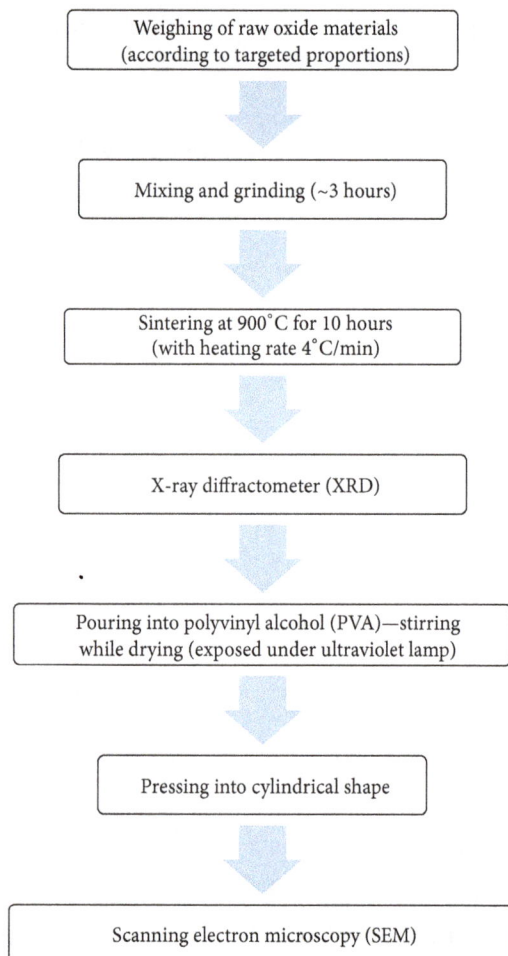

FIGURE 1: Flowchart of sample preparation to characterization.

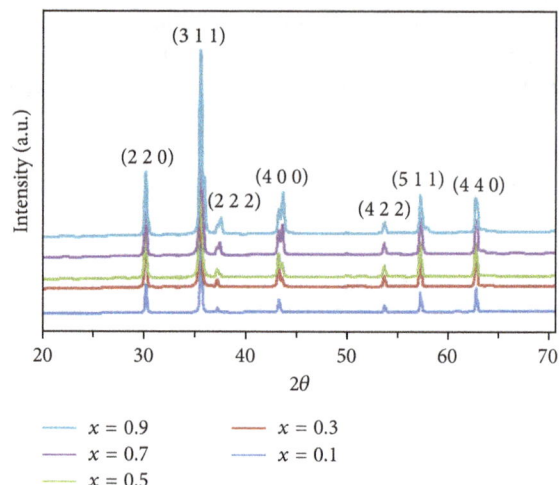

FIGURE 2: XRD profile of $Ni_xZn_{1-x}Fe_2O_4$ samples sintered at 900°C.

TABLE 1: d-spacing for $Ni_xZn_{1-x}Fe_2O_4$ samples at the main (3 1 1) hkl plane.

x	2θ (°)	d-spacing (Å)
0.1	35.317	2.543
0.3	35.319	2.536
0.5	35.412	2.526
0.7	35.426	2.518
0.9	35.778	2.511

3. Method

3.1. Sample Preparation and Structural and Morphological Characterization. The materials required for preparing $Ni_xZn_{1-x}Fe_2O_4$ samples were obtained from Alfa Aesar: Iron(III) Oxide (99.500%), Nickel(II) Oxide (99.000%), and Zinc Oxide (99.900%). The sample preparation procedures are roughly illustrated in the flowchart as in Figure 1.

3.2. Complex Permittivity Measurement. In this work, the permittivities of the samples were measured using Agilent 85070B Dielectric Probe Kit. Air and short and distilled water were used as standard materials for calibration as recommended by the manufacturer. The Dielectric Probe Kit automatically determined the complex permittivity of the materials under test by measuring both the magnitude and phase of the reflection coefficients. The measurement was started with the standard test materials which consist of air, Teflon, RT-duroid 5880, and Perspex. Then, the measurement continued with the Iron(III) Oxide (Fe_2O_3), Nickel(II) Oxide (NiO), and Zinc Oxide (ZnO) materials. The materials already prepared by the manufacturer (Alfa Aesar) in powder form were pressed into cylindrical mold at 4 tons

using mechanical pressing machine. The measurement was continued with $Ni_xZn_{1-x}Fe_2O_4$ samples that have different fractional compositions of x. The sintered mixture powder of $Ni_xZn_{1-x}Fe_2O_4$ samples was pressed into cylindrical mold at 4 tons using mechanical pressing machine as well.

4. Results and Discussion

4.1. Structure Characterization and Morphology of $Ni_xZn_{1-x}Fe_2O_4$

4.1.1. XRD Profiles. Figure 2 presents the XRD patterns of $Ni_xZn_{1-x}Fe_2O_4$ samples after sintering at 900°C for 10 hours with heating rate of 4°C/min. The patterns showed distinct diffraction lines with the highest peaks at 35.317, 35.319, 35.412, 35.426, and 35.778 of the 2θ (°) for all samples with an increment of x which in turn decrease the d-spacing accordingly (Table 1). The d-spacing was linearly decreased as the fractional composition of x increased as shown in Figure 3. The distinct diffraction lines could be observed for the powders sintered at 900°C meaning that the intensity of XRD peaks increased as the amorphous phase transformed into the crystalline phase for $Ni_{0.1}Zn_{0.9}Fe_2O_4$ sample. This could be related to the development of crystal growth of the entire particles. The peaks for (2 2 0), (3 1 1), (2 2 2), (4 0 0), (4 2 2), (5 1 1), and (4 4 0) occurred at the reflections planes originated at the 2θ (°) values 30.003, 35.317,

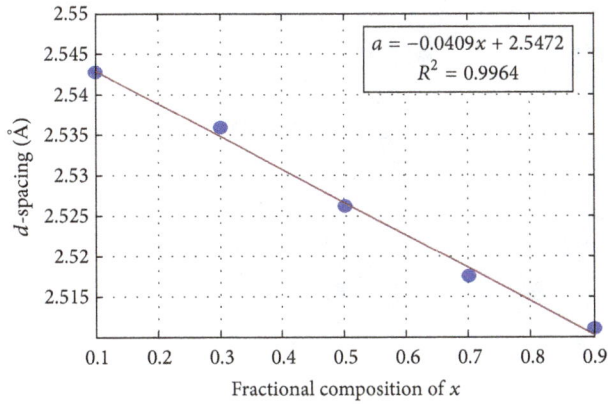

FIGURE 3: d-spacing against fractional composition of x for $Ni_xZn_{1-x}Fe_2O_4$.

36.931, 42.905, 53.172, 56.680, and 62.228, yielding to the d-spacing [Å] values of 2.978, 2.543, 2.434, 2.108, 1.723, 1.624, and 1.492 consecutively thus indicating that a pure cubic ferrite phase formed according to the reference spectrum of Ni-Zn ferrite (Joint Committee of Powder Diffraction Standards). The XRD profiles of different x are also presented in Figure 2 that showed the same behaviors as described above for $Ni_{0.1}Zn_{0.9}Fe_2O_4$ sample with slight difference in the intensity of 2θ (°) and decreased pattern for d-spacing [Å] as x increased in the fractional composition.

4.1.2. Lattice Constant. The lattice constant a was obtained as a function of fractional composition of x substitution in $Ni_xZn_{1-x}Fe_2O_4$ calculated from the combination of Bragg's equation and d-spacing expression:

$$\frac{1}{d_{hkl}^2} = \frac{h^2}{a^2} + \frac{k^2}{a^2} + \frac{l^2}{a^2} \qquad (5)$$

for cubic system equation. The calculation of lattice constant for all samples was considered at the single phase crystallite (3 1 1) hkl planes and thus the value of lattice constant was established. A linear relationship with negative sensitivity could be obtained between lattice constant and x for $Ni_xZn_{1-x}Fe_2O_4$ sample as shown in Figure 4. Other studies also found that the lattice constant a decreased with the increasing of x concentration [17, 18].

4.1.3. Density. The X-ray densities (D_x) of the $Ni_xZn_{1-x}Fe_2O_4$ samples were calculated using

$$D_x = \frac{8 \times M}{N \times a^3} \, g \cdot cm^{-3}. \qquad (6)$$

It was found that the density of the $Ni_xZn_{1-x}Fe_2O_4$ samples increased linearly with increasing of the substituted amount of x inside the Ni-Zn ferrite sample (Figure 5). Every reduction in number of molecular masses for all compositions gave a higher density value (Table 2).

TABLE 2: Calculated true X-ray density of $Ni_xZn_{1-x}Fe_2O_4$ samples.

Composition (x)	Molecular mass (g/mole)	Lattice constant a (Å)	X-ray density (g/cm^3)
0.1	240.4234	8.4334	5.3250
0.3	239.0802	8.4105	5.3386
0.5	237.7370	8.3789	5.3689
0.7	236.3938	8.3500	5.3942
0.9	235.0506	8.3280	5.4062

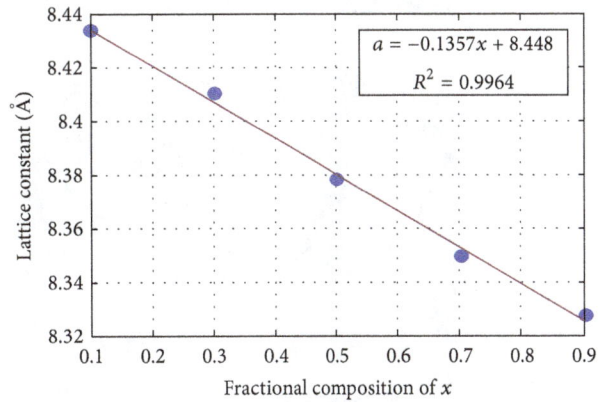

FIGURE 4: Lattice constant against fractional composition of x for $Ni_xZn_{1-x}Fe_2O_4$.

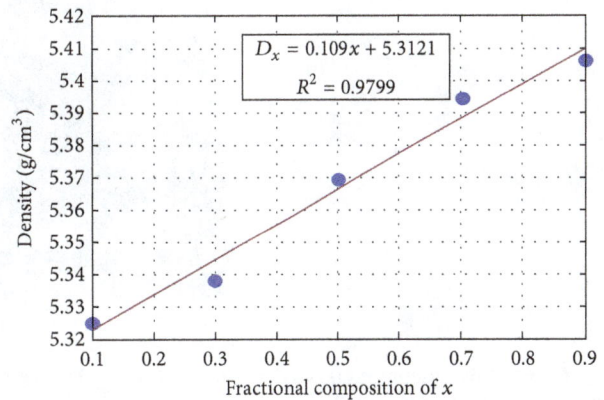

FIGURE 5: X-ray densities against fractional composition of x for $Ni_xZn_{1-x}Fe_2O_4$.

4.1.4. SEM Morphologies. The microstructural properties of the molded $Ni_xZn_{1-x}Fe_2O_4$ samples were obtained by scanning electron microscope as in Figure 6. The raw mixture in the form of powder was first sintered at 900°C for 10 hours before being poured into mold and compacted using mechanical pressing machine. The measurement was conducted on the plane surfaces of the molded samples which gave information in terms of grain morphology, grain boundaries, and porosity. The grain size of each sample was randomly selected through 60000 magnifications from the morphology picture so that the grain size could be seen clearly. The tabulated grain size for all samples was in the range of 62 nm–175 nm. Lots of pores could be seen from the morphology

(a)

(b)

(c)

(d)

(e)

FIGURE 6: SEM micrograph of (a) $Ni_{0.1}Zn_{0.9}Fe_2O_4$, (b) $Ni_{0.3}Zn_{0.7}Fe_2O_4$, (c) $Ni_{0.5}Zn_{0.5}Fe_2O_4$, (d) $Ni_{0.7}Zn_{0.3}Fe_2O_4$, and (e) $Ni_{0.9}Zn_{0.1}Fe_2O_4$ sample at 60000 magnifications.

and that was probably due to inhomogeneous size of particle; thus there would be air gaps between the particles. If the sintering time is increased, the pores will reduce because of the formation of strong bonds between the adjacent particles [19].

4.2. Permittivity Results

4.2.1. Standard Material. The measurement procedure to determine complex permittivity using the Agilent Dielectric Probe Kit 85070B was described above. The permittivity values also provide insights into the effect of the fractional composition of x on the bulk permittivity values $Ni_xZn_{1-x}Fe_2O_4$. Vector Network Analyzer 8720B (VNA) was connected via coaxial cable to the Agilent Dielectric Probe Kit 85070B.

The technique was done by pressing the dielectric probe against the sample material. The microwave signal launched by the VNA was reflected by the sample. The reflected wave was received by the VNA which then used the wave to calculate the dielectric constant and loss factor.

The dielectric constant and loss factor of air and several standard materials including Teflon, RT-duroid 5880, and Perspex with thickness of 20 mm, 19.05 mm, and 20 mm were measured in the frequency range between 1 GHz and 10 GHz as shown in Figure 7. The dielectric constant values for all the samples were almost constant for the whole frequency range with slight dispersion toward the higher end of the frequency range except for air which was lossless. The slight dispersion for all the samples at the higher frequency end was due to the increase of the loss factor because of higher

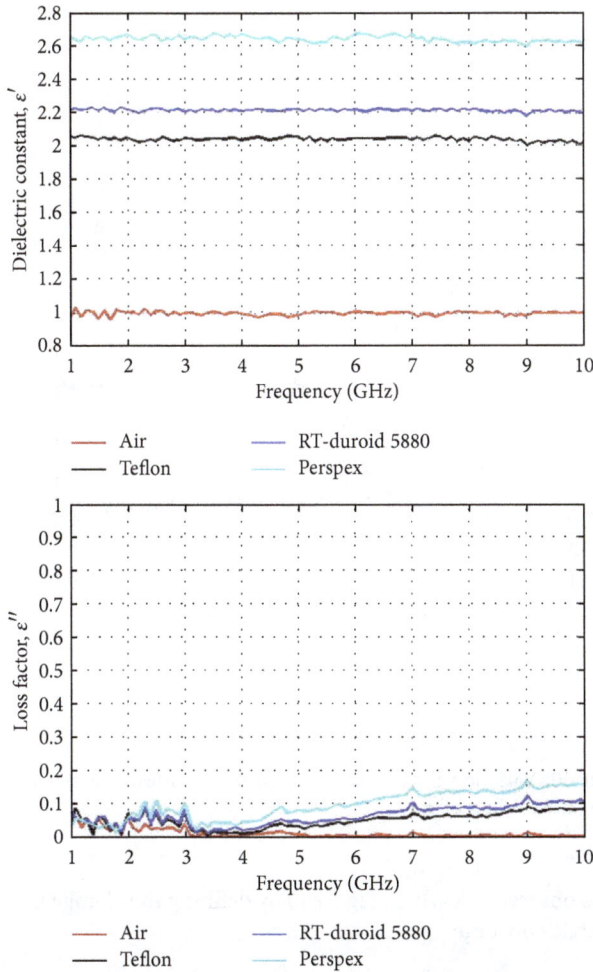

FIGURE 7: Complex permittivity of standard samples measured with Agilent 85070B Dielectric Probe Kit.

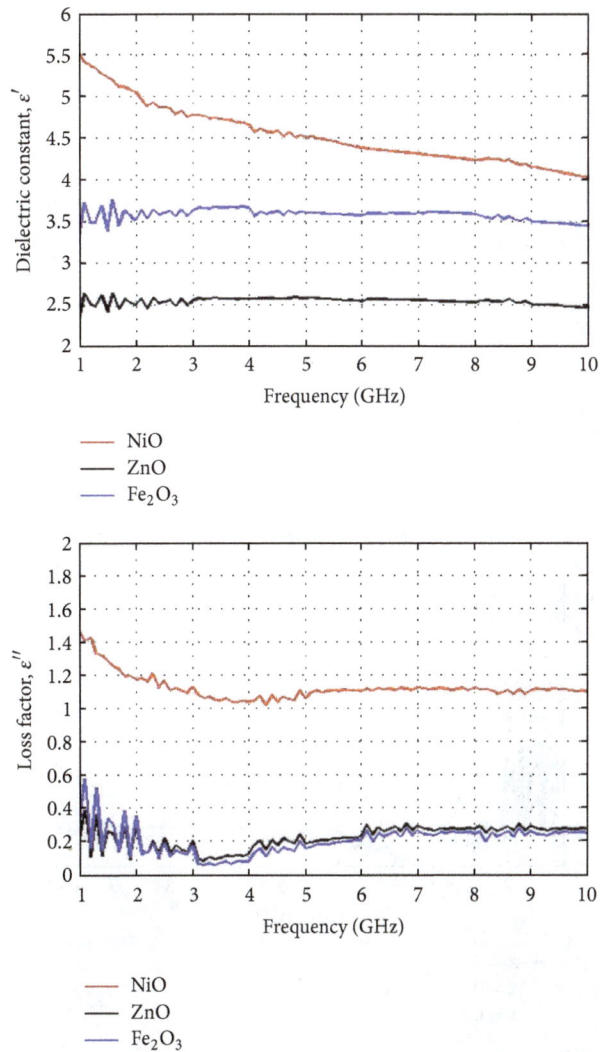

FIGURE 8: Complex permittivity of raw materials measured with Agilent 85070B Dielectric Probe Kit.

absorption loss. The dielectric constants of air, Teflon, RT-duroid 5880, and Perspex at 10^5 Hz to 1 MHz were found to be 1, 2.1 (Tecaflon PTFE, Technical Datasheet), 2.2 (Rogers Corporation, Technical Datasheet), and 2.6 (Goodfellow Group, Technical Information-Polymethylmethacrylate) which were in very good agreement with available data.

The slight dispersion for all the samples at the higher frequency end was probably due to several factors. Firstly, the minimum sample thickness recommended by the manufacturer ($d = 30\,\text{mm}/\sqrt{\varepsilon'}$) for the 85070B Dielectric Probe Kit should be more than 20 mm for $\varepsilon' = 2.05$. The higher the dielectric constant is, the lower the required minimum thickness shall be based on the higher dielectric. Small errors could be attributed to the fact that Dielectric Probe Kit 85070B was designed for liquid materials. The permittivity computation for the Dielectric Probe Kit 85070B was a simplified version of Debye model obtained from empirical fitting of several known liquids [20]; thus the permittivity calculations were less accurate for solid materials.

4.2.2. Pure NiO, ZnO, and Fe_2O_3. The dielectric constant and loss factor consisting of Nickel(II) Oxide (NiO), Zinc Oxide (ZnO), and Iron(III) Oxide (Fe_2O_3) are shown in Figure 8. It could be clearly observed from the graph that NiO had both higher dielectric constant and loss factor compared to ZnO and Fe_2O_3. The dielectric constants for both ZnO and Fe_2O_3 were almost stable for the whole frequency range. However the dielectric constant of NiO was gradually decreased from 5.5 at 1 GHz to 4 at 10 GHz. Interestingly, it could be observed clearly that NiO had loss factor approximately 5 times larger than ZnO and Fe_2O_3 thus qualifying it to be categorized as a highly loss material.

4.2.3. $Ni_xZn_{1-x}Fe_2O_4$. The measurement of complex permittivity of $Ni_xZn_{1-x}Fe_2O_4$ samples using open ended coaxial probe with different fractional composition of x was also performed. The thickness of the all samples was 8 mm. Figure 9 shows the results for each $Ni_xZn_{1-x}Fe_2O_4$ sample, where $x = 0.1, 0.3, 0.5, 0.7,$ and 0.9.

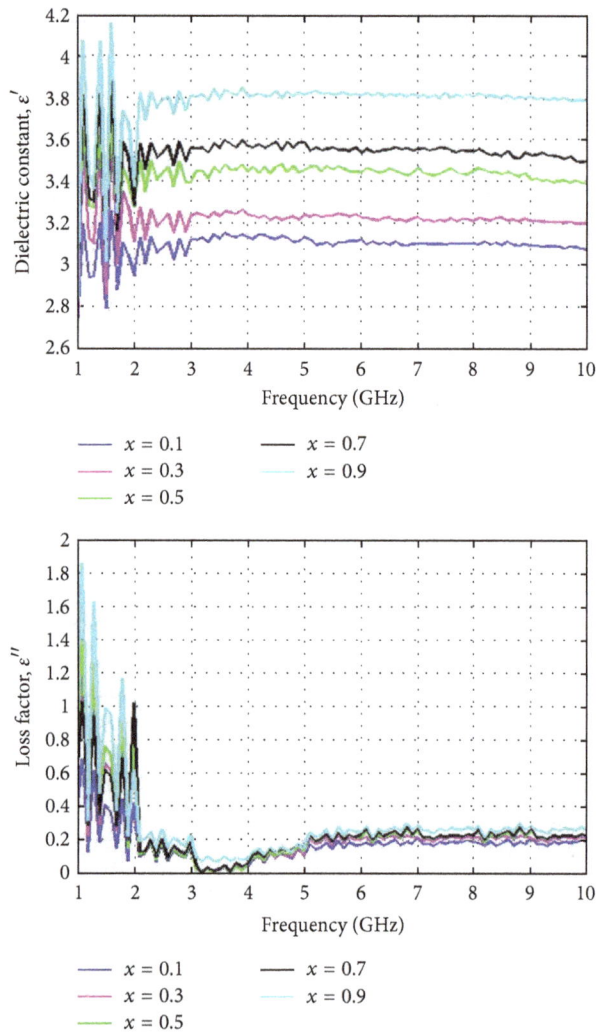

FIGURE 9: Complex permittivity of $Ni_xZn_{1-x}Fe_2O_4$ samples measured with Agilent 85070B Dielectric Probe Kit.

The high uncertainties in both ε' and ε'' at frequencies below 2 GHz were due to multiple reflection effect within the sample. The samples must be infinitely thick to avoid reflection from the end face of the sample. The lower the operating frequencies, the longer the wavelengths and thus the higher the uncertainties due to incident wave reflected at the end surface of the sample. These effects were reduced at higher frequencies especially beyond 3 GHz due to shorter probing wavelength. Generally Figure 9 suggests higher fractional composition of x would result in higher values of the dielectric constant of $Ni_xZn_{1-x}Fe_2O_4$. At 5 GHz, the value of ε' increased from approximately 3.1 to 3.8 for $x = 0.1$ to 0.9. This was expected as Figure 8 showed the dielectric constant of NiO was much higher than both ZnO and Fe_2O_3. Similarly, the loss factor ε'' values for all $Ni_xZn_{1-x}Fe_2O_4$ samples increased with increasing values of fractional composition of x especially at frequencies above 3 GHz.

The effect of fractional composition of x on the dielectric constant in the frequency range between 3 GHz and 10 GHz

TABLE 3: Mean value of dielectric constant $\Delta\varepsilon'$ in the frequency range from 3 GHz to 10 GHz.

Symbol, x	Mean value of $\Delta\varepsilon'$
0.2	0.1145
0.4	0.3332
0.6	0.4395
0.8	0.7020

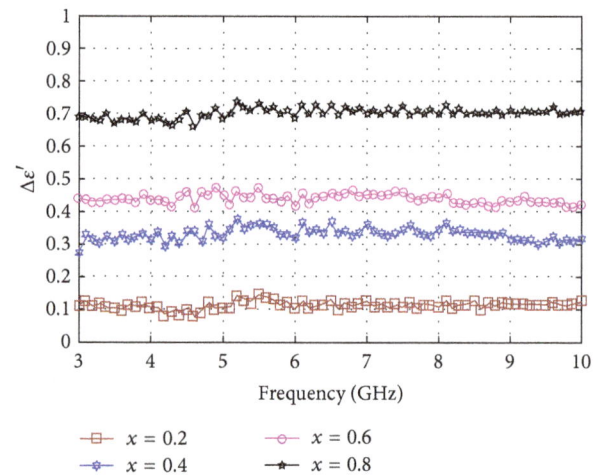

FIGURE 10: Variation in $\Delta\varepsilon'$ with frequency for various fractional composition values of x.

can be observed clearly in Figure 10 by defining the change in dielectric constant:

$$\Delta\varepsilon' = \varepsilon'_{x(i)} - \varepsilon'_{x=0.1}, \qquad (7)$$

where $\varepsilon'_{x(i)}$ is the dielectric constant of $Ni_xZn_{1-x}Fe_2O_4$ with $x(i) = 0.3, 0.5, 0.7,$ and 0.9.

The mean values $\Delta\varepsilon'$ for the whole frequency range from 3 GHz to 10 GHz are summarized in Table 3. A slight change from $x = 0.1$ to 0.3 would give a change of approximately 0.11 in the value of ε' and could be as high as 0.70 if x increased from 0.1 to 0.9. The higher the NiO content is, the higher the dielectric constant and loss factor of $Ni_xZn_{1-x}Fe_2O_4$ will be.

5. Conclusion

The permittivity of $Ni_xZn_{1-x}Fe_2O_4$ in the frequency range between 1 GHz and 10 GHz was successfully determined using an open ended coaxial probe technique as higher fractional composition of x would result in higher values of the dielectric constant of $Ni_xZn_{1-x}Fe_2O_4$. It was found that the lattice constant of the $Ni_xZn_{1-x}Fe_2O_4$ samples decreased linearly with increasing of the substituted amount of x inside the Ni-Zn ferrite sample. The tabulated grain size for all samples was in the range of 62 nm–175 nm.

Conflict of Interests

The authors declare that there is no conflict of interests regarding the publication of this paper.

Acknowledgments

The authors wish to thank Malaysia Education Ministry (Department of Higher Education) and Universiti Tun Hussein Onn Malaysia (UTHM) for their financial support and Universiti Putra Malaysia (UPM) for the provision of enabling environment to carry out this work.

References

[1] K. M. Lim, M. C. Kim, K. A. Lee, and C. G. Park, "Electromagnetic wave absorption properties of amorphous alloy-ferrite-epoxy composites in quasi-microwave band," *IEEE Transactions on Magnetics*, vol. 39, no. 3, pp. 1836–1841, 2003.

[2] K. Lakshmi, H. John, K. T. Mathew, R. Joseph, and K. E. George, "Microwave absorption, reflection and EMI shielding of PU-PANI composite," *Acta Materialia*, vol. 57, no. 2, pp. 371–375, 2009.

[3] C. A. Grimes and D. M. Grimes, "A brief discussion of EMI shielding materials," in *Proceedings of the 14th IEEE Aerospace Applications Conference*, pp. 217–226, IEEE, Steamboat Springs, Colo, USA, January 1993.

[4] M. N. Afsar, J. R. Birch, R. N. Clarke, and G. W. Chantry, "The measurement of the properties of materials," *Proceedings of the IEEE*, vol. 74, no. 1, pp. 183–199, 1986.

[5] R. Ma, Y. Wang, Y. Tian, C. Zhang, and X. Li, "Synthesis, characterization and electromagnetic studies on nanocrystalline nickel zinc ferrite by polyacrylamide gel," *Journal of Materials Science and Technology*, vol. 24, no. 3, pp. 419–422, 2008.

[6] A. Verma, T. C. Goel, R. G. Mendiratta, and R. G. Gupta, "High-resistivity nickel-zinc ferrites by the citrate precursor method," *Journal of Magnetism and Magnetic Materials*, vol. 192, no. 2, pp. 271–276, 1999.

[7] G. H. Jonker and A. L. Stuijts, "Controlling the properties of electroceramic materials through their microstructure," *Philips Technical Review*, vol. 32, no. 3-4, pp. 79–95, 1971.

[8] S. S. Stuchly, C. L. Sibbald, and J. M. Anderson, "New aperture admittance model for open-ended waveguides," *IEEE Transactions on Microwave Theory and Techniques*, vol. 42, no. 2, pp. 192–198, 1994.

[9] P. de Langhe, K. Blomme, L. Martens, and D. de Zutter, "Measurement of low-permittivity materials based on a spectral-domain analysis for the open-ended coaxial probe," *IEEE Transactions on Instrumentation and Measurement*, vol. 42, no. 5, pp. 879–886, 1993.

[10] D. Misra, M. Chabbra, B. R. Epstein, M. Mirotznik, and K. R. Foster, "Noninvasive electrical characterization of materials at microwave frequencies using an open-ended coaxial line: test of an improved calibration technique," *IEEE Transactions on Microwave Theory and Techniques*, vol. 38, no. 1, pp. 8–14, 1990.

[11] D. Xu, L. Liu, and Z. Jiang, "Measurement of the dielectric properties of biological substances using an improved open-ended coaxial line resonator method," *IEEE Transactions on Microwave Theory and Techniques*, vol. 35, no. 12, pp. 1424–1428, 1987.

[12] Y. Xu, R. G. Bosisio, and T. K. Bose, "Some calculation methods and universal diagrams for measurement of dielectric constants using open-ended coaxial probes," *IEE Proceedings H—Microwaves, Antennas and Propagation*, vol. 138, no. 4, pp. 356–360, 1991.

[13] N. Marcuvitz, *Waveguide Handbook*, McGraw-Hill, New York, NY, USA, 1951.

[14] P. D. Domich, J. Baker-Jarvis, and R. G. Geyer, "Optimization techniques for permittivity and permeability determination," *Journal of Research of the National Institute of Standards and Technology*, vol. 96, no. 5, pp. 565–575, 1991.

[15] D. F. Kelley, T. J. Destan, and R. J. Luebbers, "Debye function expansions of complex permittivity using a hybrid particle swarm-least squares optimization approach," *IEEE Transactions on Antennas and Propagation*, vol. 55, no. 7, pp. 1999–2005, 2007.

[16] M. Y. Koledintseva, K. N. Rozanov, A. Orlandi, and J. L. Drewniak, "Extraction of the Lorentzian and Debye parameters of dielectric and magnetic dispersive materials for FDTD modeling," *Journal of Electrical Engineering*, vol. 53, no. 9, pp. 97–100, 2002.

[17] Q. Lv, L. Liu, J. P. Zhou, X. M. Chen, X. B. Bian, and P. Liu, "Influence of nickel-zinc ratio on microstructure, magnetic and dielectric properties of $Ni_{(1-x)}Zn_xFe_2O_4$ ferrites," *Journal of Ceramic Processing Research*, vol. 13, no. 2, pp. 110–116, 2012.

[18] K. R. Krishna, K. V. Kumar, and D. Ravinder, "Structural and electrical conductivity studies in nickel-zinc ferrite," *Advances in Materials Physics and Chemistry*, vol. 2, no. 3, pp. 185–191, 2012.

[19] W. E. Lee and W. M. Rainforth, *Ceramic Microstructures: Property Control by Processing*, Chapman & Hall, London, UK, 1994.

[20] D. V. Blackham and R. D. Pollard, "An improved technique for permittivity measurements using a coaxial probe," *IEEE Transactions on Instrumentation and Measurement*, vol. 46, no. 5, pp. 1093–1099, 1997.

7

Ultrawideband Noise Radar Imaging of Impenetrable Cylindrical Objects Using Diffraction Tomography

Hee Jung Shin,[1] Ram M. Narayanan,[1] and Muralidhar Rangaswamy[2]

[1]*The Pennsylvania State University, University Park, PA 16802, USA*
[2]*Air Force Research Laboratory, Wright-Patterson Air Force Base, OH 45433, USA*

Correspondence should be addressed to Ram M. Narayanan; ram@ee.psu.edu

Academic Editor: Gian Luigi Gragnani

Ultrawideband (UWB) waveforms achieve excellent spatial resolution for better characterization of targets in tomographic imaging applications compared to narrowband waveforms. In this paper, two-dimensional tomographic images of multiple scattering objects are successfully obtained using the diffraction tomography approach by transmitting multiple independent and identically distributed (iid) UWB random noise waveforms. The feasibility of using a random noise waveform for tomography is investigated by formulating a white Gaussian noise (WGN) model using spectral estimation. The analytical formulation of object image formation using random noise waveforms is established based on the backward scattering, and several numerical diffraction tomography simulations are performed in the spatial frequency domain to validate the analytical results by reconstructing the tomographic images of scattering objects. The final image of the object based on multiple transmitted noise waveforms is reconstructed by averaging individually formed images which compares very well with the image created using the traditional Gaussian pulse. Pixel difference-based measure is used to analyze and estimate the image quality of the final reconstructed tomographic image under various signal-to-noise ratio (SNR) conditions. Also, preliminary experiment setup and measurement results are presented to assess the validation of simulation results.

1. Introduction

Research on the use of random or pseudorandom noise transmit signals in radar has been conducted since the 1950s [1, 2]. Noise radar has been considered a promising technique for the covert identification of target objects due to several advantages, such as excellent electronic countermeasure (ECM), low probability of detection (LPD), low probability of interception (LPI) features, and relatively simple hardware architectures [3–5]. Also, advances in signal and imaging processing techniques in radar systems have progressed so that multidimensional representations of the target object can be obtained [6].

In general, radar imaging tends to be formulated in the time domain to exploit efficient back-projection algorithms, generate accurate shape features of the target object, and provide location data [7]. For multistatic radar systems, the images of a target are reconstructed based on range profiles obtained from the distributed sensor elements. When a transmitter radiates a waveform, spatially distributed receivers collect samples of the scattered field which are related to the electrical parameters of the target object. For the next iteration, a different transmitter is activated, and the scattered field collection process is repeated. Finally, all collected scattered field data are relayed for signal processing and subsequent image formation algorithms.

Tomography-based radar imaging algorithms have been developed based on microwave image reconstruction method [9], characterizing the material property profiles of the target object in the frequency domain and reconstructing specific scattering features inside the interrogation medium by solving the inverse scattering problem. The capability of microwave imaging techniques has been found attractive in malignant breast cancer detection [10–13], civil infrastructure assessment [14–16], and homeland security [17–19] applications due to the advantages of nondestructive diagnosis and evaluation of obscured objects. The quality of the reconstructed image for different values of the electrical contrast for a UWB imaging system was investigated and published for both low-contrast and high-contrast object cases. For low-contrast objects, the obtained target image using a single frequency achieves a good reconstruction of the electrical contrast that is almost equivalent to the one obtained with the entire UWB frequency range. For the high-contrast case, while the formation of a Moiré pattern affects the single frequency reconstruction, this artifact does not appear in the UWB frequency image [20]. Thus, UWB radar tomography is expected to provide advantages over the single or narrow band frequency operation in terms of resolution and accuracy for any target object.

The goal of this paper is to demonstrate successful image reconstruction of the cylindrical conducting objects using the diffraction tomography theorem for bistatic UWB noise radar systems. The paper is organized as follows. First, the paper defines the characteristics of UWB random noise signal and discusses the shortcomings of using such noise signal as a radar transmit waveform in tomographic image reconstruction process in Section 2. The empirical solution to bypass the shortcoming of using UWB random noise waveform is also proposed. The formulations of the image reconstruction of two-dimensional scattering geometry of a bistatic imaging radar system using Fourier diffraction theorem under the assumption of plane wave illumination are presented in Section 3. In Section 4, the numerical simulation results of diffraction tomography using UWB random noise waveforms show that the tomographic image of the target is successfully reconstructed. The image quality measures of the reconstructed images, SNR effects for multiple transmissions of UWB random noise waveforms, and preliminary experimental validation are discussed in Section 5. Conclusions are presented in Section 6.

2. Analysis of White Gaussian Noise Model

The main advantage of transmitting a random noise waveform is to covertly detect and image a target without alerting others about the presence of radar system. Such LPI characteristics of the noise radar are guaranteed because the transmitted random noise waveform is constantly varying and never repeats itself exactly [21]. The random noise waveform can be experimentally generated simply by amplifying the thermal noise generated in resistors or noise diodes while maintaining relatively flat spectral density versus frequency [22]. Hence, relatively simple hardware designs can be achieved for noise radars compared to the conventional radar systems using complicated signal modulation schemes.

For a random noise waveform model, let $x[n]$ be a discrete time WSS and ergodic random process and a sequence of iid random variable drawn from a Gaussian distribution, $\mathcal{N}(0, \sigma^2)$. $x[n]$ defined herein is white Gaussian noise; that is, its probability density function follows a Gaussian distribution and its power spectral density is ideally a nonzero constant for all frequencies. However, the finite number of random noise amplitude samples must be chosen for

waveform generation for any numerical simulations and practical experiments.

Assume that a sequence of only l samples of $x[n]$ is selected for generating a white Gaussian noise. In this case, the estimate for the power spectral density, $\widehat{S}_l(\omega)$, is given by [23]

$$I_l(\omega) = \widehat{S}_l(\omega) = \sum_{m=-(l-1)}^{l-1} \widehat{r}_l[m] e^{-j\omega m}, \tag{1}$$

where $\widehat{r}_l[m]$ is the estimate for the autocorrelation sequence. $I_l(\omega)$ is defined as the periodogram estimate, and the rigorous analysis of the expected value and variance of the periodogram estimate for any arbitrary ω is described in [23–25]. The expected value of the periodogram estimate is [23, 24]

$$\begin{aligned} \mathrm{E}\left[I_l(\omega)\right] &= \sum_{m=-(l-1)}^{l-1} \mathrm{E}\left[\widehat{r}_l[m]\right] e^{-j\omega m} \\ &= \sum_{m=-(l-1)}^{l-1} \left(1 - \frac{|m|}{l}\right) r_l[m] e^{-j\omega m}, \end{aligned} \tag{2}$$

which suggests that $I_l(\omega)$ is a biased estimator. However, it is considered to be asymptotically unbiased as l approaches infinity. In this case, the expected value of $I_l(\omega)$ becomes a constant such that

$$\mathrm{E}\left[I_l(\omega)\right] \simeq S_x(\omega) = \sigma_x^2 \quad \text{as } l \longrightarrow \infty. \tag{3}$$

The variance of the periodogram estimate of the white Gaussian noise waveform formed by a sequence of l samples is given by [23, 24]

$$\mathrm{VAR}\left[I_l(\omega)\right] = S_x(\omega)^2 \left(1 + \left(\frac{\sin(\omega l)}{l \sin(\omega)}\right)^2\right), \tag{4}$$

which is proportional to the square of the power spectrum density and does not approach zero as l increases. In order to decrease the variance of $I_l(\omega)$, the periodogram averaging method has been proposed by Bartlett [26]. The average of K independent and identically distributed periodograms on samples of size l is given by

$$\overline{I}_{l,K}(\omega) = \frac{1}{K} \sum_{i=0}^{K-1} I_{l,i}(\omega), \tag{5}$$

and the expected value of the average with K iid periodogram estimate is written as [23]

$$\begin{aligned} \mathrm{E}\left[\overline{I}_{l,K}(\omega)\right] &= \mathrm{E}\left[I_l(\omega)\right] \\ &= \sum_{m=-(l-1)}^{l-1} \left(1 - \frac{|m|}{l}\right) r_l[m] e^{-j\omega m}, \end{aligned} \tag{6}$$

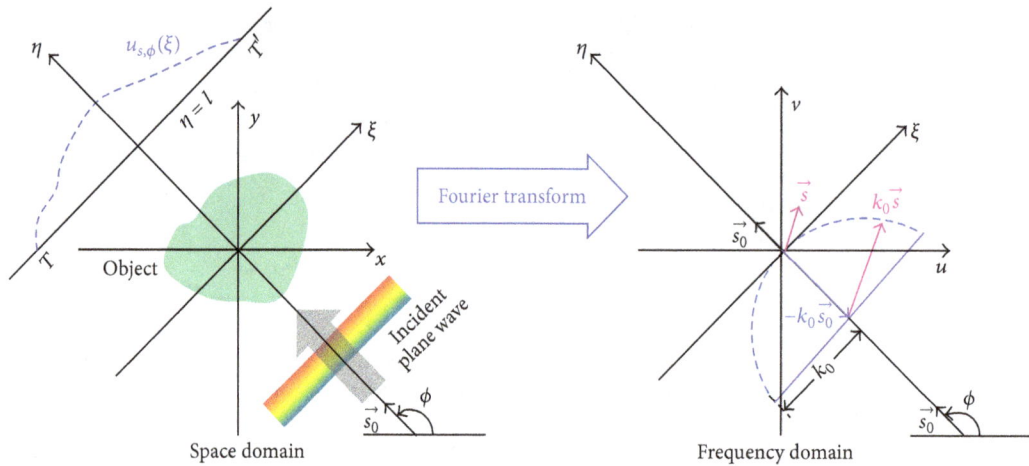

FIGURE 1: The Fourier diffraction theorem relates the Fourier transform of a diffracted projection to the Fourier transform of the object along a semicircular arc. An arbitrary object is illuminated by a plane wave propagating along the unit vector, and the coordinate system is rotated [8].

which is considered to be asymptotically unbiased as l approaches infinity. Also the variance of the averaged periodogram estimate is given by [23]

$$\mathrm{VAR}\left[\bar{I}_{l,K}\left(\omega\right)\right] = \frac{1}{K}\mathrm{VAR}\left[I_{l}\left(\omega\right)\right]$$
$$\simeq \frac{1}{K}S_{x}\left(\omega\right)^{2}. \tag{7}$$

The variance of the averaged periodogram estimate is inversely proportional to the number of iid periodograms K, and consequently the variance approaches zero as K approaches infinity. We use (6) and (7) to conclude that the expected value remains unchanged, but only the variance of white Gaussian noise decreases for averaging K iid periodogram estimates. Increasing the number of K in averaging periodogram estimate truly flattens the spectral density, and the successful tomographic image can be achieved by transmitting K multiple random noise waveforms with a large sequence size l. For the numerical simulations performed in this paper, a total of 10 iid random noise waveforms are transmitted, and each iid noise waveform is generated with 500 random amplitude samples drawn from $\mathcal{N}(0,\sigma^{2})$. The tomographic image is formed based on the dataset from 41 discrete frequencies chosen uniformly within X-band from 8 GHz to 10 GHz in steps of 50 MHz.

3. Formulation

In this section, the scattering properties for two-dimensional cylindrical impenetrable conducting object in the bistatic scattering arrangement are discussed, and the Fourier diffraction tomography algorithm is applied to reconstruct the image of the object based on the bistatic scattering properties. The Fourier diffraction theorem has been extensively applied in the area of acoustical imaging [27–29]. The goal of diffraction tomography is to reconstruct the properties of a slice of an object from the scattered field. For planar geometry, an object is illuminated with a plane wave, and the scattered

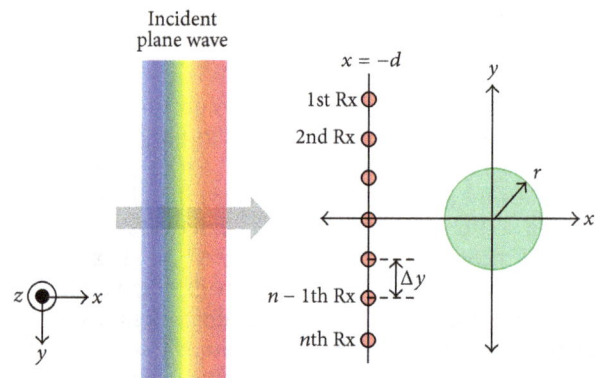

FIGURE 2: Two-dimensional backward scattering geometry for a cylindrical conducting object. Red dots and green circle represent a linear receiving array and the PEC cylinder, respectively.

fields are calculated or measured over a straight line parallel to the incident plane wave. The mathematical formulation and proof of validity of the Fourier diffraction theorem shown in Figure 1 are not presented in this section since they were already stated and published [30].

Two-dimensional backward scattering geometry for a cylindrical conducting object is shown in Figure 2. Prior research on microwave imaging has already proven that the image reconstructed in the backward scattering case is better than that obtained in the forward scattering case, based on numerical results [31].

As shown in Figure 2, a single cylindrical perfect electric conductor (PEC) object is located at the center of the simulation scene. The height of the cylindrical object is assumed to be infinitely long along the z-axis and the incident z-polarized plane wave is illuminated from $-x$ direction. The receiver spacing, Δy, is calculated to avoid aliasing effect and is given by

$$\Delta y \leq \frac{\lambda_{\min}}{2}, \tag{8}$$

where $\lambda_{\min}$ is the minimum wavelength. A linear receiving array is located at $x = -d$ from the center of the cylinder object, collecting scattered field for reconstruction of the object image. The frequency sampling interval, Δf, is also considered to ensure an image that is free of false aliases and is given by

$$\Delta f \leq \frac{c}{2 \cdot r}, \qquad (9)$$

where c is the speed of wave propagation in free space, and r is the radius of the cylindrical PEC object.

As shown in the previous section, the frequency response of the transmitted white Gaussian noise waveform shows that the field amplitude is a constant nonzero amplitude value for all possible frequencies. Thus, the z-polarized incident plane wave of a single transmitted white Gaussian noise waveform for N discrete frequencies takes the form

$$E_{\text{inc}}(\vec{r}) = \hat{z} \cdot \left(E_1 e^{-jk_1 \hat{x} \cdot \vec{r}} + E_1 e^{-jk_2 \hat{x} \cdot \vec{r}} + \cdots + E_N e^{-jk_N \hat{x} \cdot \vec{r}} \right)$$

$$= \hat{z} \cdot \sum_{n=1}^{N} E_n e^{-jk_n \hat{x} \cdot \vec{r}}, \qquad (10)$$

where $k_n = \omega_n/c$ is the wavenumber, and E_n is the field amplitude of the transmitted white Gaussian noise waveform at each discrete frequency of interest. The field amplitude of the noise waveform E_n becomes the nonzero constant value as the sequence of the noise waveform approaches infinity.

We start the analysis from a single frequency and develop the process for N multiple frequencies by summing up the analysis results for N. Also the entire analysis must be repeated for K times when K multiple iid noise waveforms are transmitted. The z-polarized incident waveform is defined as

$$E_{\text{inc}}(\vec{r}) = \hat{z} \cdot E_n e^{-jk_n \hat{x} \cdot \vec{r}},$$

$$H_{\text{inc}}(\vec{r}) = -\hat{y} \cdot \frac{1}{\eta} \cdot E_n e^{-jk_n \hat{x} \cdot \vec{r}}, \qquad (11)$$

where $\eta = \sqrt{\mu_0/\varepsilon_0}$ is the intrinsic impedance in free space. If the object consists of a material having a certain dielectric constant value, the equivalent electric current distribution, J_{eq}, is calculated for the scattered field. The object is defined as PEC so that the scattered field observed at the linear receiving array in the y-direction located at $x = -d$ is calculated by applying the physical optics approximation and is given by

$$E_{\text{scat,single}}(k_n, x = -d, y)$$

$$= -j\omega\mu_0 \int_S J_{\text{eq}} \cdot G(\vec{r} - \vec{r}') d\vec{r}' \qquad (12)$$

$$= -jk_n\eta \int_S \left(2\hat{n}(\vec{r}') \times H_{\text{inc}}(\vec{r}') \right) \cdot G(\vec{r} - \vec{r}') d\vec{r}',$$

where S is the boundary of scatterer, $\hat{n}(\vec{r}')$ is the outward unit normal vector to S, and $G(\vec{r} - \vec{r}')$ is Green's function for two-dimensional geometry defined as

$$G(\vec{r} - \vec{r}') = -\frac{j}{4} H_0^{(2)} \left(k_n (\vec{r} - \vec{r}') \right), \qquad (13)$$

where $H_0^{(2)}$ is the zeroth-order Hankel function of the second kind.

Similarly, for the case of discrete scattering objects, the scattered field of p objects over a linear array in the y-direction located at $x = -d$ is expressed as

$$E_{\text{scat,multiple}}(k_n, x = -d, y) = \sum_p E_{\text{scat,single}}(p). \qquad (14)$$

On assuming the polarization in the z-direction, the general form of scattered field, u_{scat}, obtained by the receivers at $x = -d$ becomes

$$u_{\text{scat}}(k_n, x = -d, y)$$

$$= \hat{z} \cdot E_{\text{scat}}(k_n, x = -d, y) \qquad (15)$$

$$= -jk_n E_n \iint o_{\text{scat}}(\vec{r}') e^{-jk_n \hat{x} \cdot \vec{r}'} G(\vec{r} - \vec{r}') d^2\vec{r}',$$

where

$$o_{\text{scat,single}}(\vec{r}') = -2\hat{n}(\vec{r}') \cdot \hat{x}\delta(S(\vec{r}')) \qquad (16)$$

is defined as the scattering object function of a single PEC object which is related to the object shape, and $\delta(S(\vec{r}'))$ is a Dirac delta function defined as

$$\delta(S(\vec{r}')) \begin{cases} = 0 & \text{as } \vec{r}' \in S \\ \neq 0 & \text{elsewhere.} \end{cases} \qquad (17)$$

Similarly, the scattering function shown in (16) for discrete scattering objects can be written as

$$o_{\text{scat,multiple}}(\vec{r}') = \sum_p o_{\text{scat,single}}(p). \qquad (18)$$

By using the plane wave expansion of Green's function [32], the one-dimensional Fourier transform of u_{scat}, defined in (15), in y-direction can be written as

$$\widetilde{U}_{\text{scat}}(k_n, x = -d, k_y) = \frac{k_n^2 E_n}{j2\gamma} e^{-j\gamma d} \widetilde{O}_{\text{scat}}(-\gamma - k_n, k_y), \qquad (19)$$

where

$$\gamma = \begin{cases} \sqrt{k_n^2 - k_y^2} & \text{as } |k_y| \leq k_n \\ -j\sqrt{k_y^2 - k_n^2} & \text{as } |k_y| > k_n. \end{cases} \qquad (20)$$

If k_x is defined as

$$k_x = -\gamma - k_n, \qquad (21)$$

the Fourier transform of the two-dimensional scattering object function, $\widetilde{O}_{\text{scat}}(k_x, k_y)$, defined in (19) is given by

$$\widetilde{O}_{\text{scat}}(k_x, k_y) = \iint o_{\text{scat}}(x, y) e^{-j(k_x x + k_y y)} dx \, dy. \qquad (22)$$

In this case, the arguments of $\widetilde{O}_{\text{scat}}(k_x, k_y)$ are related by

$$(k_x + k_n)^2 + k_y^2 = k_n^2. \qquad (23)$$

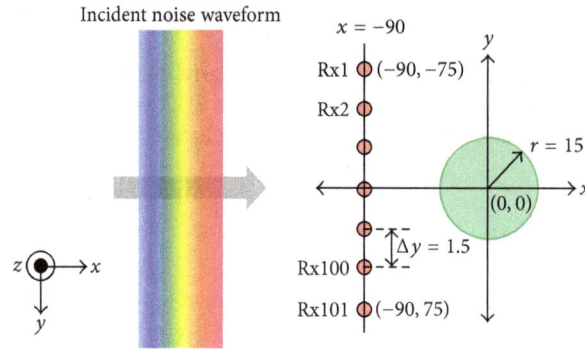

FIGURE 3: Two-dimensional backward scattering simulation geometry for a cylindrical conducting object. A single PEC cylinder with a radius of 15 cm is located at $(0,0)$, and a linear receiving array is located 90 cm away from the origin in $-x$ direction.

Equations (19) and (23) show that as a two-dimensional scattering object is illuminated by a plane wave at the frequency ω_n, the one-dimensional Fourier transform of the scattered field yields a semicircle centered at $(0, -k_n)$ with radius k_n in the two-dimensional Fourier space, which relates to the Fourier diffraction theorem depicted in Figure 1. By using N number of frequencies, the radius in the two-dimensional Fourier space changes, which enhances the resolution and accuracy of the image. Also the variance of the frequency response is reduced by taking average of the frequency responses of K multiple transmissions of iid UWB noise waveforms [8, 33].

4. Numerical Simulation Results

Based on the formulation described in previous section, this section discusses the numerical simulation results of diffraction tomography using band-limited iid UWB WGN waveforms for various scattering target geometries.

4.1. Diffraction Tomography with a Single Transmitted WGN Waveform for a Single Scattering Object. As shown in Figure 3, two-dimensional backward scattering geometry with a single cylindrical conducting object is simulated with two band-limited iid UWB WGN waveforms. The cylindrical PEC object with a radius of 15 cm is located at the origin, and the cylinder is assumed to be infinitely long along the z-axis.

The scattered field is uniformly sampled at receiving array Rx1 through Rx101 with frequency swept within X-band from 8 GHz to 10 GHz in 41 steps of 50 MHz. The locations of Rx1 and Rx101 are at $(-90\,\text{cm}, -75\,\text{cm})$ and $(-90\,\text{cm}, 75\,\text{cm})$, respectively, and the receiver spacing, Δy, is set to 1.5 cm based on (8) when $f_{\max}$ is 10 GHz. The maximum frequency stepping interval, Δf, is also calculated using (9), which yields a value of 1 GHz; however, 50 MHz frequency stepping interval enhances the quality of tomographic image compared to the maximum frequency stepping interval of 1 GHz.

For two band-limited iid UWB WGN transmitted waveforms, each iid WGN waveform is generated with 500 random amplitude samples drawn from $\mathcal{N}(0, \sigma^2)$, which are

shown in Figures 4 and 5, to reconstruct tomographic images of the cylindrical PEC object based on the scattered field observed at Rx1 through Rx101.

A block diagram shown in Figure 6 displays the tomographic image reconstruction method using diffraction tomography. The one-dimensional Fourier transformed scattered field data collected at the receiving array Rx1 through Rx101 is Fourier transformed into two-dimensional object Fourier space data, $\widetilde{O}(k_x, k_y)$, by using (19). Such Fourier space data is two-dimensional inverse Fourier transformed to obtain the scattering object function, $o(x, y)$.

Fourier space data and the tomographic images of the single cylindrical conducting object, using the first UWB WGN waveform shown in Figure 4(a), are calculated and displayed in Figures 7(a) and 7(b), respectively. The tomographic image of the single PEC cylinder appears to be successfully reconstructed using the first UWB WGN transmitted waveform.

Similarly, both Fourier space data and the tomographic images of the single PEC cylinder using the second UWB WGN waveform shown in Figure 5(a) are also shown in Figures 8(a) and 8(b), respectively. In this case, the tomographic image is significantly affected by the unexpected notch observed in the frequency spectrum at 9.1 GHz in Figure 5(b), so that the tomographic image of the object cannot be achieved correctly compared to the previous case where the first WGN waveform is transmitted. Based on the simulation results with two band-limited iid UWB WGN waveforms, a single transmission of WGN waveform may or may not be sufficient to reconstruct a successful tomographic image of the object with diffraction tomography algorithm due to the undesired and unexpected notches in the spectral density of the practical UWB WGN transmitted waveforms.

4.2. Diffraction Tomography with Multiple Transmitted iid WGN Waveforms for a Single Scattering Object. Figure 9 shows K iid UWB WGN waveforms being transmitted to reconstruct a final image of the object in order to bypass a shortcoming of the single transmission of WGN waveform displayed in Figure 8(b). The proposed imaging method with multiple iid WGN transmitted waveforms is established based on the method of the WGN periodogram averaging described in Section 2.

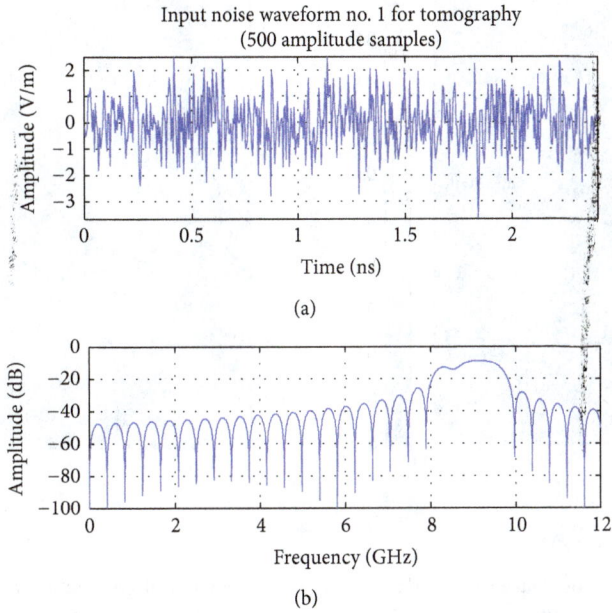

FIGURE 4: (a) The first band-limited UWB WGN transmitted waveform generated with 500 amplitude samples ($l = 500$) drawn from $\mathcal{N}(0, \sigma^2)$. Pulse duration is 2.4 ns. (b) The frequency spectrum of the time domain WGN waveform shown in Figure 4(a). The frequency ranges are shown from DC to 12 GHz only [8].

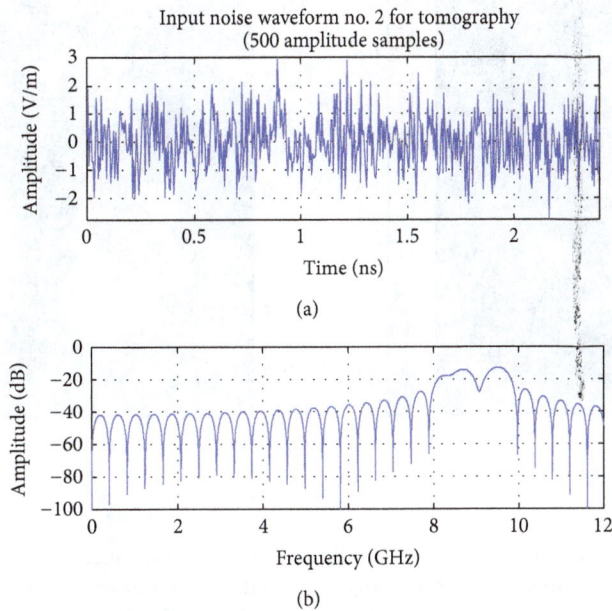

FIGURE 5: (a) The second band-limited UWB WGN transmitted waveform generated with 500 amplitude samples ($l = 500$) drawn from $\mathcal{N}(0, \sigma^2)$. Pulse duration is 2.4 ns. (b) The frequency spectrum of the time domain WGN waveform is shown in Figure 5(a). The frequency ranges are shown from DC to 12 GHz only [8].

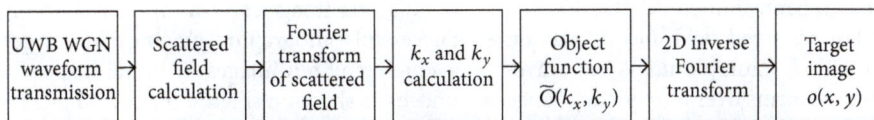

FIGURE 6: The image reconstruction method using diffraction tomography [8].

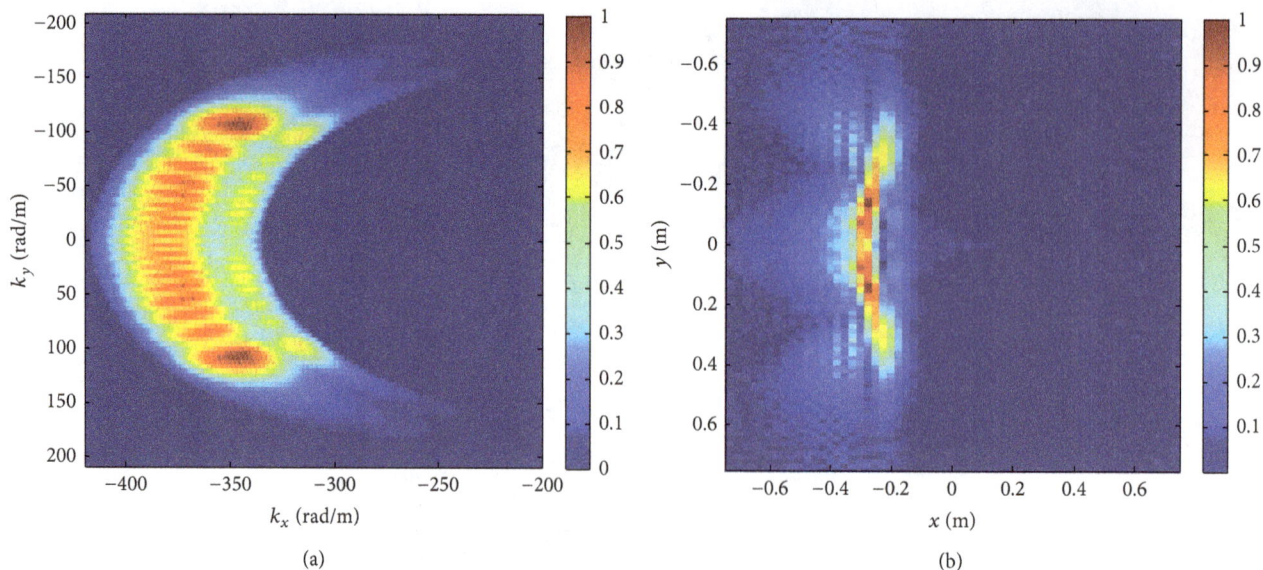

FIGURE 7: (a) The normalized magnitude of Fourier space data of the single cylindrical conducting object. The colorbar indicates the normalized magnitude of $\widetilde{O}(k_x, k_y)$. (b) The tomographic image using the first UWB WGN waveform shown in Figure 4(a). The colorbar indicates the normalized magnitude of scattering object function, $o(x, y)$.

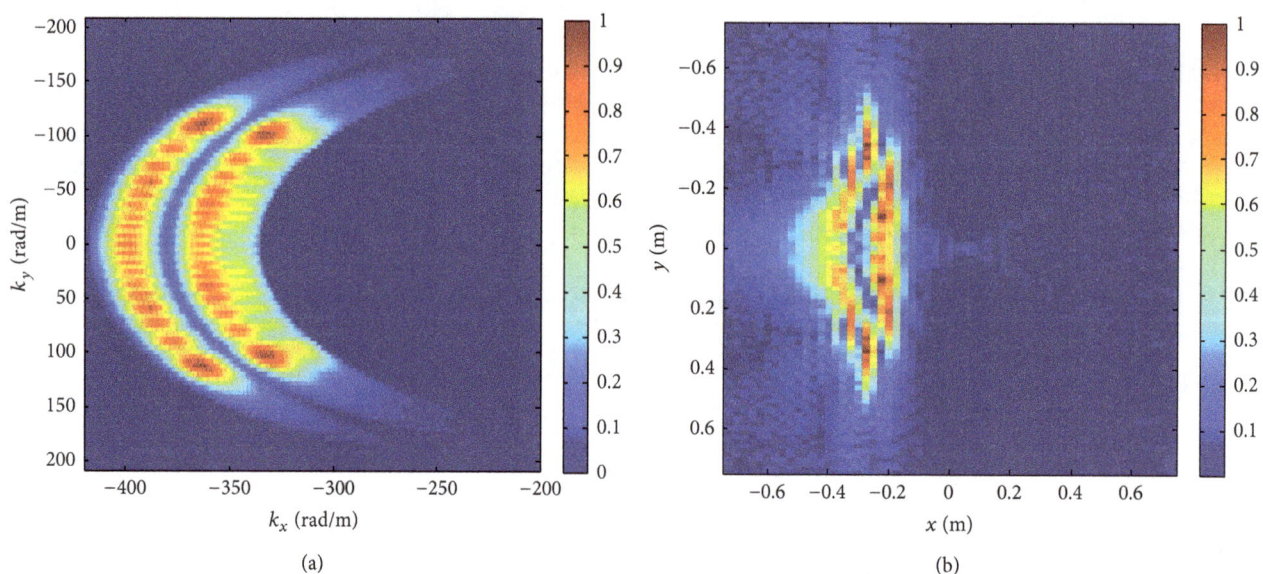

FIGURE 8: (a) The normalized magnitude of Fourier space data of the single cylindrical conducting object. The colorbar indicates the normalized magnitude of $\widetilde{O}(k_x, k_y)$. (b) The tomographic image using the second UWB WGN waveform shown in Figure 5(a). The colorbar indicates the normalized magnitude of scattering object function, $o(x, y)$.

As shown in Figure 9, the final tomographic image of the scattered object is reconstructed via sum and average of all K discrete images for Kth band-limited iid UWB WGN transmitted waveforms. For scattered field and diffraction tomography simulations with K multiple iid WGN waveforms, 10 iid UWB WGN waveforms over a frequency range from 8 to 10 GHz are generated with 500 random amplitude samples drawn from $\mathcal{N}(0, \sigma^2)$ and transmitted for backward scattering field data.

Figures 10(a), 10(b), 10(c), and 10(d) display the four final tomographic images when one, three, seven and all ten discrete images are summed and averaged, respectively. Successful tomographic imaging of the target is achieved after averaging all ten images by visual inspection of the formed images as shown in Figure 10(d), and increasing the number of transmissions of the iid UWB WGN waveform tends to enhance the quality of final tomographic image of the object by reducing the variance of the spectral response of WGN.

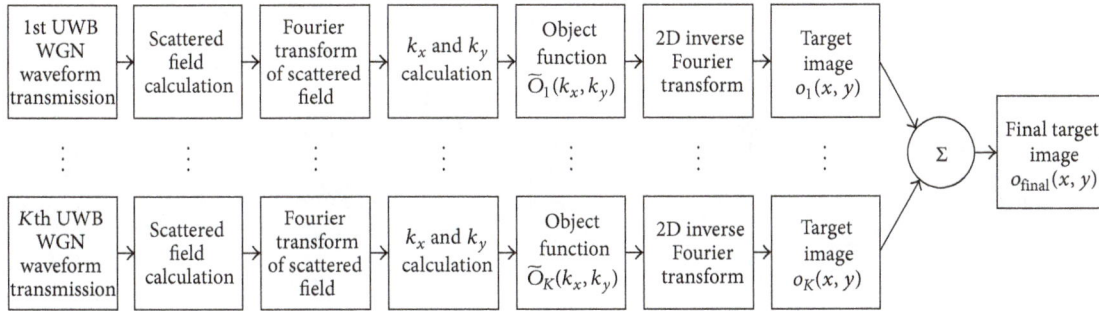

FIGURE 9: The image reconstruction method with K multiple iid WGN transmitted waveforms using diffraction tomography.

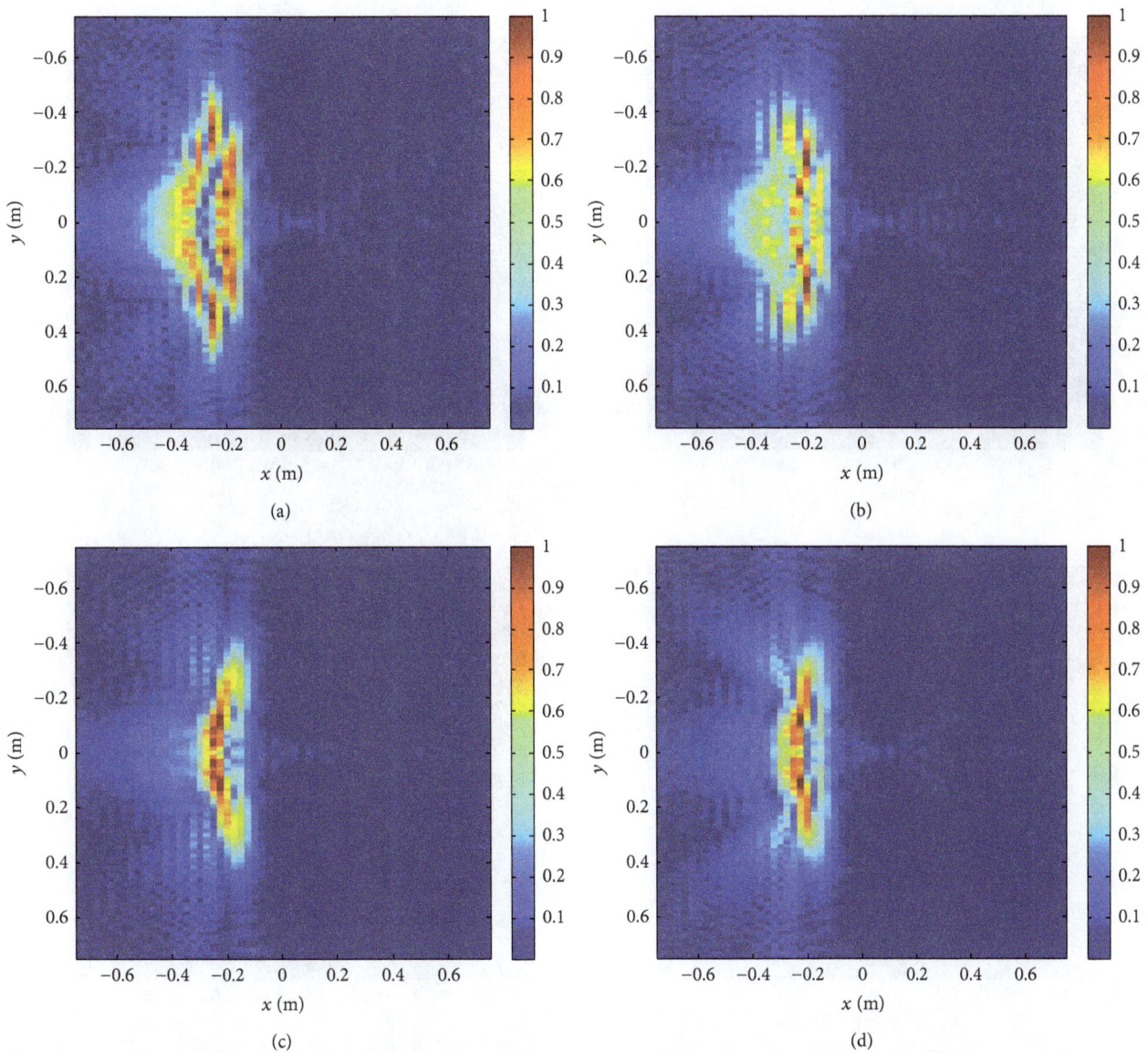

FIGURE 10: The final tomographic image of a single PEC cylinder located at $(0, 0)$ after summing and averaging process with the (a) one transmitted WGN waveform image, (b) three transmitted WGN waveform images, (c) seven transmitted WGN waveform images, and (d) all ten transmitted WGN waveform images. The colorbar indicates the normalized magnitude of scattering object function, $o(x, y)$.

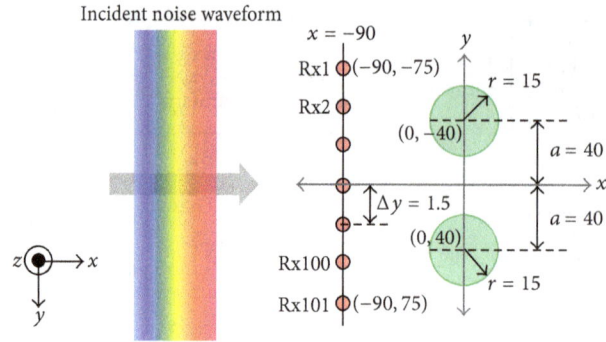

FIGURE 11: Two-dimensional backward scattering simulation geometry for two symmetrically distributed cylindrical conducting objects. PEC cylinders with radii of 15 cm are located at (0 cm, −40 cm) and (0 cm, 40 cm), and a linear receiving array is located 90 cm away from the origin in −x direction.

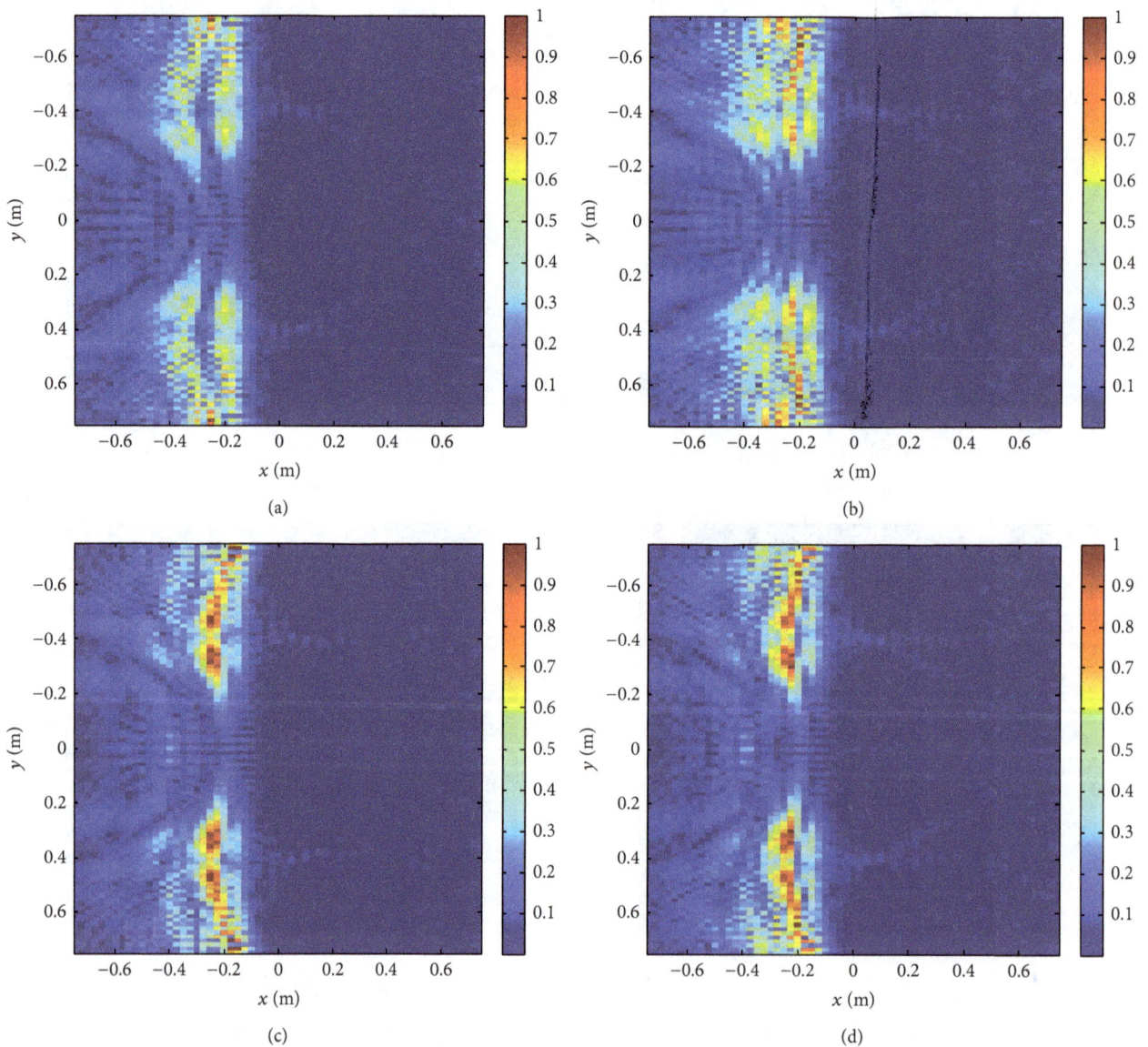

FIGURE 12: The final tomographic image of two symmetrically distributed PEC cylinders located at (0 cm, −40 cm) and (0 cm, 40 cm) after summing and averaging process with the (a) one transmitted WGN waveform image, (b) three transmitted WGN waveform images, (c) seven transmitted WGN waveform images, and (d) all ten transmitted WGN waveform images.

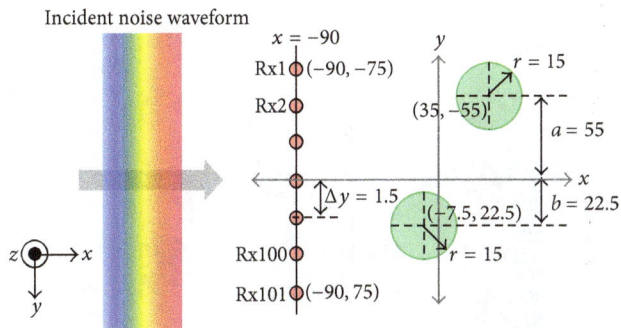

FIGURE 13: Two-dimensional backward scattering simulation geometry for two randomly distributed cylindrical conducting objects. PEC cylinders with radii of 15 cm are located at (−7.5 cm, 22.5 cm) and (35 cm, −55 cm).

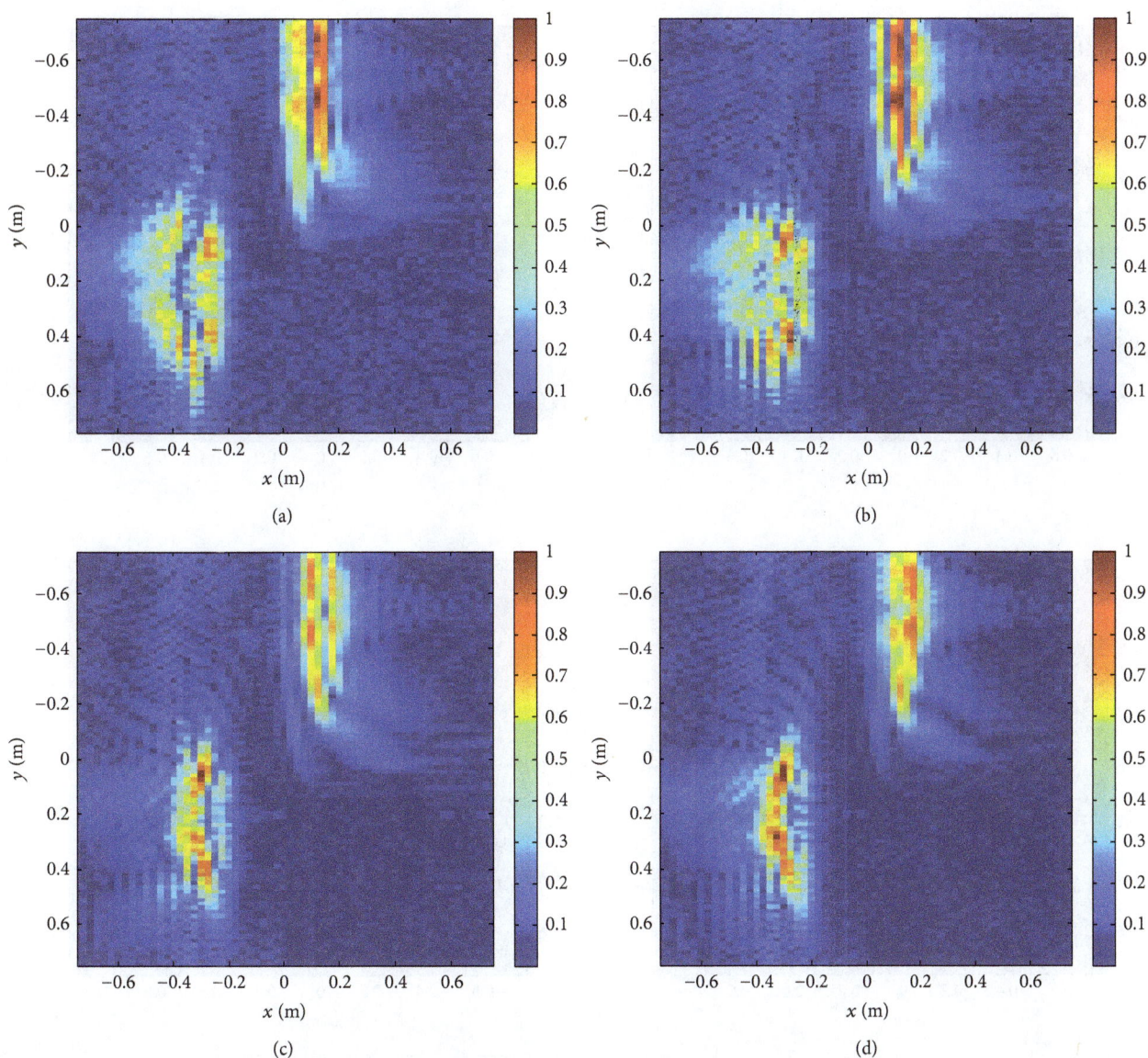

FIGURE 14: The final tomographic image of two randomly distributed PEC cylinders located at (−7.5 cm, 22.5 cm) and (35 cm, −55 cm) after summing and averaging process with the (a) one transmitted WGN waveform image, (b) three transmitted WGN waveform images, (c) seven transmitted WGN waveform images, and (d) all ten transmitted WGN waveform images.

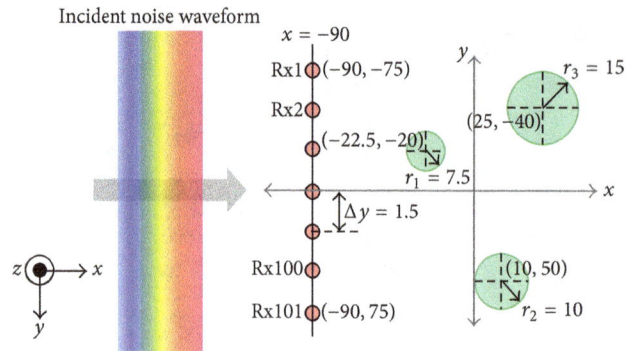

FIGURE 15: Two-dimensional backward scattering simulation geometry for three randomly distributed cylindrical conducting objects in different sizes. Three PEC cylinders with radii of 7.5 cm, 10 cm, and 15 cm are located at (−22.5 cm, −20 cm), (10 cm, 50 cm), and (25 cm, −40 cm), respectively.

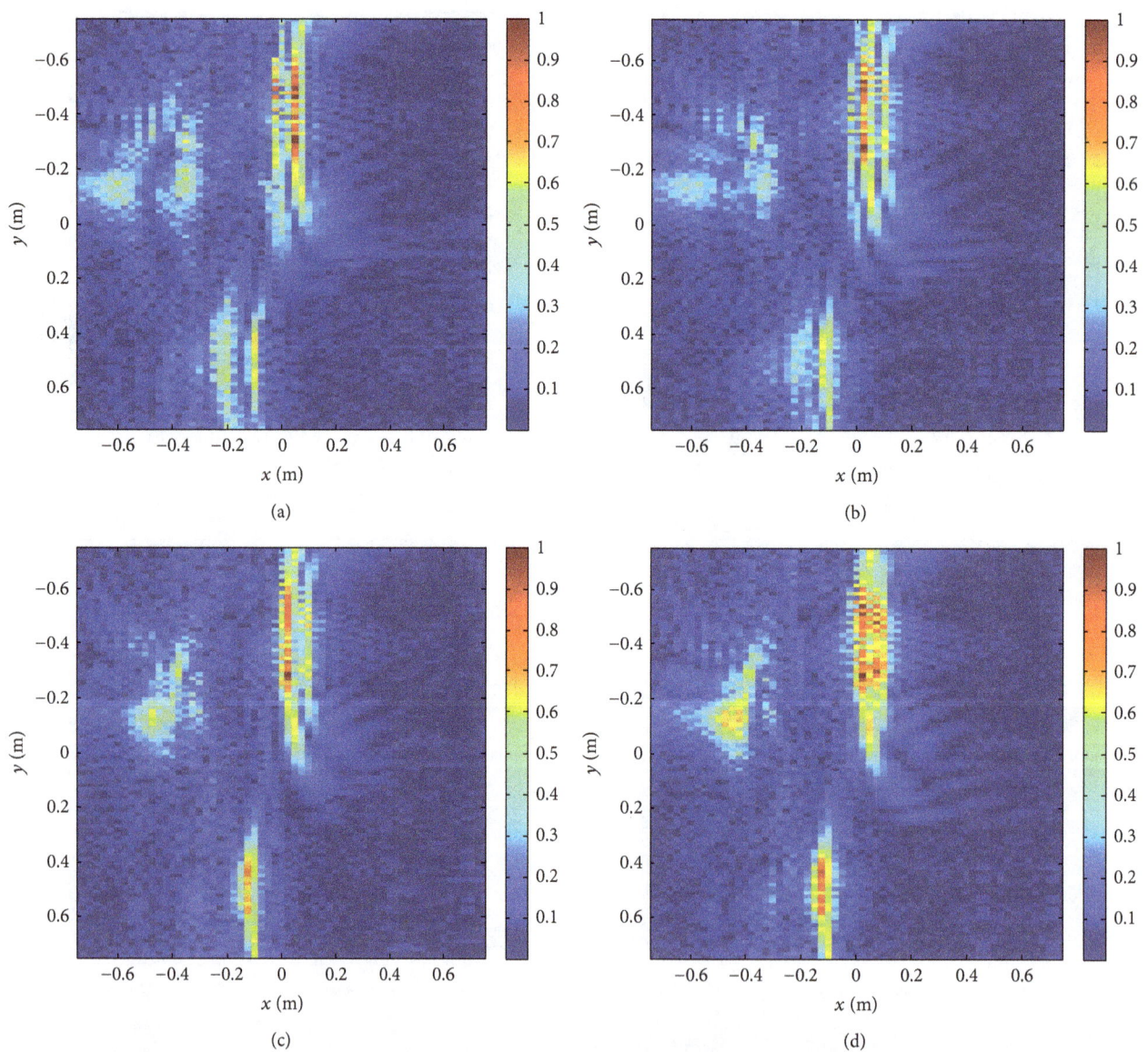

(a)

(b)

(c)

(d)

FIGURE 16: The final tomographic image of three randomly distributed PEC cylinders in different sizes after summing and averaging process with the (a) one transmitted WGN waveform image, (b) three transmitted WGN waveform images, (c) seven transmitted WGN waveform images, and (d) all ten transmitted WGN waveform images.

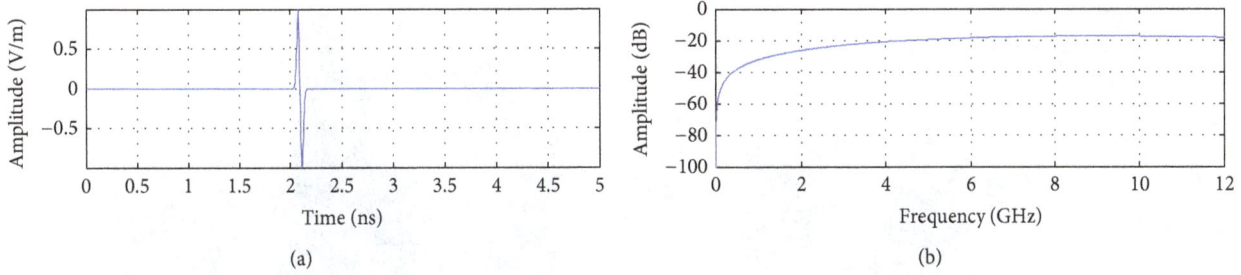

FIGURE 17: (a) The first derivative Gaussian input waveform with a pulse width of 0.1 ns. (b) The frequency spectrum of the first derivative Gaussian input waveform shown in Figure 17(a). The frequency spectrum ranges are displayed from DC to 12 GHz only.

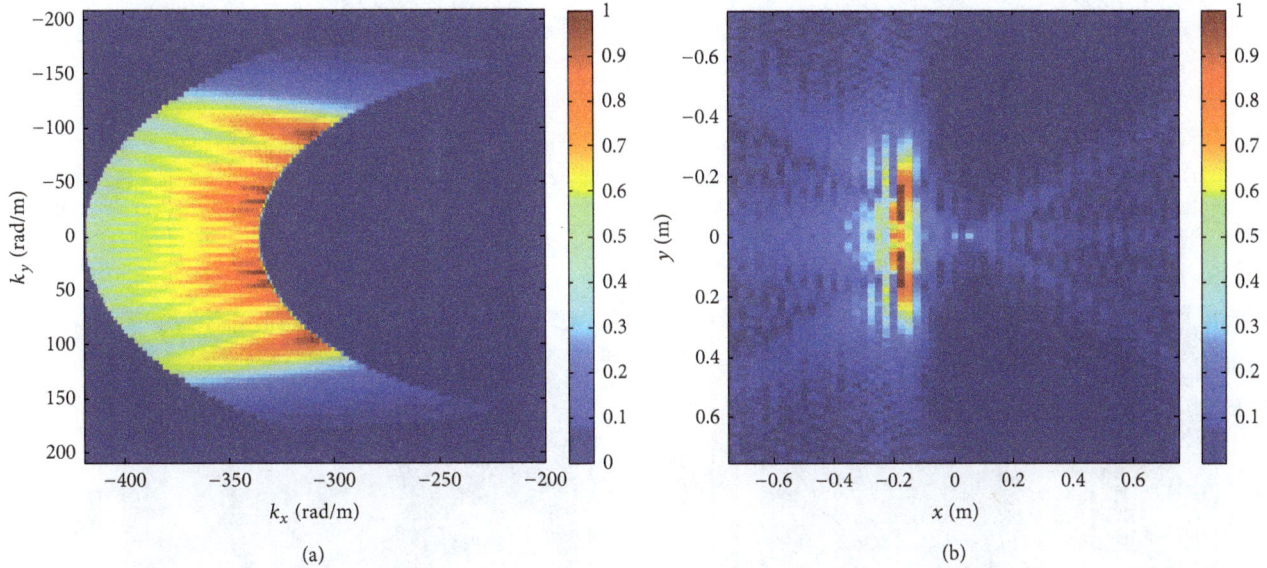

FIGURE 18: (a) The magnitude of Fourier space data based on the transmission of the first derivative Gaussian waveform shown in Figure 17(a) for the single cylindrical PEC object shown in Figure 3. The colorbar indicates the normalized magnitude of $\widetilde{O}(k_x, k_y)$. (b) The tomographic image of the single cylindrical conducting object located at $(0, 0)$. The colorbar indicates the normalized magnitude of scattering object function, $o(x, y)$.

4.3. Diffraction Tomography with Multiple Transmitted iid WGN Waveforms for Two Symmetrically Distributed Scattering Objects.

Based on the simulation result of the single PEC cylinder case with multiple transmitted noise waveforms, we can point out that transmitting multiple iid UWB WGN waveforms delivers the acceptable quality of image of the object for diffraction tomography.

Two-dimensional backward scattering geometry for two cylindrical conducting objects, which is shown in Figure 11, is also simulated with 10 iid UWB WGN waveforms. Two cylindrical PEC objects with radii of 15 cm are located at (0 cm, −40 cm) and (0 cm, 40 cm) such that they are symmetrically positioned with respect to the x-axis. The distance from the center of each PEC cylinder to the x-axis, a, is 40 cm, and they are also assumed to be infinitely long along the z-axis. The coordinates of all 101 receivers, and the receiver spacing, Δy, remained the same as shown in Figure 3. The scattered field is uniformly sampled at Rx1 through Rx101 with

same frequency swept within X-band over 8–10 GHz in 41 steps as well.

For scattered field and diffraction tomography simulations for multiple scattering objects with multiple iid WGN waveforms, the same 10 iid UWB WGN waveforms, which are used for the simulation with a single PEC object in the previous section, are transmitted in the same sequence. The scattered field due to two symmetrically positioned PEC objects is calculated based on the equations derived in Section 3, and the final tomographic image of objects is formed via the proposed image reconstruction method shown in Figure 9.

Figures 12(a), 12(b), 12(c), and 12(d) display four final tomographic images for two symmetrically distributed PEC cylinders cases when one, three, seven and all ten discrete images are summed and averaged, respectively. As shown in Figure 12(d), tomographic image of the two PEC cylinders is successfully reconstructed with multiple transmitted noise waveforms as expected. However, the mutual coupling effects

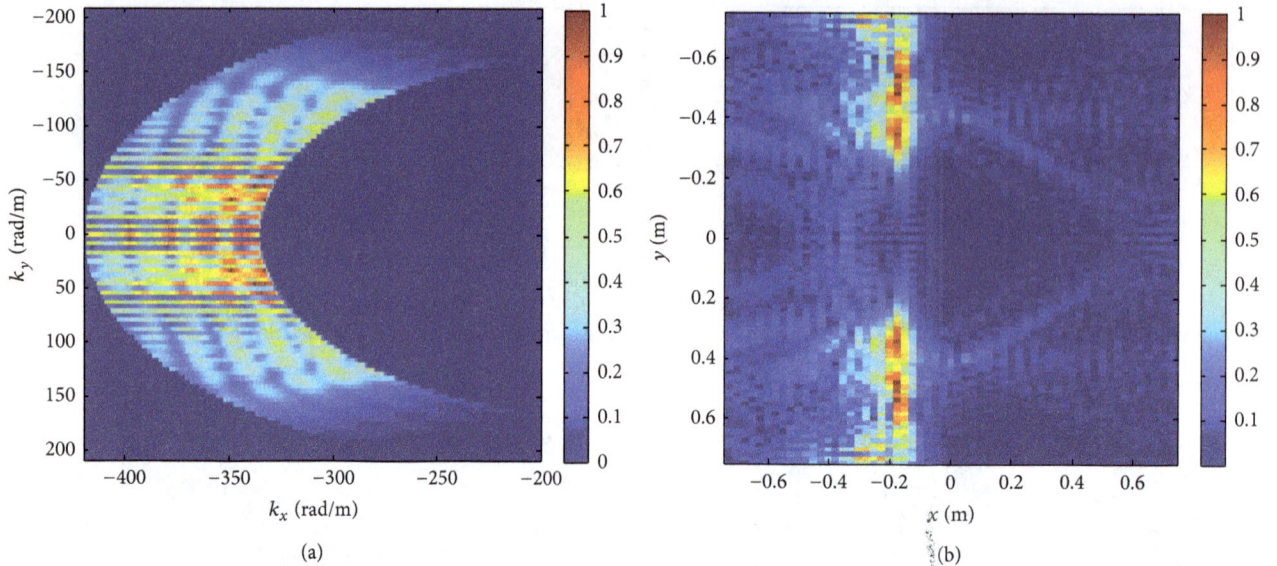

FIGURE 19: (a) The magnitude of Fourier space data based on the transmission of the first derivative Gaussian waveform shown in Figure 17(a) for two symmetrically distributed cylindrical PEC objects shown in Figure 11. (b) The tomographic image of two symmetrically distributed cylindrical conducting objects located at (0 cm, −40 cm) and (0 cm, 40 cm).

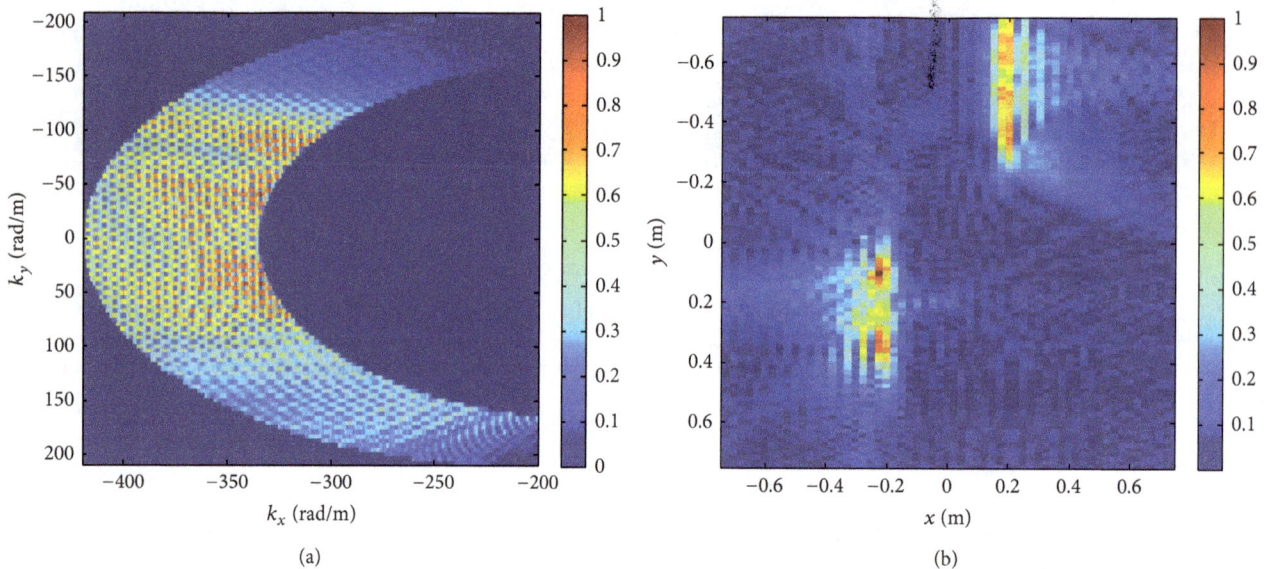

FIGURE 20: (a) The magnitude of Fourier space data based on the transmission of the first derivative Gaussian waveform shown in Figure 17(a) for two randomly distributed cylindrical PEC objects shown in Figure 13. (b) The tomographic image of two randomly distributed cylindrical conducting objects located at (−7.5 cm, 22.5 cm) and (35 cm, −55 cm).

due to the multiple PEC objects are also imaged in all tomographic images.

4.4. Diffraction Tomography with Multiple Transmitted iid WGN Waveforms for Two Randomly Distributed Scattering Objects.

Two-dimensional backward scattering geometry for two randomly distributed cylindrical conducting objects is shown in Figure 13. The radii of both cylinders are 15 cm, and the positions of two cylinders are (−7.5 cm, 22.5 cm) and (35 cm, −55 cm). The general configuration of simulation geometry, such as number of transmitted UWB WGN waveforms, frequency swept ranges, coordinates of the linear receiving array, and the receiver spacing, is identical to the previous cases.

Four final tomographic images for two randomly distributed PEC cylinders are displayed in Figures 14(a), 14(b), 14(c), and 14(d) when one, three, seven and all ten discrete images are summed and averaged, respectively. Again, the mutual coupling effects are shown in the tomographic images as expected. The reconstructed images are shown to be in good agreement with the simulation geometry given in Figure 13.

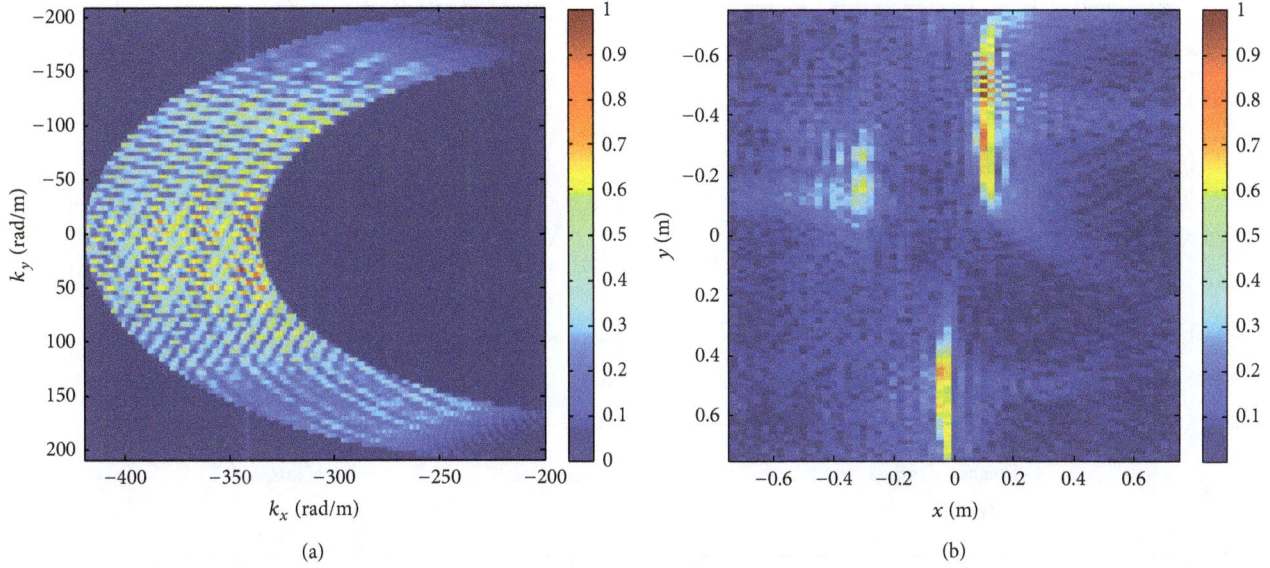

(a)   (b)

FIGURE 21: (a) The magnitude of Fourier space data based on the transmission of the first derivative Gaussian waveform shown in Figure 17(a) for three randomly distributed cylindrical PEC objects in different sizes shown in Figure 15. (b) The tomographic image of three randomly distributed cylindrical conducting objects in different sizes located at (−22.5 cm, −20 cm), (10 cm, 50 cm), and (25 cm, −40 cm).

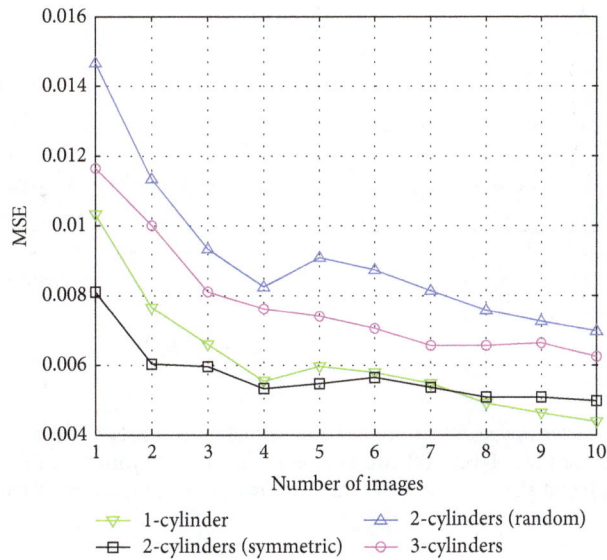

FIGURE 22: MSE versus number of reconstructed images after normalization. MSE decreases as the number of reconstructed images is overlapped.

4.5. Diffraction Tomography with Multiple Transmitted iid WGN Waveforms for Three Randomly Distributed Scattering Objects in Different Sizes. As shown in Figure 15, three PEC cylinders with radii of 7.5 cm, 10 cm, and 15 cm are located at (−22.5 cm, −20 cm), (10 cm, 50 cm), and (25 cm, −40 cm), respectively. Again, the general configurations of simulation geometry remain unchanged as defined in previous cases.

Figures 16(a), 16(b), 16(c), and 16(d) show four final tomographic images for the scattering geometry given in Figure 15 when one, three, seven and all ten discrete images are

summed and averaged, respectively. As shown in the previous cases, the image quality of reconstructed tomographic images for unevenly distributed multiple scattering objects is affected by the mutual coupling effects. Successful tomographic image of the target is reconstructed based on multiple transmitted noise waveforms as shown in Figure 16(d).

Based on the numerical simulation results of various scattering target geometries using multiple band-limited iid UWB WGN waveforms, we conclude that increasing the number of transmissions of the iid UWB WGN waveform

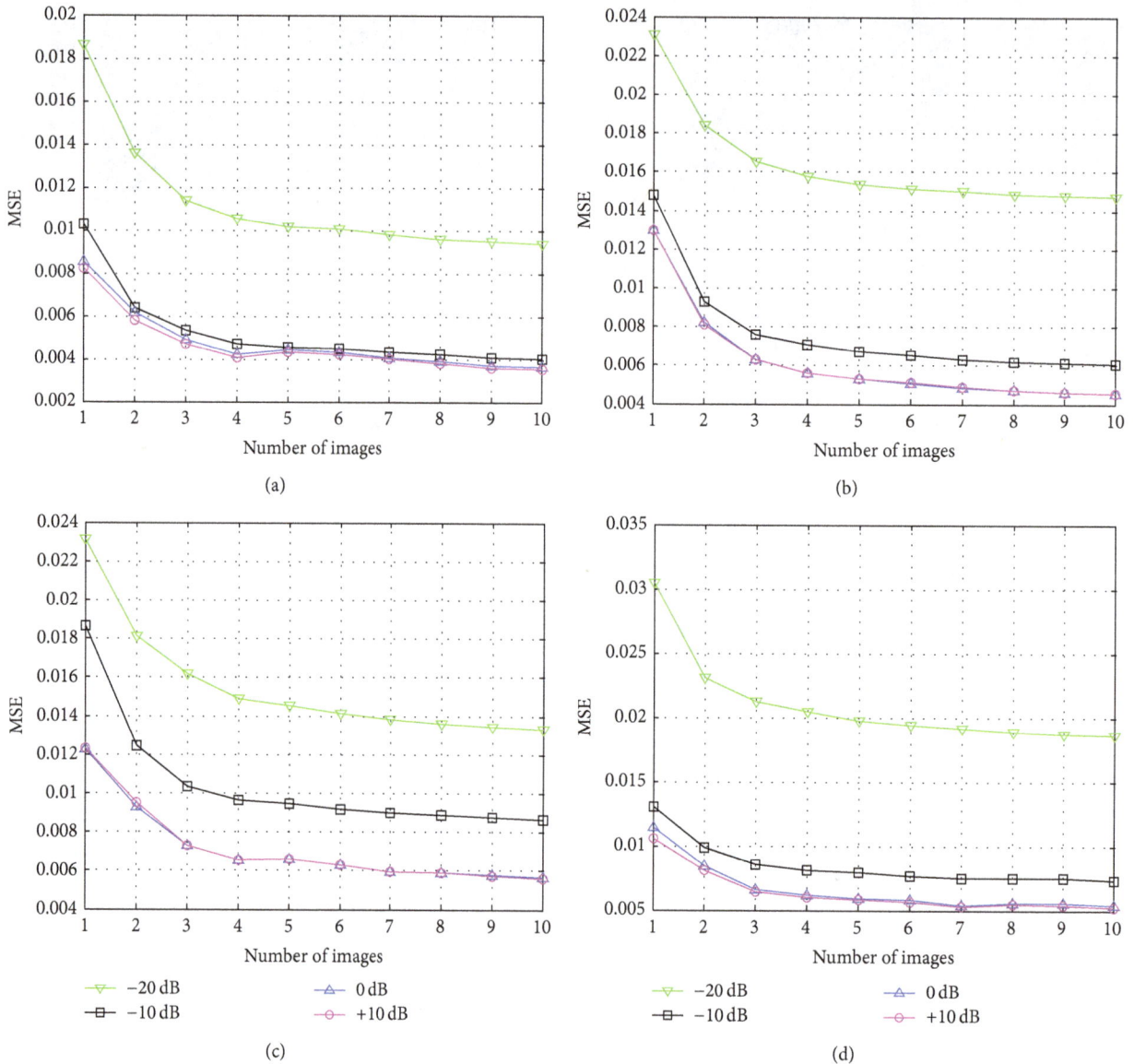

FIGURE 23: MSE versus number of reconstructed images after normalization with 4 different SNR values for (a) single cylindrical PEC object, (b) two symmetrically distributed cylindrical PEC objects, (c) two randomly distributed cylindrical PEC objects, and (d) three randomly distributed cylindrical PEC objects in different sizes. For all cases, MSE decreases as the number of reconstructed images is overlapped regardless of SNRs.

improves the quality of tomographic image by reducing the variance of the spectral response of WGN as stated in Section 2.

5. Image Quality Measure

As concluded in Section 4, increasing the number of transmissions of the iid WGN waveform tends to enhance the formed tomographic image. The image quality measures (IQMs) are discussed to determine the quality of tomographic images based on quantitative analysis. A good subjective assessment is required to evaluate the image quality and the performance of imaging systems. Various methods

to measure the image quality and investigate their statistical performance have been studied [34]. The most frequently used measures are deviations between the reference and reconstructed image with varieties of the mean square error (MSE) [35, 36]. The reasons for the widespread popularity of the analysis based on MSE calculations are their mathematical tractability and the fact that it is often straightforward to design systems that minimize the MSE.

5.1. Reference Image Generation. Prior to measuring the image quality of reconstructed tomographic images shown in the previous section, a "reference" image is obtained via diffraction tomography proposed in Figure 6 using the first

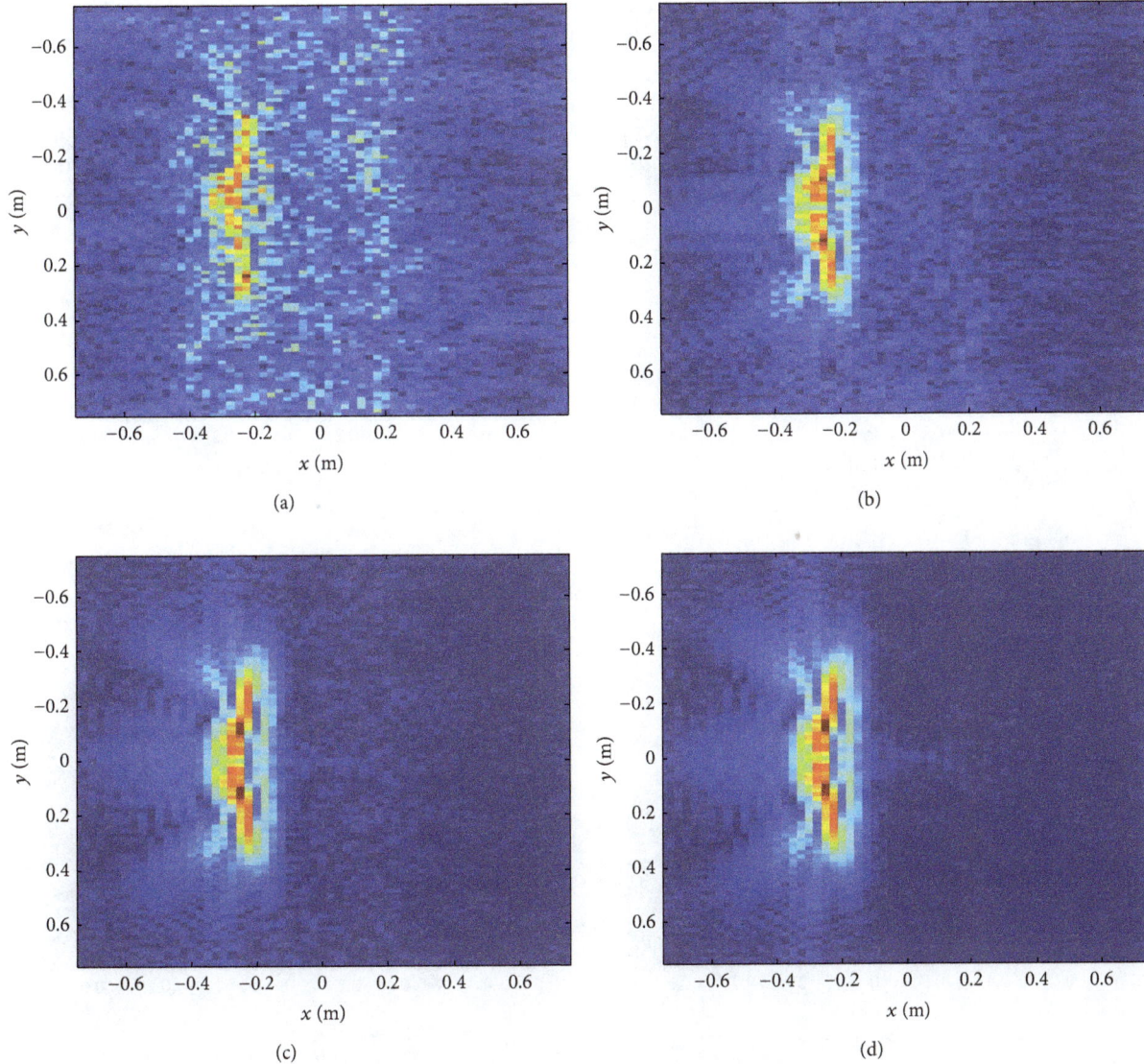

FIGURE 24: The final tomographic images after summing and averaging process with the 10 iid noise waveforms with various SNRs for a single PEC cylinder: (a) SNR = −20 dB, (b) SNR = −10 dB, (c) SNR = 0 dB, and (d) SNR = +10 dB.

derivative Gaussian waveform. The first derivative Gaussian waveform possesses the desirable property in that it has zero average value. Since it has no DC component, the first derivative Gaussian waveform is the most popular pulse shape for various UWB applications [37, 38]; thus, it can be considered to be an appropriate waveform for generating reference images to compare with our reconstructed images. Figure 17(a) displays the first derivative Gaussian input waveform with a pulse width of 0.1 ns, and the frequency spectrum of the pulse is relatively flat in the range of 8–10 GHz as shown in Figure 17(b).

The Fourier space data and the tomographic image of four scattering target geometries in Section 4 are generated with the first derivative Gaussian input waveform shown in Figure 17(a), and they are displayed in Figures 18, 19, 20, and 21. These tomographic images are considered to be the reference images for IQM analysis.

5.2. Pixel Difference-Based Measure. The tomographic images shown in Section 4 are obtained based on 10 iid UWB WGN transmitted waveforms, and the image enhancement technique is not implemented in diffraction tomography algorithm. Every pixel in all images is generated based on the scattering object function defined in (22). MSE is the cumulative mean squared error between the corresponding pixels of the reference images and the reconstructed images based on multiple iid UWB WGN transmitted waveforms. MSE is defined as

$$\text{MSE} = \frac{1}{MN} \sum_{m=0}^{M-1} \sum_{n=0}^{N-1} \left[R\left(x_m, y_n\right) - S\left(x_m, y_n\right) \right]^2, \quad (24)$$

where $R(x_m, y_n)$ and $S(x_m, y_n)$ represent the value of each pixel of the reference image and reconstructed tomographic

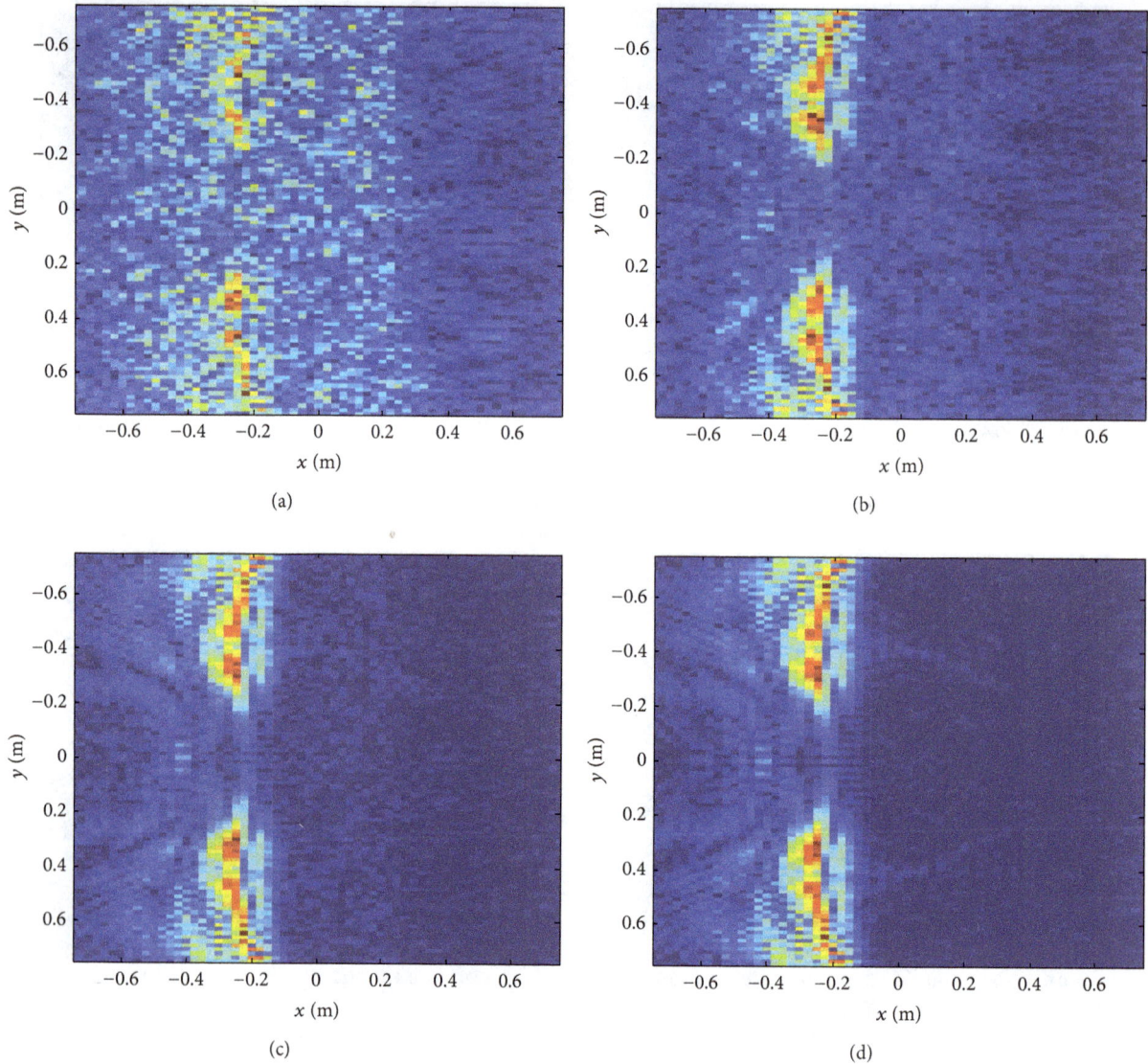

FIGURE 25: The final tomographic images after summing and averaging process with the 10 iid noise waveforms with various SNRs for two symmetrically distributed PEC cylinders: (a) SNR = −20 dB, (b) SNR = −10 dB, (c) SNR = 0 dB, and (d) SNR = +10 dB.

images with multiple iid UWB WGN transmitted waveforms, respectively, and M and N are the dimensions of the images.

Each pixel of tomographic images represents the normalized magnitude of scattering object function, $o(x, y)$, and the dimensions of images are equivalent to the number of pixels across the x-axis and y-axis. By using (24), MSE values are calculated to evaluate the deviation of the pixel values from those of the corresponding reference image, as the iid noise images are summed, averaged, and normalized. Figure 22 displays the MSE versus number of overlapping reconstructed images. Such quantitative analysis shows that the tomographic images of the objects become clear and distinctive as the number of reconstructed images based on iid WGN waveforms increases. To be more specific, as shown in Figure 22, MSE decreases as the number of overlapped reconstructed images based on iid UWB WGN transmitted waveforms increases.

5.3. *Signal-to-Noise Ratio Effects on Image Quality.* All tomographic images shown above are formed in the absence of noise in the system; that is, signal-to-noise ratio (SNR) is equal to infinity. In practical situations, the received signal shows unpredictable perturbations due to the contributions of various noise sources in communication and imaging systems [39, 40]. For example, thermal noise is present in all electronic devices and transmission media and is uniformly distributed across the frequency spectrum. When backscattering data acquisition is performed with a signal analyzer, the recorded data consists of scattered field information with some thermal noise from the equipment and cables. However, the collected data can be viewed as the sum of the scattered field and the additive white Gaussian noise since they are uncorrelated.

In this section, the white Gaussian noise is added accordingly to the collected scattered field dataset, establishing

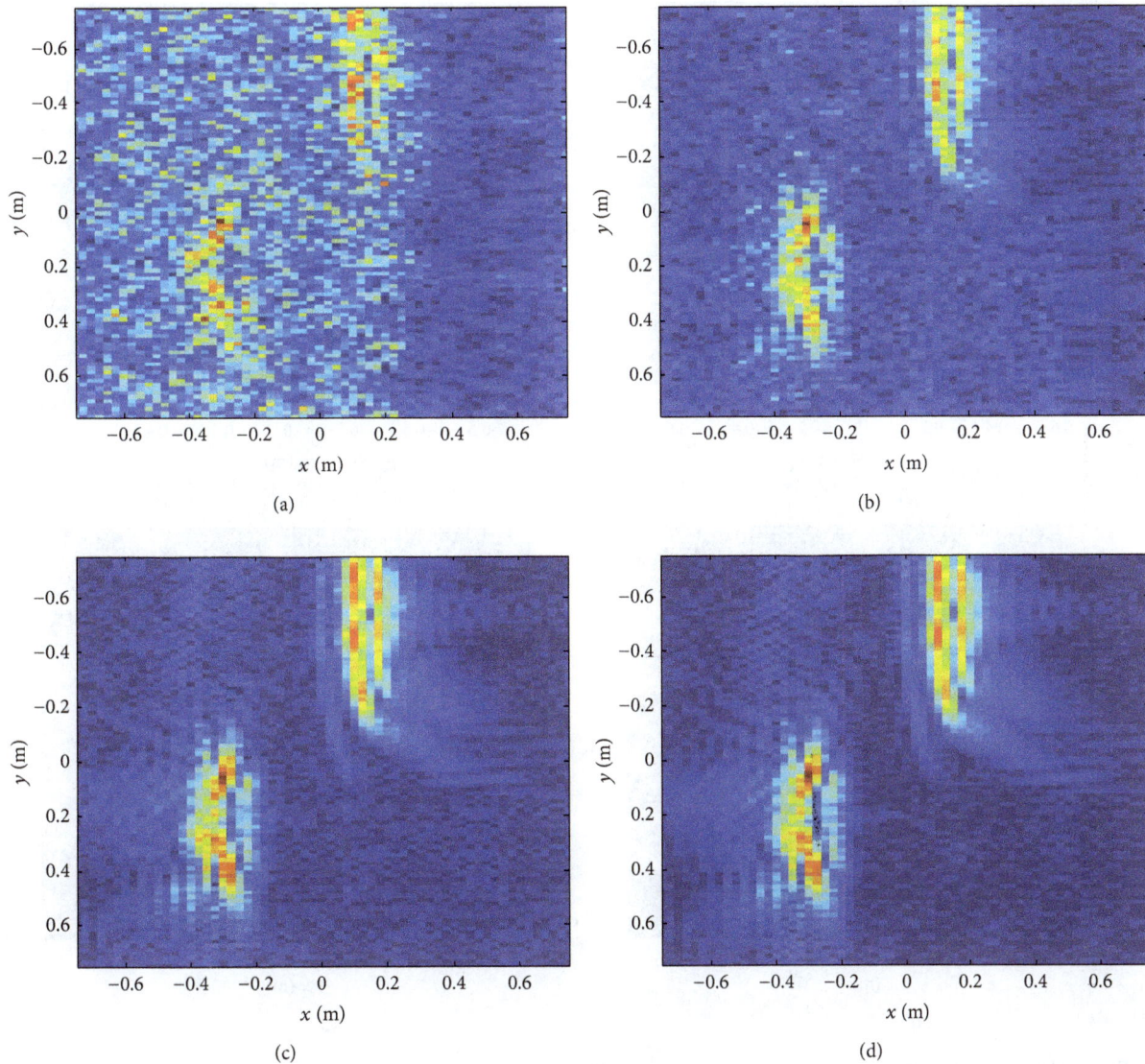

FIGURE 26: The final tomographic images after summing and averaging process with the 10 iid noise waveforms with various SNRs for two randomly distributed PEC cylinders: (a) SNR = −20 dB, (b) SNR = −10 dB, (c) SNR = 0 dB, and (d) SNR = +10 dB.

simulation environment with 4 different SNR values: −20 dB, −10 dB, 0 dB, and +10 dB. Prior to measuring the quality of the final tomographic images with various SNRs, MSE values are calculated to determine the deviation of the pixel values from that of the corresponding reference image, as the iid noise images generated with the corresponding SNR values are summed, averaged, and normalized. Based on the previous quantitative analysis results shown in Figure 22, the image quality of the tomographic image is expected to be enhanced as the number of reconstructed images increases regardless of SNR values. Figure 23 displays the MSE versus number of overlapping reconstructed images with 4 different SNRs for four scattering target geometries shown in Section 4.

After completion of total 10 transmissions of iid noise waveforms for 4 different SNR values defined above, the final tomographic images of all four scattering geometries are achieved and displayed in Figures 24, 25, 26, and 27.

As shown in Figure 24 through Figure 27, the image quality of the final tomographic image after summing, averaging, and normalizing process with transmitting 10 iid noise waveforms is truly affected by the additive white Gaussian noise. By visual inspection of the formed images, the image degradation due to the additive Gaussian noise is clearly displayed when SNR is set to −20 dB. In contrast, no such image degradation is observed at relatively higher SNRs, that is, SNR of +10 dB.

In order to evaluate the overall impact of noise effect on the formed tomographic images, MSE is calculated at the SNR range from −20 dB to +10 dB. Figure 28 explains the relationship between the image quality degradation and SNR. The rate of change in MSE is the maximum at the SNR range between −20 dB and −10 dB, and this is the SNR range where the final tomographic image is severely degraded by unwanted noise. However, the rate of change

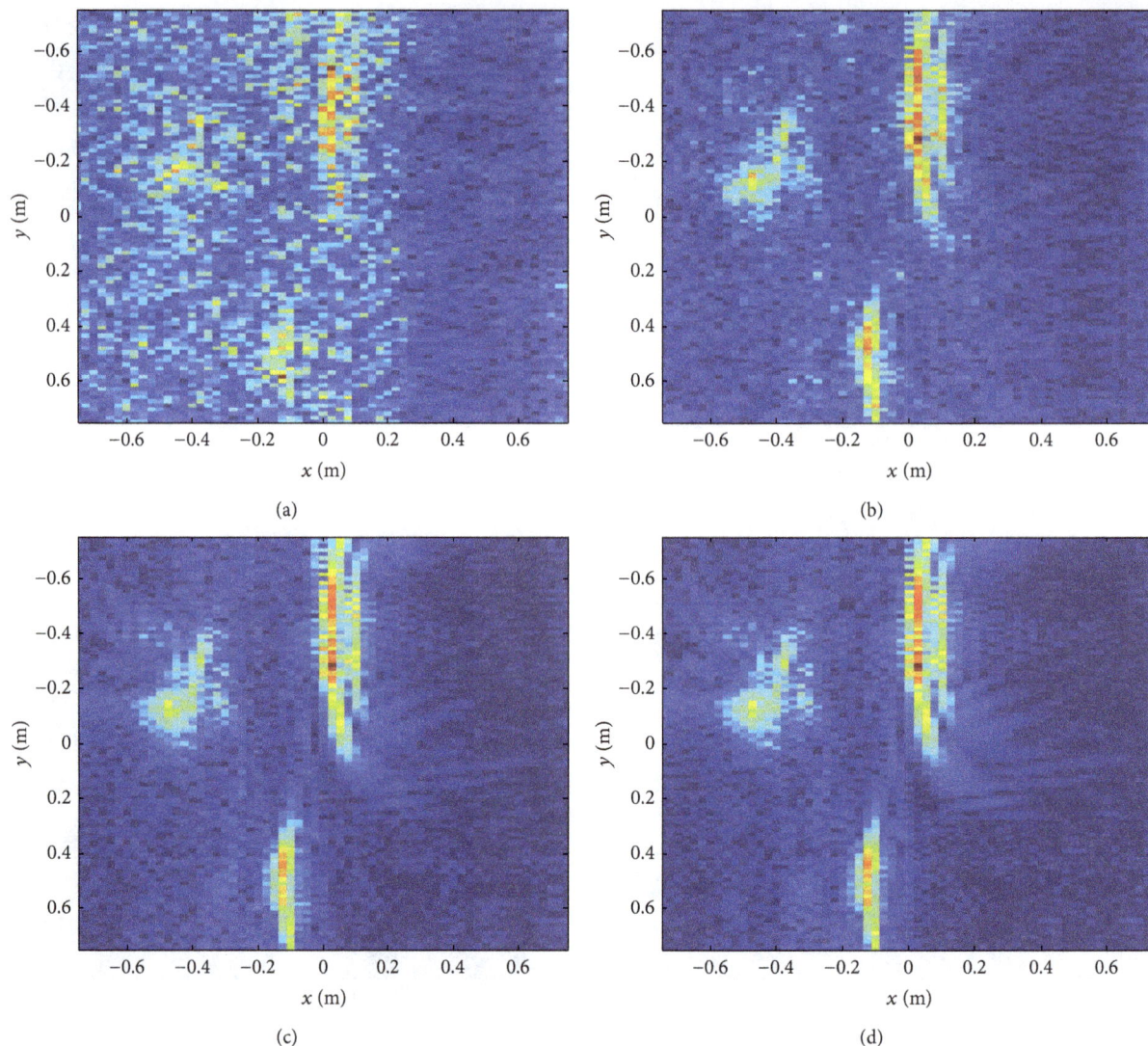

FIGURE 27: The final tomographic images after summing and averaging process with the 10 iid noise waveforms with various SNRs for three randomly distributed PEC cylinders in different sizes: (a) SNR = −20 dB, (b) SNR = −10 dB, (c) SNR = 0 dB, and (d) SNR = +10 dB.

in MSE starts falling as SNR increases, which indicates that the tomographic image is less affected by the noise contributions at high SNR. The results shown in Figure 28 are the rough measure to estimate the image quality on the final tomographic images in practical implementation of the imaging system and designs for actual experiment.

5.4. Preliminary Experimental Validation. The two-dimensional backward scattering geometry with a single cylindrical conducting object shown in Figure 3 is established for validating simulation results. The basic hardware configuration for the preliminary experiment is shown in Figure 29. For data acquisition, a scanner is configured to transmit the noise waveform via Tx antenna and collect the scattered field via Rx antenna at the same location on the y-axis, and both Tx and Rx antennas move along the scanning direction after collecting scattering data. The data acquisition process repeats for 101 positions along the scanning

direction. The first and last scanning positions are (−89 cm, −40 cm) and (−89 cm, 40 cm), respectively. In the transmit chain, the arbitrary waveform generator (AWG) generates a white Gaussian noise waveform with 500 random amplitude samples over the 2 GHz to 4 GHz frequency range, and a low pass filter (LPF) for frequencies from DC to 4 GHz is implemented to reject undesired high frequency signals coming from the AWG before frequency mixing. A signal generator provides the 6 GHz signal for upconversion, and a mixer is used to upconvert the generated noise waveform frequency from 8 GHz to 10 GHz. The receiver is designed to collect the scattering data with a signal analyzer, and the final tomographic image of the target is obtained after postprocessing with a computer.

For validation purposes, only a single noise waveform is transmitted to validate the simulation results by generating the tomographic image of the target. Experimental results of Fourier space data and the tomographic images

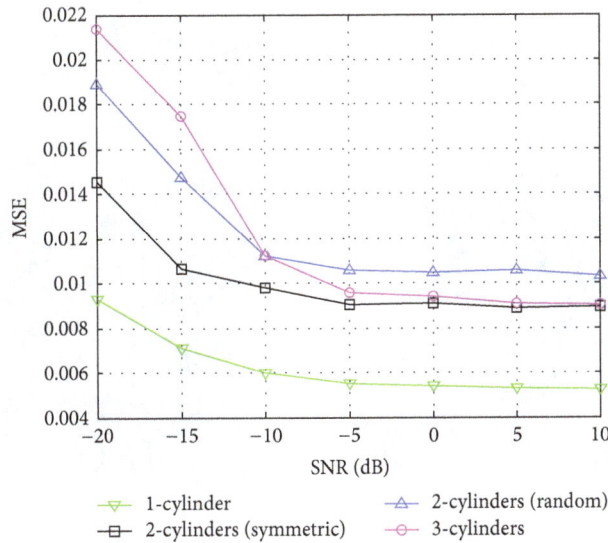

FIGURE 28: MSE versus SNR values for all final tomographic images shown in Figure 24 through Figure 27. MSE decreases as the level of SNR increases.

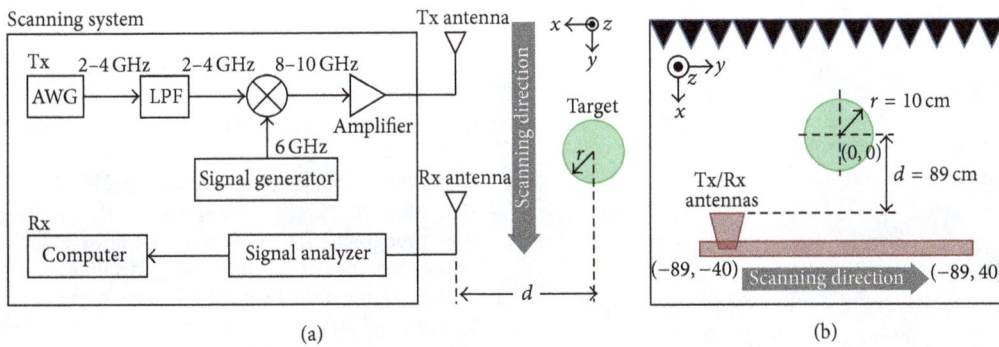

FIGURE 29: (a) Experiment setup for the two-dimensional backward scattering. The scanner is designed to transmit and receive the signals along the scanning direction. (b) Top view of experiment configuration for a single scattering geometry. The radius (r) of the cylindrical target is 10 cm, and the distance (d) from the target to the antennas is 89 cm.

of the single cylindrical conducting object are displayed in Figures 30(a) and 30(b), respectively. As shown in Figure 30, the experimental results compare well with the numerical simulation results shown in Figures 7 and 8. A more detailed experimental configuration and comprehensive results for various scattering geometries with multiple transmissions of iid noise waveforms will be presented and published soon.

6. Conclusion

This paper shows that tomographic images of scattering objects are successfully achieved with multiple transmissions of random noise waveforms. From the simulation results, we conclude that a single transmitted UWB WGN may not be sufficient to generate the correct tomographic image due to the practical implementation of a transmitted random noise waveform. However, multiple iid UWB WGN transmitted

waveforms can bypass the shortcoming of a single transmission of UWB WGN waveform, forming a correct image of target objects by summing and averaging discrete object images based on each iid WGN waveform.

Also the image quality of the tomographic image after completion of multiple transmissions of iid noise waveforms is analyzed. MSE is used to measure the image quality of all tomographic images in this paper. Image quality of the tomographic images based on the random noise waveform is enhanced as the number of iid noise waveform transmissions increases. The presence of white Gaussian noise degrades the image quality; however, the suppression of unwanted noise contributions by controlling SNR can help to achieve successful tomographic images in practical radar imaging systems. Also, numerical simulation results for the single scattering object scenario are validated with the preliminary experiment results.

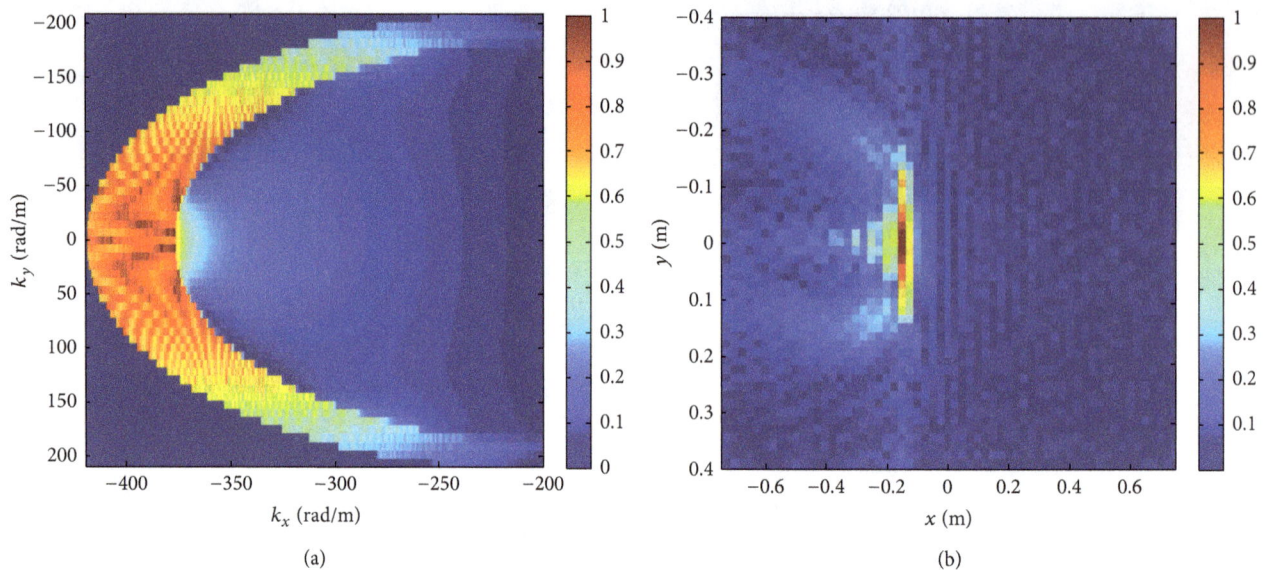

FIGURE 30: Experimental results of the single cylindrical conducting object geometry shown in Figure 3. (a) The magnitude of Fourier space data of the single cylindrical conducting object. The colorbar indicates the normalized magnitude of $\widetilde{O}(k_x, k_y)$. (b) The obtained tomographic image after transmitting a single UWB WGN waveform. The colorbar indicates the normalized magnitude of scattering object function, $o(x, y)$.

Conflict of Interests

The authors declare that there is no conflict of interests regarding the publication of this paper.

Acknowledgment

This work was supported by the Air Force Office of Scientific Research (AFOSR) Contract no. FA9550-12-1-0164.

References

[1] B. M. Horton, "Noise-modulated distance measuring systems," *Proceedings of the IRE*, vol. 47, no. 5, pp. 821–828, 1959.

[2] M. P. Grant, G. R. Cooper, and A. K. Kamal, "A class of noise radar systems," *Proceedings of the IEEE*, vol. 51, no. 7, pp. 1060–1061, 1963.

[3] M. Dawood and R. M. Narayanan, "Multipath and ground clutter analysis for a UWB noise radar," *IEEE Transactions on Aerospace and Electronic Systems*, vol. 38, no. 3, pp. 838–853, 2002.

[4] K. Kulpa, J. Misiurewicz, Z. Gajo, and M. Malanowski, "A simple robust detection of weak target in noise radars," in *Proceedings of the 4th European Radar Conference (EURAD '07)*, pp. 275–278, Munich, Germany, October 2007.

[5] Y. Zhang and R. M. Narayanan, "Design consideration for a real-time random-noise tracking radar," *IEEE Transactions on Aerospace and Electronic Systems*, vol. 40, no. 2, pp. 434–445, 2004.

[6] D. A. Ausherman, A. Kozma, J. L. Walker, H. M. Jones, and E. C. Poggio, "Developments in radar imaging," *IEEE Transactions on Aerospace and Electronic Systems*, vol. 20, no. 4, pp. 363–400, 1984.

[7] H. J. Shin, R. M. Narayanan, and M. Rangaswamy, "Tomographic imaging with ultra-wideband noise radar using time-domain data," in *Radar Sensor Technology XVII*, vol. 8714 of *Proceedings of SPIE*, pp. 1–9, Baltimore, Md, USA, April 2013.

[8] H. J. Shin, R. M. Narayanan, and M. Rangaswamy, "Diffraction tomography for ultra-wideband noise radar and imaging quality measure of a cylindrical perfectly conducting object," in *Proceedings of the 2014 IEEE Radar Conference*, pp. 702–707, Cincinnati, Ohio, USA, May 2014.

[9] L. Jofre, A. Broquetas, J. Romeu et al., "UWB tomographie radar imaging of penetrable and impenetrable objects," *Proceedings of the IEEE*, vol. 97, no. 2, pp. 451–464, 2009.

[10] X. Li, E. J. Bond, B. D. van Veen, and S. C. Hagness, "An overview of ultra-wideband microwave imaging via space-time beamforming for early-stage breast-cancer detection," *IEEE Antennas and Propagation Magazine*, vol. 47, no. 1, pp. 19–34, 2005.

[11] T. M. Grzegorczyk, P. M. Meaney, P. A. Kaufman, R. M. Diflorio-Alexander, and K. D. Paulsen, "Fast 3-D tomographic microwave imaging for breast cancer detection," *IEEE Transactions on Medical Imaging*, vol. 31, no. 8, pp. 1584–1592, 2012.

[12] M. H. Khalil, W. Shahzad, and J. D. Xu, "In the medical field detection of breast cancer by microwave imaging is a robust tool," in *Proceedings of the 25th International Vacuum Nanoelectronics Conference (IVNC '12)*, pp. 228–229, Jeju Island, Republic of Korea, July 2012.

[13] Z. Wang, E. G. Lim, Y. Tang, and M. Leach, "Medical applications of microwave imaging," *The Scientific World Journal*, vol. 2014, Article ID 147016, 7 pages, 2014.

[14] Y. J. Kim, L. Jofre, F. De Flaviis, and M. Q. Feng, "Microwave reflection tomographic array for damage detection of civil structures," *IEEE Transactions on Antennas and Propagation*, vol. 51, no. 11, pp. 3022–3032, 2003.

[15] S. Kharkovsky, J. Case, M. Ghasr, R. Zoughi, S. Bae, and A. Belarbi, "Application of microwave 3D SAR imaging technique

for evaluation of corrosion in steel rebars embedded in cement-based structures," in *Review of Progress in Quantitative Nondestructive Evaluation*, vol. 31, pp. 1516–1523, 2012.

[16] O. Güneş and O. Büyüköztürk, "Microwave imaging of plain and reinforced concrete for NDT using backpropagation algorithm," in *Nondestructive Testing of Materials and Structures*, O. Güneş and Y. Akkaya, Eds., vol. 6 of *RILEM Bookseries*, pp. 703–709, Springer, 2013.

[17] D. Zimdars and J. S. White, "Terahertz reflection imaging for package and personnel inspection," in *Terahertz for Military and Security Applications II*, vol. 5411 of *Proceedings of the SPIE*, pp. 78–83, Orlando, Fla, USA, April 2004.

[18] S. Almazroui and W. Wang, "Microwave tomography for security applications," in *Proceedings of the International Conference on Information Technology and e-Services (ICITeS '12)*, pp. 1–3, Sousse, Tunisia, March 2012.

[19] O. Yurduseven, "Indirect microwave holographic imaging of concealed ordnance for airport security imaging systems," *Progress in Electromagnetics Research*, vol. 146, pp. 7–13, 2014.

[20] L. Jofre, A. P. Toda, J. M. J. Montana et al., "UWB short-range bifocusing tomographic imaging," *IEEE Transactions on Instrumentation and Measurement*, vol. 57, no. 11, pp. 2414–2420, 2008.

[21] C.-P. Lai and R. M. Narayanan, "Ultrawideband random noise radar design for through-wall surveillance," *IEEE Transactions on Aerospace and Electronic Systems*, vol. 46, no. 4, pp. 1716–1730, 2010.

[22] R. Vela, R. M. Narayanan, K. A. Gallagher, and M. Rangaswamy, "Noise radar tomography," in *Proceedings of the IEEE Radar Conference: Ubiquitous Radar (RADAR '12)*, pp. 720–724, Atlanta, Ga, USA, May 2012.

[23] A. Leon-Garcia, *Probability and Random Processes for Electrical Engineering*, Addison-Wesley, Reading, Mass, USA, 2nd edition, 1994.

[24] G. Jenkins and D. Watts, *Watts, Spectral Analysis and Its Applications*, Holden-Day, San Francisco, Calif, USA, 1968.

[25] A. V. Oppenheim, R. W. Schafer, and J. R. Buck, *Discrete-Time Signal Processing*, Prentice Hall, Upper Saddle River, NJ, USA, 2nd edition, 1999.

[26] M. S. Bartlett, "Periodogram analysis and continuous spectra," *Biometrika*, vol. 37, pp. 1–16, 1950.

[27] S. K. Kenue and J. F. Greenleaf, "Limited angle multifrequency diffraction tomography," *IEEE Transactions on Sonics and Ultrasonics*, vol. 29, no. 4, pp. 213–217, 1982.

[28] B. A. Roberts and A. C. Kak, "Reflection mode diffraction tomography," *Ultrasonic Imaging*, vol. 7, no. 4, pp. 300–320, 1985.

[29] M. Soumekh, "Surface imaging via wave equation inversion," in *Acoustical Imaging*, L. W. Kessler, Ed., pp. 383–393, Springer, 1988.

[30] S. X. Pan and A. C. Kak, "A computational study of reconstruction algorithms for diffraction tomography: interpolation versus filtered-backpropagation," *IEEE Transactions on Acoustics, Speech, and Signal Processing*, vol. 31, no. 5, pp. 1262–1275, 1983.

[31] T.-H. Chu and K.-Y. Lee, "Wide-band microwave diffraction tomography under Born approximation," *IEEE Transactions on Antennas and Propagation*, vol. 37, pp. 515–519, 1992.

[32] J. C. Bolomey and C. Pichot, "Microwave tomography: from theory to practical imaging systems," *International Journal of Imaging Systems and Technology*, vol. 2, pp. 144–156, 1990.

[33] H. J. Shin, R. M. Narayanan, and M. Rangaswamy, "Ultrawideband noise radar imaging of cylindrical PEC objects using diffraction tomography," in *Radar Sensor Technology XVIII*, vol. 9077 of *Proceedings of SPIE*, pp. 1–10, Baltimore, Md, USA, May 2014.

[34] I. Avcibaş, B. Sankur, and K. Sayood, "Statistical evaluation of image quality measures," *Journal of Electronic Imaging*, vol. 11, no. 2, pp. 206–223, 2002.

[35] A. M. Eskicioglu and P. S. Fisher, "Image quality measures and their performance," *IEEE Transactions on Communications*, vol. 43, no. 12, pp. 2959–2965, 1995.

[36] A. M. Eskicioglu, "Application of multidimensional quality measures to reconstructed medical images," *Optical Engineering*, vol. 35, no. 3, pp. 778–785, 1996.

[37] J. Hu, T. Jiang, Z. Cui, and Y. Hou, "Design of UWB pulses based on Gaussian pulse," in *Proceedings of the 3rd IEEE International Conference on Nano/Micro Engineered and Molecular Systems (NEMS '08)*, pp. 651–655, Sanya, China, January 2008.

[38] A. Thakre and A. Dhenge, "Selection of pulse for ultra wide band communication (UWB) system," *International Journal of Advanced Research in Computer and Communication Engineering*, vol. 1, pp. 683–686, 2012.

[39] C. E. Shannon, "Communication in the presence of noise," *Proceedings of the IRE*, vol. 37, pp. 10–21, 1949.

[40] M. Bertero and P. Boccacci, *Introduction to Inverse Problems in Imaging*, Institute of Physics Publishing, Bristol, UK, 1998.

8

Implementation of a Cross-Spectrum FFT Analyzer for a Phase-Noise Test System in a Low-Cost FPGA

Patrick Fleischmann,[1] **Heinz Mathis,**[1] **Jakub Kucera,**[2] **and Stefan Dahinden**[2]

[1]*Institute for Communication Systems ICOM, University of Applied Sciences of Eastern Switzerland, 8640 Rapperswil, Switzerland*
[2]*Anapico Ltd., 8152 Glattbrugg, Switzerland*

Correspondence should be addressed to Heinz Mathis; heinz.mathis@hsr.ch

Academic Editor: Giovanni Ghione

The cross-correlation method allows phase-noise measurements of high-quality devices with very low noise levels, using reference sources with higher noise levels than the device under test. To implement this method, a phase-noise analyzer needs to compute the cross-spectral density, that is, the Fourier transform of the cross-correlation, of two time series over a wide frequency range, from fractions of Hz to tens of MHz. Furthermore, the analyzer requires a high dynamic range to accommodate the phase noise of high-quality oscillators that may fall off by more than 100 dB from close-in noise to the noise floor at large frequency offsets. This paper describes the efficient implementation of a cross-spectrum analyzer in a low-cost FPGA, as part of a modern phase-noise analyzer with very fast measurement time.

1. Introduction

Phase noise, the random phase fluctuations of a periodic signal, is an important parameter to characterize high-frequency devices, in particular reference oscillators and microwave synthesizers. Phase noise is important because it has a large impact on the performance of many applications [1], such as high-speed communications [2], radar, and precision navigation [3].

There exist various methods for phase-noise measurement, of which the cross-correlation method achieves the best sensitivity and the widest frequency range, at the expense of a relatively complex setup [4–6]. Various fully automated integrated phase-noise analyzers that implement this method are available on the market, for example, the Anapico APPH6040/20G (7 or 26 GHz), the Agilent E5052B, or the Rohde & Schwarz FSUP.

In this paper, we will first review the basics of modelling phase noise and give an outline of the associated terminology. After that, the basics of phase-noise measurement methods will be discussed. Finally, the design and implementation of a novel cross-spectrum FFT analyzer in a low-cost FPGA are described in detail.

2. Phase-Noise Modelling and Terminology

A perfect fixed-frequency oscillator without noise would produce a perfect sine wave. In reality, any oscillator is affected by internal random noise processes, such as thermal and flicker noise, as well as aging and external influences, such as temperature and vibrations. To be able to characterize the phase noise of a real oscillator, its output signal can be modelled by

$$V(t) = (V_0 + \epsilon(t)) \sin(2\pi\nu_0 t + \phi(t)), \qquad (1)$$

where $\phi(t)$ denotes the random phase fluctuations, V_0 the nominal amplitude, and ν_0 the nominal frequency. The random amplitude fluctuations $\epsilon(t)$ can generally be neglected for high-quality oscillators [7].

Phase fluctuations are characterized in the frequency-domain by their one-sided power spectral density $S_\phi(f)$, defined as

$$S_\phi(f) = \phi^2(f) \frac{1}{\text{BW}} \left[\text{rad}^2/\text{Hz}\right], \qquad (2)$$

where $\phi(f)$ is the root mean squared (rms) phase fluctuation and BW is the measurement bandwidth. The Fourier frequency f ranges from 0 to ∞ in this one-sided spectrum which contains the power of both sidebands around the nominal frequency [8].

The standard measure of phase noise in the frequency-domain is the single-sideband phase-noise $\mathscr{L}(f)$ defined by the IEEE standard 1139 [8] as

$$\mathscr{L}(f) \equiv \frac{1}{2} S_\phi(f). \qquad (3)$$

$\mathscr{L}(f)$ is usually specified in dBc/Hz, that is, dB below the carrier in a 1 Hz bandwidth.

3. Measuring Phase Noise

The simplest phase-noise measurement method is the direct spectrum measurement using a spectrum analyzer. However, this method is only suitable for sources with relatively high noise because the phase noise of the spectrum analyzer must be significantly lower than the noise of the device under test. Furthermore, the dynamic range of this method is very limited because the carrier signal is not suppressed.

Another class of measurement methods are the frequency-discriminator methods. The advantage of these methods is that they do not require a reference oscillator. However, these methods cannot achieve the sensitivity of the phase detector methods described below [6].

The method whose implementation is described in Section 4 is the cross-correlation method. Therefore, the remainder of this section will first describe the single-channel quadrature method, which is the basis of the cross-correlation method, before the cross-correlation method is explained.

3.1. Quadrature Method Phase-Noise Measurement.
The basic principle of the quadrature method is depicted in Figure 1. The device under test (DUT) signal is mixed with a reference (REF) signal at the same frequency using a phase detector mixer. A phase locked loop (PLL) ensures that the DUT and REF signals stay in phase quadrature during the measurement, so that the output of the mixer (after low-pass filtering) will be approximately proportional to the phase fluctuations of the input signals. Thus, the mixer operates as a phase detector. The output voltage can then be measured using a baseband FFT spectrum analyzer [6].

The main disadvantage and the limiting factor for the measurement accuracy of this method is that the reference source must exhibit significantly lower phase noise than the DUT because any noise on the reference is added to the DUT noise. One possible solution of this problem is to use two identical sources as DUT and reference, so that the two sources contribute the same amount of noise to the output. The measured noise power is then twice the noise power of a single source, assuming the phase noise of the two sources is uncorrelated.

Another disadvantage of the quadrature method is that the PLL forms a high-pass filter for the phase noise, as it inherently tries to compensate for phase fluctuations. Therefore, the PLL loop bandwidth must be made substantially lower than the lowest required noise frequency. Depending

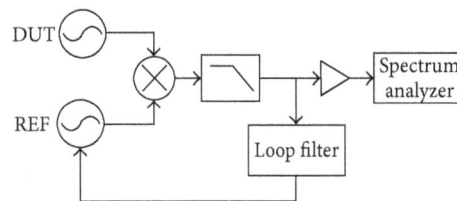

FIGURE 1: Block diagram of the quadrature method. The reference oscillator (REF) is phase-locked to the device under test (DUT). The output of the phase detector (mixer) is first amplified with a low noise amplifier and then measured with a baseband spectrum analyzer.

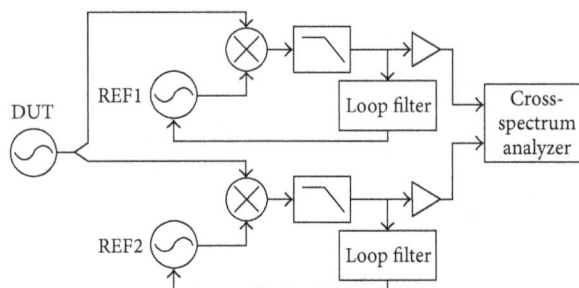

FIGURE 2: Block diagram of the cross-correlation method. This method uses two independent reference sources (REF1, REF2) and phase detectors (mixers). The two baseband outputs are measured with a cross-spectrum analyzer.

on the frequency stability of the DUT and reference, the loop bandwidth cannot be made arbitrarily small because the PLL might lose lock [6]. Therefore, the high-pass effect of the PLL is often canceled after the measurement, using signal processing.

3.2. The Cross-Correlation Method.
The cross-correlation method solves the problem of the reference source noise by using two independent reference sources and phase-detector circuits (see Figure 2). The basic reasoning behind this method is that the noise of the reference sources can be averaged away by cross-correlating and averaging the outputs of the two mixers [4, 5]. In practice, the noise floor can be improved by about 20 dB over the single-channel quadrature method [9].

In a cross-spectrum FFT analyzer, the discrete Fourier transforms (DFTs) of the two input signals are computed and the DFTs are multiplied pointwise, taking the complex conjugate of one signal, to obtain an estimate of the cross-spectrum. Several of these cross-spectra can then be averaged.

The uncorrelated noise products will have random amplitude and phase in the DFT and will therefore be eliminated by the averaging; they will decrease proportionally to $1/\sqrt{m}$, where m denotes the number of averages, if they are completely uncorrelated. This means that the measurement sensitivity will be increased by at most 5 dB for a tenfold increase in the number of averages [9].

However, for the correlated part of the noise, the product equals the squared magnitude. Therefore, more averages

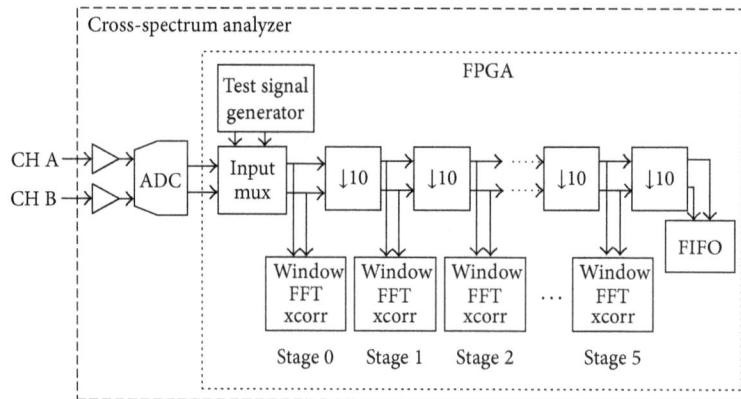

FIGURE 3: Architecture of the cross-spectrum analyzer and the signal processing inside the FPGA.

will improve the estimation of the correlated noise, that is, the phase noise of the DUT. Once the uncorrelated noise is averaged away, the variance of the power estimate will decrease proportionally to $1/m$ [5, 10].

In effect, we can accurately measure a noise source that has a lower noise level than the noise floor of a single measurement channel by using the cross-correlation method. An extensive tutorial of the cross-correlation method can be found in [5].

4. FPGA Cross-Spectrum Analyzer Implementation

This section describes the implementation of a wideband cross-spectrum analyzer for phase-noise measurement in a low-cost Spartan-6 FPGA (Field Programmable Gate Array) from Xilinx.

Figure 3 shows an overview of the complete cross-spectrum analyzer. The two analog input channels A and B are connected to two independent quadrature measurement systems, as depicted in Figure 2. The inputs are digitized using a high dynamic range ADC (Analog-to-Digital Converter) operating at a sampling rate of 100 to 150 MHz. Sufficient ADC resolution is required to cope with largely changing phase-noise profiles.

Inside the FPGA, the two channels are processed by a cascade of decimators with downsampling factors of 10 and then fed to the signal processing stages to compute the cross-spectral density of the two channels. Decimating by a factor of 10 has the advantage that the resulting plot has a constant number of samples per decade. Multiple correlations are summed up in an accumulator memory for averaging in every stage. The accumulated correlations are read out via a softcore microprocessor that provides the output interface. Figure 4 shows the architecture of one signal processing stage. The first signal processing stage operates at the full ADC sampling rate of 125 MHz and each following stage operates at a sampling rate ten times smaller than the preceding stage. This architecture simultaneously produces an estimation of the cross-spectral density in multiple, logarithmically spaced frequency ranges. The samples with the lowest sampling rate,

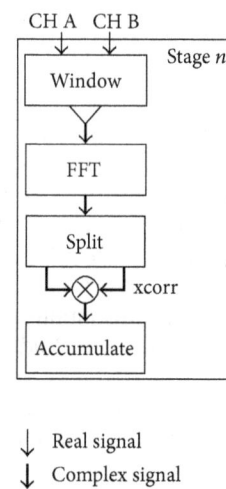

FIGURE 4: Architecture of one signal processing stage. One complex FFT block, in combination with a split block, is used to compute the DFTs of the two real input channels.

which come out of the last downsampling stage, are not processed in a hardware signal processing block but are stored in a FIFO memory (first in, first out). The FIFO memory is periodically read out by the microprocessor and the signal processing for these low-frequency samples is performed in software.

The logarithmically spaced frequency ranges are important because phase-noise power spectral densities are always plotted in a log-log scale. In the low-frequency range, the frequency resolution therefore needs to be much smaller than in the high-frequency range. When computing the DFT (Discrete Fourier Transform) of a signal, the frequency resolution is proportional to f_s/N_{DFT}, where f_s is the sampling rate and N_{DFT} is the length of the DFT in samples. To obtain a frequency resolution of 1 Hz at a sampling rate of 125 MHz, a length of $N_{DFT} = 125 \cdot 10^6$ samples would be required. This is clearly infeasible on a platform with limited memory resources. However, if the FFT block for the lowest frequency range operates at a sampling rate of only 125 Hz, the frequency resolution with $N_{DFT} = 1024$ is 0.122 Hz.

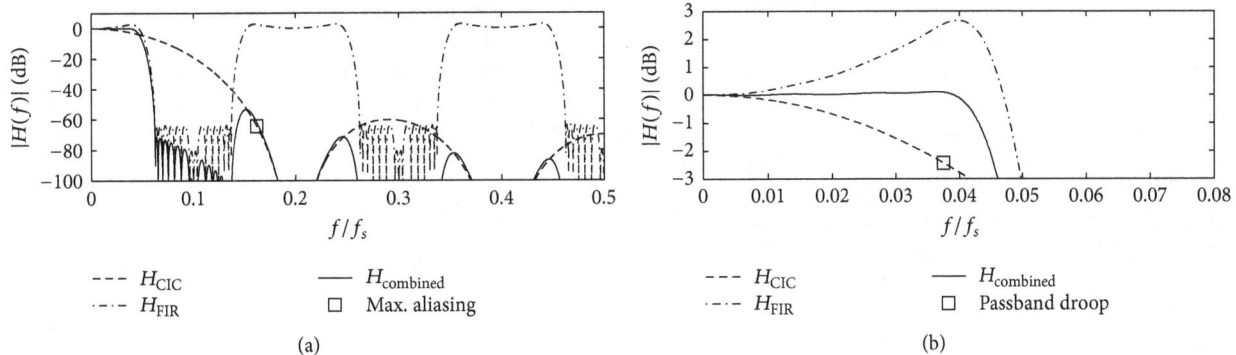

FIGURE 5: Frequency responses of the CIC filter, FIR filter, and combined filter. The CIC filter and FIR filter have downsampling rates of 5 and 2, respectively. The frequency axis is relative to the input sampling rate.

TABLE 1: Number of required multipliers for optimal FIR filter implementation, depending on the decimation-rate allocation.

R_{CIC}	R_{FIR}	N_{coeff}	N_{mult}
1	10	119	6
2	5	72	4
5	2	31	2

Another benefit of this architecture is that the measurement time to obtain one estimate of the power spectral density is dramatically reduced in the high-frequency ranges. If we again assume a frequency-resolution requirement of 1 Hz and want to estimate the spectral density using one FFT, we would require a measurement time of 1 s, because the measurement time is inversely proportional to the frequency resolution. Using the implemented architecture, we can perform multiple FFTs and correlations in the high-frequency stages in parallel, while the lowest-frequency stage may only be able to perform one correlation in the given measurement time.

4.1. Decimation. The cascade of decimators provides antialiasing filtering and downsampling for the FFT stages. The first decimator stage operates at an input sampling rate of 125 MHz and is downsampled to 12.5 MHz. It was implemented as a combination of a cascaded integrator comb (CIC) filter [11] and a subsequent finite impulse response (FIR) filter. This configuration was chosen to minimize the number of required multipliers, which are a limited resource in low-cost FPGAs.

The required specification for the decimation filter was alias suppression of at least 60 dB and passband flatness of 0.1 dB. The transition bandwidth should not exceed 50% of the output bandwidth. This means that the usable bandwidth is 75% of the total output bandwidth.

For example, if such a filter was implemented as a single FIR filter, the number of required coefficients would be over a hundred; see Table 1. The number of multipliers required for the implementation of an FIR filter in an FPGA can be estimated by dividing the number of coefficients N_{coeff} by the hardware-oversampling rate. The hardware-oversampling

rate is just the hardware clock rate f_{clk}, divided by the output sampling rate of the filter $f_{s,out}$. In this example, the number of multipliers is

$$N_{mult} \approx \left\lceil \frac{N_{coeff} f_{s,out}}{f_{clk}} \right\rceil = \left\lceil \frac{119 \cdot 12.5\,\text{MHz}}{125\,\text{MHz}} \right\rceil = 12. \quad (4)$$

If the filter coefficients are symmetric, the number of required multipliers can be approximately halved. Consequently, this filter could theoretically be implemented using six multipliers.

To reduce the number of required multipliers, it is often beneficial to choose a filter configuration consisting of a CIC filter followed by an FIR filter and distribute the overall downsampling between the two filters [12]. The advantage of CIC filters is that they do not require multipliers but only adders. Their main drawback is that the passband is not flat (passband droop). However, an FIR filter following the CIC can be used to compensate for the passband droop and also sharpen the transition from the passband to the stopband.

Theoretically there are four different possibilities to distribute the downsampling rate of 10 between two filters: $(1, 10)$, $(2, 5)$, $(5, 2)$, and $(10, 1)$. The last theoretical option would only use a CIC decimation filter with a downsampling factor of 10. However, such a filter cannot meet the requirements. The remaining three options were tested for the number of multipliers they require; see Table 1. The filter configuration with 31 symmetric FIR coefficients, which theoretically only requires two multipliers, was chosen for the implementation. Figure 5 shows the frequency response of the CIC filter, the FIR filter, and the combined filter with the points of maximum aliasing and passband droop.

The number of actually used multipliers depends on how well the algorithm that synthesizes the netlist for the FPGA can exploit the hardware oversampling and the coefficient symmetry. Furthermore, it also depends on the bit-width of the data and coefficients. For this example, the Xilinx tools generated a filter using three multipliers.

4.2. Windowing and Overlapping. Windowing is performed before the FFT to prevent spectral leaking. The implemented window is a 4-term minimum-sidelobe Blackman-Harris window that has large sidelobe suppression of 92 dB but a relatively large equivalent noise bandwidth (ENBW) of 2 bins [13]. The ENBW has to be taken into account to correct the estimated power spectral density.

To increase the number of available FFT samples for averaging, stage 3 to stage 5 employ an overlapping block in front of the windowing. This technique accelerates the convergence of the averaged power spectral densities, which is important to reduce the measurement time at the lower sampling rates.

When the squared magnitude of K independent and identically distributed samples are averaged, the variance of the average σ_{avg}^2 will decrease by

$$\frac{\sigma_{\text{avg}}^2}{\sigma_{\text{samp}}^2} = \frac{1}{K}. \tag{5}$$

If overlapping of 50% is used, the individual FFT samples are correlated. This means that the variance will decrease more slowly. A good approximation of the variance reduction (for more than ten averages) is

$$\frac{\sigma_{\text{avg}}^2}{\sigma_{\text{samp}}^2} \approx \frac{1}{K}\left[1 + 2c^2\left(0.5\right)\right], \tag{6}$$

where $c(0.5)$ is the correlation coefficient of the window for 50% overlap [10]. The overlap correlation coefficient of the window used is only $c(0.5) = 3.8\%$. This means that the variance of the average will approximately be reduced as if the individual FFT samples were uncorrelated.

4.3. Two-Channel FFT. The FFT blocks compute the Discrete Fourier Transforms (DFT) of the two windowed input sequences using intellectual property (IP) cores from Xilinx. The length of the DFT at all stages is 1024 samples; the bit-width of the input and output sequences increases substantially from the first stage to the last stages to accommodate the larger dynamic range in the low-frequency stages.

The FFT IP cores compute the standard DFT, which takes one complex-valued input sequence to produce one complex-valued output sequence. In this application, however, we need to compute the DFTs of two real-valued input sequences at the same time. The simplest solution to this problem would be to use two independent FFT cores but this would be a waste of FPGA resources.

To save resources, we use the well-known trick for computing the DFTs of two real sequences using only one complex DFT [14]. To explain how this works, we start by defining the DFT of a sequence $x(n)$ of length N as

$$X(k) = \sum_{n=0}^{N-1} x(n)e^{-j2\pi kn/N}, \quad k = 0, 1, \ldots, N-1. \tag{7}$$

If $x(n)$ is real-valued, $X(k)$ has a complex conjugate symmetry about $N/2$; that is,

$$X(N-k) = X^*(k), \quad k = 1, 2, \ldots, N-1. \tag{8}$$

In other words, the real component of $X(k)$ exhibits even symmetry and the complex component odd symmetry.

We can exploit this symmetry to compute the DFTs of two real sequences $x(n)$ and $y(n)$ by computing the DFT of the complex input sequence $z(n) = x(n) + jy(n)$. Because the DFT is linear, the transformed sequence is simply

$$\begin{aligned} Z(k) &= \text{DFT}\left[x(n) + jy(n)\right] \\ &= \left\{X_r(k) - Y_i(k)\right\} + j\left\{X_i(k) + Y_r(k)\right\} \quad (9) \\ &= Z_r(k) + jZ_i(k), \end{aligned}$$

where we used the indices r and i to denote the real and imaginary components. Now we can split $Z(k)$ by separating the even and odd parts of its real and complex components to get the DFTs of $x(n)$ and $y(n)$ as follows:

$$\begin{aligned} X(k) &= \frac{1}{2}\left\{Z_r(k) + Z_r(N-k)\right\} \\ &\quad + j\frac{1}{2}\left\{Z_i(k) - Z_i(N-k)\right\}, \\ Y(k) &= \frac{1}{2}\left\{Z_i(N-k) + Z_i(k)\right\} \\ &\quad + j\frac{1}{2}\left\{Z_r(N-k) - Z_r(k)\right\}. \end{aligned} \tag{10}$$

Because of their symmetry, we only need to compute $X(k)$ and $Y(k)$ for $k = 0, 1, \ldots, N/2$.

Using this algorithm, the cost of computing the DFT of two real, length-N sequences is equal to the cost of one complex length-N FFT plus two additions per complex output sample. The memory cost is also increased, compared to one complex FFT, because the complete FFT output sequence has to be stored before splitting can be performed.

4.4. Cross-Correlation and Vector-Averaging. After the splitting block, the two transformed sequences are multiplied using a complex multiplier block to compute the complex-valued cross-spectrum. Before multiplication, the imaginary part of the sequence from channel b is inverted to obtain the complex conjugate. Finally, multiple correlations are summed up in an accumulator memory which can accommodate more than 10,000 correlations. The accumulator memories of all stages can then be read out and the averaged cross-spectrum can be displayed to the user.

5. Measurements

The cross-spectrum analyzer described above was successfully implemented in the commercially available APPH6040 signal source analyzer from Anapico [15]. Figures 6 and 7 show example phase-noise measurements with the APPH6040, demonstrating the excellent sensitivity achieved by the cross-correlation method. To that end, two traces are plotted in both figures. One trace shows the result of only one correlation, whereas the other trace shows the average of many correlations, which results in a significantly lower noise floor.

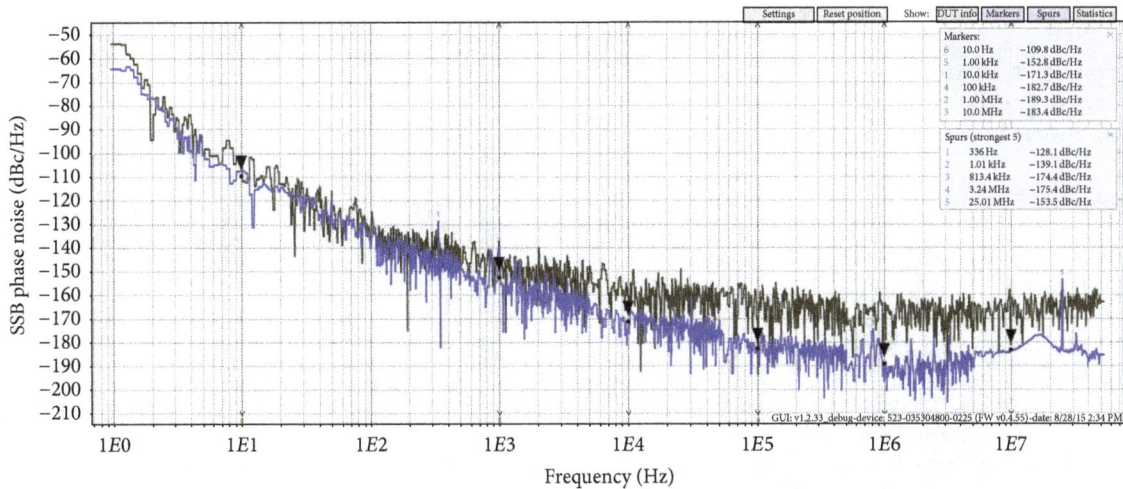

FIGURE 6: Cross-correlation phase-noise measurement with the APPH6040. The DUT was ultra-low-phase-noise 100 MHz OCXO. The grey trace shows a measurement with only one correlation. The blue trace shows a measurement with 10^6 correlations in stage 0 (5–50 MHz), 10^5 correlations in stage 1 (0.5–5 MHz), and so on. Both measurements were performed with the internal reference sources of the APPH6040. Same measurement time of ca. 11 sec for both measurements.

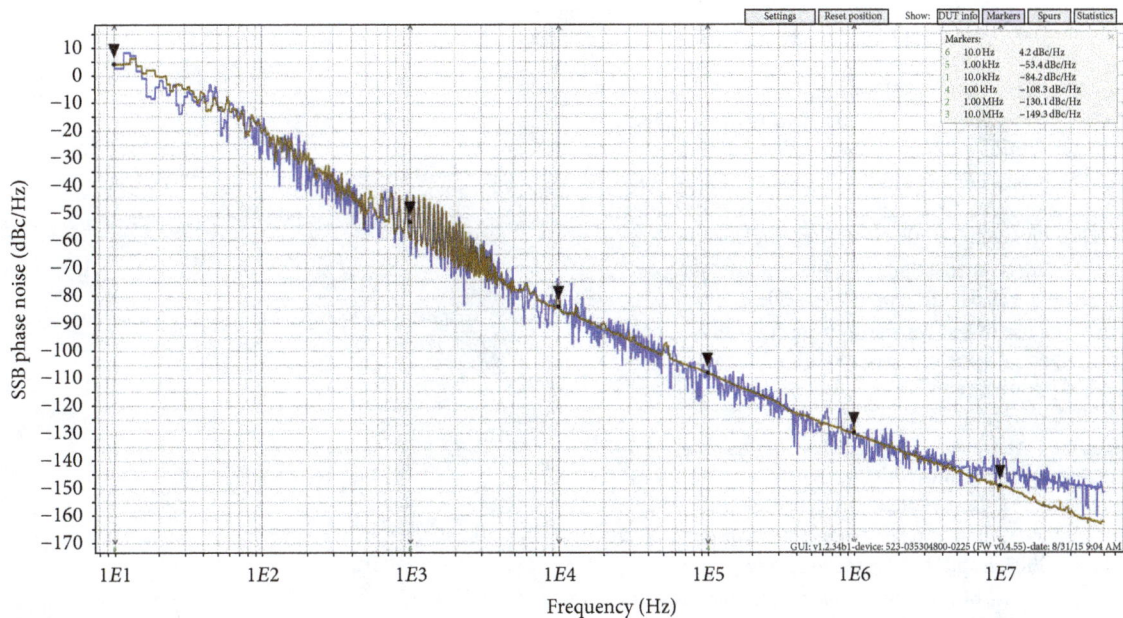

FIGURE 7: Cross-correlation phase-noise measurement of a free-running wideband VCO at 2.16 GHz with the APPH6040. Single correlation (blue) and maximum correlation (brown) per individual decades. Same measurement time of 1.2 sec for both traces.

6. Conclusion

We have presented the implementation of a novel cross-spectrum FFT analyzer architecture for phase-noise measurements, which was successfully integrated into the commercially available APPH6040 signal source analyzer from Anapico.

Several efficient signal processing techniques had to be employed to enable integration of the FFT analyzer into a low-cost FPGA. For example, the use of CIC filters for signals with high sample rates dramatically reduces the number of required hardware multipliers, a scarce resource in low-cost FPGAs. Furthermore, the successive downsampling architecture uses FFT blocks with small length to cover a large measurement bandwidth, which saves memory inside the FPGA.

Conflict of Interests

The authors declare that there is no conflict of interests regarding the publication of this paper.

Acknowledgment

Financial support by the Swiss Commission for Technology and Innovation is gratefully acknowledged (CTI projects 11904.2 PFNM-NM and 13461.1 PFLE-NM).

References

[1] W. P. Robins, *Phase Noise in Signal Sources: Theory and Applications*, vol. 9, IET, 1984.

[2] S. Wu and Y. Bar-Ness, "OFDM systems in the presence of phase noise: consequences and solutions," *IEEE Transactions on Communications*, vol. 52, no. 11, pp. 1988–1996, 2004.

[3] W. Yu, G. Lachapelle, and S. Skone, "PLL performance for signals in the presence of thermal noise, phase noise, and ionospheric scintillation," in *Proceedings of the 19th International Technical Meeting of the Satellite Division (ION GNSS '06)*, pp. 1–17, Fort Worth, Tex, USA, September 2006.

[4] W. F. Walls, "Cross-correlation phase noise measurements," in *Proceedings of the IEEE Frequency Control Symposium*, pp. 257–261, Hershey, Pa, USA, May 1992.

[5] E. Rubiola and F. Vernotte, "The cross-spectrum experimental method," http://arxiv.org/abs/1003.0113.

[6] U. L. Rohde, A. K. Poddar, and A. M. Apte, "Getting its measure," *IEEE Microwave Magazine*, vol. 14, no. 6, pp. 73–86, 2013.

[7] J. Rutman, "Characterization of phase and frequency instabilities in precision frequency sources: fifteen years of progress," *Proceedings of the IEEE*, vol. 66, no. 9, pp. 1048–1075, 1978.

[8] J. Vig, *IEEE Standard Definitions of Physical Quantities for Fundamental Frequency and Time Metrology-Random Instabilities (IEEE Std 1139-1999)*, vol. 1, IEEE, New York, NY, USA, 1999.

[9] J. Breitbarth, "Cross correlation in phase noise analysis," *Microwave Journal*, vol. 54, no. 2, pp. 78–86, 2011.

[10] P. D. Welch, "The use of fast Fourier transform for the estimation of power spectra: a method based on time averaging over short, modified periodograms," *IEEE Transactions on Audio and Electroacoustics*, vol. 15, no. 2, pp. 70–73, 1967.

[11] E. B. Hogenauer, "An economical class of digital filters for decimation and interpolation," *IEEE Transactions on Acoustics, Speech, and Signal Processing*, vol. 29, no. 2, pp. 155–162, 1981.

[12] R. Lyons, "Understanding cascaded integrator comb filters," embedded.com, 2005, http://www.embedded.com/.

[13] F. J. Harris, "On the use of windows for harmonic analysis with the discrete Fourier transform," *Proceedings of the IEEE*, vol. 66, no. 1, pp. 51–83, 1978.

[14] H. V. Sorensen, D. L. Jones, M. T. Heideman, and C. S. Burrus, "Realvalued fast Fourier transform algorithms," *IEEE Transactions on Acoustics, Speech, and Signal Processing*, vol. 35, no. 6, pp. 849–863, 1987.

[15] *APPH6040/APPH20G Specification V2.11*, AnaPico, 2015, http://www.anapico.com/.

9

Design of UWB Planar Monopole Antennas with Etched Spiral Slot on the Patch for Multiple Band-Notched Characteristics

Swarup Das, Debasis Mitra, and Sekhar Ranjan Bhadra Chaudhuri

Department of Electronics & Telecommunication Engineering, Indian Institute of Engineering Science and Technology, Shibpur, Howrah 711 103, India

Correspondence should be addressed to Swarup Das; dasswarup08@yahoo.in

Academic Editor: Mustapha C. E. Yagoub

Three types of Ultrawideband (UWB) antennas with single, double, and triple notched bands are proposed and investigated for UWB communication applications. The proposed antennas consist of CPW fed monopole with spiral slot etched on the patch. In this paper single, double, and also triple band notches with central frequency of 3.57, 5.12, and 8.21 GHz have been generated by varying the length of a single spiral slot. The proposed antenna is low-profile and of compact size. A stable gain is obtained throughout the operation band except the three notched frequencies. The antennas have omnidirectional and stable radiation patterns across all the relevant bands. Moreover, relatively consistent group delays across the UWB frequencies are noticed for the triple notched band antenna. A prototype of the UWB antenna with triple notched bands is fabricated and the measured results of the antenna are compared with the simulated results.

1. Introduction

In modern communication there has been increasing demand in designing Ultrawideband (UWB) systems [1]. The UWB radio system occupies UWB frequency band, that is, 3.1–10.6 GHz, approved by Federal Communications Commission (FCC) [2], in which there might potentially exist several narrow band interferences caused by other wireless communication systems, such as IEEE 802.11a wireless local area network (WLAN) in the frequency band of 5.15–5.35 GHz and 5.725–5.825 GHz and WiMAX mainly around 3.5 GHz. Therefore, it is necessary for UWB antennas to perform band-notched function in those frequency bands to avoid potential interferences. Recently, a number of antennas with band-notched property have been discussed in [3–15] and various methods have been used to achieve the function. The widely used methods are etching slots on the patch or on the ground plane, that is, C-shaped, H-shaped, L-shaped, U-shaped, V-shaped, arc-shaped, and pie-shaped slot [3–9]. Slot-type split ring resonators (SRRs) etched on the patch

were found to have better performance in this regard [10, 11]. Adding L-shaped and ring-shaped parasitic elements with suitable designs on the bottom of the substrate was another method to generate notched bands [12, 13]. Band-notched property has been realized in Ultrawideband monopole antennas by using a strip bar and a folded strip [14, 15].

Ultrawideband antenna with single notched band was reported in [16–18]; then different methods were applied to produce double band-notched function in Ultrawideband antennas [19–21]. Lately a number of recent techniques have been proposed to generate triple notched bands [22–25]. In [22] triple band notches are realized by adding closed-loop ring in three different layers of the substrate. In [23] three open ended quarter-wavelength slots are used to obtain band-notched characteristics at three frequencies 3.5, 5.5, and 7.5 GHz. The triple band-notched characteristic is obtained by etching a complementary meander line split ring resonator inside the radiation patch and ground plane of a rectangular antenna in [24]. It is shown that the triple band-notched performance at 3.31, 5.81, and 8.53 GHz can be obtained. In [25] the antenna consists of a modified stair cased V-shaped radiating element and partial ground plane. The triple band-notched characteristics are achieved by embedding two different vertical up C-shaped slots with a vertical down C-shaped slot in the radiating patch and in the ground plane, respectively. In [26] four notched bands were observed using four different metallic strips. Further in this structure using only three metallic strips four notched bands were obtained with some modification in the antenna geometry. Therefore to generate multiple band notches the above designs are complicated structures leading to increased fabrication costs, antenna size, and difficulty in the integration with microwave integrated circuits.

In this paper a single spiral slot has been used to generate single, double, and also triple notched bands by varying spiral slot length with central frequency of 3.57, 5.12, and 8.21 GHz, respectively. The main objective of this paper is to present a simple and compact realization with stable radiation performance of a triple band-notched planar antenna suitable for UWB applications. The notched characteristic is achieved in antenna using spiral-shaped slot etched on the radiating patch.

In the proposed structure UWB operation was obtained by using a simple rectangular patch. But in some other structures many complex techniques were used like beveling of patch and ground plane, using slit with matching steps and Defected Ground Structure (DGS) in [23, 24, 26], respectively. In [22–26] multiple notched bands were obtained using several metallic resonators or various types of slots while in the proposed structure multiple notched bands are found using a single spiral slot. In the proposed structure the spiral slot is etched on the single layer of patch which is much simpler to realize compared to the multilayered structure as described in [22]. Slots were etched on ground plane in [24, 25] to produce notched bands whereas ground plane is unaffected in the proposed structure. Some of the notched bandwidths are controlled to make them sharper in the proposed structure than the notched bandwidths given in [23–25].

The paper is organized into three main sections. The first section is concerned with the antenna that has single turn spiral slot etched on the patch to generate a single notched band. The second section deals with the spiral slot with two turns to achieve double notch bands and in the last section it is described that the spiral slot with three turns etched on the patch is used to achieve triple notch bands. All of these antennas with multiple notched bands are fabricated and experimentally verified.

2. Spiral Slot Loaded Antenna with Single Notch

The proposed structure of spiral slot loaded antenna with single notch is shown in Figure 1. The antenna is printed on the top of a lightweight FR4 (ε_r = 4.4, tan δ = 0.02) substrate of thickness h = 1.6 mm. The length (L) and width (W) of the substrate are 30 mm and 30 mm, respectively. We optimize the antenna geometry for S_{11} < −10 dB over the wideband frequency range. A symmetric slot has been etched on the patch which produced a single notch. The feed line is a 50 Ω CPW

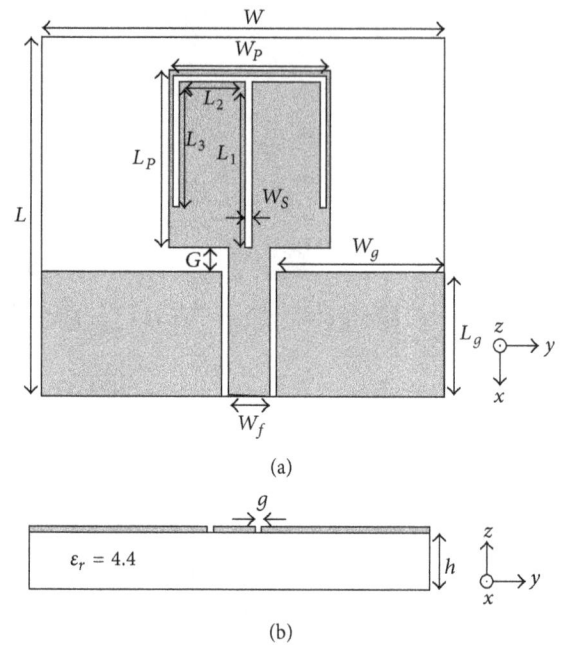

Figure 1: CPW fed planar monopole antenna with single notch: (a) top view, (b) side view.

line which is connected to the radiating element. For the design of the 50 Ω CPW feed line, the dimensions are chosen to be W_f = 3 mm and g = 0.3 mm, where W_f is the width of the feed line and g is the gap between feed line and ground plane. The length (L_P) and width (W_P) of the radiating patch element are 12 mm and 10 mm, respectively. The optimized value of gap between patch and ground plane is kept at G = 1.7 mm. Each of the ground planes has a size of ($L_g \times W_g$) mm² where L_g = 12.3 mm and W_g = 13.2 mm. The spiral-shaped slot etched on patch has a width of W_S = 0.5 mm. The total length of the slot can be calculated as $L_1 + 2(L_2 + L_3)$ = 11.25 + 2(4 + 8) mm = 27.25 mm. To reject the interference with existing wireless band a thin spiral slot has been printed on the radiating patch as a half-guided wavelength resonator to generate the notched band. In our design, the spiral slot is etched with a width of 0.5 mm to produce stronger resonance that guarantees better band-rejected performance.

The simulated and measured return loss of the structure is shown in Figure 2. The simulation has been conducted using High Frequency Structure Simulator 11 (HFSS11). The reference antenna (without slot) exhibits a bandwidth of 7.84 GHz (10.80 GHz–2.96 GHz). By using the Agilent N5230A network analyzer, S_{11} has been measured. Measured result shows that a notched band is generated from 3.4 GHz to 3.65 GHz.

3. Spiral Slot Loaded Antenna with Double Notch

On the same structure as described previously slot length has been increased 14 mm on both sides which gives dual notch. The gap between two slots is kept at M_S = 0.25 mm. This structure has been shown in Figure 3.

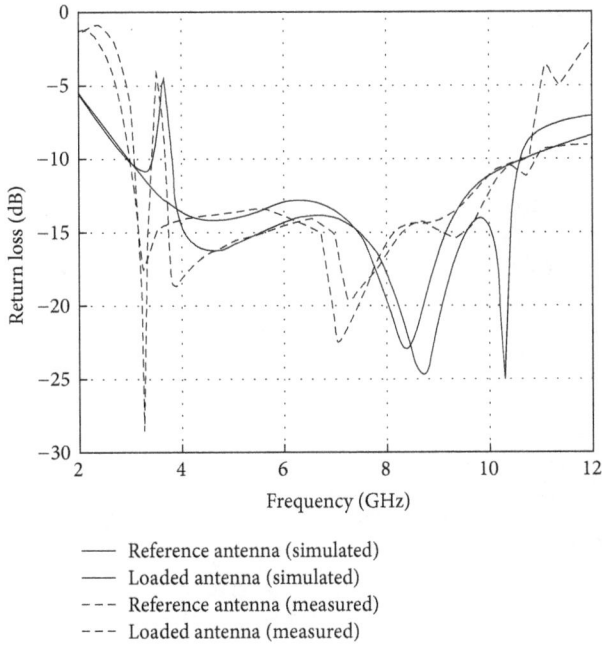

FIGURE 2: Return loss plot of the monopole antenna with single notch.

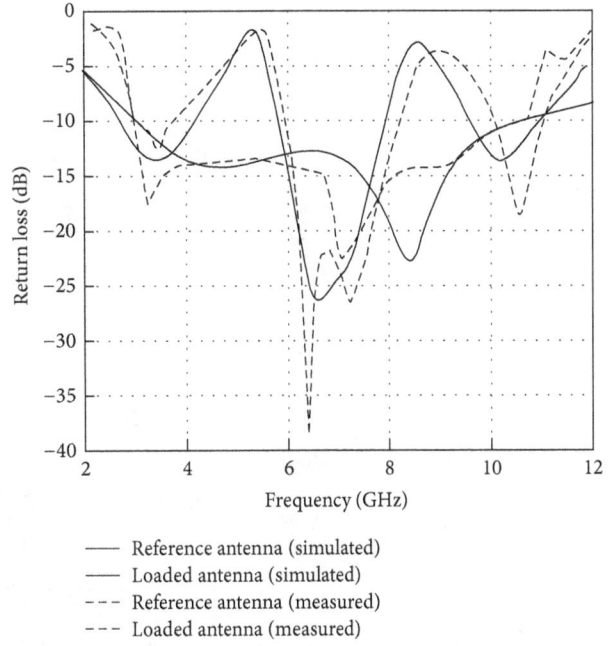

FIGURE 4: Return loss plot of the monopole antenna with double notch.

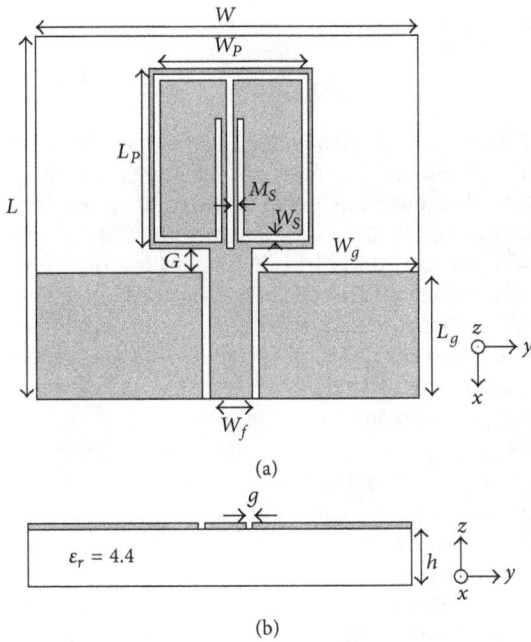

FIGURE 3: CPW fed planar monopole antenna with double notch: (a) top view, (b) side view.

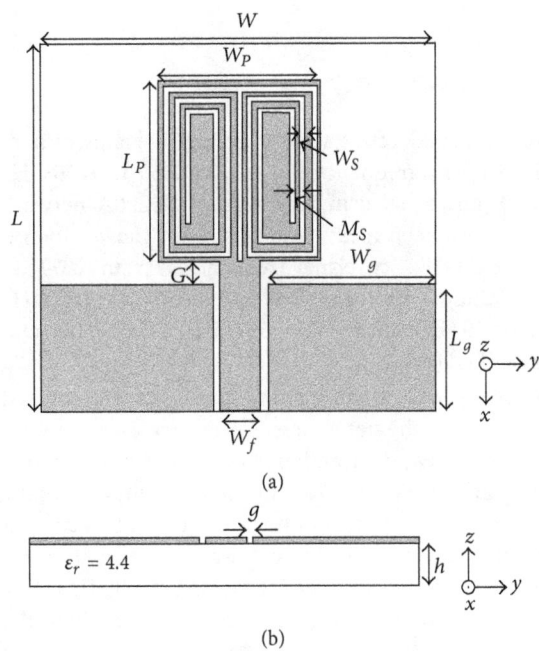

FIGURE 5: CPW fed planar monopole antenna with triple notch: (a) top view, (b) side view.

The simulated and measured return loss of the structure is shown in Figure 4. The simulation has been conducted using High Frequency Structure Simulator 11 (HFSS11). By increasing the slot length two notched bands are obtained. By using the Agilent N5230A network analyzer, S_{11} has been measured. Measured result shows that it has a bandwidth of 8.11 GHz (11.05 GHz–2.94 GHz) with two notched bands of 2.17 GHz (5.97 GHz–3.8 GHz) and 1.87 GHz (10.07 GHz–8.2 GHz).

4. Spiral Slot Loaded Antenna with Triple Notch

Figure 5 shows that on the same structure as described previously (Figure 3) slot length has been increased 35.4375 mm on both sides. This increased slot length gives triple notch at desired frequencies. The gap between two slots is kept $M_S = 0.25$ mm.

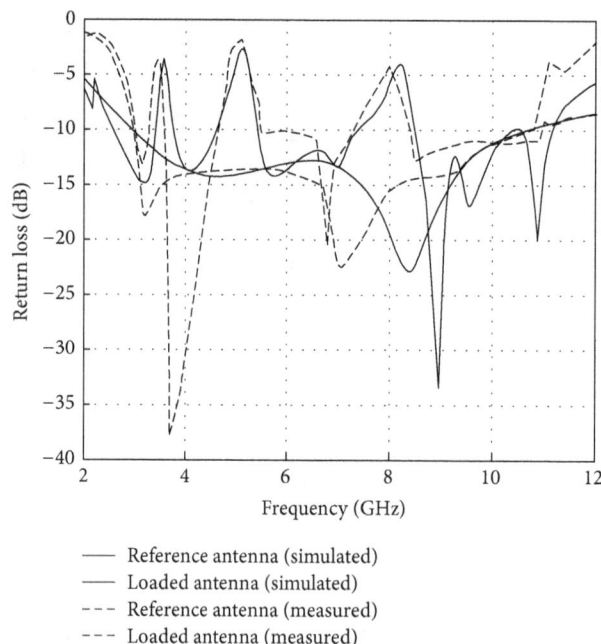

FIGURE 6: Return loss plot of monopole antenna with triple notch.

From simulated result we are getting a bandwidth of 8.57 GHz covering frequency range from 2.59 to 11.16 GHz shown in Figure 6. By using the Agilent N5230A network analyzer, S_{11} has been measured. This result shows a bandwidth of 8.11 GHz, covering frequencies from 3.04 to 10.90 GHz along with three notched bands of 0.32 GHz (3.61 GHz–3.29 GHz), 0.84 GHz (5.49 GHz–4.65 GHz), and 1.11 GHz (8.41 GHz–7.3 GHz).

Generally speaking for getting the required notched bands the length of the slot is increased by increasing number of turns. The current distribution is not uniform on the surface of the patch; that is, it varies from center to the edge of the patch. By increasing number of turns of the slot this current distribution is affected in different way and also it introduces new capacitive and inductive loading effect. Therefore this slot is resonating at different frequencies for which the notched bands are obtained.

In order to further understand the behavior of the resonating structure, especially in the notched bands, surface current distribution at five different frequencies 3.57 GHz, 5.12 GHz, 6 GHz, 8.21 GHz, and 9 GHz is simulated and displayed in Figures 7(a), 7(b), 7(c), 7(d), and 7(e), respectively. It is seen that the current distribution around spiral slot resonating structure increases drastically at 3.57 GHz, 5.12, and 8.21 GHz which implies that the spiral slot resonates near 3.57 GHz, 5.12, and 8.21 GHz. Thus, from both the return loss characteristic and the simulated surface current distribution, it can be concluded that the spiral resonator generates the frequency notched function.

The E plane ($\varphi = 0°$ plane) and H plane ($\varphi = 90°$ plane) radiation patterns of the antenna structures are shown in Figure 8. The radiation patterns were measured in an anechoic chamber in the entire bandwidth. The radiation patterns in two planes at six different frequencies are shown in Figure 8. It is seen that this antenna has the nearly monopole-like, omnidirectional radiation pattern. It is observed that at higher frequencies the radiation pattern has tilted because of the fact that at higher frequency surface current distribution increases at ground plane.

The antenna gain is simulated and measured in the entire band. From Figure 9(a) it is found that gain decreases sharply in the notched frequency band. For other frequencies out of the notched frequency band, the antenna exhibits moderate gain. Sharp decrease in gain is observed at the three notched frequencies. The same type of result is obtained for radiation efficiency as shown in Figure 9(b). The efficiency decreases sharply at the notched bands and for other frequencies out of the notched band it shows reasonable values. Hence it can be concluded that the antenna is radiating effectively outside notched bands without great amount of losses due to surface waves.

The simulated and measured results of all the three structures are summarized in Table 1.

5. Parametric Variation

Figure 10(a) shows the variation of notch band frequency with respect to the gap between two slots (M_S) and slot width (W_S). Figure 10(b) shows the effects of the gap (G) between

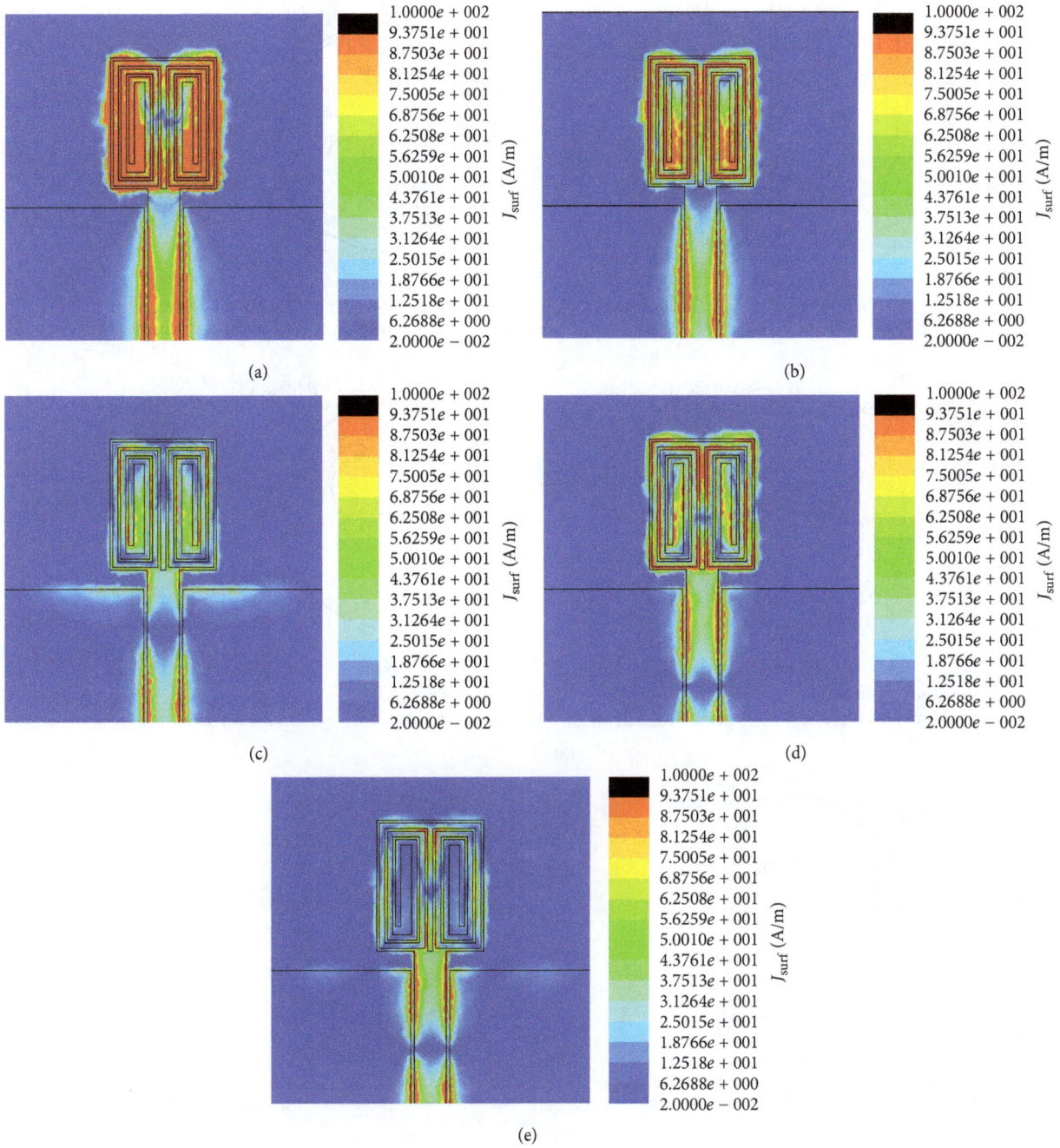

FIGURE 7: Surface current distribution: (a) 3.57 GHz, (b) 5.12 GHz, (c) 6 GHz, (d) 8.21 GHz, and (e) 9 GHz.

TABLE 1: Simulated and measured notch frequency and notch bandwidth for three types of antenna.

Antenna type	Notch frequency		Notch bandwidth	
	Simulated	Measured	Simulated	Measured
Single notch	3.66 GHz	3.52 GHz	3.40–3.80 GHz	3.40–3.65 GHz
Double notch	5.30 GHz	5.48 GHz	4.18–5.82 GHz	3.80–5.97 GHz
	8.58 GHz	9.04 GHz	7.92–9.61 GHz	8.20–10.07 GHz
Triple notch	3.57 GHz	3.48 GHz	3.40–3.76 GHz	3.29–3.61 GHz
	5.12 GHz	5.07 GHz	4.60–5.40 GHz	4.65–5.49 GHz
	8.21 GHz	7.98 GHz	7.30–8.53 GHz	7.30–8.41 GHz

the monopole and ground plane. This gap can control the bandwidth of third notch. In this case, results show that the notched bandwidth for proposed antenna becomes wider when G increases from 1.4 mm to 2 mm.

In order to verify the capability of the proposed antenna to operate as UWB antenna, it is necessary to achieve a consistent group delay. The group delay properties of the proposed multiband antenna have been studied and results have been shown in Figure 11. The results show that the simulated group delay is flat with variations below 0.05 ns whereas measured group delay variations are below 1 ns which is acceptable. The fabricated proposed UWB antenna is shown in Figure 12.

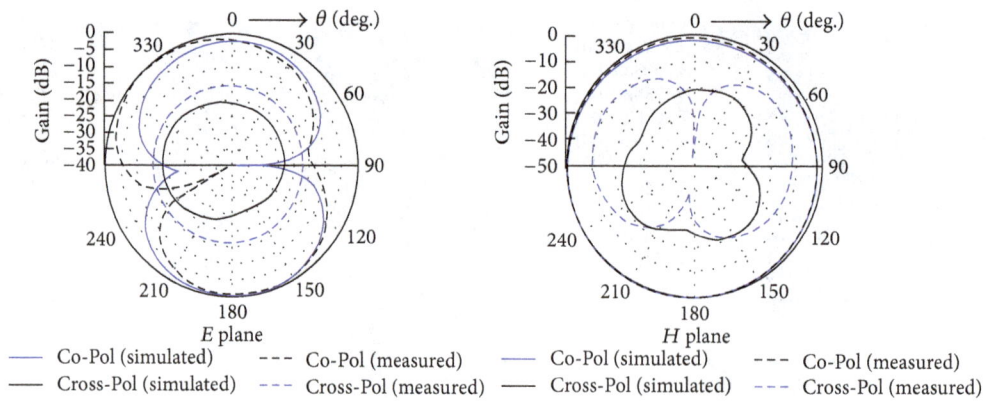

(a)

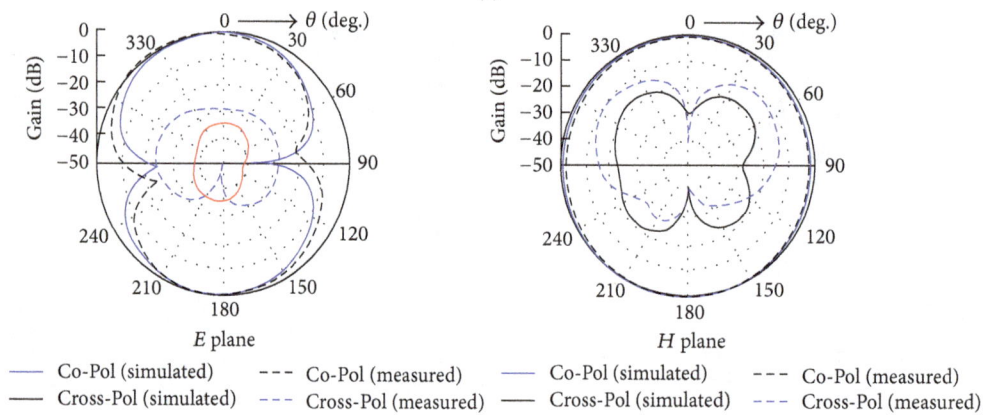

(b)

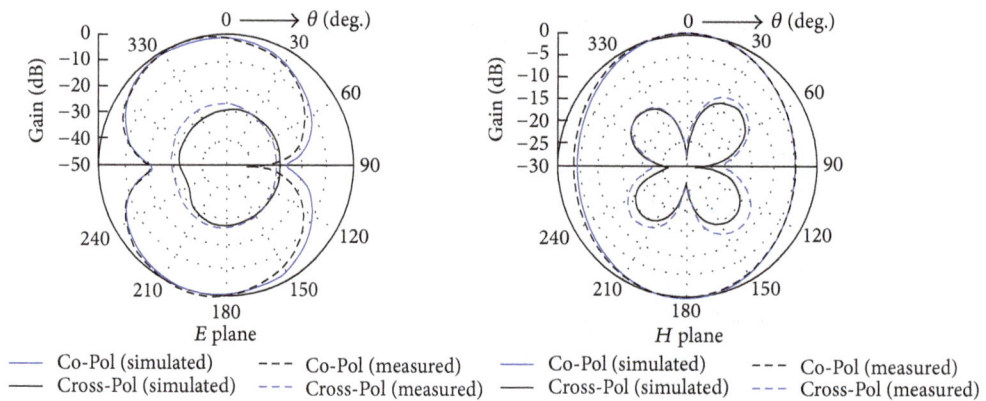

(c)

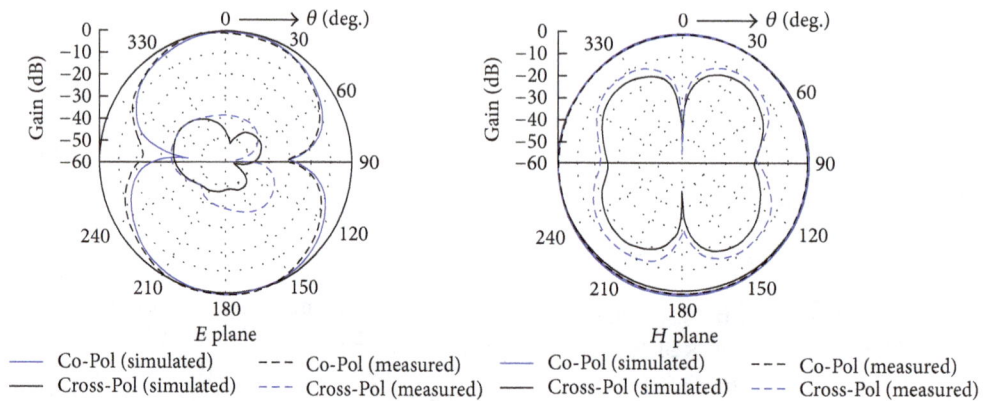

(d)

FIGURE 8: Continued.

(e)

(f)

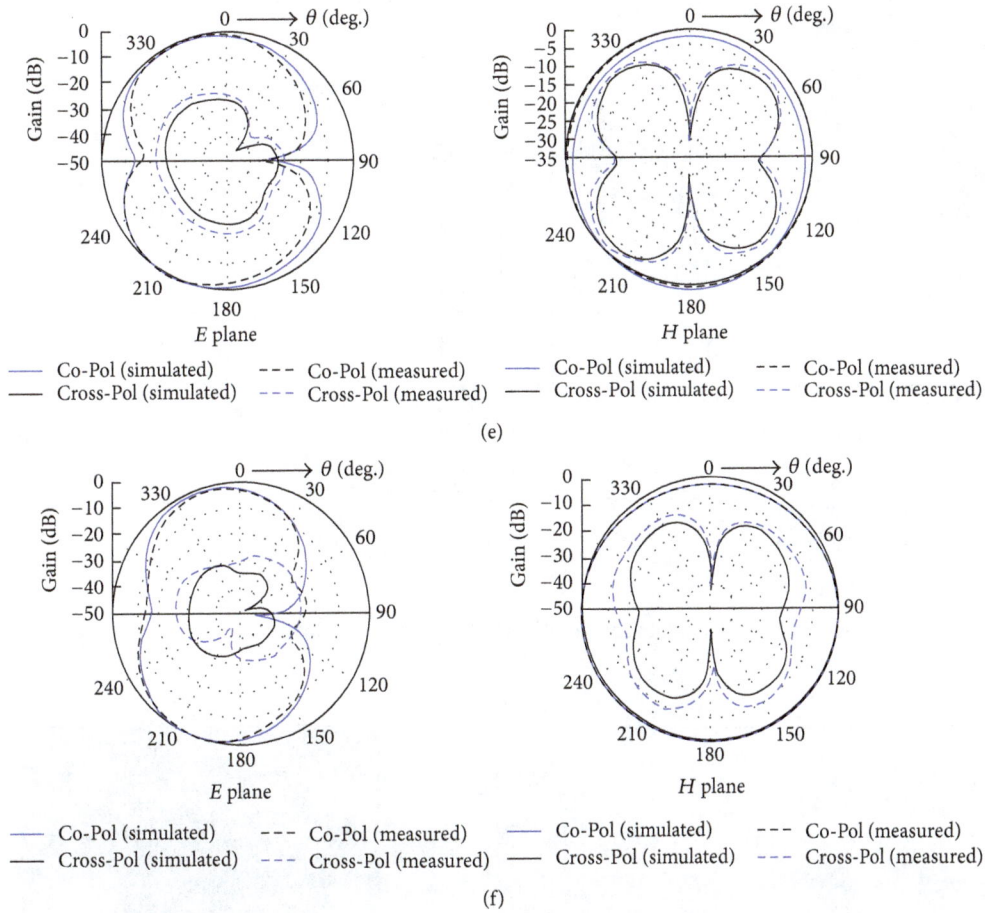

FIGURE 8: Simulated and measured normalized radiation pattern at various frequencies of (a) 3 GHz, (b) 3.57 GHz, (c) 5.12 GHz, (d) 6 GHz, (e) 8.21 GHz, and (f) 9 GHz.

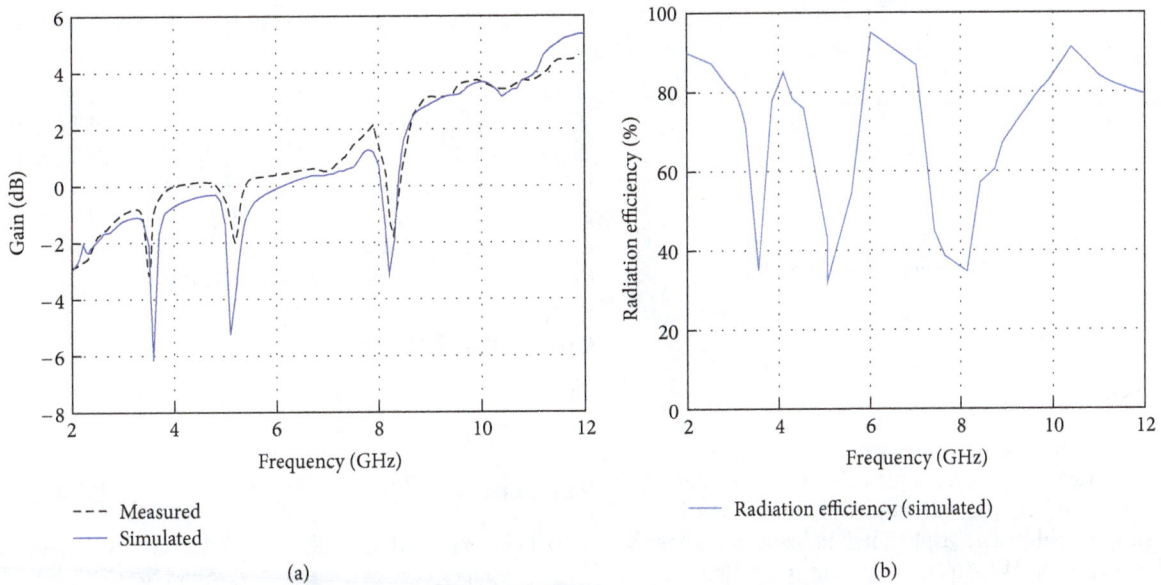

FIGURE 9: (a) Gain versus frequency plot of proposed UWB antenna with triple notch bands. (b) Radiation efficiency versus frequency plot of proposed UWB antenna with triple notch bands.

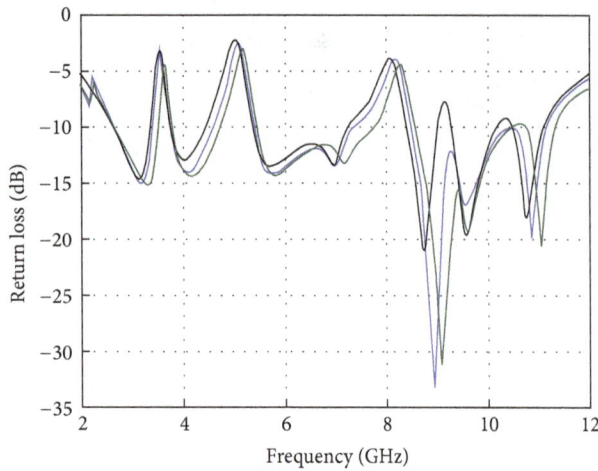

$W_S = 0.5\,\text{mm}, M_S = 0.25\,\text{mm}$
$W_S = 0.6\,\text{mm}, M_S = 0.15\,\text{mm}$
$W_S = 0.4\,\text{mm}, M_S = 0.35\,\text{mm}$

(a)

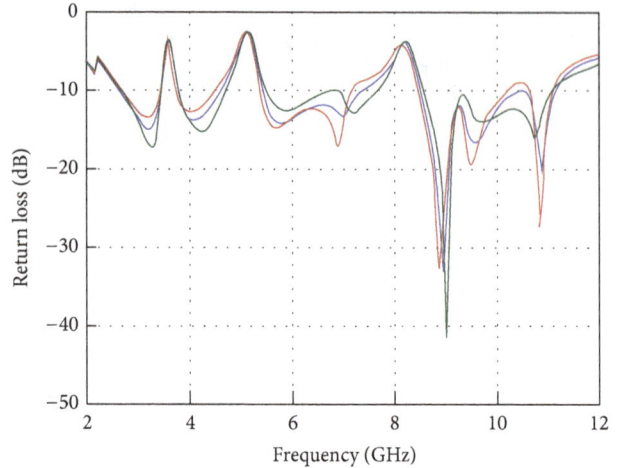

$G = 1.7\,\text{mm}$
$G = 2\,\text{mm}$
$G = 1.4\,\text{mm}$

(b)

Figure 10: (a) Return loss plot for different values of W_S, M_S. (b) Return loss plot for different values of gap (G).

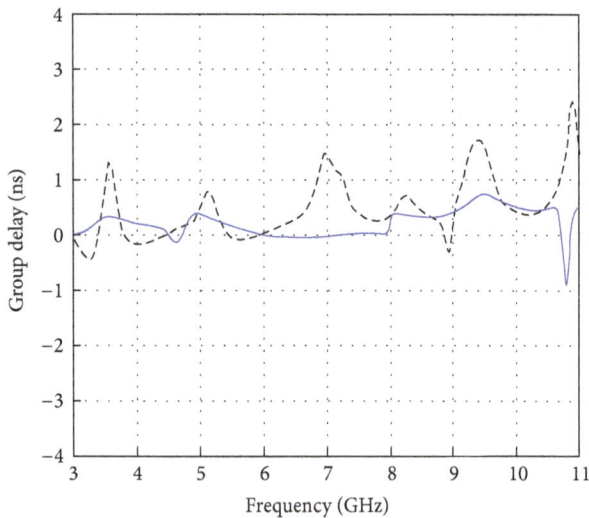

- - - Measured
— Simulated

Figure 11: Group delay versus frequency plot.

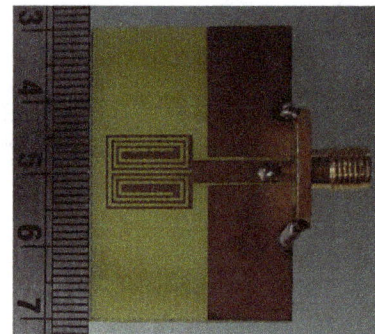

Figure 12: Fabricated photograph of the proposed antenna.

gain is obtained throughout the operation band except the three notched frequencies. The simulated results have been verified with the experimental results and are found to be in good agreement. The proposed multiple band-notched UWB antenna is low-profile, compact having simple structure, and easily compatible with microwave integrated circuits.

Conflict of Interests

The authors declare that there is no conflict of interests regarding the publication of this paper.

6. Conclusion

A novel concept for design of compact UWB monopole antennas with variable single, double, and triple filtering function is proposed. By properly adjusting the length of the spiral slot single, double, and triple notched bands have been generated at WiMAX, WLAN, and satellite communication frequencies. It is also observed that the radiation patterns of the proposed antennas are nearly omnidirectional over the entire operating bandwidth as well as relatively consistent group delay across the UWB frequencies. Moreover, a stable

References

[1] I. Oppermann, M. Hämäläinen, and J. Iinatti, *UWB Theory and Applications*, chapter 1, John Wiley & Sons, New York, NY, USA, 2004.

[2] Federal Communications Commission, "Revision of part 15 of the commission's rules regarding ultra-wideband transmission systems," First Report and Order FCC 02.V48, FCC, 2002.

[3] Y. Kim and D.-H. Kwon, "CPW-fed planar ultra wideband antenna having a frequency band notch function," *Electronics Letters*, vol. 40, no. 7, pp. 403–405, 2004.

[4] K.-L. Wong, Y.-W. Chi, C.-M. Su, and F.-S. Chang, "Band-notched ultra-wideband circular-disk monopole antenna with an arc-shaped slot," *Microwave and Optical Technology Letters*, vol. 45, no. 3, pp. 188–191, 2005.

[5] C.-Y. Huang and W.-C. Hsia, "Planar ultra-wideband antenna with a frequency notch characteristic," *Microwave and Optical Technology Letters*, vol. 49, no. 2, pp. 316–320, 2007.

[6] S. Barbarino and F. Consoli, "UWB circular slot antenna provided with an inverted-L notch filter for the 5 GHz WLAN band," *Progress in Electromagnetics Research*, vol. 104, pp. 1–13, 2010.

[7] K. Chung, J. Kim, and J. Choi, "Wideband microstrip-fed monopole antenna having frequency band-notch function," *IEEE Microwave and Wireless Components Letters*, vol. 15, no. 11, pp. 766–768, 2005.

[8] W. Choi, K. Chung, J. Jung, and J. Choi, "Compact ultra-wideband printed antenna with band-rejection characteristic," *Electronics Letters*, vol. 41, no. 18, pp. 990–991, 2005.

[9] W.-S. Lee, D.-Z. Kim, K.-J. Kim, and J.-W. Yu, "Wideband planar monopole antennas with dual band-notched characteristics," *IEEE Transactions on Microwave Theory and Techniques*, vol. 54, no. 6, pp. 2800–2805, 2006.

[10] J. C. Ding, Z. L. Lin, Z. N. Ying, and S. L. He, "A compact ultra-wideband slot antenna with multiple notch frequency bands," *Microwave and Optical Technology Letters*, vol. 49, no. 12, pp. 3056–3060, 2007.

[11] J. Kim, C. S. Cho, and J. W. Lee, "5.2 GHz notched ultra-wideband antenna using slot-type SRR," *Electronics Letters*, vol. 42, no. 6, pp. 315–316, 2006.

[12] S.-H. Lee, J.-W. Baik, and Y.-S. Kim, "A coplanar waveguide fed monopole ultra-wideband antenna having band-notched frequency function by two folded-striplines," *Microwave and Optical Technology Letters*, vol. 49, no. 11, pp. 2747–2750, 2007.

[13] K.-H. Kim and S.-O. Park, "Design of the band-rejected UWB antenna with the ring-shaped parasitic patch," *Microwave and Optical Technology Letters*, vol. 48, no. 7, pp. 1310–1313, 2006.

[14] K. Chung, S. Hong, and J. Choi, "Ultrawide-band printed monopole antenna with band-notch filter," *IET Microwaves, Antennas & Propagation*, vol. 1, no. 2, pp. 518–522, 2007.

[15] T.-G. Ma and S.-J. Wu, "Ultrawideband band-notched folded strip monopole antenna," *IEEE Transactions on Antennas and Propagation*, vol. 55, no. 9, pp. 2473–2479, 2007.

[16] M. Zhang, Y.-Z. Yin, J. Ma, Y. Wang, W.-C. Xiao, and X.-J. Liu, "A racket-shaped slot UWB antenna coupled with parasitic strips for band-notched application," *Progress in Electromagnetics Research Letters*, vol. 16, pp. 35–44, 2010.

[17] L. Lizzi, G. Oliveri, P. Rocca, and A. Massa, "Planar monopole UWB antenna with UNII1/UNII2 WLAN-band notched characteristics," *Progress In Electromagnetics Research B*, vol. 25, pp. 277–292, 2010.

[18] M. Xie, Q. Guo, and Y. Wu, "Design of a miniaturized UWB antenna with band-notched and high frequency rejection capability," *Journal of Electromagnetic Waves and Applications*, vol. 25, no. 8-9, pp. 1103–1112, 2011.

[19] G.-P. Gao, Z.-L. Mei, and B.-N. Li, "Novel circular slot UWB antenna with dual band-notched characteristic," *Progress in Electromagnetics Research C*, vol. 15, pp. 49–63, 2010.

[20] Y.-Q. Xia, J. Luo, and D.-J. Edwards, "Novel miniature printed monopole antenna with dual tunable band-notched characteristics for UWB applications," *Journal of Electromagnetic Waves and Applications*, vol. 24, no. 13, pp. 1783–1793, 2010.

[21] R. Shi, X. Xu, J. Dong, and Q. Luo, "Design and analysis of a novel dual band-notched UWB antenna," *International Journal of Antennas and Propagation*, vol. 2014, Article ID 531959, 10 pages, 2014.

[22] M. Almalkawi and V. Devabhaktuni, "Ultrawideband antenna with triple band-notched characteristics using closed-loop ring resonators," *IEEE Antennas and Wireless Propagation Letters*, vol. 10, pp. 959–962, 2011.

[23] D. T. Nguyen, D. H. Lee, and H. C. Park, "Very compact printed triple band-notched UWB antenna with quarter-wavelength slots," *IEEE Antennas and Wireless Propagation Letters*, vol. 11, pp. 411–414, 2012.

[24] J.-Y. Kim, B.-C. Oh, N. Kim, and S. Lee, "Triple band-notched UWB antenna based on complementary meander line SRR," *Electronics Letters*, vol. 48, no. 15, pp. 896–897, 2012.

[25] C. Abdelhalim and D. Farid, "A compact planar UWB antenna with triple controllable band-notched characteristics," *International Journal of Antennas and Propagation*, vol. 2014, Article ID 848062, 10 pages, 2014.

[26] V. M. Nangare and V. G. Kasabegoudar, "Ultra-wideband monopole antenna with multiple notch characteristics," *International Journal of Electromagnetics and Applications*, vol. 4, no. 3, pp. 70–76, 2014.

10

Logarithmic Slots Antennas Using Substrate Integrated Waveguide

Jahnavi Kachhia, Amit Patel, Alpesh Vala, Romil Patel, and Keyur Mahant

Department of Electronics and Communication Engineering, Charotar University of Science & Technology, Changa, Anand, Gujarat, India

Correspondence should be addressed to Amit Patel; amitvpatel.ec@charusat.ac.in

Academic Editor: Xianming Qing

This paper represents new generation of slotted antennas for satellite application where the loss can be compensated in terms of power or gain of antenna. First option is very crucial because it totally depends on size of satellite so we have proposed the high gain antenna creating number of rectangular, trapezoidal, and I shape slots in logarithm size in Substrate Integrated Waveguide (SIW) structure. The structure consists of an array of various shape slots antenna designed to operate in C and X band applications. The basic structures have been designed over a RT duroid substrate with dielectric constant of 2.2 and with a thickness of 0.508 mm. Multiple slots array and shape of slot effects have been studied and analyzed using HFSS (High Frequency Structure Simulator). The designs have been supported with its return loss, gain plot, VSWR, and radiation pattern characteristics to validate multiband operation. All the proposed antennas give gain more than 9 dB and return loss better than −10 dB. However, the proposed structures have been very sensitive to their physical dimensions.

1. Introduction

Rapid development in the field of wireless communication system that operates in the microwave range is payed more attention from industry and academia [1–3]. In the present, rectangular waveguides are widely used in microwave engineering for antenna [4], filters [5, 6], couplers [7], and so forth due to their advantages of low losses, high power handling, and high isolation [8]. In addition, slot array antennas based on waveguides that feature favorable antenna performance such as high directivity, low cross-polarization, and low crosstalk have been presented [9, 10, 15].

In 1943, the slot array was invented at McGill University in Montreal by Watson. One of the best features of this antenna is horizontal polarization and omnidirectional gain around the azimuth. A slot along the length is cut into the wall of a waveguide that disrupts the transverse current flowing in the wall, which enforces the current to travel at border of the slot and induces an electric field in the slot [11, 12]. The location of the slot in the rectangular waveguide decides the current flow. Thus, the pose determines the impedance introduced to the transmission line and the amount of energy coupled and radiated from the slot. Slotted waveguide arrays (SWA) have some

advantages over microstrip antennas such as having low loss, high isolation, and high power handling [11]. Due to these advantages, SWA antennas are widely used in communication and radar systems particularly at microwave wave frequencies. Regarding conformal array applications, microstrip antennas are easier to implement compared to SWA antennas; however, excitation of antenna elements by a waveguide feed network is advantageous compared to microstrip feed network in terms of eliminating radiation losses and cross coupling problems. Moreover, complex feed network structure is not needed in SWA to excite the slots in the same waveguide. Furthermore, the array of waveguides can be formed without cross coupling problems.

However, the manufacturing of these waveguides needs to be accomplished with sufficient accuracy so as to allow for operation at millimeter wave frequencies. On the other hand, antennas and microwave components used at lower frequencies typically rely on planar designs which are mostly realized with the Printed Circuit Board (PCB) processing technique [13, 14]. Moreover, conformal arrays have a specific shape determined by the parameters other than radiation pattern and input match requirements and they can easily be implemented using microstrip technology using a flexible substrate or multifaceted surfaces [15].

This mature technology leads to, not only low-cost designs, but also the possibility to easily integrate them with common electronic components. However, these planar designs are by nature not fully shielded and thus subject to radiation, cross-talk, and packaging problems [16]. It has also complex feed networks such as probes feeding patch elements. Strip line can be used to eliminate the radiation losses; however, cross coupling problems are encountered in the feed network with strip line structure. These drawbacks added to the potentially high conductor losses make this technology not feasible to implement high frequency complex structures, such as large feeding networks [17]. It is then obvious that a large performance gap exists between components based on metallic waveguides and the ones based on PCBs.

A very promising candidate to fill this gap and to provide widespread commercial solutions is the SIW technology or, more in general, the Substrate Integrated Circuit (SIC) architecture [18]. This technology allows building high performance, low cost, and reliable waveguide-like components using planar processing techniques, such as the PCB or the Low-Temperature Cofired Ceramic (LTCC) [19]. It has also unique features such as compact size in comparison with waveguide antennas and also simpler structure in comparison with reflector antennas [1].

On the basis of the above two, SIW slot array antennas have been proposed with high gain and ultrawide bandwidth specifically for C band, X band, and Ku band applications. The proposed structures contain three different shapes of slots like rectangle, trapezoidal, and I shape and its size varies according to logarithm value. All the structures generate quite better results and respond to resonant frequencies according to their size. One of the unique features of them is that they generate sharp radiation pattern and isolation between bands is much more high. Due to these advantages, the slotted SIW antennas are good candidates for the conformal array applications especially when SIW is implemented using a flexible substrate. There is a limited study in the literature on the conformal array applications with slotted waveguide arrays.

With the aim of the above, the structure of paper is as follows: Section 2 describes the theory of basic SIW and based on it is the design of slotted linear arrays of antenna. Then, Section 3 shows the design and calculation of various proposed antennas and their simulation results are shown in Section 4. Section 5 contains comparison of all the proposed antennas and discussion of results. Finally paper is ended with conclusion.

2. SIW Antenna Design

SIW is described as two conducting layers are connected by cylindrical vias row at both sides which is shown in Figure 1 [20]. In Figure 1, d is the diameter of vias, w is the equivalent width of SIW, p is the spacing between two vias, and h is the thickness of substrate. Structure of SIW contains low cost, high Q-factor, low radiation, and high density integration which makes it preferable.

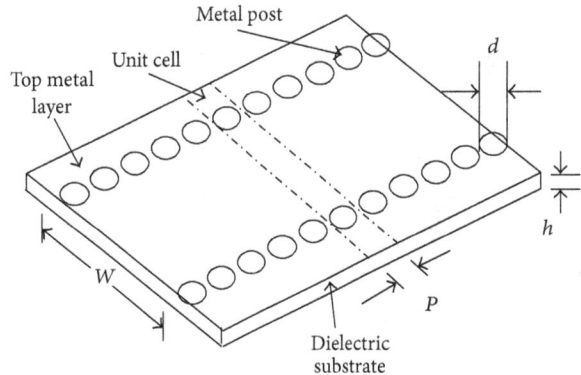

FIGURE 1: Basic structure of SIW.

The cutoff frequency of SIW is defined as $f_c = c/2a$, where a is width of waveguide; it can change by varying the width in conciliation of degrading the overall characteristics of its components.

As we know that, the current in the walls of the waveguide must be comparative to the difference in the electric field between two points (so the selection of slot in SIW is very important) [21]. Therefore, to make a slot in the correct center of the broad wall of the waveguide will not radiate at all, since the electric field is not asymmetrical around the center of the guide and thus is indistinguishable at both edges of the slot. If the position of slot is moved from the centerline, the difference in field concentration between the rims of the slot is larger, so that more current is interrupted and more energy is coupled to the slot that ultimately increases radiated power. In other sides of the waveguide, the field strength is very weak, since the sidewalls are short circuited for the electric field. The produced current must also be small; longitudinal slots which are far from the center or created in the sidewall will not radiate significantly.

In this paper, we keep the distance between slots as $\lambda_g/2$ such that the slots will be fed in the same phase (spacing between the slots at $\lambda_g/2$ intervals in the waveguide is an equivalent electrical spacing of 180°. Therefore, each slot is exactly out of phase with its neighbors, so their radiation cancelled each other. On the other side, slots on opposite sides of the center axis of the guide are out of phase (180°), so we can swap the slot displacement around the center axis and have a total phase difference of 360° between slots, putting them back in phase) and the beam will not be inclined [22]. For the position of the last slot, the center of slot is kept at guided quarter wavelength away from the closed end of the waveguide. As we know that a short circuited quarter wavelength stub of transmission line works as an open-circuit, the closed end does not impact on the impedance. Another reason to keep the closed end is space $3\lambda_g/4$ for mechanical fabrication; the extra half-wavelength is crystal clear.

3. SIW Slots Array Antenna Parameters

The SIW slots array antenna is shown in Figure 2. Here, prime important parameters in the design are distance and diameter of posts which controlled the flow of electric field. Slots are

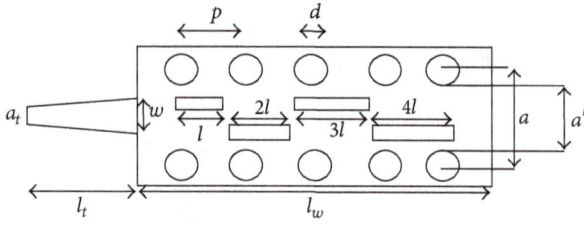

FIGURE 2: Schematic of proposed antenna.

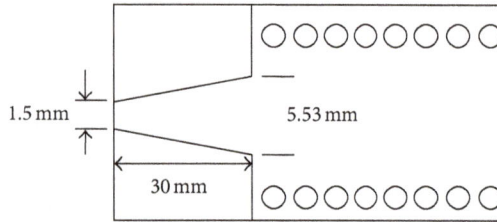

FIGURE 3: Microstrip to SIW transition.

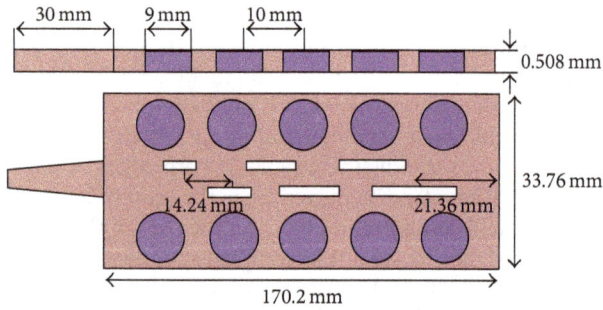

FIGURE 4: Side and top view of 1st proposed logarithmic rectangular slots array antenna.

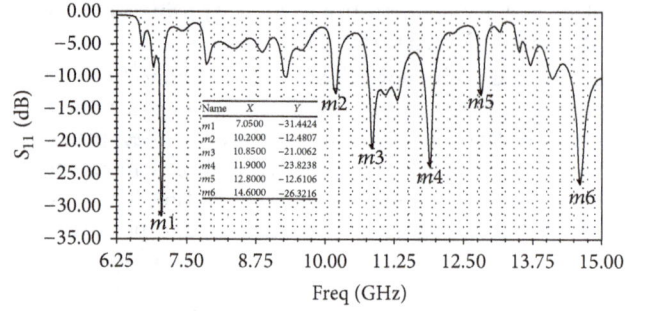

Curve info
—— S_{11} (dB)
Setup1: sweep

FIGURE 5: S_{11} of 1st proposed logarithmic rectangular slots array antenna.

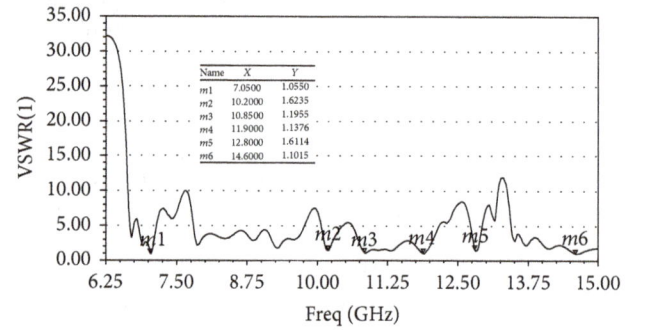

Curve info
—— VSWR(1)
Setup1: sweep

FIGURE 6: VSWR of 1st proposed logarithmic rectangular slots array antenna.

printed on a 0.508 mm thick Roger RT duroid 5880 substrate (the relative dielectric constant is 2.2) with the size of $W \times L = (33.76 \times 170.2)$ mm. The selection of feed is also very important and here we have selected feeding using tapered slot which is shown in Figure 3 that provide transition from microstrip to SIW [23]. A tapered microstrip line has the following parameters: the width of the feed line $a_t = 1.5$ mm, $w = 5.53$ mm and the length of feed line $l_t = 30$ mm (calculation of all the parameters are shown below). The finite ground plane has an area of $W \times H = (33.76 \times 170.2)$ mm.

Parameters Calculation [21]

(1) For designing SIW based antenna define first substrate dimensions which are given by $a \times b$ for that a' and a must be known as shown below (a' inversely propositional to cutoff frequency):

$$a' = \frac{c}{2 f_c \sqrt{\varepsilon_r}}. \tag{1}$$

(2) Now obtain center-to-center distance with the help of a' as

$$a = a' + \frac{d^2}{0.95 p}. \tag{2}$$

(3) To select the diameter of post and also distance between two posts, we have to consider the following condition which minimizes the losses [1]:

$$p \le 2d$$

$$0.05 < \left(\frac{p}{\lambda_c} \right) < 0.25 \tag{3}$$

$$\frac{\lambda_g}{5} < d,$$

where λ_g is given by

$$\lambda_g = \frac{\lambda_c}{\sqrt{\varepsilon_r}}. \tag{4}$$

Here, p is the center-to-center distance between two posts; d is the diameter of the post; λ_g is guided wavelength; λ_c is cutoff wavelength; ε_r is relative permittivity of dielectric medium.

(4) To apply tapper feed at antenna, first find the tapper width by using traditional microstrip calculation which is denoted by (w_1) and after that feeding width at dielectric side.

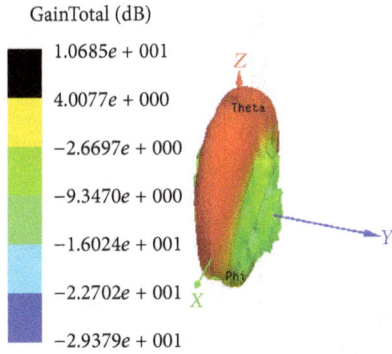

Antenna gain plot for 7.05 GHz

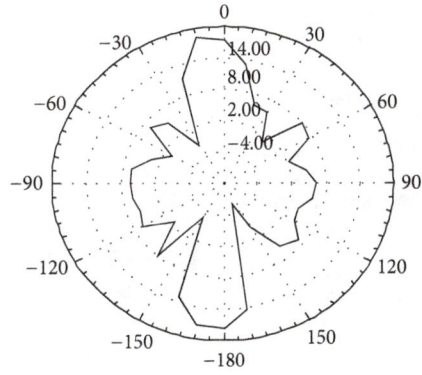

Curve info
—— rETotal (dB)
Setup1: LastAdaptive
Freq = "7.05 GHz" Theta = "90 deg"

Radiation pattern for solution frequency 7.05 GHz

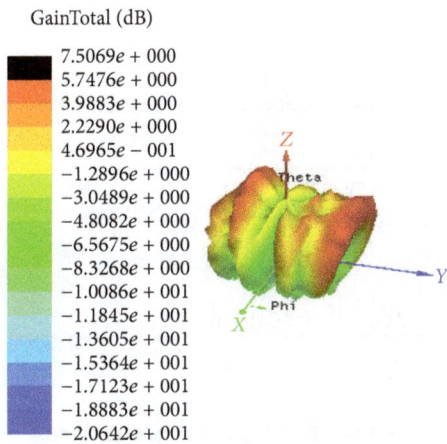

Antenna gain plot for 11.9 GHz

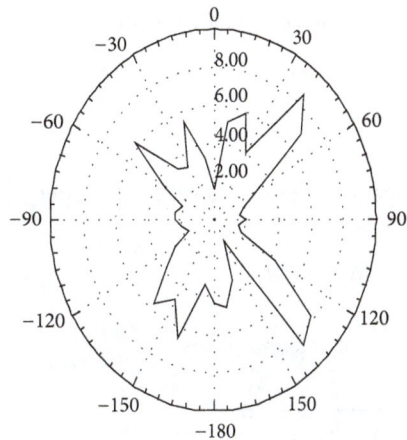

Curve info
—— rETotal
Setup1: LastAdaptive
Freq = "10.85 GHz" Theta = "90 deg"

Radiation pattern for solution frequency 10.8 GHz

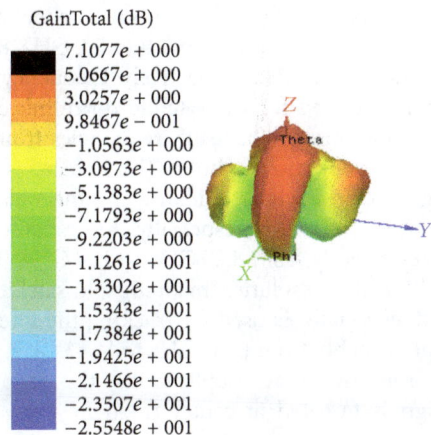

Antenna gain plot for 10.8 GHz

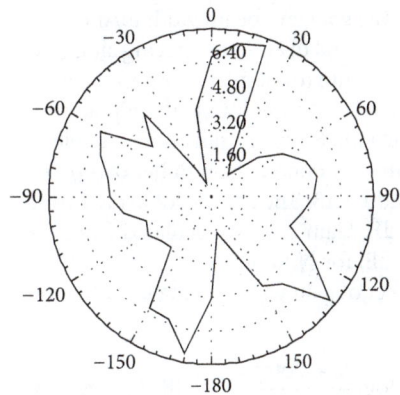

Curve info
—— rETotal
Setup1: LastAdaptive
Freq = "11.9 GHz" Theta = "90 deg"

Radiation pattern for solution frequency 11.9 GHz

FIGURE 7: Continued.

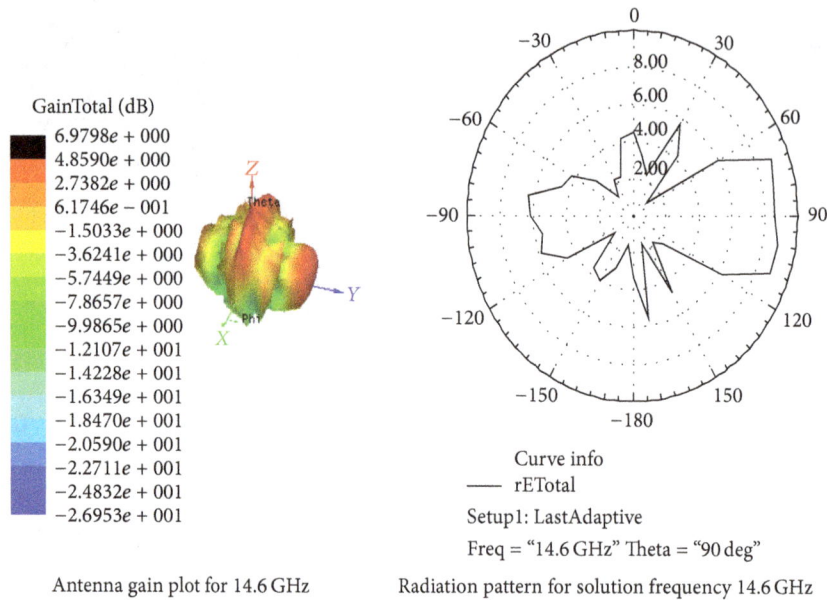

GainTotal (dB)

6.9798e + 000
4.8590e + 000
2.7382e + 000
6.1746e − 001
−1.5033e + 000
−3.6241e + 000
−5.7449e + 000
−7.8657e + 000
−9.9865e + 000
−1.2107e + 001
−1.4228e + 001
−1.6349e + 001
−1.8470e + 001
−2.0590e + 001
−2.2711e + 001
−2.4832e + 001
−2.6953e + 001

Curve info
—— rETotal
Setup1: LastAdaptive
Freq = "14.6 GHz" Theta = "90 deg"

Antenna gain plot for 14.6 GHz

Radiation pattern for solution frequency 14.6 GHz

FIGURE 7: Gain plots and radiation patterns at different resonant frequencies.

The width of tapper at feed side is 0.4 times the opening of patch antenna:

$$a_t = 0.4 \left(a - d\right). \tag{5}$$

The width of tapper at antenna side is

$$w = \frac{c}{2f\sqrt{(\varepsilon_r + 1)/2}}. \tag{6}$$

Finally, length of tapper for impedance match is given by

$$l_t = \frac{n * \lambda_g}{4} \quad \text{where } n = 1, 2, 3, \ldots. \tag{7}$$

It is already derived that the gain and beamwidth formula for the slots have equal distance end to end and spacing along the waveguide. A simple procedure to calculate the gain of a slot antenna is on array of dipoles. As we double the dipole slots they increase the double gain. So for 16 slots it is possible to get gain of 12 dB. Each time we double the number of dipoles, we double the gain, or add 3 dB. Thus, a 16-slot array would have a gain of about 12 dB. Gain can be calculated from the equation $G = 10 \log(N)$ dB, for N total slots.

Now, include the effect of gain with spacing of slot better to be described as

$$\text{Gain} = 10 \log \frac{N * \text{slotspacing}}{\lambda_0} \text{ dB.} \tag{8}$$

Calculated gain is always equal to average gain; now beamwidth is given by

$$\text{Beamwidth} = 50.7 \frac{\lambda_0}{(N/2) * \text{slotspacing}} \text{ Degree.} \tag{9}$$

TABLE 1: Specification of the proposed structure.

Parameters	Value
Center frequency	7 GHz
Return loss	>−10 dB
VSWR	1
Gain	>5 dB
Dielectric constant (RT duroid)	2.2

4. Design of Proposed Structures

Here, we have considered the design of proposed structure based on the specification given in Table 1. These specifications are considered based on the satellite and RADAR applications in which we are targeting C, X, and Ku band frequencies. The C band, X band, and Ku Band defined by an IEEE standard for radio waves and radar engineering with frequencies that range from 4.0 to 8.0 GHz, 8.0 to 12.0 GHz, and 12.0 to 18.0 GHz, respectively. Frequency range of 3.7 to 4.2 GHz is used for the satellite downlink communication in C band and the band of frequencies from 5.925 GHz to 6.425 GHz for their uplinks. The X band is used for short range tracking, missile guidance, marine, radar, and airborne intercept. It is used, especially, for radar communication ranges roughly from 8.29 GHz to 11.4 GHz. The Ku band is used for high resolution mapping and satellite altimetry. Ku band, especially, is used for tracking the satellite within the ranges roughly from 12.87 GHz to 14.43 GHz.

From the above applications, the proposed structure design is targeted at center frequency of 7 GHz. Physical dimensions of the first proposed structure are calculated by the above formula and summarized in Table 2. The wire line structure of proposed antenna is shown in Figure 3.

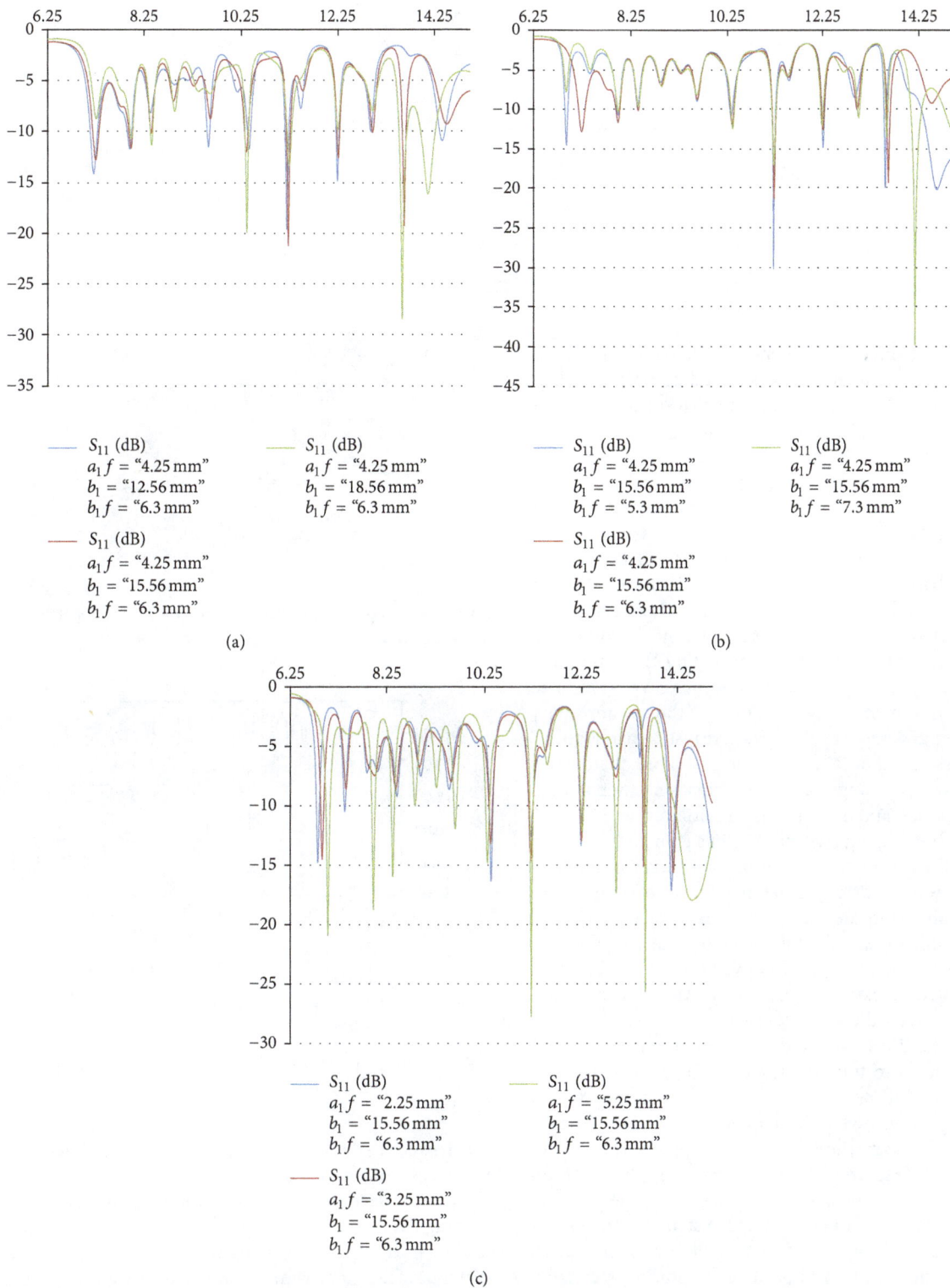

FIGURE 8: (a) S_{11} of logarithmic rectangular slots array antenna with respect to variations in 1st slot position. (b) S_{11} of logarithmic rectangular slots array antenna with respect to variations in 1st slot position. (c) S_{11} of logarithmic rectangular slots array antenna with respect to variations in width of 1st slot.

TABLE 2: Calculated parameters.

Parameter	f_c	a^l	D	P	a_s	ε_r	λ_c	λ_g	w	a_t	l_w	l_t
Value	7.1 GHz	14.2 mm	9 mm	10 mm	22.76 mm	2.2	42.25 mm	28.48 mm	5.504 mm	1.5 mm	170 mm	30 mm

TABLE 3: Summaries of simulated parameters.

Resonant frequency	Gain in dB	Bandwidth in MHz	VSWR	Peak return loss in dB
7.05 GHz	10.85	100	1.05	−31.43
10.85 GHz	7.5	200	1.19	−21.00
11.9 GHz	7.1	250	1.13	−23.71
14.6 GHz	6.9	400	1.097	−26.64

As shown in Figure 4, it contains six rectangular slots array whose longitudinal length is varied according to logarithmic manner. We are keeping center-to-center distance between two slots as 14.24 mm ($\lambda_g/2$) and center of last slot to broadside wall as 21.36 mm.

The structure has been simulated using HFSS (High Frequency Structure Simulator) and it generates return loss shown in Figure 5. As shown in Figure 5, the structure generates six resonant frequencies one in C band (7.05 GHz), three in X band (10.2, 10.55, and 11.5 GHz), and two in Ku band (12.8 and 14.6 GHz), respectively. These results are totally expected since the proposed antenna is composed of different resonant slot lengths providing the multiband performance. At all the resonant frequencies VSWR is also good which is shown in Figure 6. One of the extraordinary performances generated by this structure is shown in Figure 7 which represents the gain value that is more than 10 dB in single layer and single array combination compared to other traditional antennas having multilayer and 2 to 3 arrays for achieving higher gain. Various gain and radiation patterns of the proposed logarithmic slots array of antenna for different resonant frequencies are shown in Figure 6. All the simulation results are summarized in Table 3.

It is possible to tune the frequency by changing the geometrical position with respect to wave propagation direction and physical dimensions of the slots. All these methods ultimately changed the value of reactance of resonators and coupling of the EM field. Here we have demonstrated three possible methods to tune the response: (i) pose of slot; (ii) length of slot; (iii) width of slot. For demonstration and analysis purpose only we applied the above tuning methods to the first slot. These methods can also be applicable for other remaining slots. Figure 8(a) demonstrates simulation result of reflection coefficient obtained by changing the position of first slot which was kept at $\lambda_g/2$ distance with respect to second slot in longitude direction. It shows that when we move the slot position in positive longitude direction, it improves the value of return loss and increment in opposite direction, it reduces the value of return loss. Figure 8(b) shows the simulation result with increasing the length of the first slot. It decreases the values of inductance and increases the capacitance value. Variation in length gave minor effect on a reflection coefficient value. Similarly Figure 8(c) is the result of reflection coefficient with the change in width of

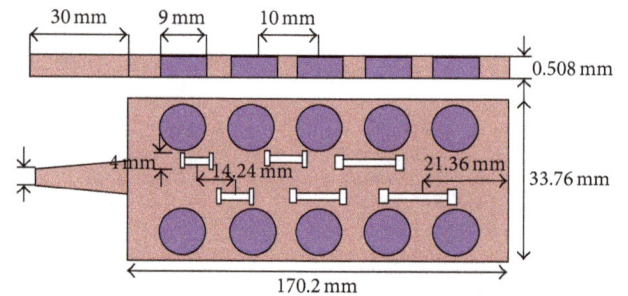

FIGURE 9: 2nd proposed structure contains I shape logarithmic slots array antenna.

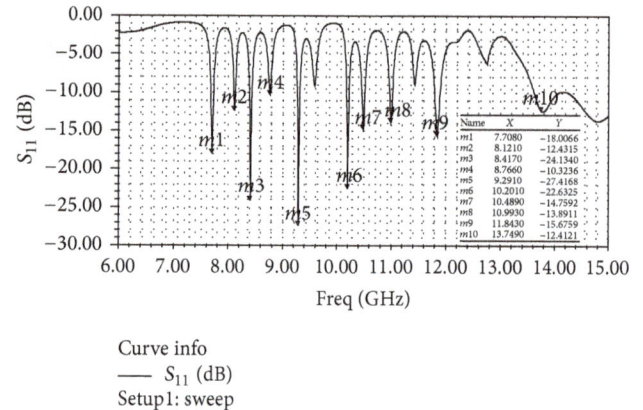

FIGURE 10: S_{11} plot for I shaped logarithmic slots array antenna.

the first slot that tunes resonant frequency effectively and also produces very sharp bandwidth.

In the second proposed structure, instead of taking rectangular shape slots, longitudinal I shape is selected which is shown in Figure 9. Spacing between two slots and center distance of last slots from the end boadsie wall are the same as 1st proposed design. In first I shape slot, vertical height of the slot is 4 mm and length of slot is 10 mm which progressively increases for other slots according to logarthmic nature.

This structure is also simulated by using HFSS and its return loss is shwon in Figure 10. This structure radiates at six resonant frequencies: one in C band (7.7 GHz), three in X band (8.12, 8.41, 8.76, 9.29, 10.2, 10.48, 10.99, 11.45, and

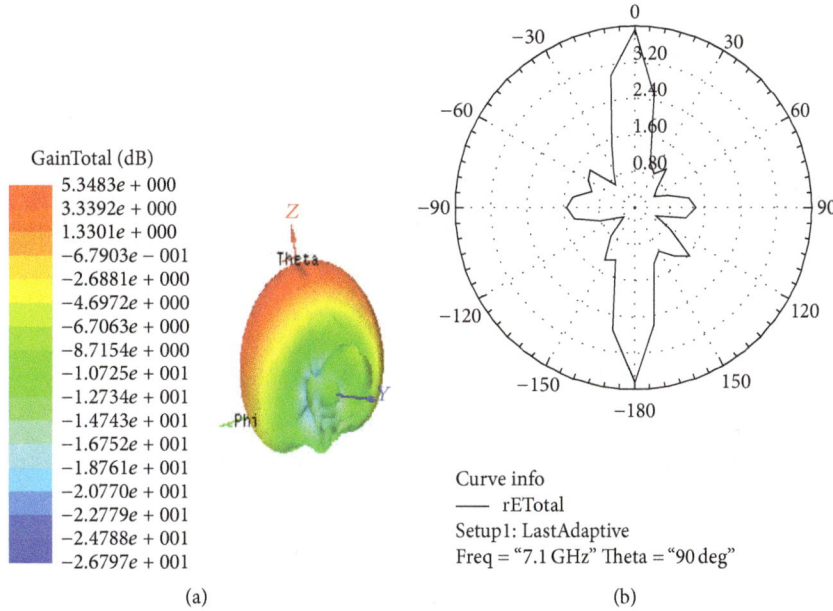

GainTotal (dB)

5.3483e + 000
3.3392e + 000
1.3301e + 000
−6.7903e − 001
−2.6881e + 000
−4.6972e + 000
−6.7063e + 000
−8.7154e + 000
−1.0725e + 001
−1.2734e + 001
−1.4743e + 001
−1.6752e + 001
−1.8761e + 001
−2.0770e + 001
−2.2779e + 001
−2.4788e + 001
−2.6797e + 001

Curve info
——— rETotal
Setup1: LastAdaptive
Freq = "7.1 GHz" Theta = "90 deg"

(a)   (b)

FIGURE 11: (a) Gain plot and (b) radiation pattern for I shape logarithmic slots array antenna.

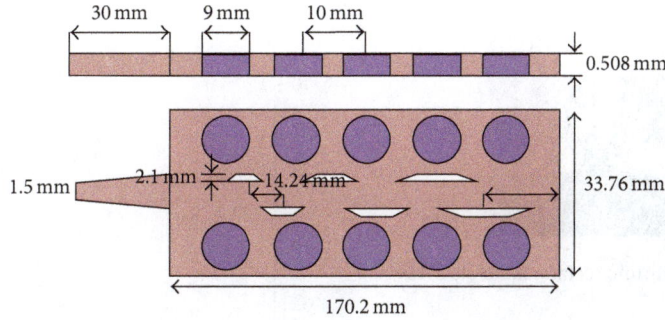

FIGURE 12: 3rd proposed structure contains trapezoidal logarithmic slots array antenna.

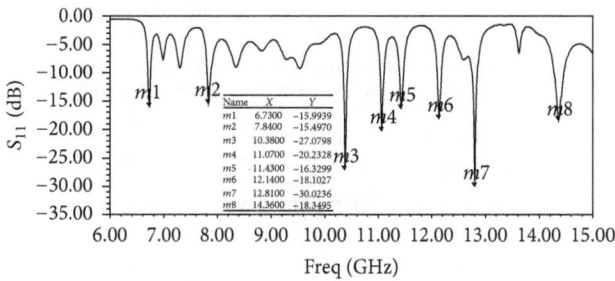

Curve info
——— S_{11} (dB)
Setup1: sweep

FIGURE 13: S_{11} plot for trapezoidal logarithmic slots array antenna.

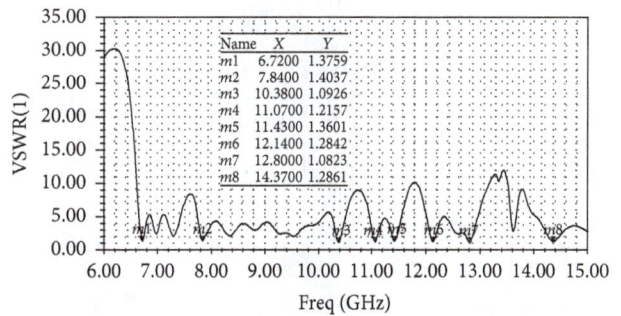

Curve info
——— VSWR(1)
Setup1: sweep

FIGURE 14: VSWR of 3rd proposed trapezoidal logarithmic slots array antenna.

11.9 GHz), and two in Ku band (13.95 and 14.5 GHz). The gain plot and radiation pattern for 6.7 GHz frequency are shown in Figures 11(a) and 11(b), which shows that it generates gain of 5 dB which is half compared to first proposed structure.

Third proposed structure contains trapezoidal logarithmic slots array as shown in Figure 12. Distances of all the slots are the same as 1st proposed design. Simulated result of the proposed structure is shown in Figure 13, which shows higher return loss occurring at eight different frequencies: two in C band (6.7 and 7.54 GHz), four in X band (10.25, 11, 11.43, and

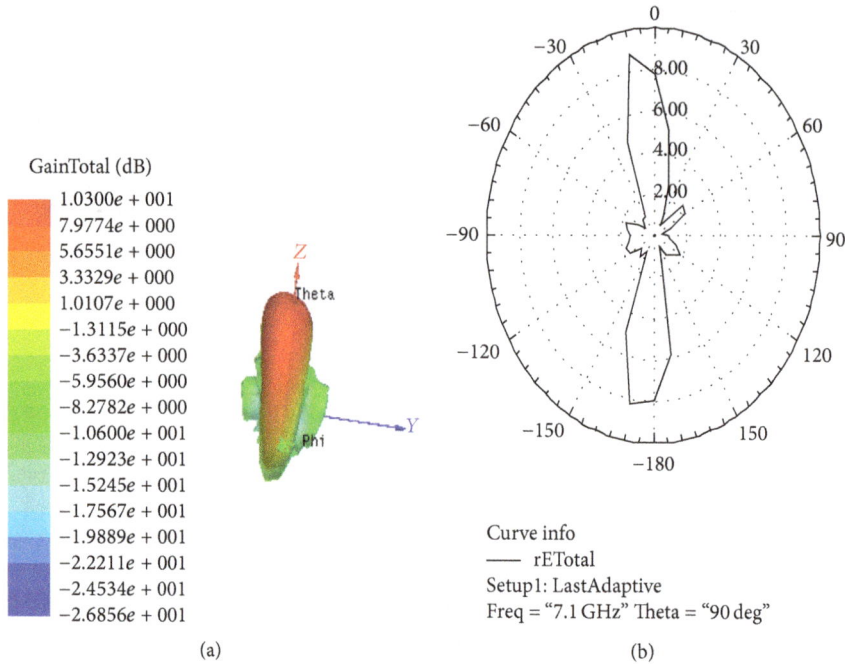

(a)   (b)

FIGURE 15: (a) Gain plot and (b) radiation pattern for trapezoidal logarithmic slots array antenna.

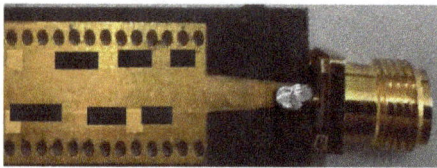

FIGURE 16: Fabricated model of logarithmic rectangular slots array antenna.

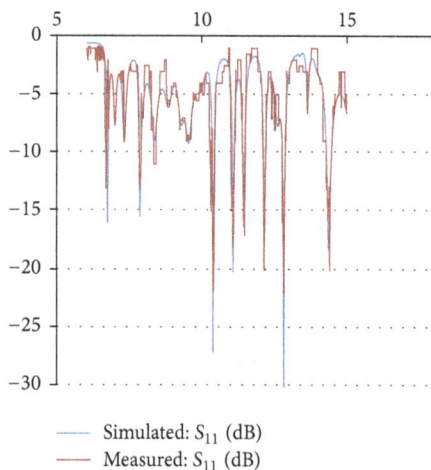

FIGURE 17: S_{11} of logarithmic rectangular slots array antenna.

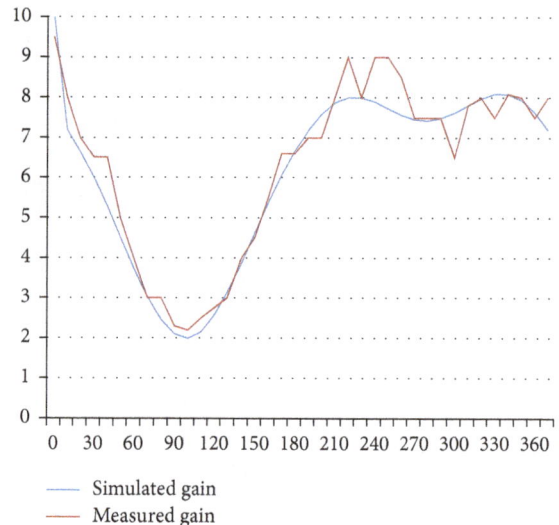

FIGURE 18: Gain of logarithmic rectangular slots array antenna.

5. Measured Results

Logarithmic rectangular slots array antenna has been fabricated on Rogers RT Duroid 5880 high frequency substrate with a thickness of 0.787 mm, relative permittivity of 2.2, and relative permeability of 1 and loss tangent of 0.0009. The top side of the antenna has logarithmic slots and the other side is ground plane. The photograph of fabricated antenna is shown in Figure 16.

The antenna S_{11} parameter and gain are measured using the Vector Network Analyzer MVNA-8-350 with probe station. The simulated and measured S_{11} parameter for antenna is shown in Figure 17. A slight difference is observed between

12.14 GHz), and Ku band (12.81 and 14.36 GHz). Its VSWR is shown in Figure 14. Gain plot and radiation pattern are shown in Figures 15(a) and 15(b), respectively. It gives more than 10 dB gain at 6.7 GHz frequency.

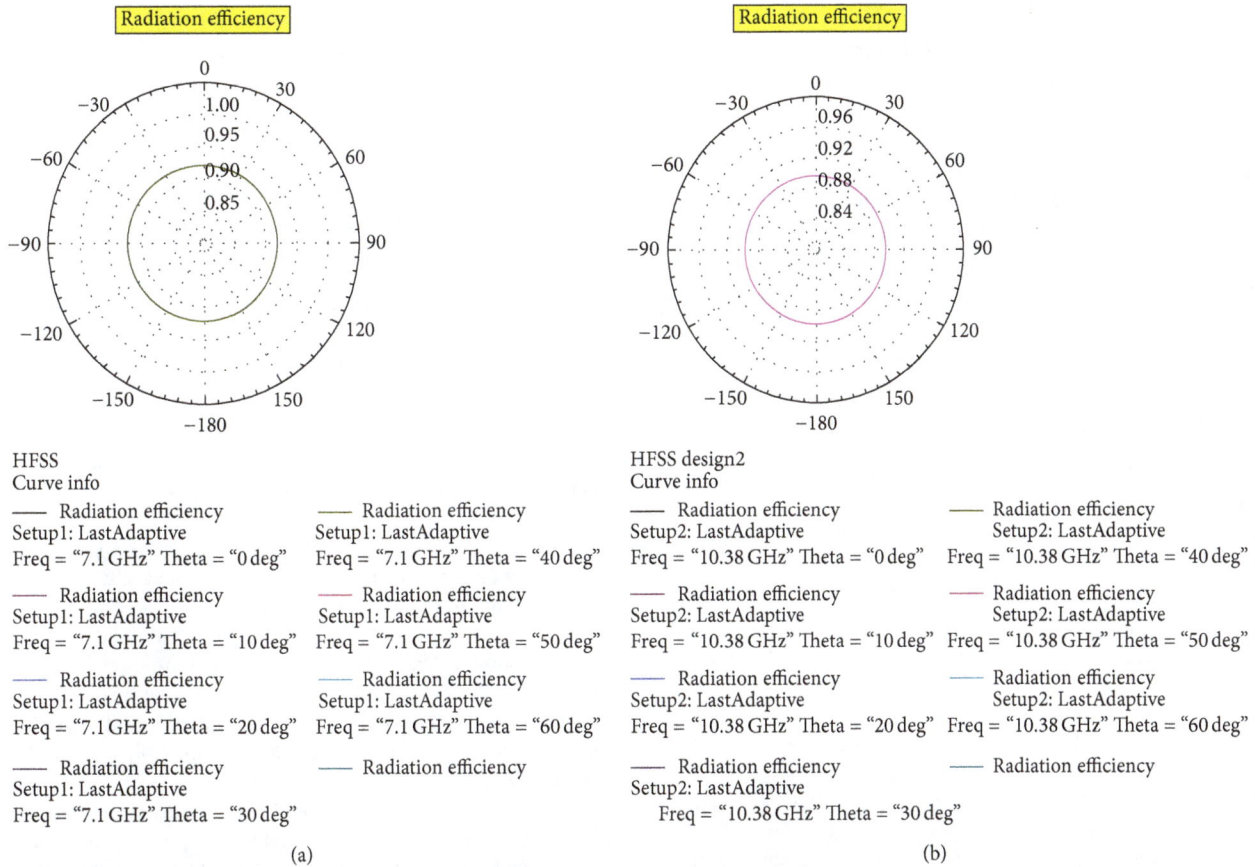

Radiation efficiency

Radiation efficiency

HFSS
Curve info

—— Radiation efficiency
Setup1: LastAdaptive
Freq = "7.1 GHz" Theta = "0 deg"

—— Radiation efficiency
Setup1: LastAdaptive
Freq = "7.1 GHz" Theta = "40 deg"

—— Radiation efficiency
Setup1: LastAdaptive
Freq = "7.1 GHz" Theta = "10 deg"

—— Radiation efficiency
Setup1: LastAdaptive
Freq = "7.1 GHz" Theta = "50 deg"

—— Radiation efficiency
Setup1: LastAdaptive
Freq = "7.1 GHz" Theta = "20 deg"

—— Radiation efficiency
Setup1: LastAdaptive
Freq = "7.1 GHz" Theta = "60 deg"

—— Radiation efficiency
Setup1: LastAdaptive
Freq = "7.1 GHz" Theta = "30 deg"

—— Radiation efficiency

HFSS design2
Curve info

—— Radiation efficiency
Setup2: LastAdaptive
Freq = "10.38 GHz" Theta = "0 deg"

—— Radiation efficiency
Setup2: LastAdaptive
Freq = "10.38 GHz" Theta = "40 deg"

—— Radiation efficiency
Setup2: LastAdaptive
Freq = "10.38 GHz" Theta = "10 deg"

—— Radiation efficiency
Setup2: LastAdaptive
Freq = "10.38 GHz" Theta = "50 deg"

—— Radiation efficiency
Setup2: LastAdaptive
Freq = "10.38 GHz" Theta = "20 deg"

—— Radiation efficiency
Setup2: LastAdaptive
Freq = "10.38 GHz" Theta = "60 deg"

—— Radiation efficiency
Setup2: LastAdaptive
Freq = "10.38 GHz" Theta = "30 deg"

—— Radiation efficiency

(a)

(b)

FIGURE 19: Radiation efficiency of logarithmic rectangular slots array antenna in (a) C band and (b) X band.

the measured value and simulated value. The difference between the measured and simulated S_{11} of the antenna is caused by the microstrip to SIW transition.

The simulated and measured gain of the antenna are shown in Figure 18. The maximum measured gain is also very close to 10 dB at 7 GHz for the antenna which is max. The measured result shows that the bandwidth of the antenna covers over 300 MHz while the gain of the antenna is kept almost constant within such a wide bandwidth of the antenna. Simulation results specify that metallic and dielectric losses do not have a substantial effect on the bandwidth and matching condition of the antenna, while they decrease the measured gain of antenna by 0.5 dB. The apparent difference between the simulation and measured gains might be due to the calibration linked tolerance range of the antenna reference in anechoic chamber.

The radiation pattern of the antenna at resonant frequency is shown in Figure 18. The radiation pattern measurement is carried out in a conventional far field anechoic chamber which uses a V connector to connect the antenna. Due to the size of the connector compared to the antenna, the rear radiation patterns were not incorporated in the results. However, the similar effects were observed in the simulation results and most significantly, the radiating behavior of the antenna is very similar to simulation results. As shown in Figures 19(a) and 19(b), they represent the radiation efficiency in C band and X band, respectively, which is 92% and 88%.

6. Observation and Discussion

From the simulation results we have observed the following things:

(i) All the structure gives six or more than six resonant frequencies in C band, X band, and Ku band.

(ii) Gain generated by rectangular and trapezoidal logarithmic slots array antenna is more than 10 dB compared to I shape logarithmic slots array antenna.

(iii) Isolation between two radiation bands is much higher in trapezoidal logarithmic slots array antenna and also generates low leakage loss in stop band.

(iv) Size of entire proposed antennas is compact compared to traditional antennas which are proposed for high gain applications.

(v) Bandwidth produced by all the antennas at resonant frequencies is more than 100 MHz.

7. Conclusion

From the design and simulation results, it is clear that all the antennas are compact in size and they are capable to generate multiband frequencies which are very important factors for the proposed design process of antennas. Antennas based on SIW technology have been proposed in this paper.

These structures can find many applications in C band and X band for radar and remote sensing mechanism and are one of the major areas in docking satellite where communication is done in C band while ranging is carried out in X band. So this antenna fulfilled both requirements instead of using two different antennas. Significant increment of gain parameter has been obtained for introduction of more number and variety of shapes of slots. Compared to identical slots, logarithmic slots give higher gain. The effect has been extensively studied unlike any other recent publication in this field. The structure is very simple and development of the prototype is easy in presence of advanced PCB fabrication technology.

Conflict of Interests

The authors declare that there is no conflict of interests regarding the publication of this paper.

References

[1] P. N. Richardson and H. Y. Lee, "Design and analysis of slotted waveguide arrays," *Microwave Journal*, vol. 31, pp. 109–125, 1988.

[2] S. Ravish, A. Patel, V. V. Dwivedi, and H. B. Pandya, "Design, development and fabrication of post coupled band pass waveguide filter at 11.2 GHz for radiometer," *Microwave Journal*, vol. 51, 2008.

[3] A. Patel, Y. Kosta, N. Chhasatia, and K. Pandya, "Multiple band waveguide based microwave resonator," in *Proceedings of the 1st International Conference on Advances in Engineering, Science and Management (ICAESM '12)*, pp. 84–87, Nagapattinam, India, March 2012.

[4] S. R. Rengarajan, "Compound radiating slots in a broad wall of a rectangular waveguide," *IEEE Transactions on Antennas and Propagation*, vol. 37, no. 9, pp. 1116–1123, 1989.

[5] A. Patel and Y. P. Kosta, "Multiple-band waveguide based bandstop filter," *International Journal of Applied Electromagnetics and Mechanics*, vol. 47, no. 2, pp. 563–581, 2015.

[6] S. Fallahzadeh, H. Bahrami, and M. Tayarani, "A novel dual-band bandstop waveguide filter using split ring resonators," *Progress in Electromagnetics Research Letters*, vol. 12, pp. 133–139, 2009.

[7] I. Ohta, Y. Yumita, K. Toda, and M. Kishihara, "Cruciform directional couplers in H-plane rectangular waveguide," in *Proceedings of the Asia-Pacific Microwave Conference (APMC '05)*, vol. 2, December 2005.

[8] A. Patel, Y. Kosta, N. Chhasatia, and F. Raval, "Design and fabrication of microwave waveguide resonator: with improved characteristic response," *European Journal of Scientific Research*, vol. 102, no. 2, pp. 163–174, 2013.

[9] R. S. Elliott and L. A. Kurtz, "The design of small slot arrays," *IEEE Transactions on Antennas and Propagation*, vol. 26, no. 2, pp. 214–219, 1978.

[10] R. S. Elliott, "An improved design procedure for small arrays of shunt slots," *IEEE Transactions on Antennas and Propagation*, vol. 31, no. 1, pp. 48–53, 1983.

[11] A. F. Stevenson, "Theory of slots in rectangular wave-guides," *Journal of Applied Physics*, vol. 19, pp. 24–38, 1948.

[12] W. Coburn, M. Litz, J. Miletta, N. Tesny, L. Dilks, and B. King, "A slotted-waveguide array for high-power microwave transmission," Progress Report for FY 2000, US Army Research Laboratory, Adelphi, Md, USA, 2001.

[13] F. Raval, Y. P. Kosta, J. Makwana, and A. V. Patel, "Design & implementation of reduced size microstrip patch antenna with metamaterial defected ground plane," in *Proceedings of the 2nd International Conference on Communication and Signal Processing (ICCSP '13)*, pp. 186–190, IEEE, Melmaruvathur, India, April 2013.

[14] M. Kahrizi, T. K. Sarkar, and Z. A. Maricevic, "Analysis of a wide radiating slot in the ground plane of a microstrip line," *IEEE Transactions on Microwave Theory and Techniques*, vol. 41, no. 1, pp. 29–37, 1993.

[15] J. Galejst, "Admittance of a rectangular slot which is backed by a rectangular cavity," *IEEE Transactions on Antennas and Propagation*, vol. 11, no. 2, pp. 119–126, 1963.

[16] K.-L. Wong, *Compact and Broadband Microstrip Antennas*, John Wiley & Sons, New York, NY, USA, 2002.

[17] W.-S. Chen, "Single-feed dual-frequency rectangular microstrip antenna with square slot," *Electronics Letters*, vol. 34, no. 3, pp. 231–232, 1998.

[18] D. Busuioc, M. Shahabadi, A. Borji, G. Shaker, and S. Safavi-Naeini, "Substrate integrated waveguide antenna feed—design methodology and validation," in *Proceedings of the IEEE Antennas and Propagation Society International Symposium (AP-S '07)*, pp. 2666–2669, Honolulu, Hawaii, USA, June 2007.

[19] H. Kumar, R. Jadhav, and S. Ranade, "A review on substrate integrated waveguide and its microstrip interconnect," *Journal of Electronics and Communication Engineering*, vol. 3, no. 5, pp. 36–40, 2012.

[20] M. Bozzi, F. Xu, D. Deslandes, and K. Wu, "Modeling and design considerations for substrate integrated waveguide circuits and components," in *Proceedings of the 8th International Conference on Telecommunications in Modern Satellite, Cable and Broadcasting Services (TELSIKS '07)*, pp. P-7–P-16, IEEE, Niš, Serbia, September 2007.

[21] S. Moitra, A. Mukhopadhyay, and A. Bhattacharjee, "Ku-band substrate integrated waveguide (SIW) slot array antenna for next generation networks," *Global Journal of Computer Science and Technology E: Network, Web & Security*, vol. 13, no. 5, pp. 11–16, 2013.

[22] A. J. Farrall and P. R. Young, "Integrated waveguide slot antennas," *Electronics Letters*, vol. 40, no. 16, pp. 974–975, 2004.

[23] D. Pozar, *Microwave Engineering*, John Wiley & Sons, Hoboken, NJ, USA, 3rd edition, 2005.

11

Application of Defected Ground Structure to Suppress Out-of-Band Harmonics for WLAN Microstrip Antenna

Pravin Ratilal Prajapati

A. D. Patel Institute of Technology, Department of Electronics and Communication Engineering, Gujarat 388121, India

Correspondence should be addressed to Pravin Ratilal Prajapati; pravinprajapati05@gmail.com

Academic Editor: Dmitry Kholodnyak

An application of defected ground structure (DGS) to reduce out-of-band harmonics has been presented. A compact, proximity feed fractal slotted microstrip antenna for wireless local area network (WLAN) applications has been designed. The proposed 3rd iteration reduces antenna size by 43% as compared to rectangular conventional antenna and by introducing H shape DGS, the size of an antenna is further reduced by 3%. The DGS introduces stop band characteristics and suppresses higher harmonics, which are out of the band generated by 1st, 2nd, and 3rd iterations. H shape DGS is etched below the 50 Ω feed line and transmission coefficient parameters (S_{21}) are obtained by CST Microwave Studio software. The values of equivalent L and C model have been extracted using a trial version of the diplexer filter design software. The stop band characteristic of the equivalent LC model also has been simulated by the Advance Digital System software, which gives almost the same response as compared to the simulation of CST Microwave Studio V. 12. The proposed antenna operates from 2.4 GHz to 2.49 GHz, which covers WLAN band and has a gain of 4.46 dB at 2.45 GHz resonance frequency.

1. Introduction

Printed antennas have been widely used because of their advantages like low profile, easy fabrication, low cost, small size, and so forth [1, 2]. To reduce the size of the antenna without much more adverse effect on bandwidth and the gain of an antenna, various methods have been proposed like using dielectric substrates with high permittivity [3], applying magneto inductive waveguide loading [4], and using notches or slots on patch antenna [5]. By embedding specific slot (fractal slot) on microstrip antenna, surface current path increases, which lowers the resonant frequency of an antenna, and thus antenna size reduction can be possible [6]. To improve radiation efficiency of the antenna, it becomes necessary to suppress higher harmonics, which cause loss of power. Higher harmonics also produce spurious radiation. To suppress higher harmonics, various techniques like PBG (photonic band gap structure) [7–10], Filter, and EBG (electromagnetic band gap structure) [11, 12] have been proposed. The conducting metal etched off in specific shape from the ground plane provides wide rejection band covering some frequency range [13–17]. The structure of

this type is known as *defected ground structure* (DGS). The main advantages of DGS is that it introduces slow wave effect. This effect produced because of the DGS equivalent L and C components. The transmission line witH-DGS gives a higher effective impedance and also introduces high slow wave effect, which provides rejection band in some frequency range. The microstrip line witH-DGS has a large electrical length as compared to conventional microstrip for the same physical length. Thus, DGS helps to lower resonance frequency and therefore to reduce the size of an antenna [18].

In this paper, an application of the DGS to suppress higher harmonics and thus to improve radiation efficiency of the planar antenna is demonstrated. Moreover, this paper also shows the significant role of the fractal geometry for size miniaturization of the radiating element of the patch antenna.

2. Antenna Geometry and Operational Mechanism

The geometry of the proximity coupled patch antenna is presented in Figure 1. The argon material, which has a relative

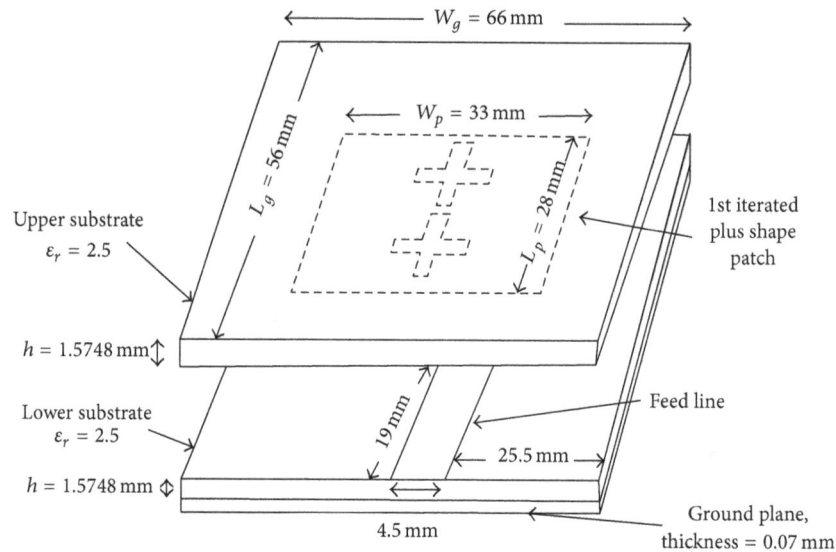

FIGURE 1: Geometry of proximity coupled plus shape fractal slot antenna.

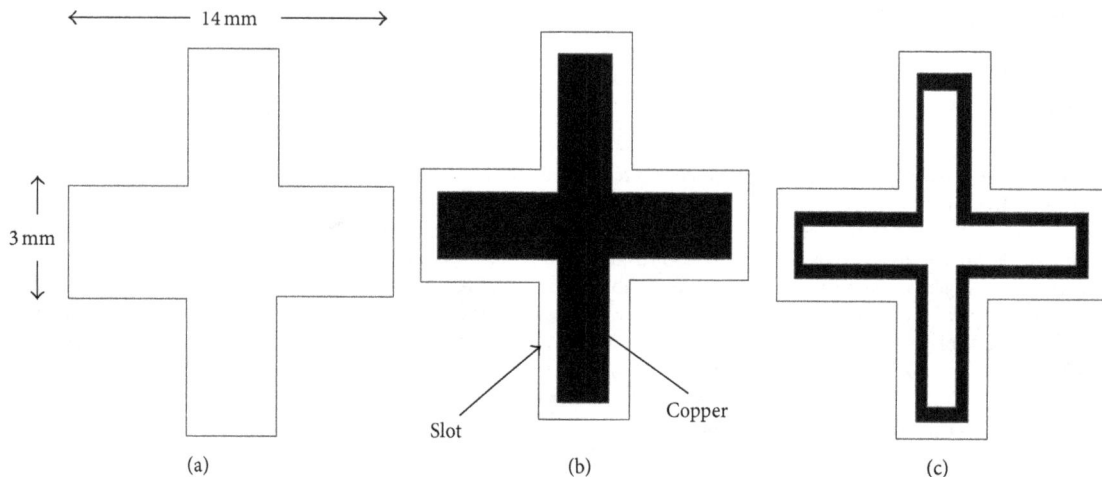

FIGURE 2: Geometry of plus shape fractal slot: (a) 1st iteration (b) 2nd iteration, (c) 3rd iteration.

permittivity ε_r = 2.5, thickness 1.5748 mm, and loss tangent 0.0025, is used for both substrates. Low dielectric constant material has been selected to get good radiation efficiency. The ground plane dimensions are $W_g \times L_g = 66 \times 56 \, \text{mm}^2$. The microstrip line with 50 Ω impedance, width W_f = 4.5 mm, and length L_f = 19 mm is fabricated on upper side of the lower substrate and H shape DGS is etched out from the lower side (ground plane) of lower substrate. The number of DGS shapes such as hook shape [19], arc shape [20], concentric ring shape [21], and spiral shape [22] was reported. The reason of selection of H shape DGS is that, as per parametric simulation, it was concluded that the effect of changing of length arm of H shape DGS (L_1), width of arm (L_2), distance between the two arms (C_1), and so forth on cut-off frequency of DGS is almost linear. So it is easy to get specific bandstop region by selecting appropriate dimensions of H shape. The rectangular patch with size $W_p \times L_p = 33 \times 26 \, \text{mm}^2$ is fabricated on upper side of upper substrate.

From that patch, plus shape slot is taken out. This procedure is repeated for next two iterations as shown in Figure 2. Figure 3 shows inner dimensions of the plus shape slot. H shape DGS is created in the ground plane of the antenna. The dimensions and location of H shaped DGS are mentioned in Figure 4.

3. Behavior of DGS as Band Stop Filter

Initially, H-DGS is considered below the 50 Ω microstrip line as shown in Figure 5. The microstrip line has a width of 4.5 mm considered from the calculation obtained by CST Microwave Studio, V. 12. The DGS cell is simulated by the same software and from S_{21} parameters, it was concluded that H-DGS has characteristics of a one-pole band stop filter. For getting the desired value of upper edge and lower edge of band stop frequencies, specific dimension of H-DGS was varied by keeping other dimensions constant. The effect of H-DGS dimensions on upper edge and lower edge band stop

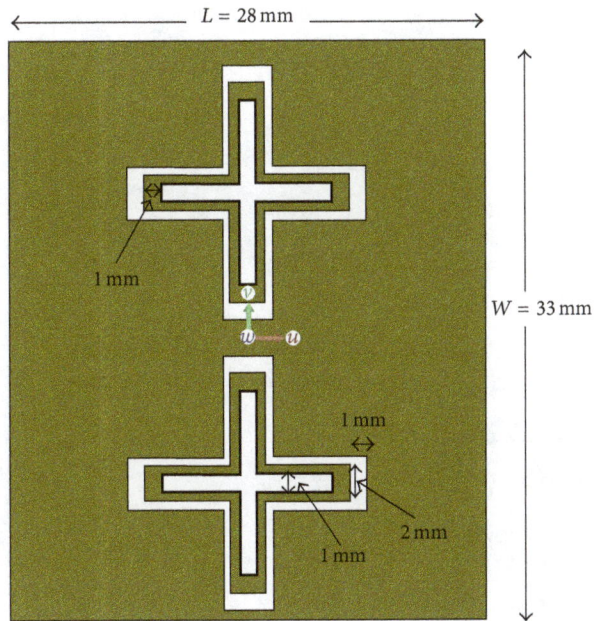

FIGURE 3: Design geometry of 3rd iterated plus shape patch.

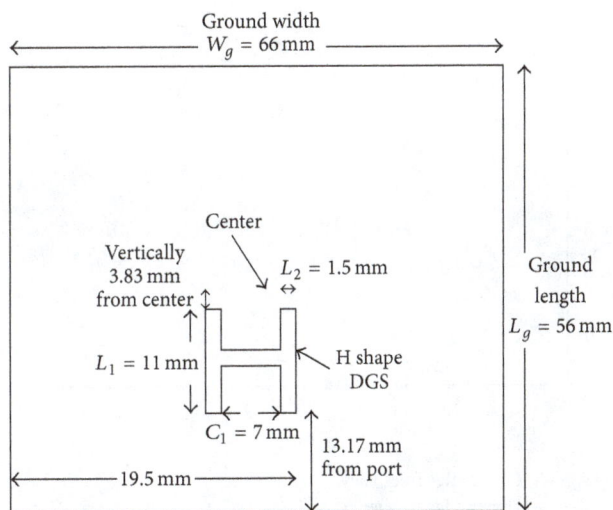

FIGURE 4: Design geometry of DGS on the ground plane of the proposed antenna.

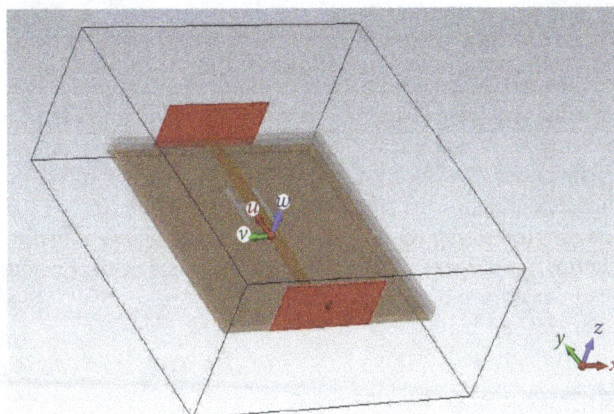

FIGURE 5: Setup of H-DGS under 50 Ω strip line in CST Microwave Studio to extract S_{21} parameters.

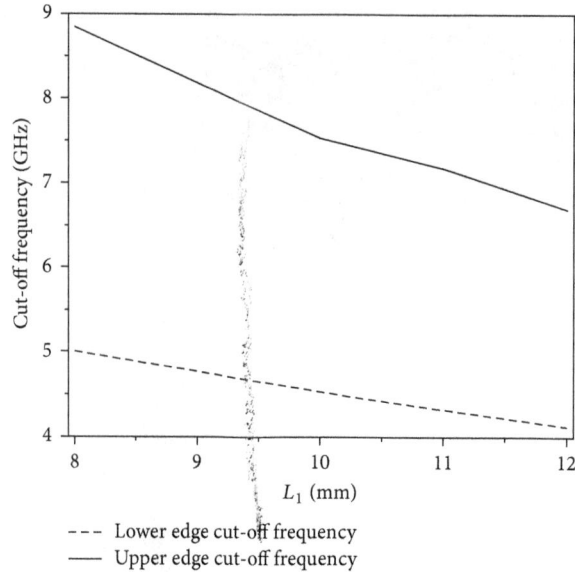

FIGURE 6: Effect of variation of L_1 on cut-off frequency of DGS as band stop filter.

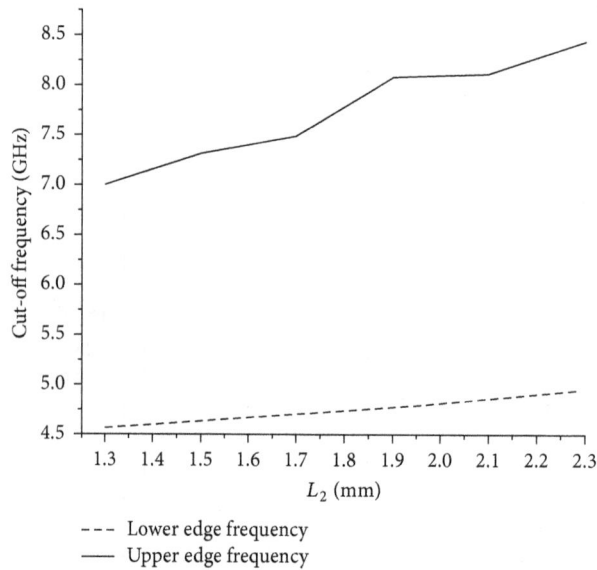

FIGURE 7: Effect of variation of L_2 on cut-off frequency of DGS as band stop filter.

frequencies is shown in Figures 6, 7, and 8. These figures show that as the arm length (L_1) of DGS and the gap (C_1) between two arms increase, frequency of upper edge and lower edge decreases and as thickness of both arms (L_2) increases, frequency of upper edge and lower edge increases. For getting the desired value of stop band frequencies, dimensions of H-DGS (L_1, L_2, and C_1) have been optimized by CST Microwave Studio software and $L_1 = 11$ mm, $L_2 = 1.5$ mm, and $C_1 = 5$ mm have been considered for upper edge 4.62 GHz and lower edge 7.19 GHz frequency. For desired band stop characteristics, the values of the LC equivalent model have been extracted by diplexer filter software.

The LC equivalent model of DGS has been shown in Figure 9. The equivalent LC model of DGS has been simulated by Advance Digital System (ADS) software and its S_{21}

response has been compared with that of CST Microwave Studio, as shown in Figure 10. Even though there is a difference between the attenuation level at attenuation pole frequency, the 3 dB bandstop frequency is almost same for ADS and CST Microwave simulation. The reason of mismatch result of attenuation level is that ADS simulate with consideration of infinite ground plane, while in CST Microwave Studio finite ground plane has been considered.

4. Results and Discussion

The geometry of the proposed microstrip antenna has been optimized and simulated with CST Microwave Studio. Figure 11 shows the simulated return losses of the

TABLE 1: Comparison of conversional rectangular patch antenna with various fractal slot iterations of antenna.

Iteration	Resonating frequency (GHz)	Number of higher harmonics	Return loss (dB)	VSWR	Gain (dB)	Radiation efficiency (%)	Bandwidth of fundamental resonance frequency (%)	higher harmonics	Bandwidth of 2nd harmonics (%)	Patch size (mm²)	Size reduction (%)
0	2.42	1	−12.53	1.61	6.3	67.6	2.46	2.90	—	1725	—
1st	2.43	2	−13.44	1.54	3.3	55.4	3.27	2.64	5.08	983.25	43
2nd	2.42	2	−22	1.17	3.8	51.2	3.30	9.19	4.34	983.25	43
3rd	2.42	2	−27	1.2	4.16	51.5	3.27	2.29	4.93	983.25	43
3rd with H-DGS	2.45	0	−23.26	1.14	4.46	63.8	3.68	—	—	932	46

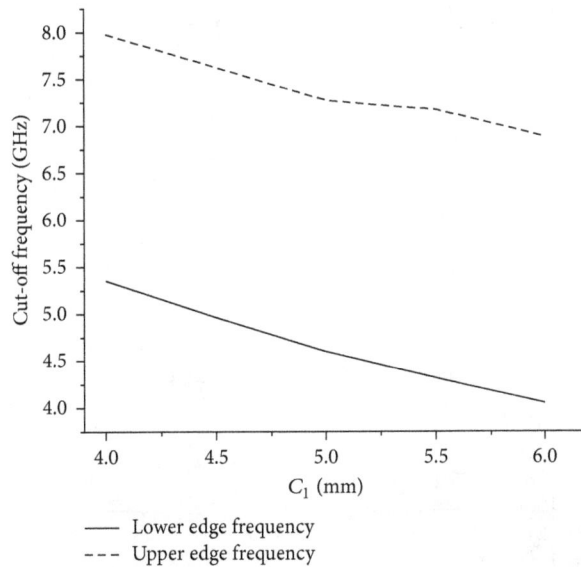

FIGURE 8: Effect of variation of C_1 on cut-off frequency of DGS as band stop filter.

conventional antenna and all iteration antennas. It can be observed that, besides fundamental mode, the conventional rectangular antenna gives higher harmonics, which are not useful. Here, to reduce the size of the radiating element, slot technique is used. In simulation, by taking different slots, we found that plus shape slot gives the maximum size reduction factor and also does not have any adverse effect on the desired characteristics of the patch antenna. To increase the gain of the antenna, the space of the plus shape slot used with an innovative method and copper layer is etched in the space of the slot. To further increment in the gain, again the same method is repeated (2nd iteration). Table 1 shows that 1st, 2nd, and 3rd iterations reduce the size of the conventional antenna by 43%. By etching H-DGS at the ground plane, the size of fractal slot antenna further reduced by 3% and the bandwidth is enhanced from 2.46% to 3.68%. Table 1 shows that as iterations increase from 1st to 3rd, the gain of the antenna at resonance frequency increases. This is according to babinet's principle [2], which states that each slot acts as a radiator and gives contribution in gain of the antenna.

Table 1 shows that, after 1st iteration, if one more iteration is introduced, that is, in 2nd iteration antenna, VSWR improves due to matching and thus radiation loss also

improves. The gain of the 2nd iteration antenna also increases according to babinet's principle, but radiation efficiency decreases due to increment in the bandwidth of 1st harmonics. Similarly, 3rd iteration antenna gives more gain at a fundamental resonance frequency as compared to 2nd iteration antenna. The 3rd iteration antenna has two higher harmonics, by introducing H-DGS at the ground plane, the size of an antenna further reduced by 3% as compared to 3rd iterated antenna without DGS, and due to the reduction of the size of an antenna, the 2nd harmonics are suppressed by about 14.75 dB. The 1st harmonics generated in 3rd iterated antenna are suppressed by about 6.89 dB as shown in Figure 12. This suppression occurred due to band stop characteristics of H-DGS. Due to suppression of higher harmonics by H-DGS, the power wastage reduced and thus radiation efficiency increased as compared to 1st, 2nd, and 3rd iteration antennas without DGS, as shown in Table 1.

Figure 13 shows simulated current distribution on plus shaped slot, which shows that most of the current density concentrates on the joints and edges. Figure 14 plots measured and simulated H-plane and E-plane radiation patterns of the 3rd iteration plus the slot antenna witH-DGS at 2.42 GHz resonance frequency.

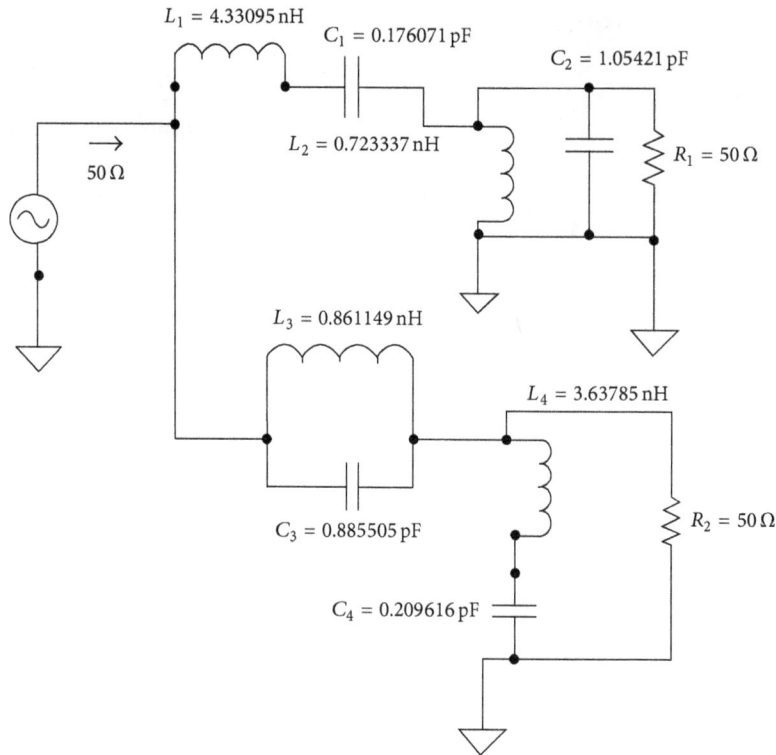

FIGURE 9: Equivalent LC model of H shape DGS.

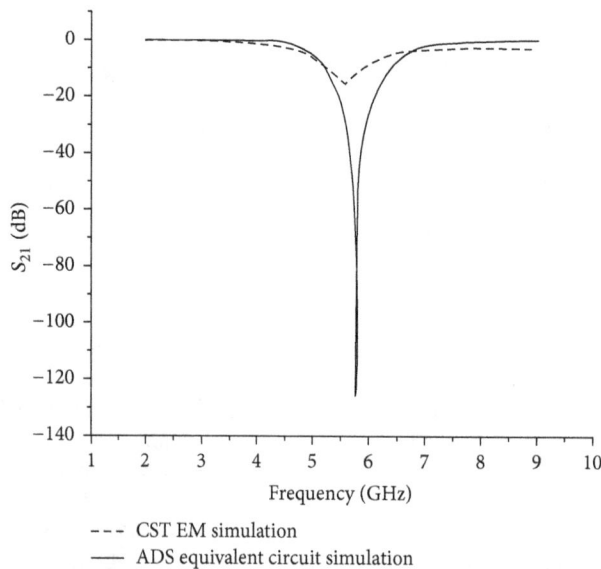

--- CST EM simulation
— ADS equivalent circuit simulation

FIGURE 10: S_{21} parameter of LC equivalent model compared with that of H-DGS.

5. Conclusion

To obtain an impedance matching, high radiation efficiency, higher harmonic suppression, and the size reduction, a novel type of 3rd iteration plus shape fractal slot antenna has been proposed. By introducing fractal slots, VSWR improves and the size of an antenna reduces, but it also generates higher

harmonics. To suppress higher harmonics, H shape DGS and its equivalent circuit have been proposed. The parameter extraction method for the proposed H-DGS has also been explained. Furthermore, by employing the extracted parameters, the band stop characteristics of H-DGS are explained, which suppressed higher harmonics. The radiated power of the proposed antenna in 1st and 2nd harmonic frequency is

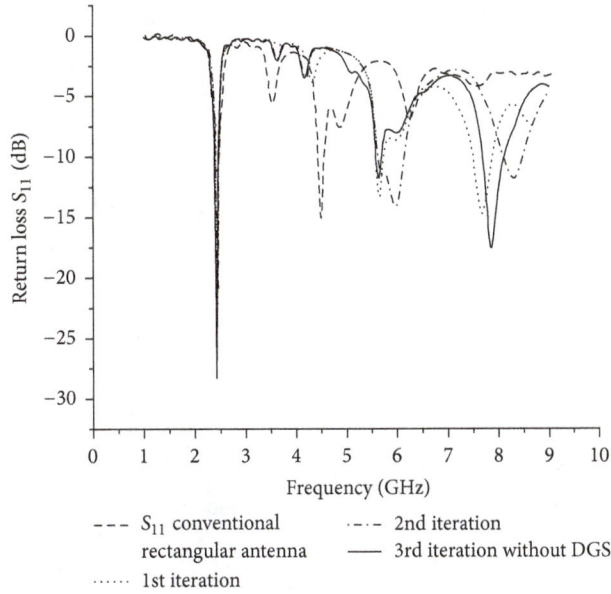

FIGURE 11: Simulated return losses S_{11} of conventional rectangular antenna and 1st to 3rd iterated plus fractal slot antennas.

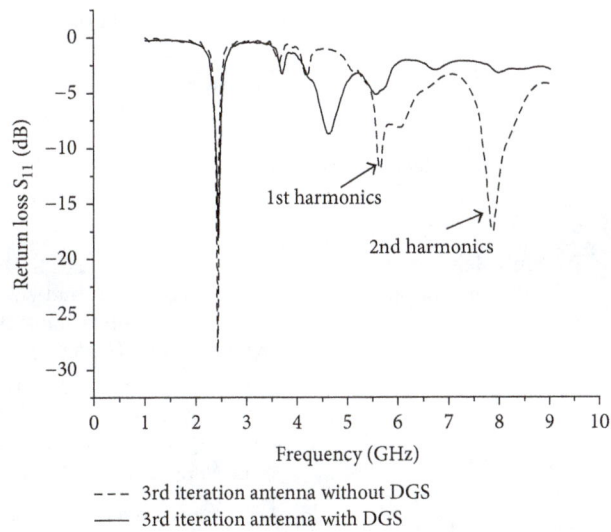

FIGURE 12: Comparison of return loss of 3rd iterated antenna with and without DGS.

FIGURE 13: Current distribution in 3rd iteration fractal slot antenna witH-DGS.

Gain Abs ($\phi = 90$)   Far field ($f = 2.42$) [1]

θ (°) versus (dB)

Frequency = 2.42
Main lobe magnitude = 4.8 dB
Main lobe direction = −5.0 deg.
Angular width (3 dB) = 88.9 deg.
Side lobe level = −11.5 dB

(a)

Gain Abs ($\phi = 0$)   Far field ($f = 2.42$) [1]

θ (°) versus (dB)

Frequency = 2.42
Main lobe magnitude = 4.8 dB
Main lobe direction = 0.0 deg.
Angular width (3 dB) = 85.0 deg.
Side lobe level = −11.6 dB

(b)

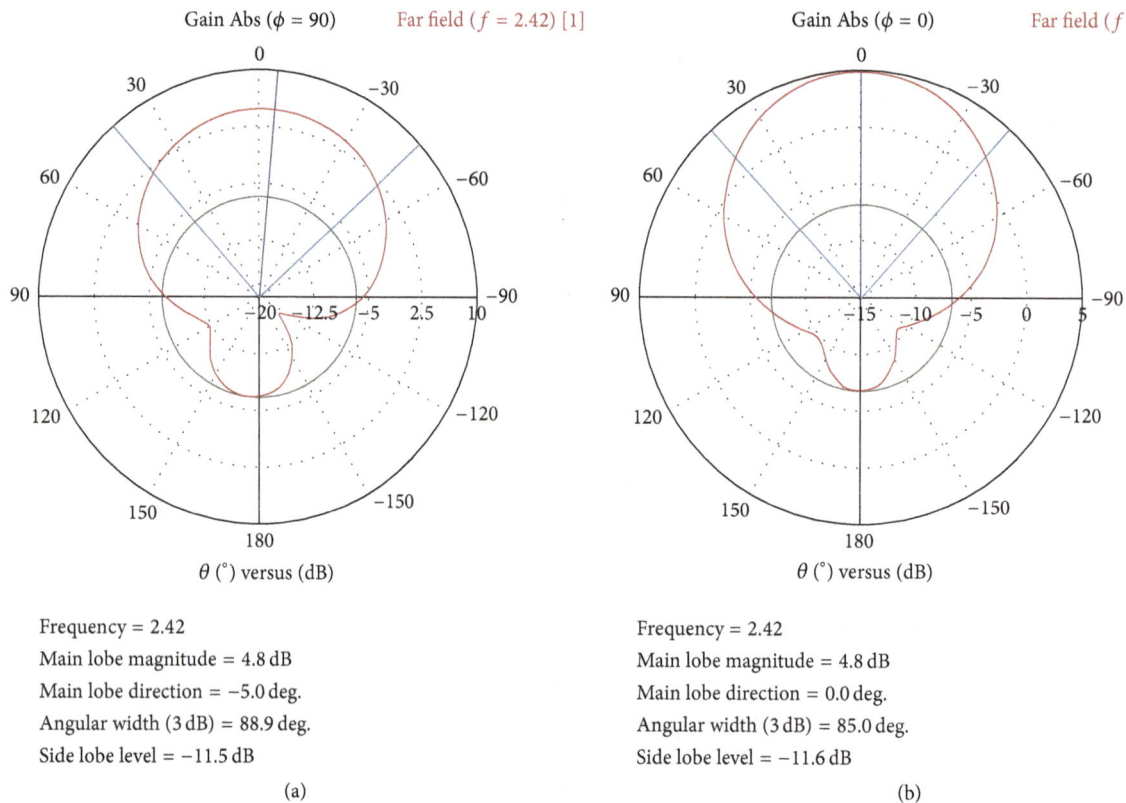

FIGURE 14: Simulated radiation pattern of 3rd iteration fractal slot antenna witH-DGS: (a) E-plane, (b) H-plane.

very low. The proposed DGS unit and its equivalent circuit could also find applications like microwave filter, coupler, power divider, and so forth.

Conflict of Interests

The author declares that there is no conflict of interests regarding the publication of this paper.

Acknowledgments

The author is thankful to G. H. Patel College of Engineering and Technology, Gujarat, for providing access of ADS Simulation software and the management of A. D. Patel Institute of Technology and Charutar Vidyamandal, Gujarat, India, for motivation and support for the research work.

References

[1] G. Kumar and K. P. Ray, *Broadband Microstrip Antennas*, Artech House, 2003.

[2] C. A. Balanis, *Antenna Theory: Analysis and Design*, Wiley Publication, 2nd edition, 2007.

[3] X. Tang, H. Wong, Y. Long, Q. Xue, and K. L. Lau, "Circularly polarized shorted patch antenna on high permittivity substrate with wideband," *IEEE Transactions on Antennas and Propagation*, vol. 60, no. 3, pp. 1588–1592, 2012.

[4] J. G. Joshi, S. S. Pattnaik, S. Devi, and M. R. Lohokare, "Bandwidth enhancement and size reduction of microstrip patch antenna by magneto inductive waveguide loading," *Wireless Engineering and Technology*, vol. 2, no. 2, pp. 37–44, 2011.

[5] R. Chair, C.-L. Mak, K.-F. Lee, K.-M. Luk, and A. A. Kishk, "Miniature wide-band half U-slot and half E-shaped patch antennas," *IEEE Transactions on Antennas and Propagation*, vol. 53, no. 8, pp. 2645–2652, 2005.

[6] A. Kordzadeh and F. H. Kashani, "A new reduced size microstrip patch antenna with fractal shaped defects," *Progress in Electromagnetics Research B*, vol. 11, pp. 29–37, 2009.

[7] Z. Harouni, L. Osman, and A. Gharsallah, "Efficient 2.45 GHz proximity coupled microstrip patch antenna design including harmonic rejecting device for microwave energy transfer," in *Proceedings of the International Renewable Energy Congress (IREC '10)*, pp. 73–75, Sousse, Tunisia, November 2010.

[8] Y. Horri and M. Tsutsumi, "Harmonic control by photonic bandgap on microstrip patch antenna," *IEEE Microwave and Guided Wave Letters*, vol. 9, no. 1, pp. 13–15, 1999.

[9] H. Liu, Z. Li, X. Sun, and J. Mao, "Harmonic suppression with photonic bandgap And defected ground structure for a microstrip patch antenna," *IEEE Microwave and Wireless Components Letters*, vol. 15, no. 2, pp. 55–56, 2005.

[10] X. Lin, L. Wang, and J. Sun, "Harmonic suppression by photonic bandgap on CPW fed loop slot antenna," *Microwave and Optical Technology Letters*, vol. 41, pp. 154–156, 2004.

[11] Z. Zakaria, W. Y. Sam, M. Z. A. A. Aziz, A. A. M. Isa, and F. M. Johar, "Design of integrated rectangular SIW filter and

microstrip patch antenna," in *Proceedings of the 5th IEEE Asia-Pacific Conference on Applied Electromagnetics (APACE '12)*, pp. 137–141, Melaka, Malaysia, December 2012.

[12] O. A. Nova, J. C. Bohórquez, N. M. Peña, G. E. Bridges, L. Shafai, and C. Shafai, "Filter-antenna module using substrate integrated waveguide cavities," *IEEE Antennas and Wireless Propagation Letters*, vol. 10, pp. 59–62, 2011.

[13] J.-S. Lim, J.-S. Park, Y.-T. Lee, D. Ahn, and S. Nam, "Application of defected ground structure in reducing the size of amplifiers," *IEEE Microwave and Wireless Components Letters*, vol. 12, no. 7, pp. 261–263, 2002.

[14] A. K. Arya, M. V. Kartikeyan, and A. Patnaik, "Defected ground structure in the perspective of microstrip antennas: a review," *Frequenz*, vol. 64, no. 5-6, pp. 79–84, 2010.

[15] M. K. Mandal and S. Sanyal, "A novel defected ground structure for planar circuits," *IEEE Microwave and Wireless Components Letters*, vol. 16, no. 2, pp. 93–95, 2006.

[16] A. Boutejdar, A. Omar, E. P. Burte, and R. Mikuta, "An improvement of defected ground structure lowpass/bandpass filters using H-slot resonators and coupling matrix method," *Journal of Microwaves, Optoelectronics and Electromagnetic Applications*, vol. 10, no. 2, pp. 295–307, 2011.

[17] S. K. Parui and S. Das, "Modeling of modified split ring type defected ground structure and its application as bandstop filter," *Radioengineering*, vol. 18, no. 2, pp. 149–154, 2009.

[18] P. R. Prajapati, A. Patnaik, and M. V. Kartikeyan, "Design and characterization of an efficient multi-layered circularly polarized microstrip antenna," *International Journal of Microwave and Wireless Technologies*, 2015.

[19] W. T. Li, X. W. Shi, and O. Q. Hei, "Novel planar UWB monopole antenna with triple band-notched characteristics," *IEEE Antennas and Wireless Propagation Letters*, vol. 8, pp. 1094–1098, 2009.

[20] D. Guha, C. Kumar, and S. Pal, "Improved cross-polarization characteristics of circular microstrip antenna employing arc-shaped Defected Ground Structure (DGS)," *IEEE Antennas and Wireless Propagation Letters*, vol. 8, pp. 1367–1369, 2009.

[21] D. Guha, S. Biswas, M. Biswas, J. Y. Siddiqui, and Y. M. M. Antar, "Concentric ring-shaped defected ground structures for microstrip applications," *IEEE Antennas and Wireless Propagation Letters*, vol. 6, no. 1, pp. 402–405, 2006.

[22] D. Nashaat, H. A. Elsadek, E. Abdallah, H. Elhenawy, and M. F. Iskander, "Multiband and miniaturized inset feed microstrip patch antenna using multiple spiral-shaped defect ground structure (DGS)," in *Proceedings of the Antennas and Propagation Society International Symposium (APSURSI '09)*, pp. 1–4, IEEE, Charleston, SC, USA, June 2009.

12

Development of a Novel Switched-Mode 2.45 GHz Microwave Multiapplicator Ablation System

Guido Biffi Gentili,[1] **Cosimo Ignesti,**[2] **and Vasco Tesi**[3]

[1] *Department of Information Engineering, University of Florence, Via di S. Marta 3, 50139 Florence, Italy*
[2] *Biomedical Srl, Via G.B. Lulli 43, 50144 Florence, Italy*
[3] *WaveComm S.r.l., Loc. Belvedere, Ingresso 2, 53034 Colle Val d'Elsa, Siena, Italy*

Correspondence should be addressed to Cosimo Ignesti; cosimo.ignesti@gmail.com

Academic Editor: Gian Luigi Gragnani

The development of a novel switched-mode 2.45 GHz microwave (MW) multiapplicator system intended for laparoscopic and open surgical thermoablative treatments is presented. The system differs from the other synchronous and asynchronous commercially available equipments because it employs a fast sequential switching (FSS) technique for feeding an array of up to four high efficiency MW applicators. FSS technology, if properly engineered, allows improving system compactness, modularity, overall efficiency, and operational flexibility. Full-wave electromagnetic (EM) and thermal (TH) simulations have been made to confirm the expected performances of the FSS technology. Here we provide an overview of technical details and early *ex-vivo* experiments carried out with a full functional β-prototype of the system.

1. Introduction

In cancer treatment, open surgery and chemotherapy are still the physicians' first choices. The gold standard treatment for most of the tumors in the liver, lung, and kidney is surgical resection. However, up to 80% of liver cancer patients and 50% of lung cancer patients are refractory to surgery due to multifocal disease, poor baseline health, or comorbidities such as cirrhosis and emphysema [1, 2]: for some patients, removal of tumors with open surgery is not possible or involves a too high risk due to the poor condition of the patient himself. Therefore their success rate largely depends on the type of malignancy treated and on the progress of the disease.

Minimally invasive surgery or percutaneous interventions may in these cases be adequate to increase safety, reduce trauma, and shorten operative time [3–5].

Microwave ablation (MWA) is a relatively new technology in continuous development because it offers some advantages when compared to the radiofrequency ablation (RFA) [6–8], the technology which currently represents the prevailing clinical focal therapy. MWA can generate higher temperatures in less time since tissue charring does not hinder the radiation of MW fields and it is less susceptible to the *heat-sink effect* of peritumoral vessels [9]. For these reasons, MWA and their minimally invasive approach have fertile ground for innovation through future systems and technological developments. Clinical MWA equipment operate at 915 MHz or 2.45 GHz since these bands, which are allowed for medical use and relatively high-power devices, are readily available providing a balance between localized heating and sufficient energy penetration to treat most focal tumors [10].

The power that the coaxial structure of a microwave applicator can safely handle is proportional to its external diameter; therefore very small diameter applicators can handle proportionally smaller powers. This physical limit can be overcome by cooling the applicator shaft or by combining more than one antenna into an array of applicators in order to increase the coagulation volume [11, 12]. However, due to the high losses of the internal coaxial feeding line, a small diameter shaft cooled antenna intrinsically suffers of low radiating efficiency that rapidly falls with the decreasing of its external diameter.

A single applicator appears fundamentally unable to uniformly and efficiently heat a large volume of tissue because the emitted radiation is subjected to a very strong attenuation, due to the intrinsic high propagation losses of the biological tissue. Recent researches [10, 13] show a fundamental advantage in using multiple antennas: power distribution in a multiple-probe system is more effective, even compared to that of a single antenna providing the same amount of energy. Heating produced simultaneously by multiple nearby synchronous (coherent) or asynchronous (incoherent) arrays of radiating sources can create larger ablation areas than what we might expect from a single applicator radiating the same amount of energy, by means of the effect called *thermal synergy* [14]. This ability to perform multiple ablation simultaneously may allow the treatment of large tumors with concurrent overlapping thermal lesions or the ablation of several anatomically separate tumor lesions at once [11].

The median tumor diameter currently being targeted by thermal ablation is approximately 25 mm. Therefore, a 35–40 mm ablation zone is recommended to treat an average tumor with an appropriate 5–10 mm radial margin, aiming at a lower probability of local tumor progression [15, 16].

Furthermore properly assembled linear or conformal arrays of applicators can be used as surgical resection devices to reduce blood losses and to assist in coagulation of liver tissue during intraoperative and laparoscopic surgical procedures.

Multiprobe ablations reduce the need to repeat treatments, decrease inadequate treatments of larger tumors, and increase the speed of the therapy, thereby decreasing the complication rate.

The purpose of this paper is to present the development of a novel 2.45 GHz multiprobe modular thermoablation system conceived to reduce system complexity and cost while maintaining a very high energetic efficiency.

The simplest way to implement a MWA N-needle ablative system consists in the simultaneous use of N power generators, where each generator independently feeds an applicator of the array but does not communicate with the others.

This obvious asynchronous solution is certainly not optimal from the engineering point of view because solid-state MW power generators are very expensive and a separate control unit ought to be used because independent control of each generator is impractical in the clinical environment.

Alternatively N microwave power amplifiers (MPA) fed in parallel by a common low-power MW source could be employed, leading to a synchronous or coherent solution that allows phase control of the array. In principle this solution should offer better energy focalization inside the tumor by taking advantage of near-field array optimal phasing. In practice, however, the heat diffusion in the perfused tissue smoothes the behavior of the temperature distribution inside the treated volume (See Appendix A), thus drastically reducing the advantages of the coherent feeding.

The use of multiple generators/amplifiers can be avoided by employing a single generator and a passive power splitter as shown in Figure 1.

This straightforward solution allows reducing system complexity and costs but suffers of poor flexibility because

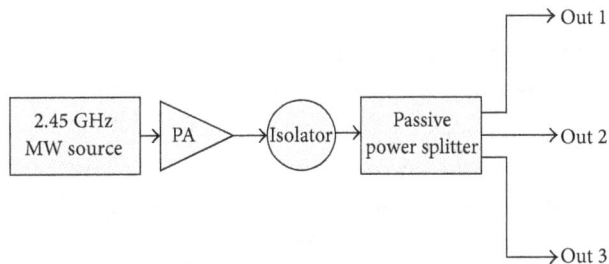

Figure 1: Block diagram of a single generator multioutputs MWA system.

the total available power P_{tot} at the generator output port is equally split among N ports; therefore when only a single applicator is used, the maximum allowable power is P_{tot}/N, unless the splitter is removed.

To overcome this limitation and to increase system flexibility and usability, the power distribution among N applicators can be done using a switched-mode approach [17] where the N applicators of the array are fed sequentially in time using a very fast solid-state high power switch, as early proposed in [18].

This approach leads to a simple and highly flexible solution called fast sequential switching (FSS), where the switching frequency must be kept high enough to prevent localized temperature drop during the applicator's off-time. By adjusting the duty-cycle of the signal that drives the switch, "on-the-fly" corrections or adaptive control of the power distribution among the N applicators can be done without any energy loss.

In Section 2 we outline the system concept and in Section 3 we describe the engineering development of the new equipment. Section 4 concerns the critical aspects of system safety and risk management; finally the conclusions are presented in Section 5.

Two appendices are included to assist the reader in recognizing the potentialities of the new multiapplicator system. Appendix A is devoted to the numerical modeling of a single and a dual applicator immersed in a muscle phantom; Appendix B reports some significant *ex vivo* experiments in swine liver and loin.

2. System Concept

The system has been conceived to evolve from the usual single applicator system to the new four-applicator FSS technology with the following objectives:

(i) addressing the perceived deficiencies in first-generation systems;

(ii) treating larger tumors, improving ablative margins and decreasing treatment time;

(iii) improving flexibility, usability, and procedural safety.

The system concept is based on the fact that the thermal time constant τ of a biological tissue is relatively high (in the order of seconds); thus if MW power is applied in a periodic pulsed mode with a short pulse period, the tissue temperature tends to be related to the average applied power. Therefore we can conceive a switched-mode regulator that uses a continuous wave (CW) source with the output power P that is periodically switched ON/OFF to a load (tissue to be ablated) with period T, thus transferring to it a controlled fraction of the power P. The lower useful limit of T is given by the commutation time of the switch while the upper limit is given by the ratio T/τ that should be <1. It was experimentally found that using a period T between 0.5 and 2 seconds the switched-mode system is practically equivalent to an ideal asynchronous system.

During a therapy session the power P is generally held constant and equal to the total power P_t that the physician wants to dispense to the tumor through a set of applicators:

$$P = P_t = \Sigma P_k \quad \text{with } k = 1 \dots 4. \tag{1}$$

The power distribution can be easily obtained using four very fast switches; the ON time of the kth switch is given by

$$T_k = \frac{T \cdot P_k}{P_t}. \tag{2}$$

When a single applicator is employed only the corresponding switch is always ON, while the others are OFF.

3. System Development

We believe that cross-collaboration is essential for the success of our project. Such partnership is also necessary to produce novel and commercially viable technologies. Furthermore, consistent inputs from a team of clinical experts are crucial to ensure that the developed technology remains clinically relevant and, therefore, can have the greatest positive impact on the rapid diffusion of minimally invasive focal treatments at MW frequencies.

With this in mind, a fully engineered prototype of the novel switched-mode system, called thermal ablation multiprobe microwave system (TAMMS), has been developed by a small multidisciplinary team at the University of Florence, in the framework of a collaborative academic-industry agreement with two private companies: Biomedical Srl, Florence (http://www.biomedical-srl.com/), that provided financial support and laboratory facilities, and WaveComm Srl, Siena (http://www.wavecomm.it/), that provided qualified engineering support.

As shown in Figure 2, TAMMS consists of the following main blocks:

(i) a single 100 W high efficiency solid-state MW power generator working at 2.45 GHz;

(ii) an FSS microwave switching unit (MSU), which operates as a programmable "active" power splitter. This unit has four IN/OUT ports that can be automatically configured as output power ports (100 W maximum) or temperature sensing input ports;

(iii) a controller that regulates the MW generator power output and produces the time sequence that commands the single pole 4 throw (SP4T) switch of the FSS unit;

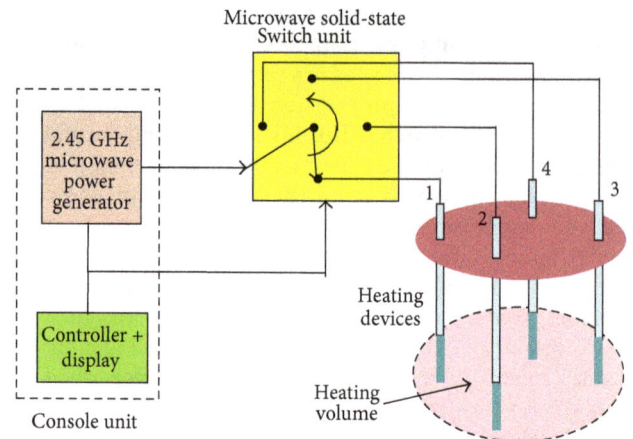

FIGURE 2: Block diagram of the four-applicator TAMMS system employing FSS technology.

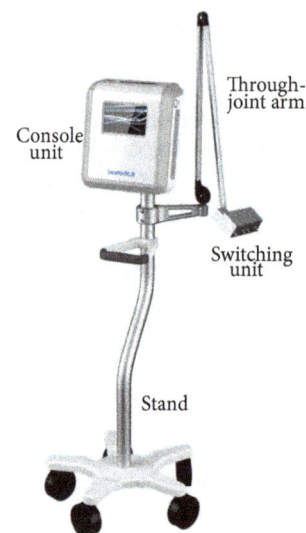

FIGURE 3: TAMMS 3D rendering.

(iv) a rugged power delivery device (interstitial applicator) in 18, 17, and 14 gauge formats preferred for percutaneous procedures and 11 gauge format for surgical applications. For high power treatments all the devices can be cryogenically cooled using CO_2.

In order to optimize the overall efficiency, the system has been designed with a two-stage power distribution concept: the first stage utilizes a large low loss coaxial cable that connects the MW power generator to the MSU, while the second stage comprises smaller more flexible cables connecting the MSU outputs to the corresponding power delivery devices.

Figure 3 shows the 3D rendering of the new system that evidences its compactness and the attention paid to the fundamental aspects of ergonomics, usability, and safety (industrial design made by Un-Real Studio, Florence, Italy— http://www.un-real.it/).

A robust and balanced mechanical arm with a double through-joint allows positioning the MSU nearby the patient to reduce the length of the cables that connect the unit to the set of interstitial applicators. Alternatively the MSU can be connected to the procedure bed in order to be fixed relative to patient. It is worth noting that if the MSU is omitted the system reduces to a conventional single applicator thermoablator with 100 W maximum CW power output.

3.1. Console Unit.

The console unit contains four main blocks: a solid-state MW generator, a switching AC/DC power supply, a digital controller, and a touch screen display.

The MW generator is capable of delivering a power output in excess of 100 W. The last-generation solid-state gallium nitride on silicon carbide (GaN-on-SiC) class AB high power amplifier (HPA), having 50% power added efficiency (PAE) at 100 W power output, can withstand 10 : 1 WSVR without damage. Furthermore automatic shut-off is provided for higher mismatching levels. The generator integrates multiple protection methodologies, including a forward power detector, a reflection power detector, and an embedded microcontroller to control alarm features. A forced air cooling system maintains the internal temperature of the whole generator below 75°C.

The 400 W medical grade AC/DC switching power supply has power efficiency in excess of 90% and automatic overload shutoff to further protect the MW generator.

The digital control unit performs core safety critical tasks by a board based on a powerful mixed signal microcontroller while the human-machine interface (HMI) is implemented by a high level program running on a single board computer (SBC) under Linux operating system.

The color touchscreen display has 800×480 pixel and the capability of showing multiple interactive screenshots dedicated to set the treatment parameters, monitor their evolution, and display devices status in real time.

The console has $35 \times 32 \times 15$ cm overall dimensions and weighs less than 8 Kg to be possibly hand carried and/or connected to a classical instrument stand.

3.2. Microwave Switching Unit.

The MSU represents a well-balanced tradeoff between system flexibility in terms of ablation volume/shape and complexity/compactness/cost of the switching MW circuitry.

The unit utilizes a matrix of high power PIN diodes as switching elements in a SP4T configuration. The design goal was to maintain the internal insertion losses <0.5 dB with resulting high system efficiency and low heat dissipation. An aluminum case provides both heat dissipation by natural convection and EM shielding. Table 1 resumes its main characteristics.

Figure 4 illustrates the realized MSU in microstrip planar technology with the cover removed.

Figure 5 depicts the unit operation in the time domain when the generator available power is equally subdivided among four applicators; in this case the driving sequence duty-cycle $D_t = T_{on}/T_{off}$ is 0.25 for all applicators.

The MSU receives commands from the console via a serial bus interface and an embedded controller locally generates the appropriate PIN diodes driving time sequences. It is

FIGURE 4: Microwave switching unit, inside view.

TABLE 1: Characteristics of the high power SP4T switching unit.

Parameter	Typical	Limit
Frequency (GHz)	2.45	ISM band
VSWR	1.13 : 1	<1.4 : 1
Insertion loss (dB)	0.4	<0.6
Power handling capability (W)	100	<150
Switching time (μS)	200	500

worth noting that anyway the total delivered power did not exceed 100 W that corresponds to the maximum allowable generator output power; this constraint is automatically verified by the console controller.

3.3. Power Delivery Device.

The power delivery device has a mechanically simple and rugged coaxial structure outlined in Figure 6 that consists of a dielectrically loaded monopole with a capacitive cap [19–21] fed by a coaxial line.

The device derives from the one described in [18], where the radiating element was a simple linear monopole. A needle shaped cap loading has been introduced to allow easy insertion inside the tissue, reduce the monopole length, and increase tip robustness while maintaining good input matching and radiating performances.

The intrinsically unbalanced design allows returning current flow on the outer conductor influencing impedance matching and producing the so-called "tail comet effect" on the radiated fields that can be reduced by inserting a "cancelling slot" [22]. An equivalent effect could be obtained using a $\lambda/4$ choke as suggested in [23]; this solution is, however, more mechanically complex and does not allow reducing shaft diameter below the 16 G dimension.

A comparison between slotted and unslotted applicators is presented in Appendix A.

It is worth noting that the applicator's structure (patent pending) is easily scalable and that the very thin commercial semirigid coax cable usually employed for feeding the radiating section of the applicator is avoided, thus simplifying design and reducing insertion losses. Stainless steel inner and outer conductors of the shaft assure high rigidity and robustness of the needle structure; in particular the radiating monopole is simply obtained by lengthening the inner conductor of the shaft, thus assuring very high breaking strength due to the absence of any mechanical junction. The whole device with its handle is shown in Figure 7; in order to reduce the adhesion of charred tissue during the thermoablative

(a)

T_on T_off

(b)

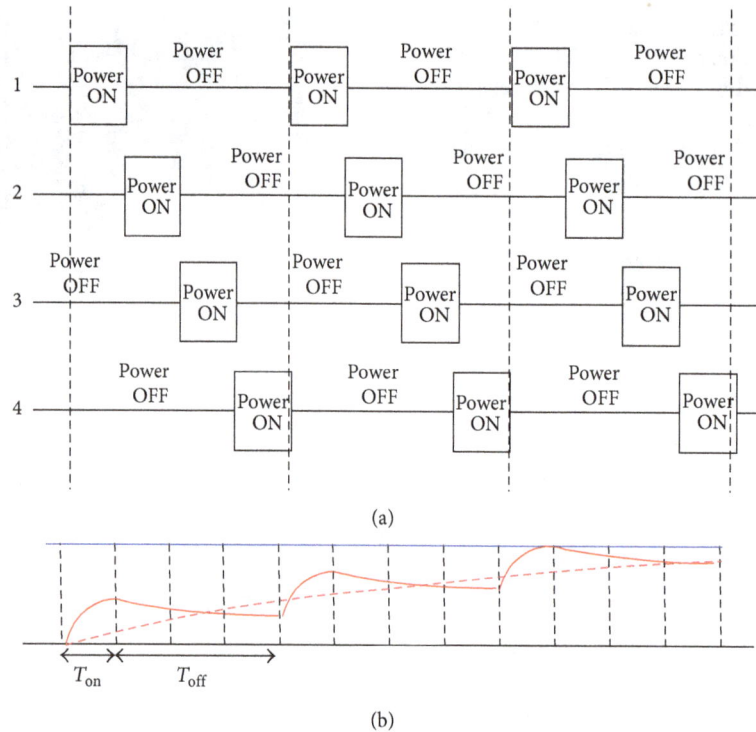

FIGURE 5: Switching applicator operation. Applicator driving time sequence (a) and temperature versus time evolution on the first applicator (b).

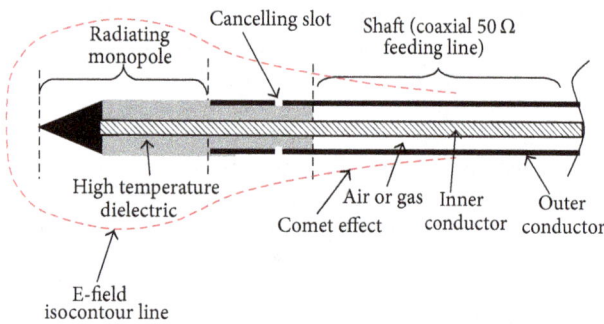

FIGURE 6: The coaxial low loss applicator structure.

treatment, the shaft can be coated with a very thin layer of Teflon or Parylene C.

High power efficiency is achieved by employing air or gas as dielectric in the coaxial shaft structure and by silver-plating the inner conductor. The applicator can be employed with and without gas cooling, depending on the applied input power. Table 2 resumes the main performances of the different sized applicators with and without CO_2 internal cooling.

Figure 8 shows the ice crystals formation on the applicator's shaft surface due to the Joule-Thompson effect when the cryogenic gas cooling is employed [24].

Very thin thermal sensors [25, 26] can be incorporated inside the applicator to monitor the temperature in correspondence with its radiating section and along the shaft.

A short circuited $\lambda/4$ stub integrated in the needle handle (shown in Figure 6) is employed to decouple the input port

FIGURE 7: The coaxial low loss applicator design.

TABLE 2: Applicators performances.

Size	11 G	14 G	17 G	18 G
Shaft diameter (mm)	3	2.1	1.52	1.27
Efficiency (20 cm shaft length)	0.99	0.96	0.94	0.92
Max CW power (W) (uncooled)	33	20	13	7
Max CW power (W) (cooled)	65	40	26	20

of the applicator from its fluidic section, thus allowing a matched connection with a $50\,\Omega$ coaxial cable that was chosen to have <1.5 dB/m loss and to be enough flexible.

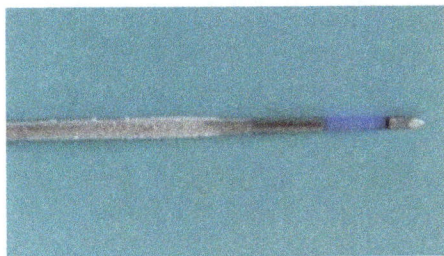

FIGURE 8: Ice crystals on the applicator's shaft surface.

Note that if we take into account the insertion losses of the whole power distribution chain (cables, connectors, and MSU) overall efficiencies practically halve and input powers double.

The applicator has been designed using the CST Microwave Studio Electromagnetic Simulation software (CST—Computer Simulation Technology—https://www.cst.com/). An optimization method has been employed in order to guarantee the maintenance of a good input matching ($S_{11} < -15$ dB at 2.45 GHz) during the entire ablative procedure, taking into account the variation of the tissue (liver, kidney, and muscle) complex permittivity with temperature and time [27].

4. System Safety and Risk Management

As other MW and RF thermal ablation equipment, TAMMS represents a great challenge from the safety standpoint. Starting from the conception phase, the first step that has been taken into account is a comprehensive risk analysis and assessment, mainly conducted with International Standard ISO 14971:2012 "*Application of Risk Management to Medical Devices.*" The lack of particular standards regarding the safety and effectiveness of surgical MW equipment forces the risk management process to be based only on the general standard EN 60601-1:2006 "*General Requirements for Basic Safety and Essential Performance*" and its corresponding relevant collateral standards (e.g., EN 60601-1-2:2007 related to EM compatibility and EN 60601-1-8 related to alarm systems).

During the risk management process we used the Failure Mode, Effects and Criticality Analysis (FMECA) mainly because of its flexibility in the definition and calibration of risk parameters (i.e., hazard likelihood, severity, and detectability). Indeed the use of the FMECA framework led to the definition of a multiattribute analysis where all the dimensions of service effectiveness were taken into account. To shape and refine the TAMMS project, an exhaustive study of the MWA process in terms of activities, information flows, tools, and different professional profiles involved has also been executed by means of a workflow analysis according to ANSI Standard ANSI/PMI 99/001/2008.

During each project step, the entire system underwent a thorough usability analysis according to the collateral standard EN 60601-1-6:2010, aimed to ensure a user-friendly and safe design.

Finally, every identified risk, including the additional ones closely related to the mitigation interventions themselves, has been brought back under the acceptability threshold. Therefore, the device is ready to go further towards a clinical evaluation and all the next steps needed for CE marking [28]. The entire system is now subject to an intensive engineering process and it will be renamed as Thermal Ablation Treatments for Oncology (TATO).

5. Conclusions

A 2.45 GHz MWA system has been developed with four output channels that can independently feed up to four high efficiency applicators.

The power distribution among the output channels is made by a state-of-the-art solid-state MSU that operates as an active programmable power splitter by employing an FSS technology.

Due to the thermal inertia of the biological tissue and to the very fast commutation of the switching unit, the new system practically performs as a pseudoasynchronous system as demonstrated by EM and TH numerical simulations and confirmed by *ex vivo* experiments (see Appendices A and B).

The novel approach used for feeding multiple applicators allows

(i) reducing system complexity and cost because only a single high power MW generator is employed;

(ii) increasing operational flexibility;

(iii) obtaining thermal lesions of up to 4 cm in diameter with optimal reproducibility and predictability, by using a couple of 17 G uncooled high efficiency applicators (see Appendix B);

(iv) reducing sensitivity to the placement of applicators and temperature sensors into the tissue.

Future studies and developments will be devoted to the following topics:

(i) radiometric temperature sensing during the OFF applicator's states;

(ii) blood flow estimation by measuring the tissue time constant τ during an ON-OFF cycle;

(iii) adaptive control of the switching duty-cycle based on the temperature feedback;

(iv) study of a radically new applicator cooling system based on heat-pipe technology;

(v) integration with robotic and EM navigation technologies.

Appendices

A. Single and Dual-Probe Numerical Simulations

Numerical EM and TH simulations represent a powerful means to estimate the performances of a single or multiapplicator system operating in a biological tissue with the purpose

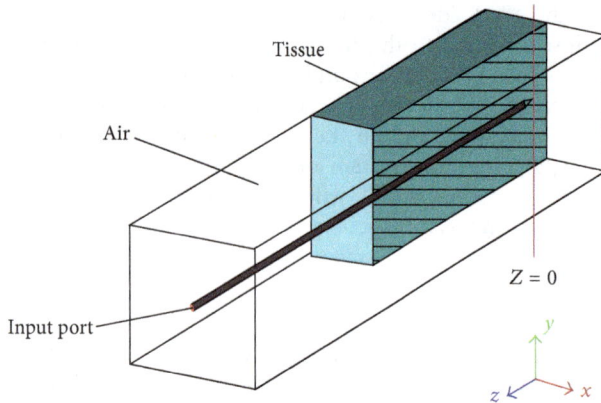

FIGURE 9: CST analysis domain.

TABLE 3: Properties of muscle tissue at 2.45 GHz frequency.

Property at 37°C	Value
Electrical and mechanical properties	
Relative permittivity ε_r	45.6
Electrical conductivity σ (S/m)	1.97
Thermal conductivity k_t (W/mK)	0.564
Specific heat c_p (J/kgK)	3400
Density ρ (kg/m³)	1050
Blood perfusion properties	
Flow coefficient (W/K/m³)	2700
Metabolic rate (W/m³)	480

of inducing hyperthermia or producing thermal ablation. In the following we analyze single and double applicator configurations that are the most frequently used in clinical practice, using the CST Microwave Studio software. The analysis domain is shown in Figure 9. Comparisons between synchronous and switched-mode feeding for a couple of applicators are also made.

Steady-state and transient thermal analyses, based on the bioheat equation [29, 30], are performed with the simplifying hypothesis that EM and TH equations are not coupled through the temperature dependence of the tissue permittivity.

The reference tissue chosen for the simulations is muscle, whose properties at 37°C and 2.45 GHz frequency are listed in Table 3.

Figure 10 shows the power density distribution produced by a 17 G applicator with (a) and without (b) cancelling slot. The corresponding steady-state temperature distributions obtained with an effective input power of 15 W and active blood perfusion are depicted in Figure 11.

Note that the "comet tail" clearly evident in the power density distribution practically disappears in the temperature behavior because of the heat sink effect of blood perfusion.

Passing to a couple of 17 G slotted applicators immersed in the same tissue at distance $D = 12$ mm, we obtain the relevant power density distributions depicted in Figure 12. Left figures (a, b, and c) refer to the xy plane ($z = 21$ mm) while right

figures (d, e, and f) refer to yz plane ($x = y = 0$). Upper figures (a, d) represent power densities in the case of in-phase feeding of both applicators, while central (b, e) and lower (c, f) figures represent the two complementary states of a switched feeding.

In both in-phase and switched feeding modalities near field interferences arise in the close proximity of the applicators' tips. In the first case the interference is "active" and produces focusing in the yz plane of symmetry while in the second case the interference is "passive" and the OFF applicator performs as a wire reflector of the ON applicator. Due to the very high attenuation of the EM fields in the medium the interference phenomena are substantially reduced at a radial distance greater than the plane wave penetration depth in the tissue, that is 20 mm in the muscle at 2.45 GHz. To evidence this phenomenon a reference circle has been traced having radius equal to the penetration depth δ.

Figure 13 refers to simulated treatments of 3 min (a, b) and 10 min (c, d) duration with 40 W total input power.

By increasing the distance D between the two applicators, similar behaviors can be observed up to the $D = 20$ mm limit. Beyond this limit some indentations begin to appear in the 60°C isotherm profile [31].

By increasing the treatment time, the 60°C isotherm that encloses the tissue volume where acute cellular necrosis instantly occurs slightly overtakes the reference circle but does not expand further for $t > 15$ min. This numerical result confirms that treatment times lasting more than 10 minutes are not advantageous at MW frequencies as it happens in RFA. Further simulations proved that the 85% of the maximum allowable size of the thermal lesion is reached only 6 minutes after the treatment starting.

From the totality of the simulations performed we observed that the isotherms for $T \leq 60°C$ are practically independent from the feeding modality (synchronous, asynchronous, or FSS) because of the heat diffusion inside the tissue.

B. *Ex Vivo* Experiments

To support numerical analysis' results, we made single applicator and multiapplicator tests using 14 G devices that were chosen because of their higher power capability with and without a cooling system. The ablations were performed on swine livers or loin pieces taken from animals euthanized less than 6 hours prior to the tests. All tissues were maintained at room temperature. During this work, we will refer to the "input effective power" P_{eff} as the net power flowing in the applicator input port that is calculated by calling off the losses of the power distribution chain. Practically this power value can be obtained by halving the power delivered by the MW generator.

Figure 14 shows ablations obtained employing a single slotted applicator with $P_{eff} = 15$ W input power and 8-minute treatment time. The thermal lesion was 4×2.5 cm and its shape and dimensions agree well with simulation results.

We have also compared the lesions obtained with and without cooling, shown in Figure 15. With $P_{eff} = 20$ W and

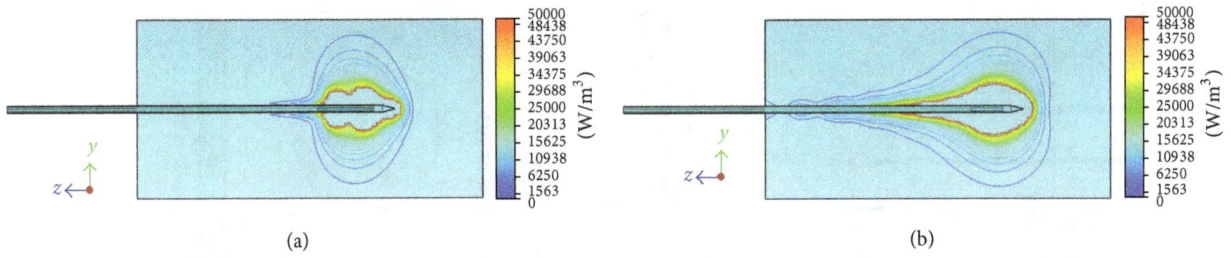

FIGURE 10: Power density distribution on the yz plane of a 17 G applicator operating inside muscle tissue with (a) and without (b) cancelling slot. Reference input power = 1 W.

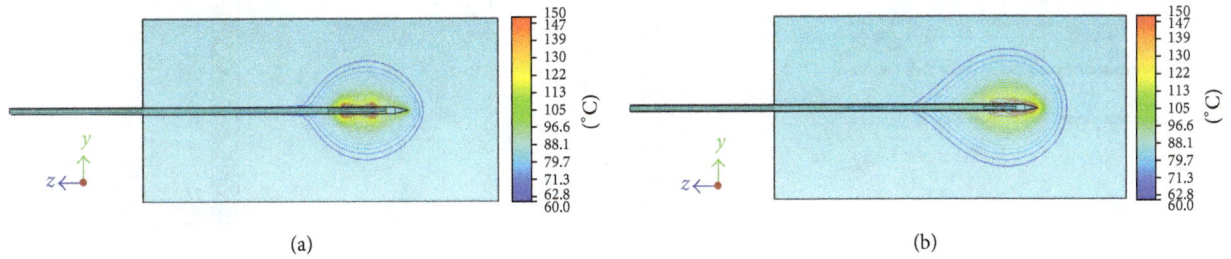

FIGURE 11: Steady-state temperature distribution on the yz plane of a 17 G applicator operating inside muscle tissue with (a) and without (b) cancelling slot. P_{in} = 15 W.

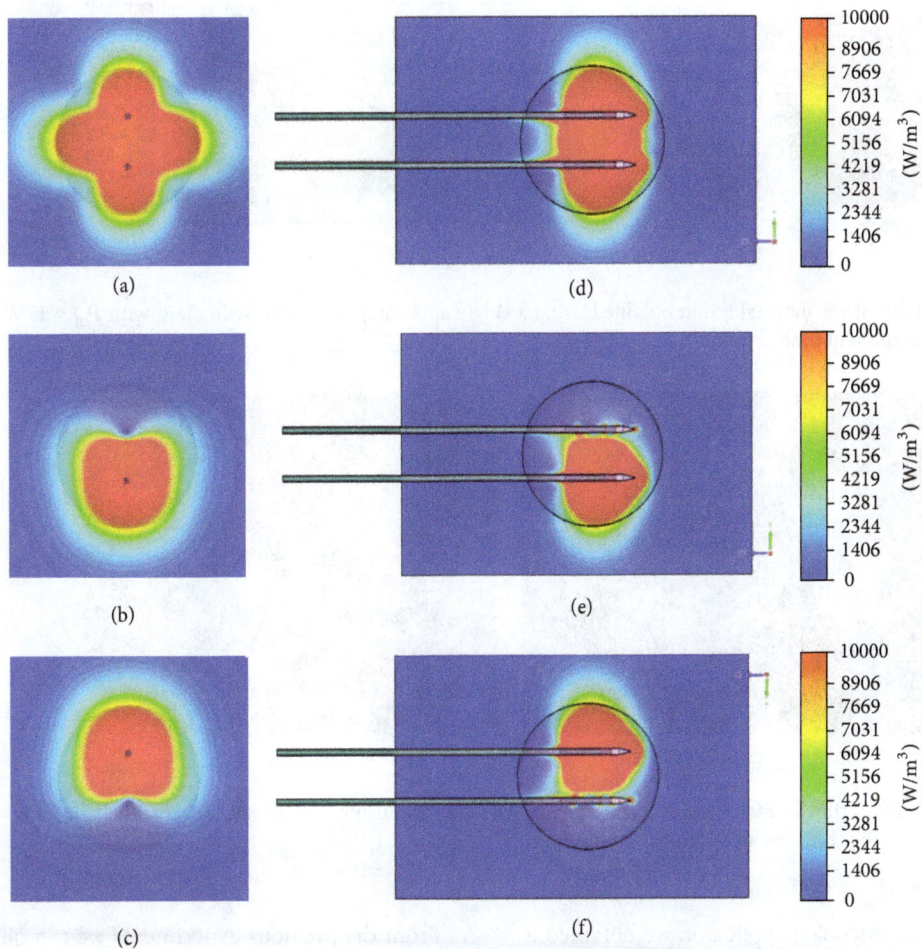

FIGURE 12: Power density distribution produced by a couple of applicators with in-phase feeding (a, d) and switched-mode feeding (b, e) and (c, f). Reference input power = 1 W.

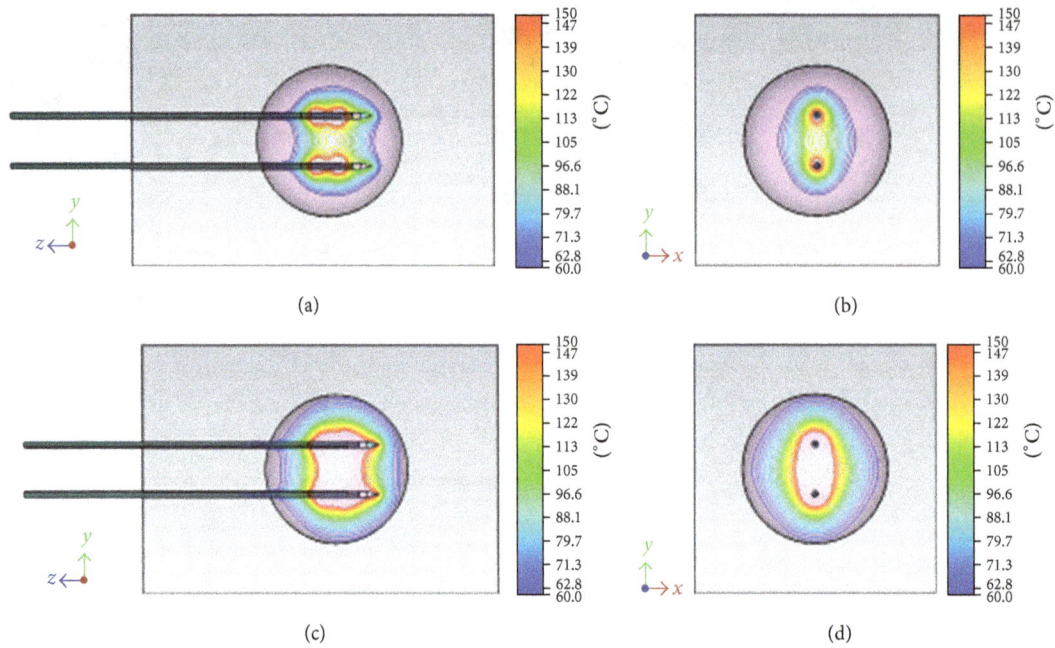

FIGURE 13: Temperature distribution after 3 min (a, b) and 10 min (c, d) simulated treatments in perfused muscle tissue. Total P_{in} = 40 W.

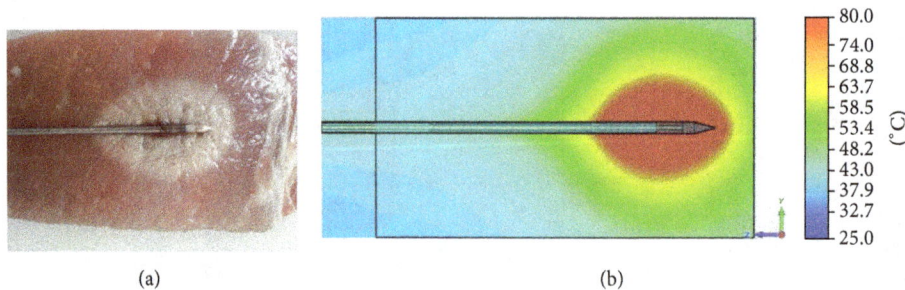

FIGURE 14: Longitudinal cut of the thermal lesion obtained using a single applicator in *ex vivo* swine loin with P_{eff} = 15 W (a) and related thermal simulation (b). Treatment time: 8 minutes.

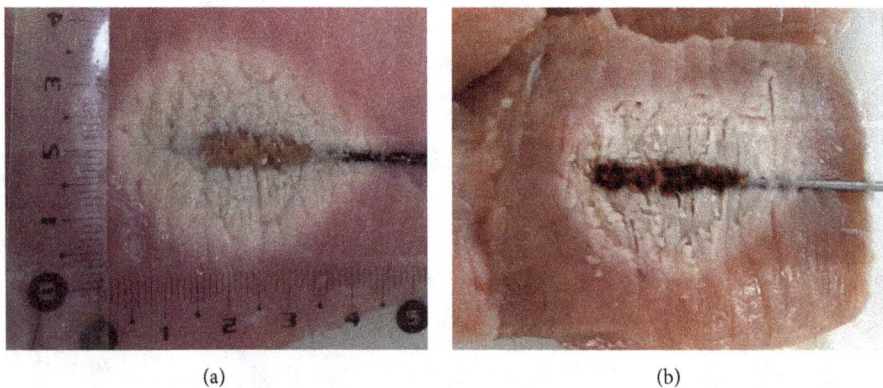

FIGURE 15: Longitudinal cuts of the thermal lesions obtained in *ex vivo* swine loin using a single applicator with P_{eff} = 20 W without cooling (a) and P_{eff} = 30 W with cooling (b). Treatment time: 10 minutes.

employing an unslotted uncooled applicator, we obtained a lesion of 4.5 × 3.4 cm with reduced charring in 10 minutes. On the other hand, with cooling active and P_{eff} = 30 W, the lesion was 5 × 4 cm.

From the previous experiments we can observe that the differences between the lesions obtained with and without gas cooling are comparable in shape and dimensions apart from the greater charring produced by the cooled applicator.

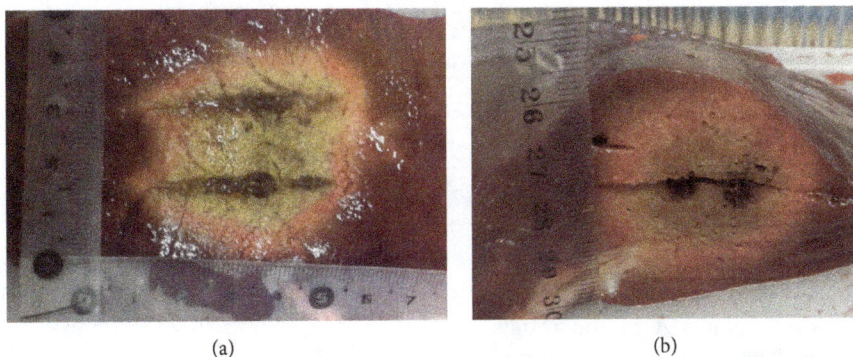

(a) (b)

FIGURE 16: Longitudinal (a) and transverse (b) cuts of the thermal lesions obtained in *ex vivo* swine liver using a couple of uncooled applicators with total P_{eff} = 40 W. Treatment time: 10 minutes.

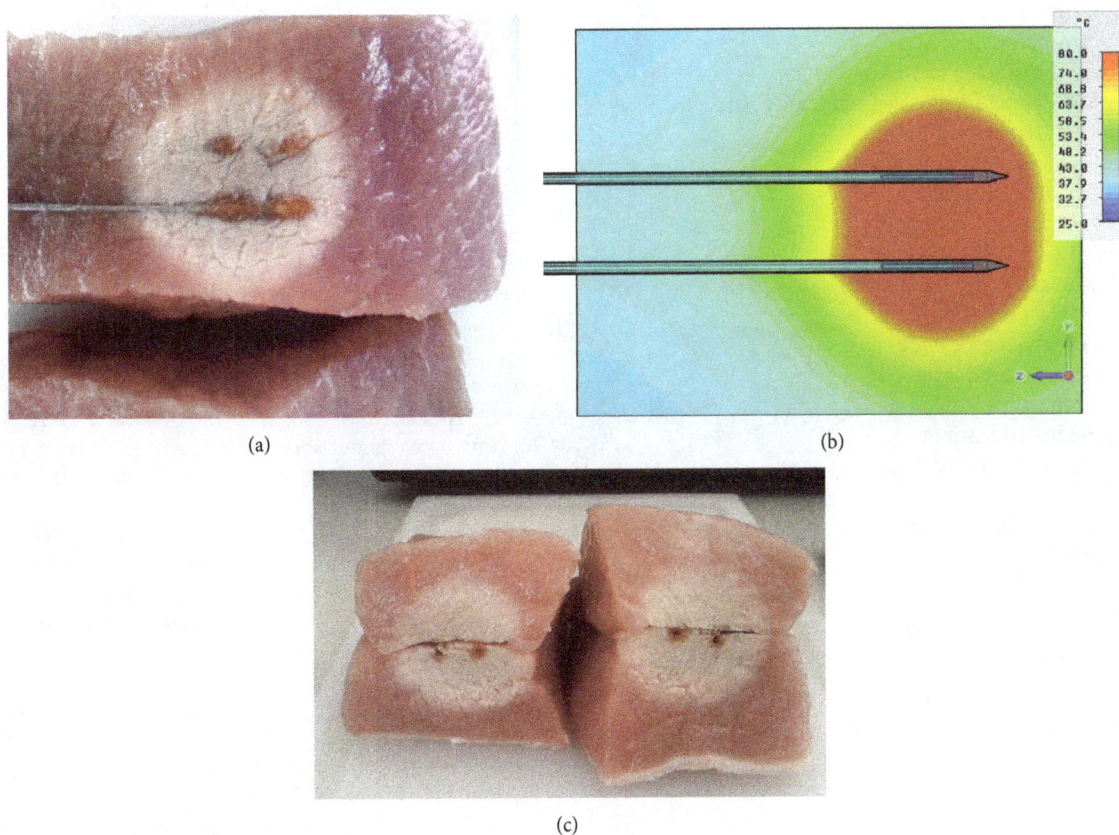

(a) (b)

(c)

FIGURE 17: Longitudinal cut (a), related thermal simulation (b), and transverse cut (c) of the thermal lesion obtained in *ex vivo* swine loin using a couple of uncooled applicators with total P_{eff} = 25 W. Treatment time: 10 min. One of the two applicators was removed to highlight higher temperature gradient zones.

It follows that, when passing from an uncooled to a cooled device, the efficiency suffers a decrease, about 20% in this case, due to the MW power wasted in the shaft far from the radiating section, where the cooling fluid subtracts heat. This drop of efficiency, that depends on shaft losses and comet effect, is influenced by several parameters such as temperature and flow velocity of the cooling fluid and does not remain constant during the treatment and furthermore cannot easily be measured.

We can ascribe the worse repeatability of the obtained results when the applicator's cooling is active to this variability and unpredictability of the total energy administered to the tissue.

For this reason we decided to perform the multiapplicator trials using only uncooled applicators. Figure 16 shows the longitudinal cut of an ablation made with two applicators on an *ex vivo* swine liver with P_{eff} = 40 W and 10-minute treatment time.

Figure 17 compares the ablation obtained using a couple of applicators immersed in swine loin with the corresponding numerical simulations. Also in this case experiments agree well with simulation results. It is worth noting that we obtained an almost spherical lesion with a 4 cm diameter and negligible charring effects, resulting in a volume of 31 cm^3. This finding confirms the preliminary results obtained in [18] using applicators without cap loading.

Referring to recent comparisons [31, 32], it can be said that the ablations shown in Figures 14–17 represent very promising findings, even when compared to the results obtainable with other commercially available systems using cooled applicators.

Abbreviations

CW: Continuous wave
EM: Electromagnetic
FIT: Finite integration technique
FMECA: Failure Mode, Effects and Criticality Analysis
FSS: Fast sequential switching
HMI: Human-machine interface
HPA: High power amplifier
MPA: Microwave power amplifier
MSU: Microwave switching unit
MW: Microwave
MWA: Microwave ablation
PAE: Power added efficiency
PML: Perfect matched layer
RFA: Radiofrequency ablation
SBC: Single board computer
SP4T: Single pole 4 throw
TH: Thermal
TAMMS: Thermal ablation multiprobe microwave system
TATO: Thermal Ablation Treatments for Oncology.

Conflict of Interests

The authors declare that there is no conflict of interests regarding the publication of this paper.

Acknowledgments

The authors would like to gratefully thank the COMIS (Centro Interdipartimentale Sviluppo di Nuove Tecnologie Mini-Invasive in Chirurgia Oncologica) for the highly skilled support in the *ex vivo* experiments. They would like to inform the reader that at the time when they originally wrote the paper, Engineer Vasco Tesi, Ph.D., and Engineer Cosimo Ignesti, they worked at the University of Florence as researchers along with Professor Biffi Gentili. At the present time they collaborate with Wavecomm and Biomedical, respectively, to continue the project and its related studies.

References

[1] C. Toso, G. Mentha, N. M. Kneteman, and P. Majno, "The place of downstaging for hepatocellular carcinoma," *Journal of Hepatology*, vol. 52, no. 6, pp. 930–936, 2010.

[2] F. J. Wolf, D. J. Grand, J. T. Machan, T. A. DiPetrillo, W. W. Mayo-Smith, and D. E. Dupuy, "Microwave ablation of lung malignancies: effectiveness, CT findings, and safety in 50 patients," *Radiology*, vol. 247, no. 3, pp. 871–879, 2008.

[3] C. J. Diederich, "Thermal ablation and high-temperature thermal therapy: overview of technology and clinical implementation," *International Journal of Hyperthermia*, vol. 21, no. 8, pp. 745–753, 2005.

[4] L. Boni, A. Benevento, F. Cantore, G. Dionigi, F. Rovera, and R. Dionigi, "Technological advances in minimally invasive surgery," *Expert Review of Medical Devices*, vol. 3, no. 2, pp. 147–153, 2006.

[5] P. Liang, Y. Wang, X. Yu, and B. Dong, "Malignant liver tumors: treatment with percutaneous microwave ablation—complications among cohort of 1136 patients," *Radiology*, vol. 251, no. 3, pp. 933–940, 2009.

[6] C. L. Brace, "Microwave tissue ablation: biophysics, technology, and applications," *Critical Reviews in Biomedical Engineering*, vol. 38, no. 1, pp. 65–78, 2010.

[7] T. P. Ryan, P. F. Turner, and B. Hamilton, "Interstitial microwave transition from hyperthermia to ablation: historical perspectives and current trends in thermal therapy," *International Journal of Hyperthermia*, vol. 26, no. 5, pp. 415–433, 2010.

[8] C. L. Brace, "Radiofrequency and microwave ablation of the liver, lung, kidney, and bone: what are the differences?" *Current Problems in Diagnostic Radiology*, vol. 38, no. 3, pp. 135–143, 2009.

[9] C. L. Brace, "Microwave ablation technology: what every user should know," *Current Problems in Diagnostic Radiology*, vol. 38, no. 2, pp. 61–67, 2009.

[10] M. G. Lubner, C. L. Brace, J. L. Hinshaw, and F. T. Lee Jr., "Microwave tumor ablation: mechanism of action, clinical results, and devices," *Journal of Vascular and Interventional Radiology*, vol. 21, no. 8, pp. S192–S203, 2010.

[11] C. L. Brace, P. F. Laeseke, L. A. Sampson, T. M. Frey, D. W. Van Der Weide, and F. T. Lee Jr., "Microwave ablation with multiple simultaneously powered small-gauge triaxial antennas: results from an in Vivo swine liver model," *Radiology*, vol. 244, no. 1, pp. 151–156, 2007.

[12] M. G. Lubner, J. L. Hinshaw, A. Andreano, L. Sampson, F. T. Lee Jr., and C. L. Brace, "High-powered microwave ablation with a small-gauge, gas-cooled antenna: initial ex vivo and in vivo results," *Journal of Vascular and Interventional Radiology*, vol. 23, no. 3, pp. 405–411, 2012.

[13] P. Laeseke, F. Lee, D. van der Weide, and C. Brace, "Multiple-antenna microwave ablation: spatially distributing power improves thermal profiles and reduces invasiveness," *Journal of Interventional Oncology*, vol. 5, no. 5, pp. 65–72, 2009.

[14] A. S. Wright, F. T. Lee Jr., and D. M. Mahvi, "Hepatic microwave ablation with multiple antennae results in synergistically larger zones of coagulation necrosis," *Annals of Surgical Oncology*, vol. 10, no. 3, pp. 275–283, 2003.

[15] C.-H. Liu, R. S. Arellano, R. N. Uppot, A. E. Samir, D. A. Gervais, and P. R. Mueller, "Radiofrequency ablation of hepatic tumours: Effect of post-ablation margin on local tumour progression," *European Radiology*, vol. 20, no. 4, pp. 877–885, 2010.

[16] T. Nakazawa, S. Kokubu, A. Shibuya et al., "Radiofrequency ablation of hepatocellular carcinoma: correlation between local tumor progression after ablation and ablative margin," *The American Journal of Roentgenology*, vol. 188, no. 2, pp. 480–488, 2007.

[17] C. L. Brace, P. F. Laeseke, L. A. Sampson, D. W. van der Weide, and F. T. Lee, "Switched-mode microwave ablation: less dependence on tissue properties leads to more consistent ablations than phased arrays," in *Proceedings of the Radiological Society of North America Meeting (RSNA '07)*, Chicago, Ill, USA, 2007.

[18] G. Biffi Gentili and M. Linari, "A novel thermo-ablative microwave multi-applicator system," in *Proceedings of the 19th Italian National Meeting on Electromagnetism (RiNEM '12)*, Rome, Italy, September 2012.

[19] M. F. Iskander and A. M. Tumeh, "Design optimization of interstitial antennas," *IEEE Transactions on Biomedical Engineering*, vol. 36, no. 2, pp. 238–246, 1989.

[20] K. Saito, Y. Hayashi, H. Yoshimura, and K. Ito, "Numerical analysis of thin coaxial antennas for microwave coagulation therapy," in *Proceedings of the IEEE Antennas and Propagation Society International Symposium*, vol. 2, pp. 992–995, Orlando, Fla, USA, July 1999.

[21] M. Cavagnaro, C. Amabile, P. Bernardi, S. Pisa, and N. Tosoratti, "A minimally invasive antenna for microwave ablation therapies: design, performances, and experimental assessment," *IEEE Transactions on Biomedical Engineering*, vol. 58, no. 4, pp. 949–959, 2011.

[22] K. Ito, K. Ueno, M. Hyodo, and H. Kasai, "Interstitial applicator composed of coaxial ring slots for microwave hyperthermia," in *Proceedings of the International Conference on Intelligent Systems Applications to Power Systems (ISAP '89)*, pp. 253–256, Tokyo, Japan, 1989.

[23] I. Longo, G. B. Gentili, M. Cerretelli, and N. Tosoratti, "A coaxial antenna with miniaturized choke for minimally invasive interstitial heating," *IEEE Transactions on Biomedical Engineering*, vol. 50, no. 1, pp. 82–88, 2003.

[24] E. M. Knavel, J. L. Hinshaw, M. G. Lubner et al., "High-powered gas-cooled microwave ablation: shaft cooling creates an effective stick function without altering the ablation zone," *The American Journal of Roentgenology*, vol. 198, no. 3, pp. W260–W265, 2012.

[25] M. Linari and G. Biffi Gentili, "Design and optimization of a miniaturized hyperthermic microwave applicator with integrated temperature sensors," in *Proceedings of the 10th International Congress on Hyperthermic Oncology (ICHO '08)*, 2008.

[26] G. B. Gentili and M. Linari, "A minimally invasive microwave hyperthermic applicator with an integrated temperature sensor," in *Proceedings of the 1st International Conference on Biomedical Electronics and Devices (BIODEVICES '08)*, pp. 113–118, January 2008.

[27] Z. Ji and C. L. Brace, "Expanded modeling of temperature-dependent dielectric properties for microwave thermal ablation," *Physics in Medicine and Biology*, vol. 56, no. 16, pp. 5249–5264, 2011.

[28] E. Iadanza, C. Ignesti, and G. B. Gentili, "Risk management process in a microwave thermal ablation system for CE marking," in *XIII Mediterranean Conference on Medical and Biological Engineering and Computing 2013*, vol. 41 of *IFMBE Proceedings*, pp. 1170–1173, 2014.

[29] H. H. Pennes, "Analysis of tissue and arterial blood temperatures in the resting human forearm," *Journal of Applied Physiology*, vol. 85, no. 1, pp. 5–34, 1998.

[30] W. L. Nyborg, "Solutions of the bio-heat transfer equation," *Physics in Medicine and Biology*, vol. 33, no. 7, pp. 785–792, 1988.

[31] F. Oshima, K. Yamakado, A. Nakatsuka, H. Takaki, M. Makita, and K. Takeda, "Simultaneous microwave ablation using multiple antennas in explanted bovine livers: relationship between ablative zone and antenna," *Radiation Medicine—Medical Imaging and Radiation Oncology*, vol. 26, no. 7, pp. 408–414, 2008.

[32] R. Hoffmann, H. Rempp, L. Erhard et al., "Comparison of four microwave ablation devices: an experimental study in ex vivo bovine liver," *Radiology*, vol. 268, no. 1, pp. 89–97, 2013.

13

The Spiral Coaxial Cable

I. M. Fabbri

Department of Physics, University of Milan, Via Celoria 16, 20133 Milan, Italy

Correspondence should be addressed to I. M. Fabbri; italomariofabbri@crfm.it

Academic Editor: Kamya Yekeh Yazdandoost

A new concept of metal spiral coaxial cable is introduced. The solution to *Maxwell's* equations for the fundamental propagating *TEM* eigenmode, using a generalization of the *Schwarz-Christoffel* conformal mapping of the spiral transverse section, is provided together with the analysis of the impedances and the Poynting vector of the line. The new cable may find application as a medium for telecommunication and networking or in the sector of the Microwave Photonics. A spiral plasmonic coaxial cable could be used to propagate subwavelength surface plasmon polaritons at optical frequencies. Furthermore, according to the present model, the myelinated nerves can be considered natural examples of spiral coaxial cables. This study suggests that a malformation of the Peters angle, which determines the power of the neural signal in the *TEM* mode, causes higher/lower power to be transmitted in the neural networks with respect to the natural level. The formulas of the myelin sheaths thickness, the diameter of the axon, and the spiral *g* factor of the lipid bilayers, which are mathematically related to the impedances of the spiral coaxial line, can make it easier to analyze the neural line impedance mismatches and the signal disconnections typical of the neurodegenerative diseases.

1. Introduction

The coaxial cable invented by Heaviside [1] is a transmission line composed of an inner conductor surrounded by an insulating layer and an outer conducting shield. The inner and the outer conductors share the same geometrical axis.

Nowadays, there exist several types of transmission lines, the coaxial cables, the hollow waveguides (rectangular, circular, elliptical, and parallel plates [2, 3]), the two wires [3, 4], and the channel waveguides (buried, strip-loaded, ridge, rib, diffused, and graded-dielectric index [5]). The two-wire transmission line used in conventional circuits is inefficient for transferring electromagnetic energy at microwave frequencies because the fields are not confined in all directions, the energy escapes by radiation, and it has large copper losses due to its relatively small surface area.

On the other hand the larger surface area of the boundary makes in general the waveguides more efficient with respect to the coaxials on reducing the copper losses.

Dielectric losses in two wires and coaxial lines, caused by the heating of the insulation between the conductors, are also lower in waveguides. The insulation behaves as the dielectric of a capacitor and the breakdown of the insulation between the conductors of a transmission line is more frequently a problem than is the dielectric loss in practical applications [3].

Waveguides are also subject to the dielectric breakdown caused by the stationary voltage spikes at the nodes of the standing waves.

In spite of the dielectric in waveguides is air, it causes arcing which decreases the efficiency of energy transfer and can severely damage the waveguide.

The spiral coaxial cable (SCC) discussed in this paper has many advantages with respect to the other types of transmission lines; first of all the elm energy can be distributed efficiently over a larger area, reducing all the undesired aforementioned effects.

The power handling versus frequency is consequently higher for the metal spiral coaxial line (MSCC) with respect to the other lines.

The Microwave Photonics (MWP) [6], a new discipline that joins together the radio-frequency engineering and optoelectronics, represents the future in many civil and defense applications like speed of the digital signal processors, cable television, and optical signal processing.

The MSCC could become one of the key objects in the field of MWP for its characteristics in terms of both power handling and energy transfer efficiency.

Recently, it has also been demonstrated experimentally that metal coaxial waveguide nanostructures perform at optical frequencies [7], opening up new research pursuits unexpected in the area of nanoscale waveguiding, field enhancement, imaging with coaxial cavities, and negative-index metamaterials [8].

Several different plasmonic waveguiding structures have been proposed such as metallic nanowires [9, 10], metal-dielectric-metal (MDM) structures [11], and metallic nanoparticle arrays [12] for achieving compact integrated photonic devices.

Most of these structures support a highly confined mode near the surface plasmon frequency [11]. This study could become the reference point to introduce the MSCC as a valid alternative to guide subwavelength surface plasmon polaritons (SPPs).

The extraordinary transmission of light through an array of subwavelength apertures, enhancement which arises from the coupling of the incident light with the SPPs through the surface grating in metal film [13], could result particularly efficiently on the spiral metal dielectric interface with periodic holes.

High sensitivity spiral biosensors and spiral photonic integrated circuits based on nonlinear surface plasmon polariton optics [14, 15] may be implemented.

The aim of this pioneering work on the SCC is to represent an initial landmark in the continuously growing sector of the microwave research.

The popularity of Video on Demand (VoD) and Over the Top Technology (OTT) services to access high definition videos over home interconnected devices of the hybrid fibre-coax (HFC) networks is driving the research toward more efficient and cost-effective cables. Particularly, the development of Converged Cable Access Platforms (CCAP) that combine video and data transmission supporting simultaneous network access of multiple users over a single coaxial cable is flourishing and sustaining the demand for new high speed transmission media.

In a metallic guide, the reflection mechanism responsible for confining the energy is due to the reflection from the conductors at the boundary [16], whose geometry is strictly related to the propagating modes.

Coaxial cables were designed to propagate high frequency radio signals. The principal constraints on performance of a coaxial are attenuation, thermal noise, and passive intermodulation noise (PIM).

In RF applications, the wave propagates essentially in the fundamental transverse electric magnetic (TEM) mode; that is, the electric and magnetic fields are both perpendicular to the direction of propagation.

In the ideal case, the conductors can be considered to have infinite conductivity and the TEM eigenmode is the basic propagating wave (see [17] page 110) along the transmission line.

Practical lines have finite conductivity, and this results in a perturbation or change of the TEM mode (see [18] page 119).

Above the cutoff frequency, transverse electric (TE) or transverse magnetic (TM) modes [19] can also propagate with different velocities within a practical cylindrical coax, interfering with each other producing distortion of the signal.

The frequency of operation for a specific outer conductor size is then limited by the highest usable cutoff frequency before undesirable modes of propagation occur.

In order to prevent higher order modes from being launched, the radiuses of the coaxial conductors must be reduced, diminishing the amount of power that can be transmitted.

On the other hand at high frequencies it is impossible to make the cylindrical coaxial line in the small size necessary to propagate the TEM mode alone.

The research described in this paper demonstrates the propagation on the fundamental TEM wave along the ideal MSCC.

Since the mode of transmission on an ideal line is the TEM wave, the relations for input impedance, reflection coefficient, return loss (RL), standing wave ratio (SWR), and so forth, given afterward in the next sections, are applicable in general to the spiral transmission lines (see [18] Chapter 3).

The metal double spiral coaxial cable or MDSCC, resulting from the superposition of two spiral conductors that share the same geometrical axis, can be made multi-turn. The amount of heat generated by the losses for heating can be distributed over a larger area and this would lower the temperature and raise the reliability of the line.

In fact, operation at higher temperatures results in a reduction in the life expectancy and reliability of the transmission line relative to the lower temperature performance.

Applications like nanoscale optical components for integration on semiconductor chips could benefit from these characteristics of the MSCC.

Where signal integrity is important, coaxial cables are needed to be shielded against radio frequency noise (RF noise). The multiturn MDSCC is naturally shielded because the highest part of the elm energy can be distributed on the inner part of the cable, which protects small signals from interference due to external electric fields.

A new class of spiral passive components, computer-aided engineering (CAE) tools as well as electromagnetic (EM) simulators, is required before new high-frequency spiral RF/microwave circuits will be implemented.

The spiral geometry occurs widely in nature; examples like the *spiral galaxies* are found at the universe level while the *myelin bundles* are common in the microcosm of the neuron cells.

Recently, a new spiral optical fibre has been proposed both in the fundamental mode [20] and in the higher order modes [21] operation.

Spirals are also of extreme interest to the field of the new metamaterials and invisible cloaking [22].

Myelinated nerve fibers are micro-spiral coaxial cables ($g \ll 1$) whose electric behaviour is still today described by neurophysiologists using *W. Thomson's* (later known as *Lord Kelvin*) *cable formula* [23] of the 1860s, which determines the velocity of the signal propagating in saltatory conduction [23, 24].

Cable theory in neurobiology has a long history, having first been applied to neurons in 1863 by *C. Matteucci* [25] who discovered that if a constant current flows through a portion of a platinum wire covered with a sheath saturated with fluid, extra-polar current can be led off which corresponds to the electrotonic current of nerves.

Since the 1950s–60s myelinated nerves have been recognized to have a spiral structure and to behave like a high loss coaxial cable [26, 27] with negligible inductance.

The mathematical model presented in this paper can be used to refine the elm theory of the myelinated nerves by taking into account their spiral geometry.

In a coaxial guide, the determination of the electromagnetic fields within any region of the guide is dependent upon one's ability to explicitly solve the *Maxwell field equations* in an appropriate coordinate system [28].

Let us consider *Maxwell's equations*

$$\nabla \times \vec{E} = -j\omega\mu\vec{H},$$

$$\nabla \times \vec{H} = j\omega\epsilon\vec{E},$$

$$\nabla \cdot \vec{E} = 0,$$

$$\nabla \cdot \vec{H} = 0,$$

$$(1)$$

where the time variation of the fields is assumed to be $\exp(j\omega t)$.

In view of the nature of the boundary surface, it is convenient to separate these field equations into components parallel and transverse to the waveguide z-axis.

This is achieved by scalar and vector multiplication of (1) with $\hat{e}_z$, a unit vector in the z direction, thus obtaining

$$\nabla_\perp \cdot \left(\hat{e}_z \times \vec{E}_\perp\right) = -j\omega\mu H_z,$$

$$\nabla_\perp \cdot \left(\hat{e}_z \times \vec{H}_\perp\right) = j\omega\epsilon E_z,$$

$$(2)$$

$$\nabla_\perp E_z - \frac{\partial E_\perp}{\partial z} = -j\omega\mu\left(\hat{e}_z \times H_\perp\right),$$

$$\nabla_\perp H_z - \frac{\partial H_\perp}{\partial z} = j\omega\epsilon\left(\hat{e}_z \times E_\perp\right).$$

$$(3)$$

Since the transmission line description of the electromagnetic field within uniform guides is independent of the particular form of the coordinate system employed to describe the cross section, no reference to cross-sectional coordinates is made on deriving the *telegrapher's equation* [28, 29].

Substituting (2) into (3) we obtain

$$\frac{\partial E_\perp}{\partial z} = -jk\xi\left(\vec{e} + \frac{1}{k^2}\nabla_\perp\nabla_\perp\right) \cdot \left(\vec{H}_\perp \times \hat{e}_z\right),$$

$$\frac{\partial H_\perp}{\partial z} = -jk\eta\left(\vec{e} + \frac{1}{k^2}\nabla_\perp\nabla_\perp\right) \cdot \left(\hat{e}_z \times \vec{E}_\perp\right).$$

$$(4)$$

Vector notation is employed with the following meanings for the symbols:

$E_\perp = E_\perp(x, y, z) =$ the rms electric field intensity transverse to the z-axis.

$H_\perp = H_\perp(x, y, z) =$ the rms magnetic field intensity transverse to the z-axis.

$\eta =$ intrinsic impedance of the medium $1/\eta = \sqrt{\mu/\epsilon}$.

$k = \omega\sqrt{\mu\epsilon} = 2\pi/\lambda =$ propagation constant in medium or the wavenumber (see [28] page 3).

$\nabla_\perp =$ gradient operator transverse to z-axis $= \nabla - \hat{e}_z(\partial/\partial z)$.

$\vec{e} =$ unit dyadic defined such that $\vec{e} \cdot \vec{A} = \vec{A} \cdot \vec{e} = \vec{A}$.

Equations (4) and (2), which are fully equivalent to the *Maxwell equations*, make evident the separate dependence of the field on the cross-sectional coordinates and on the longitudinal coordinate z. The cross-sectional dependence may be integrated out of (4) by means of a suitable set of vector orthogonal functions provided they satisfy appropriate conditions on the boundary curve or curves s of the cross section.

Such vector functions are known to be of two types: the *E-mode* functions e'_i defined by

$$e'_i = -\nabla_\perp\Phi_i,$$

$$h'_i = \hat{e}_z \times e'_i,$$

$$(5)$$

where

$$\nabla_\perp^2\Phi_i + k'^2_{ci}\Phi_i = 0,$$

$$\Phi_i = 0 \quad \text{on } s \quad \text{if } k'_{ci} \neq 0,$$

$$\frac{\partial\Phi_i}{\partial s} = 0 \quad \text{on } s \quad \text{if } k'_{ci} = 0,$$

$$(6)$$

and the *H-mode* functions e''_i defined by

$$e''_i = \hat{e}_z \times \nabla_\perp\Psi_i,$$

$$h''_i = \hat{e}_z \times e''_i,$$

$$(7)$$

where

$$\nabla_\perp^2\Psi_i + k''^2_{ci}\Psi_i = 0,$$

$$\frac{\partial\Psi_i}{\partial n} = 0 \quad \text{on } s,$$

$$(8)$$

where i denotes a double index and $\vec{n}$ is the outward normal to s in the cross-section plane.

The constants k''_{ci} and k'_{ci} are defined as the cutoff wave numbers or eigenvalues associated with the guide cross section.

The functions e_i possess the vector orthogonality properties

$$\iint e'_i \cdot e'_j dS_\perp = \iint e''_i \cdot e''_j dS_\perp = \begin{cases} 1 & \text{for } i = j \\ 0 & \text{for } i \neq j, \end{cases}$$

$$\iint e'_i \cdot e''_j dS_\perp = 0,$$

$$(9)$$

with the integration extended over the entire guide cross section with surface $S_\perp$.

The total average power flow along the guide in the z direction is

$$P_z = \frac{1}{2} \text{Re} \left(\iint E_\perp \times H_\perp^* \cdot \hat{e}_z dS_\perp \right), \tag{10}$$

where all quantities are rms and the asterisk denotes the complex conjugate.

In TEM modes, both E_z and H_z vanish, and the fields are fully transverse. Their cutoff condition $k_c^2 = 0$ or $\omega = \beta c$ (where c is the speed of the light) is equivalent to the following relation [28]:

$$\vec{H}_\perp = \frac{1}{\eta} \hat{e}_z \times \vec{E}_\perp, \tag{11}$$

between the electric and magnetic transverse fields, where $\eta = \sqrt{\mu/\epsilon}$ is the medium impedance so that $\eta/c = \mu$ and $\eta c = 1/\epsilon$.

The electric field $\vec{E}_\perp$ is determined from the rest of *Maxwell's* equations which read

$$\nabla_\perp \times \vec{E}_\perp = 0,$$
$$\nabla \cdot \vec{E}_\perp = 0. \tag{12}$$

These are recognized as the field equations of an equivalent two-dimensional electrostatic problem.

Once the electrostatic solution $\vec{E}_\perp$ is found, the magnetic field is constructed from (11).

Because of the relationship between $\vec{E}_\perp$ and $\vec{H}_\perp$, the Poynting vector S_z will be

$$S_z = \frac{1}{2} \text{Re} \left(\vec{E}_\perp \times \vec{H}_\perp^* \right) \cdot \hat{e}_z = \frac{1}{\eta} \left| \vec{E}_\perp \right|^2 = \eta \left| \vec{H}_\perp \right|^2. \tag{13}$$

2. The Spiral Differential Geometry

For the MSCC structures it is difficult to construct solutions for *Laplace's equation* with polar or cartesian coordinates.

The conformal mapping technique is a powerful method for solving two-dimensional potential problems and mapping the boundaries into a simpler configuration for which solutions to *Laplace's equation* are easily found [17, 18].

For the specific purposes of the MSCC, the following spiral coordinates based on a generalization of the *Schwarz-Christoffel* mapping (see appendix) are introduced:

$$x = e^{(\delta/g - g\theta)} \cos(\delta + \theta),$$
$$y = e^{(\delta/g - g\theta)} \sin(\delta + \theta), \tag{14}$$
$$z = z,$$

where θ, δ represent the spiral coordinates and $g > 0$ is a constant which characterizes the transformation (see appendix).

As it can be seen in Figure 1, the equation $\delta = \text{const.}$ represented by a vertical line in the δ-θ plane corresponds to a logarithmic spiral into the x-y plane and a constant coordinate line of the spiral mapping.

Observing (14) it appears clear that for $g\theta - \delta^*/g \to 0$, where δ^* is a constant, the curve in the x-y plane locally reduces (for $|\theta| \ll 1$, $g \ll 1$) to an *Archimedean spiral*.

The region between the two coaxial spirals maps into the region inside the polygon bounded by the coordinate-lines $\theta = \theta_1, \theta = \theta_2$ and $\delta = \delta_1, \delta = \delta_2, \delta = \delta_1 - 2\pi g^2/(1 + g^2)$ [18] (see Figure 1(b)).

It is also worth to observe that, if $\delta_2 - \delta_1 = 2q\pi g^2/(1 + g^2)$, $q \in \mathbb{Z}$, the two spirals $\delta = \delta_1$, $\delta = \delta_2$ are identical apart from a shift of $\Delta\theta_s = 2q\pi/(1 + g^2)$.

We then require $|\delta_2 - \delta_1| < 2\pi g^2/(1 + g^2)$ in order to avoid cyclic spirals with $\delta > \delta_2$ in the middle of the two with $\delta = \delta_1$ and $\delta = \delta_2$.

The differential one form of the spiral transformation or dual basis results in

$$dx = \frac{\partial x}{\partial \delta} d\delta + \frac{\partial x}{\partial \theta} d\theta + \frac{\partial x}{\partial z} dz,$$
$$dy = \frac{\partial y}{\partial \delta} d\delta + \frac{\partial y}{\partial \theta} d\theta + \frac{\partial y}{\partial z} dz, \tag{15}$$
$$dz = dz.$$

The arc length $d\ell$ is given by

$$d\ell^2 = dx^2 + dy^2 + dz^2$$
$$= g_{\delta\delta} d\delta^2 + g_{\theta\theta} d\theta^2 + g_{zz} dz^2, \tag{16}$$

where

$$h_\delta^2 = g_{\delta\delta} = e^{(2(\delta/g) - 2g\theta)} \left(1 + \frac{1}{g^2} \right),$$
$$h_\theta^2 = g_{\theta\theta} = e^{(2(\delta/g) - 2g\theta)} \left(1 + g^2 \right), \tag{17}$$
$$h_z^2 = g_{zz} = 1, \qquad g_{\delta\theta} = g_{\delta z} = g_{\theta z} = 0$$

are the components of the *metric tensor* and *Lamè coefficients*.

The infinitesimal volume element is given by

$$dV = \|J\| d\delta\, d\theta\, dz = e^{(2(\delta/g) - 2g\theta)} \frac{1 + g^2}{g} d\delta\, d\theta\, dz, \tag{18}$$

where the J is the *Jacobian* of the spiral transformation.

Let us now define the spiral natural basis vectors $\bar{e}_\delta, \bar{e}_\theta, \bar{e}_z$:

$$\bar{e}_\delta = \frac{\partial \vec{x}}{\partial \delta}$$
$$= e^{(\delta/g - g\theta)} \left(\frac{1}{g} \cos(\delta + \theta) - \sin(\delta + \theta) \right) \bar{e}_x$$
$$+ e^{(\delta/g - g\theta)} \left(\frac{1}{g} \sin(\delta + \theta) + \cos(\delta + \theta) \right) \bar{e}_y,$$

(a)

(b)

(c)

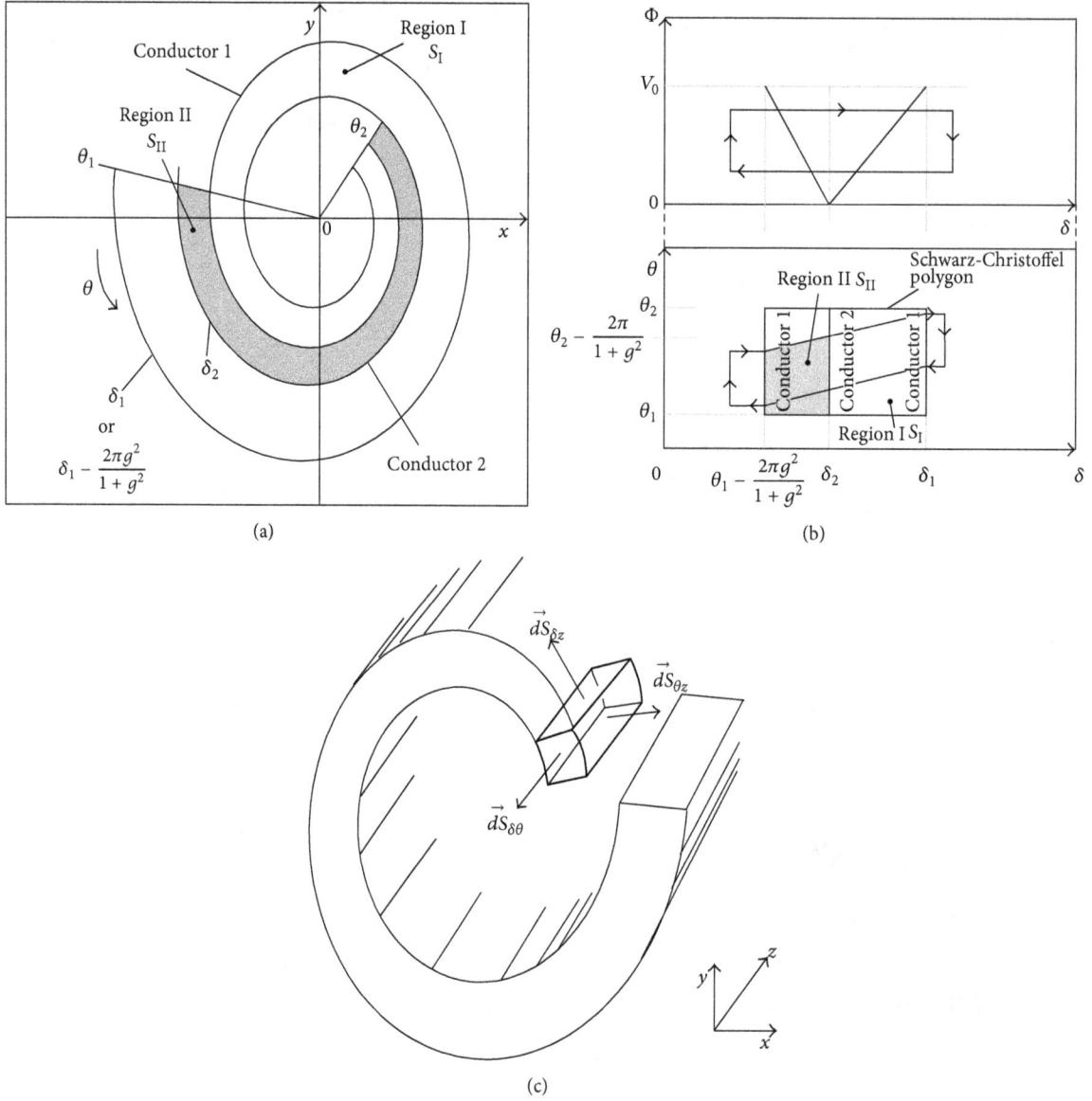

FIGURE 1: (a) The spiral coordinates lines. (b) The mapping of the spiral coaxial section and the scalar potential $\Phi(\delta, \theta)$, solution to the equivalent Laplace's equation. (c) The differential spiral surfaces.

$$\bar{e}_\theta = \frac{\partial \vec{x}}{\partial \theta}$$

$$= e^{(\delta/g - g\theta)} \left(-g\cos\left(\delta + \theta\right) - \sin\left(\delta + \theta\right) \right) \bar{e}_x$$

$$+ e^{(\delta/g - g\theta)} \left(-g\sin\left(\delta + \theta\right) + \cos\left(\delta + \theta\right) \right) \bar{e}_y,$$

$$\bar{e}_z = \frac{\partial \vec{x}}{\partial z} = \bar{e}_z \tag{19}$$

in terms of the cartesian basis vectors $\bar{e}_x, \bar{e}_y, \bar{e}_z$.

The infinitesimal surface elements transverse and longitudinal along the z-axis (see Figure 1) are given by

$$\left\| d\vec{S}_{\delta\theta} \right\| = dS_\perp = \left\| \frac{\partial \vec{x}}{\partial \delta} \times \frac{\partial \vec{x}}{\partial \theta} \right\| d\delta\, d\theta$$

$$= e^{(2\delta/g - 2g\theta)} \left(\frac{1}{g} + g \right) d\delta\, d\theta,$$

$$\left\| d\vec{S}_{\theta z} \right\| = \left\| \frac{\partial \vec{x}}{\partial \theta} \times \frac{\partial \vec{x}}{\partial z} \right\| d\theta\, dz = e^{(\delta/g - g\theta)} \sqrt{1 + g^2}\, dz\, d\theta, \tag{20}$$

$$\left\| d\vec{S}_{\delta z} \right\| = \left\| \frac{\partial \vec{x}}{\partial \delta} \times \frac{\partial \vec{x}}{\partial z} \right\| d\delta\, dz = e^{(\delta/g - g\theta)} \frac{\sqrt{1 + g^2}}{g}\, dz\, d\delta.$$

We then define the natural unitary spiral basis vectors

$$\hat{e}_\delta = \frac{\bar{e}_\delta}{h_\delta}, \qquad \hat{e}_\theta = \frac{\bar{e}_\theta}{h_\theta}, \qquad \hat{e}_z = \frac{\bar{e}_z}{h_z}. \tag{21}$$

The usual unitary relations of orthogonality hold; that is,

$$\hat{e}_\delta = \hat{e}_\theta \times \hat{e}_z, \qquad \hat{e}_\theta = \hat{e}_z \times \hat{e}_\delta, \qquad \hat{e}_z = \hat{e}_\delta \times \hat{e}_\theta,$$

$$\hat{e}_\delta \cdot \hat{e}_\theta = 0 = \hat{e}_\delta \cdot \hat{e}_z = \hat{e}_\theta \cdot \hat{e}_z = 0. \tag{22}$$

In Figure 1 a vertical segment in the θ-δ plane corresponds to a piece of spiral in the x-y plane; the circle is a particular spiral defined by the relation $\theta = \delta/g^2 - K/g$.

The radius vector in spiral coordinates becomes

$$\vec{r} = \frac{e^{(\delta/g - g\theta)}}{\sqrt{1 + g^2}} (\hat{e}_\delta - g\hat{e}_\theta) + z\hat{e}_z. \tag{23}$$

Logarithmic spirals are analogous to the straight line. The orthogonal spiral is obtained, exactly as for the straight lines by replacing the g factor (which is analogous to the slope for the straight lines) with $g_\perp = -1/g$.

It is also possible to define the orthogonal spiral coordinate mapping as follows:

$$x = e^{(-g\delta + \theta/g)} \cos(\delta + \theta),$$

$$y = e^{(-g\delta + \theta/g)} \sin(\delta + \theta), \tag{24}$$

$$z = z.$$

3. The TEM Mode for the Spiral Waveguide

Let us consider two separate perfectly conducting spiral conductors with uniform cross section, infinitely long and oriented parallel to the z-axis; for such a structure a TEM mode of propagation is possible [18].

Laplace's equation of this line transformed by means of a spiral conformal mapping [17, 18], which is the generalization of the polar conformal mapping (see appendix), is

$$\frac{e^{-2(\delta/g) + 2g\theta}}{1 + g^2} \left[g^2 \frac{\partial^2 \Phi}{\partial \delta^2} + \frac{\partial^2 \Phi}{\partial \theta^2} \right] = 0, \tag{25}$$

where the scalar electric potential $\Phi(\delta, \theta)$ represents the solution to the equivalent electrostatic problem of the transverse electromagnetic TEM mode propagating along the MSCC.

This equation has to be solved into two separate independent open regions I, II where the solution must be continuous with derivatives:

$$\Phi \in \mathscr{C}^{(0)} \left[\left[\delta_1 - \frac{2\pi g^2}{1 + g^2}, \delta_2 \right] \times (-\infty, \infty) \right]$$

$$\cap \mathscr{C}^{(2)} \left[\left[\delta_1 - \frac{2\pi g^2}{1 + g^2}, \delta_2 \right] \times (-\infty, \infty) \right], \tag{26}$$

$$\Phi \in \mathscr{C}^{(0)} \left[[\delta_2, \delta_1] \times (-\infty, \infty) \right]$$

$$\cap \mathscr{C}^{(2)} \left[[\delta_2, \delta_1] \times (-\infty, \infty) \right].$$

The derivative of the electric potential represents the electric and the magnetic fields whose values are not continuous at the two spiral metal boundary walls. In Figure 3(a) MDSCC partially composed of two infinite ideal spiral conductors filled with dielectric material having a permittivity $\epsilon = \epsilon_0 \epsilon_r$ is shown. The MDSCC has much in common with the parallel plate line [17]; the two spiral conductors are considered infinitely wide ($\theta \in [-\infty, \infty]$) and separated by $\Delta\delta = 2\pi g^2/(1 + g^2)$.

The potential $\Phi(\delta, \theta)$ is subject to the following boundary conditions in the region I (see Figure 1)

$$\Phi(\delta_1, \theta) = V_0,$$

$$\Phi(\delta_2, \theta) = 0 \quad \forall \theta \in (-\infty, \infty) \tag{27}$$

and in the region II

$$\Phi(\delta_2, \theta) = 0,$$

$$\Phi\left(\delta_1 - \frac{2\pi g^2}{1 + g^2}, \theta\right) = V_0 \quad \forall \theta \in (-\infty, \infty). \tag{28}$$

V_0 must be the same in both cases of (27) and (28) because $\delta = \delta_1$ and $\delta = \delta_1 - 2\pi g^2/(1 + g^2)$ correspond to the same conductor (see Figure 1(b); cyclic spiral) and the potential must be continuous at the spiral metal walls.

By the method of separation of variable, let $\Phi(\delta, \theta)$ be expressed in product form as

$$\Phi(\delta, \theta) = R(\delta) P(\theta). \tag{29}$$

Substituting (29) into (25) and dividing by RP give

$$\frac{g^2}{R(\delta)} \frac{\partial^2 R(\delta)}{\partial \delta^2} + \frac{1}{P(\theta)} \frac{\partial^2 P(\theta)}{\partial \theta^2} = 0. \tag{30}$$

The two terms in (30) must be equal to constants, so that

$$\frac{g^2}{R(\delta)} \frac{\partial^2 R(\delta)}{\partial \delta^2} = -k_\delta^2, \tag{31}$$

$$\frac{1}{P(\theta)} \frac{\partial^2 P(\theta)}{\partial \theta^2} = -k_\theta^2, \tag{32}$$

$$k_\delta^2 + k_\theta^2 = 0. \tag{33}$$

The general solution to (32) is

$$P(\theta) = A \cos(k_\theta \theta) + B \sin(k_\theta \theta). \tag{34}$$

Now, because the boundary conditions (27), (28) do not vary with θ, the potential $\Phi(\delta, \theta)$ should not vary with θ. Thus, k_θ must be zero. By (33), this implies that k_δ must also be zero, so that (31) for $R(\delta)$ reduces to

$$\frac{\partial^2 R(\delta)}{\partial \delta^2} = 0, \tag{35}$$

and so

$$\Phi(\delta, \theta) = C\delta + D. \tag{36}$$

The equivalent electrostatic problem in the plane (δ, θ) is the problem of finding the potential distribution between two plates [18].

Applying the boundary conditions of (27) to (36) gives two equations for the constants C and D in the region I:

$$\Phi(\delta_1, \theta) = 0 = C_I \delta_1 + D_I,$$
$$\Phi(\delta_2, \theta) = V_0 = C_I \delta_2 + D_I. \tag{37}$$

At the same time the boundary conditions of (28) into (36) give two equations for the constants C and D in the region II:

$$\Phi(\delta_2, \theta) = V_0 = C_{II} \delta_2 + D_{II},$$
$$\Phi\left(\delta_1 - \frac{2\pi g^2}{1+g^2}, \theta\right) = 0 = C_{II}\left(\delta_1 - \frac{2\pi g^2}{1+g^2}\right) + D_{II}. \tag{38}$$

After solving for $C_{I,II}$ and $D_{I,II}$, we can write the final solution for $\Phi(\delta, \theta)$:

$$\Phi(\delta, \theta) = \frac{V_0}{\delta_2 - \delta_1}(\delta - \delta_1)$$

region I $\theta \in [-\infty, \infty], \quad \delta \in [\delta_2, \delta_1],$

$$\Phi(\delta, \theta) = \frac{V_0}{\delta_2 - \delta_1 + 2\pi g^2/(1+g^2)}\left(\delta - \delta_1 + \frac{2\pi g^2}{1+g^2}\right)$$

region II $\theta \in [-\infty, \infty], \quad \delta \in \left[\delta_1 - \frac{2\pi g^2}{1+g^2}, \delta_2\right].$ (39)

The $\vec{E}$ and $\vec{H}$ fields can now be found using (5) and (39):

region I $\begin{cases} \vec{E}_\perp = E_\delta \hat{e}_\delta = -\nabla_\perp \Phi = -\frac{e^{(-\delta/g + g\theta)}}{\sqrt{1+g^2}}\frac{gV_0}{\delta_2 - \delta_1}\hat{e}_\delta, \\ E_\theta = 0, \\ \vec{H}_\perp = H_\theta \hat{e}_\theta = \frac{1}{\eta}\hat{e}_z \times E_\delta \hat{e}_\delta \\ \qquad = \frac{ge^{(-\delta/g+g\theta)}}{\eta\sqrt{1+g^2}}\frac{V_0}{\delta_2-\delta_1}\hat{e}_\theta, \\ H_\delta = 0. \end{cases}$

region II $\begin{cases} \vec{E}_\perp = E_\delta \hat{e}_\delta = -\nabla_\perp \Phi \\ \qquad = -\frac{ge^{(-\delta/g+g\theta)}}{\sqrt{1+g^2}}\frac{V_0}{\delta_2-\delta_1+2\pi g^2/(1+g^2)}\hat{e}_\delta, \\ E_\theta = 0, \\ \vec{H}_\perp = H_\theta \hat{e}_\theta = \frac{1}{\eta}\hat{e}_z \times E_\delta \hat{e}_\delta \\ \qquad = \frac{ge^{(-\delta/g+g\theta)}}{\eta\sqrt{1+g^2}}\frac{V_0}{\delta_2-\delta_1+2\pi g^2/(1+g^2)}\hat{e}_\theta, \\ H_\delta = 0. \end{cases}$ (40)

While the electric and the magnetic fields together with the surface charge and current densities vary exponentially with the spiral coordinates (δ, θ), the potential remains constant on the two conductors.

The field distribution for the TEM mode in the MSCC depicted in Figure 2 is obtained by using (40) and the quiver-MATLAB function.

As stated by the *Gauss law* [30] the whole surface density σ of charge on each of the two spiral conductors, due to the discontinuity of the electric field, is

$$\sigma(\theta) = \epsilon \vec{E}_{II} \cdot \vec{n} - \epsilon \vec{E}_I \cdot \vec{n}, \tag{41}$$

where $\vec{n} \equiv \hat{e}_\delta$ is the normal to the spiral surface of the conductors whilst $\vec{E}_I$ and $\vec{E}_{II}$ are the electric fields seen from the regions I and II, respectively.

According to (41) the electric charge distribution follows the exponential electric field.

The two spiral metal conductors are in a parallel configuration; they have the same potential difference but two different capacities and two different surface charge distributions.

At the same time the total displacement current [30] due to the discontinuity of the magnetic fields at the two conductors is

$$\vec{J}_{S_{tot}} = \vec{n} \times \vec{H}_I - \vec{n} \times \vec{H}_{II}. \tag{42}$$

The time-average stored electric energy per unit length [2, 17] in the MDSCC (see Figure 3) is

$$W_e = \frac{1}{2}\int_{S_\perp} \epsilon' \vec{E} \cdot \vec{E}^* dS_\perp, \tag{43}$$

while circuit theory gives $W_e = C'|V_0|^2/4$, resulting in the following expression for the capacitance per unit length:

$$C' = \frac{\epsilon'}{|V_0|^2}\int S_\perp \vec{E} \cdot \vec{E}^* dS_\perp, \quad [\text{F/m}]. \tag{44}$$

As in the case of the parallel plate waveguide, the MSCC is composed of finite strips.

The electric field lines at the edge of the finite spiral conductors are not perfect spirals, and the field is not entirely contained between the conductors.

The azimuthal length in real multiturn MDSCC is assumed to be much greater than the separation between the conductors ($|\theta_1 - \theta_2| \gg \Delta\delta$) with $|\theta_1|, |\theta_2|$ not too high as in the case of the myelin bundles, so that the fringing fields can be ignored [2].

Furthermore, the minimum distance between the two spiral conducting strips is chosen in such a way to avoid the dielectric voltage breakdown.

Although the MDSCC line is modeled with two capacitors, it is composed by two and not three conductors as it would be in the case of the parallel plates.

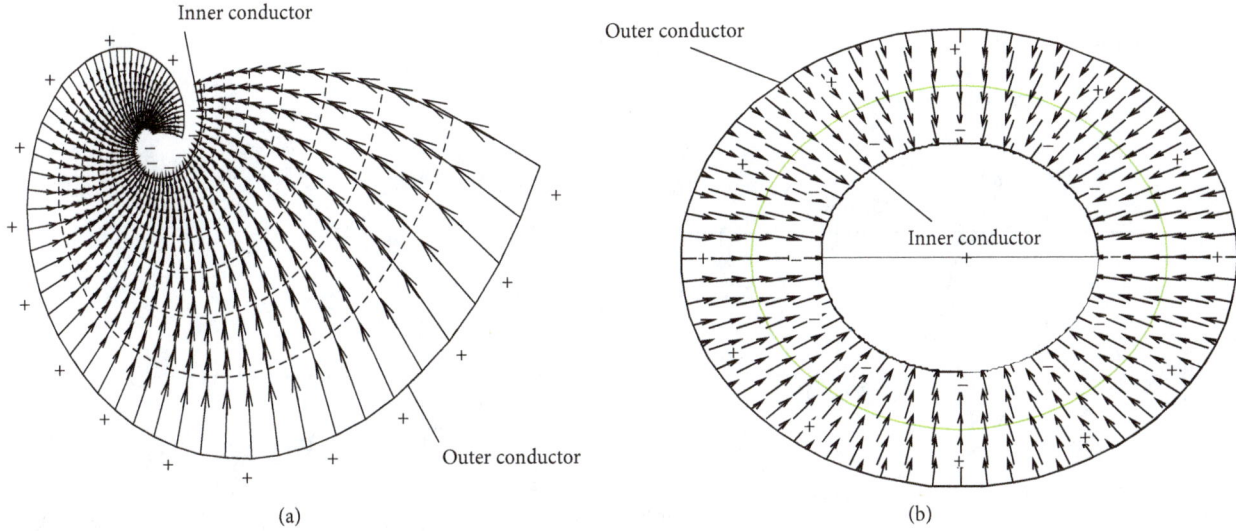

FIGURE 2: Field distribution for the TEM mode in the (a) MSCC, (b) cylindrical coax, obtained using the quiver-MATLAB function (simulations on Pentium 4, 3.2 Ghz, average CPU time: 4 min).

The two capacitors are different because their spiral dimensions are different; consequently the two capacitances are determined by

$$C_1' = \frac{\epsilon'}{|V_0|^2} \int_{S_{\mathrm{I}}} \vec{E}_{\mathrm{I}} \cdot \vec{E}_{\mathrm{I}}^* dS_\perp = \frac{g\epsilon'(\theta_2 - \theta_1)}{\delta_1 - \delta_2},$$

$$C_2' = \frac{\epsilon'}{|V_0|^2} \int_{S_{\mathrm{II}}} \vec{E}_{\mathrm{II}} \cdot \vec{E}_{\mathrm{II}}^* dS_\perp \qquad (45)$$

$$= \frac{g\epsilon'\left(\theta_2 - \theta_1 - 2\pi/\left(1+g^2\right)\right)}{\left(\delta_2 - \delta_1 + 2\pi g^2/\left(1+g^2\right)\right)}, \quad \theta_2 > \theta_1.$$

Thus,

$$C_{\mathrm{tot}}' = C_1' + C_2'$$

$$= \epsilon g W \left(\frac{\theta_2 - \theta_1}{\delta_1 - \delta_2} + \frac{\theta_2 - \theta_1 - 2\pi/\left(1+g^2\right)}{\delta_2 - \delta_1 + 2\pi g^2/\left(1+g^2\right)}\right). \qquad (46)$$

This value represents the capacitance $C_{\mathrm{tot}}' = C_{\mathrm{tot}}/W$ (e.g., farads/meter) per unit length of the spiral coaxial line with finite azimuthal dimension $\theta_1 - \theta_2$ for the first greater capacitor and $\theta_2 - \theta_1 - 2\pi/(1+g^2)$ for the smaller one (see Figure 1).

If the number of spiral turns become high enough, the difference in terms of θ between the two capacitors will be negligible.

In order to determine the inductance L' per unit length of the MDSCC, we observe that the magnetic field is orthogonal to the electric field.

The magnetic fluxes over the two infinitesimal areas $dS_{\mathrm{I}z\delta}$ and $dS_{\mathrm{II}z\delta}$ are (see Figure 4) $d\Phi_{\mathrm{I,II}} = \vec{B}_{\mathrm{I,II}_\perp} \cdot d\vec{S}_{\mathrm{I,II}z\delta}$, while the

total fluxes over the two spiral areas S_{I}, S_{II}, according to (40), are

$$\Phi_{\mathrm{I}} = \Phi_{\mathrm{II}} = WV_0\frac{\mu}{\eta}. \qquad (47)$$

The fluxes per unit length are given by

$$\Phi_{\mathrm{I,II}}' = \frac{\Phi_{\mathrm{I,II}}}{W} = L'I_0. \qquad (48)$$

Consequently

$$L' = Z_0\frac{\mu}{\eta}, \qquad (49)$$

where Z_0 and I_0 are the impedances and current of the line, respectively.

As it can be noted from (48) there is only one current I_0 flowing along the spiral coaxial cable.

The time-average stored magnetic energy for unit length (at low frequencies for nondispersive media) of the MDSCC can be written as [2, 17]

$$W_m = \frac{\mu}{2}\int_{S_\perp} \vec{H} \cdot \vec{H}^* dS_\perp. \qquad (50)$$

Circuit theory gives $W_m = LI_0^2/4$ in terms of the unique current of the line I_0 and results from the sum of two contributions $W_m = W_1 + W_2$.

Thus,

$$L' = \frac{\mu Z_0^2}{V_0^2}\left(\int_{S_{\mathrm{I}}} \vec{H} \cdot \vec{H}^* dS_\perp + \int_{S_{\mathrm{II}}} \vec{H} \cdot \vec{H}^* dS_\perp\right). \qquad (51)$$

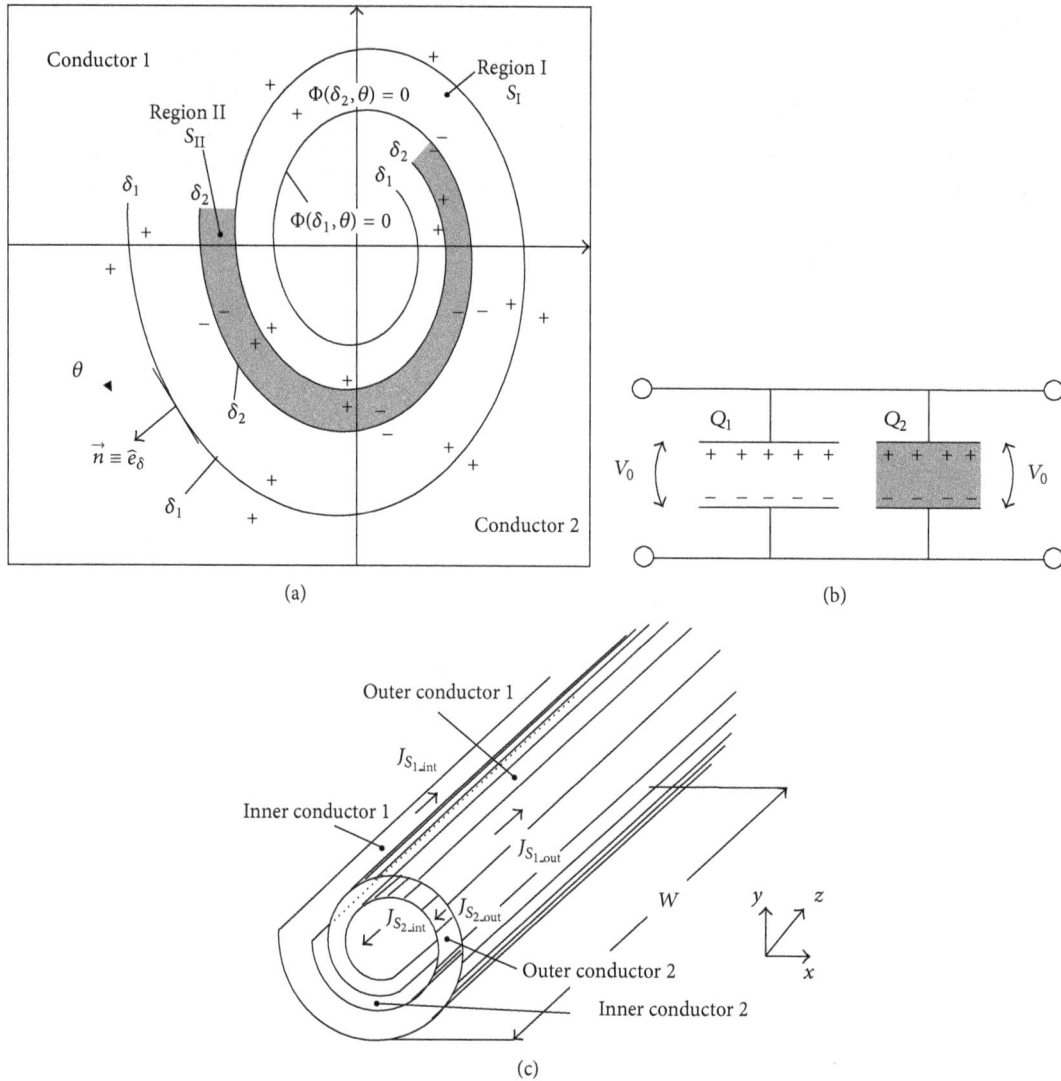

FIGURE 3: (a) Charge distributions in the electrostatic MDSCC section. (b) Parallel capacitors scheme of the electrostatic MDSCC. (c) Current distributions in the MDSCC.

Substituting (40) into (51), by considering the superposition of the two lines and using (49), gives

$$L' = \frac{\mu}{g} \cdot \left(\frac{\theta_2 - \theta_1}{\delta_1 - \delta_2} \right.$$

$$\left. + \frac{\theta_2 - \theta_1 - 2\pi / \left(1 + g^2\right)}{\delta_2 - \delta_1 + 2\pi g^2 / \left(1 + g^2\right)} \right)^{-1},$$

$$Z_0 = \frac{\eta}{g} \cdot \left(\frac{\theta_2 - \theta_1}{\delta_1 - \delta_2} \right.$$

$$\left. + \frac{\theta_2 - \theta_1 - 2\pi/(1 + g^2)}{\delta_2 - \delta_1 + 2\pi g^2/(1 + g^2)} \right)^{-1}.$$

According to the classical electromagnetism (see, e.g., [16] page 563), a periodic wave incident upon a material body gives rise to a forced oscillation of free and bound charges synchronous with the applied field, producing a secondary field both inside and outside the body; the transmitted and reflected waves have the chance to excite propagating eigenmodes solutions to *Maxwell's* equations.

From the physical point of view, the light that passes through the entrance of the spiral waveguide is subject to multiple reflections. The historical work of Mie [31, 32] for the case of the spherical topology will be the reference starting point for the analysis of the light that passes through the open MSCC section and it is scattered by the spiral surface.

Localized surface plasmon polaritons (LSPP) [15] existing on a good metal surface can be excited, propagated, and scattered on the spiral lines. The enhancement of the electromagnetic field at the metal dielectric spiral interface could be responsible for surface-enhanced optical phenomena such as Raman scattering, fluorescence, and second harmonic generation (SHG) [33].

Nevertheless, the continuity of the tangential components of the magnetic and electric fields on each spiral

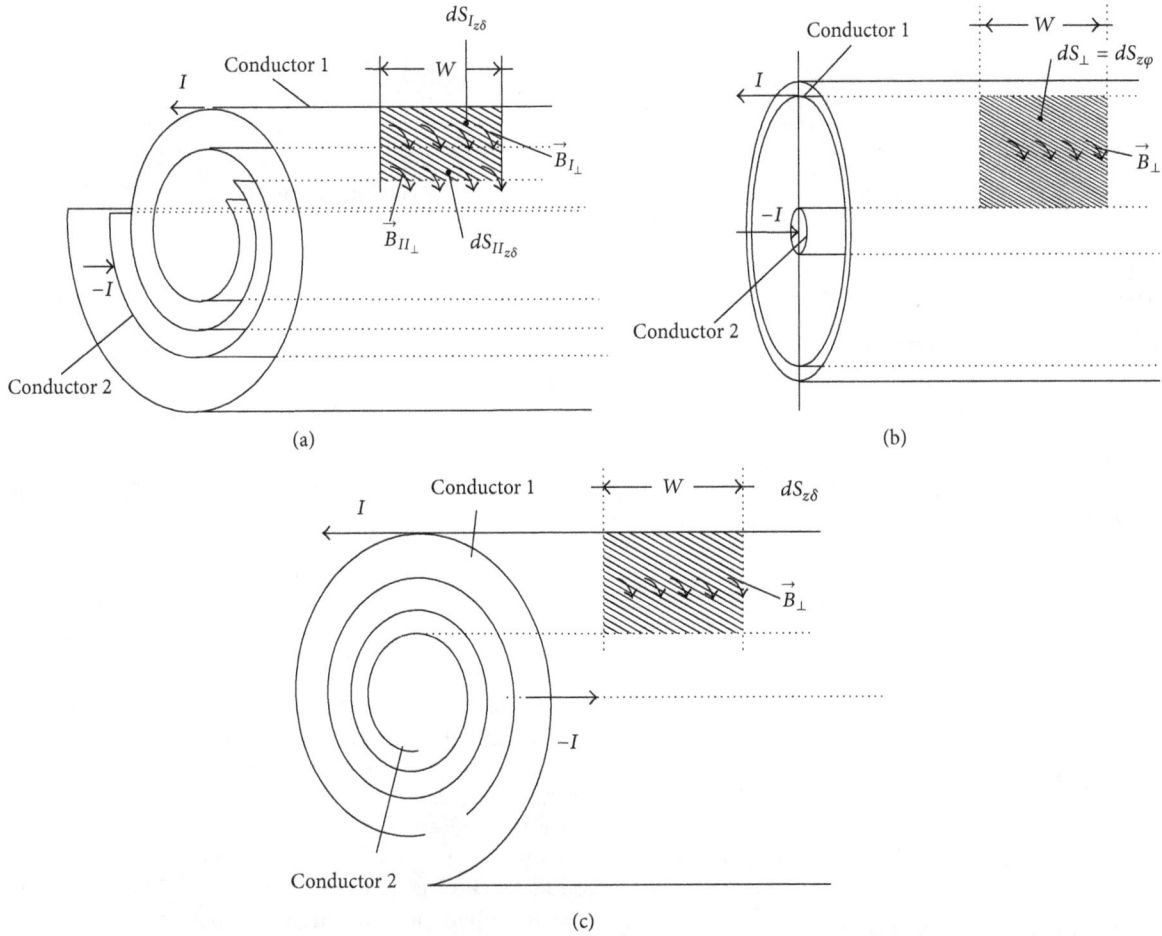

FIGURE 4: Surfaces for calculation of external inductances of (a) MDSCC, (b) cylindrical coaxial line [4], and (c) MSSCC.

metal-dielectric interface, which is essential in order to propagate the polaritons along the line [15] and includes the specific frequency-dependent dielectric constant of metals (real and imaginary parts), needs specific simulation methods [11] and dedicated mathematical analysis.

All these electromagnetic effects, which require advanced numerical techniques, validations, and comparisons in terms of CPU time, involve all the modes that pass through the waveguide. In spite of the interesting results and applications that these analyses could bring to the future of the spiral coaxial cables, their study is beyond the scope of this paper.

4. The Spiral Transmission Line

A transmission line consists of two or more conductors [2, 4, 17]. In this paper we consider two types of spiral transmission lines; their elements of line of infinitesimal length dz depicted in Figure 5 can be modeled as lumped-element circuits.

Although the MDSCC line is modeled with two capacitors, it is composed by two conductors with only one real capacitor. The series resistance R' per unit length represents the resistance due to the finite conductivity of the individual conductors, and the shunt conductance G' per unit length is due to dielectric loss in the material between the conductors.

For lossless lines, the three quantities Z, L', and C' are related as follows:

$$L' = \mu \frac{Z}{\eta},$$
$$C' = \epsilon \frac{\eta}{Z}, \tag{53}$$

where $\eta = \sqrt{\mu/\epsilon}$ is the characteristic impedance of the dielectric medium between the conductors.

The equations of the ideal spiral transmission line [4] depicted in Figure 5 are

$$\frac{\partial V}{\partial z} = -L' \frac{\partial I}{\partial t} - R'I,$$
$$\frac{\partial I}{\partial z} = -C' \frac{\partial V}{\partial t} - G'V, \tag{54}$$

where R' is the resistance per unit length of the line, expressed in [Ω/m], and G' is the conductance per unit length of the line, measured in [S/m].

The two equations (54) for $R' = 0$ and $G' = 0$ can be combined to form D'Alambert's wave equation for either

(a)   (b)

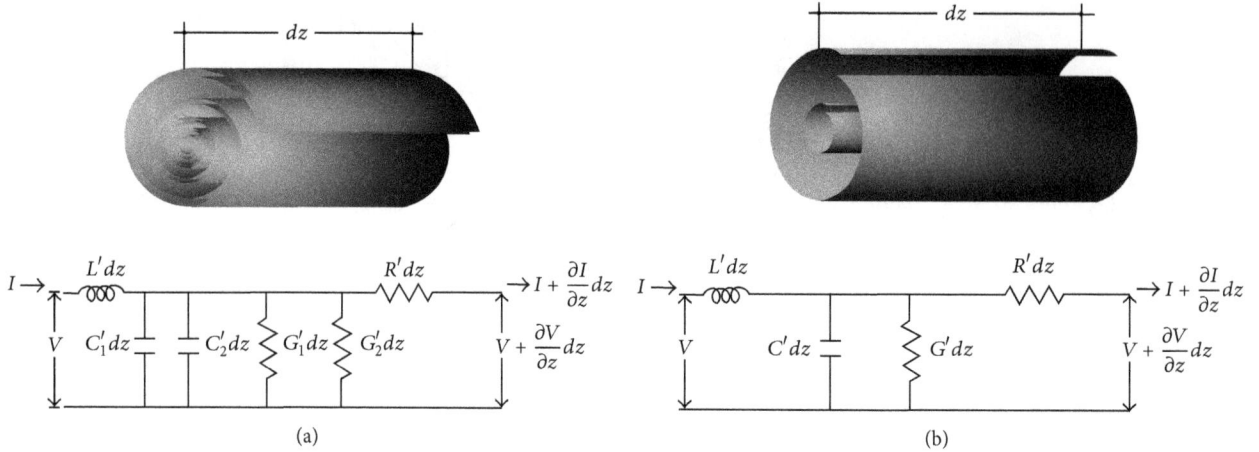

FIGURE 5: Element dz (a) MDSCC, (b) MSSCC, and their lumped-element equivalent circuits obtained using M-file with camlight programming tools (run on Pentium 4, 3.2 Ghz, average CPU time: 8 min).

variables [2], whose solutions are waves propagating along the ideal line with speed v:

$$\frac{\partial^2 V}{\partial z^2} = \frac{1}{v}\frac{\partial^2 V}{\partial t^2},$$

$$\frac{\partial^2 I}{\partial z^2} = \frac{1}{v}\frac{\partial^2 I}{\partial t^2}, \quad v = \frac{1}{\sqrt{L'C'}}. \tag{55}$$

Using the Fourier transform of the signals V, I

$$V(\omega) = \frac{1}{2\pi}\int_{-\infty}^{\infty} V(t) e^{-i\omega t} dt,$$

$$I(\omega) = \frac{1}{2\pi}\int_{-\infty}^{\infty} I(t) e^{-i\omega t} dt. \tag{56}$$

The solution to (55) may be written in terms of exponentials

$$V = V_+ e^{-\gamma z} + V_- e^{\gamma z},$$

$$I = \frac{1}{Z_0}\left(V_+ e^{-\gamma z} - V_- e^{\gamma z}\right), \tag{57}$$

$$\gamma^2 = -\omega^2 L'C'.$$

If a sinusoidal voltage is supplied to MDSCC with load impedance Z_L at $z = 0$, the reflection Γ and transmission τ coefficients will be

$$\Gamma = \frac{V_-}{V_+} = \frac{Z_L - Z_0}{Z_L + Z_0},$$

$$\tau = \frac{V_L}{V_+} = \frac{2Z_L}{Z_L + Z_0}. \tag{58}$$

If the terminating impedance is exactly equal to the characteristic impedance of the line there is no reflected wave; the line is matched with the load. According to (49) the reflected and the transmitted waves of a spiral coaxial line depend on the number of turns $n = \text{Int}(\Delta\theta/2\pi)$, on the shift $\Delta\delta$ between the spiral walls, and on the spiral g factor.

5. Waves in a Lossy Spiral Coaxial Transmission Line

Conductors used in transmission lines have finite conductivity and exhibit series resistance R which increases with an increase in the frequency of operation [17] because of the skin effect. Furthermore the two conductors are separated by a dielectric medium which have a small amount of dielectric loss due to the polarization; consequently a small shunt conductance G is added to the circuit. Differentiating the lossy transmission equation (54) we obtain

$$\frac{\partial^2 V}{\partial z^2} = R'\left(G'V + C'\frac{\partial V}{\partial t}\right) + L'\left(C'\frac{\partial V}{\partial t} + C'\frac{\partial^2 V}{\partial t^2}\right),$$

$$\frac{\partial^2 I}{\partial z^2} = R'\left(G'I + C'\frac{\partial I}{\partial t}\right) + L'\left(C'\frac{\partial I}{\partial t} + C'\frac{\partial^2 I}{\partial t^2}\right). \tag{59}$$

By using the Fourier transform of the signals V, I we obtain

$$\gamma = \left[-\omega^2 L'C' + R'G' + i\omega\left(R'C' + L'G'\right)\right]^{1/2},$$

$$Z_0 = \left(\frac{R' + i\omega L'}{G' + i\omega L'}\right)^{1/2}. \tag{60}$$

For most transmission lines the losses are very small; that is, $R' \ll \omega L'$ and $G' \ll \omega C'$; a binomial expansion of γ then holds:

$$\gamma \simeq i\omega\sqrt{L'C'} + \frac{1}{2}\sqrt{L'C'}\left(\frac{R'}{L'} + \frac{G'}{C'}\right) = \alpha + i\beta. \tag{61}$$

Thus the phase constant β remains unchanged with respect to the ideal line.

The expressions of R' reported in Table 2 can be found from the expression of the power loss per unit length due to the finite conductivity of the two metallic spiral conductors [2]; that is,

$$P_c = \frac{R_S}{2}\int_{S_{\theta z}} \vec{J}_S \cdot \vec{J}_S^* dS_{\theta z}, \tag{62}$$

where the argument of the integral is the scalar product of the displacement currents [30] flowing along the surfaces of the conductors.

In (62), $R_s = 1/(\sigma\delta_s)$ is the surface resistance of the conductors, where the skin depth, or characteristic depth of penetration, is defined as $\delta_s = \sqrt{2/(\omega\mu\sigma)}$.

The material filling the space between the conductors is assumed to have a complex permittivity $\epsilon = \epsilon' - i\epsilon''$, a permeability $\mu = \mu_0\mu_r$, and a loss tangent $\tan(\delta_{\text{mat}}) = \epsilon''/\epsilon'$.

The shunt conductance per unit length G' reported in Table 2 can be inferred from the time-average power dissipated per unit length in a lossy dielectric; that is,

$$P_d = \frac{\omega\epsilon''}{2}\int_{S_{I_\perp}}\vec{E}\cdot\vec{E}^*dS_\perp + \frac{\omega\epsilon''}{2}\int_{S_{II_\perp}}\vec{E}\cdot\vec{E}^*dS_\perp. \quad (63)$$

The total voltage and current waves on the line can then be written as a superposition of an incident and a reflected wave

$$V = V_+\left(e^{-\gamma z} + \Gamma e^{\gamma z}\right),$$
$$I = \frac{V_+}{Z_0}\left(e^{-\gamma z} - \Gamma e^{\gamma z}\right). \quad (64)$$

The time-average power flow along the line at the point z is

$$P_{\text{avg}} = \frac{1}{2}\frac{\left|V_{0_+}\right|^2}{Z_0}\left(1 - |\Gamma|^2\right). \quad (65)$$

When the load is mismatched, not all of the available power from the generator is delivered to the load, the presence of a reflected wave leads to standing waves [2], and the magnitude of the voltage on the line is not constant.

The return loss (RL) is

$$\text{RL} = -20\log|\Gamma| \quad [\text{dB}]. \quad (66)$$

A measure of the mismatch of a line is the standing wave ratio (SWR)

$$\text{SWR} = \frac{1 + |\Gamma|}{1 - |\Gamma|}. \quad (67)$$

At a distance $z = -l$ from the load, the input impedance seen looking toward the load is

$$Z_{\text{in}} = Z_0\frac{Z_L + iZ_0\tan\gamma l}{Z_0 - iZ_0\tan\gamma l}. \quad (68)$$

The power delivered to the input of the terminated line at $z = -l$ is

$$P_{\text{in}} = \frac{1}{2}\frac{\left|V_{0_+}\right|^2}{Z_0}\left(e^{2\alpha l} - |\Gamma|^2 e^{2\alpha l}\right). \quad (69)$$

The difference $P_{\text{avg}} - P_{\text{in}}$ corresponds to the power lost in the line [2].

From (58) and (49) it appears clear that $|\Gamma|$, P_{avg}, RL, SWR, Z_{in}, and the power lost depend critically on the spiral factors of the line.

Particularly it is worth to point out that the g factor acts as a "control knob" of the electromagnetic propagation along the MDSCC.

6. Single Spiral Coaxial Cable and the Myelinated Nerves

The difficulty of using a single spiral surface to construct a coaxial line is due to the constraint of having the constant potential on the conductor.

The problem can be solved by using two independent stripes of the same single spiral surface with $|\theta_f - \theta_i| \le 2\pi$ and $|\theta_1|$, $|\theta_2|$ not too high, separated by a shift $\Delta\delta = 2n\pi g^2/(1+g^2)$ to form a system of two independent faced conductors with one grounded (as depicted in Figures 5(b) and 6(a)).

The metal single spiral coaxial cable (MSSCC) does not differ geometrically too much from the cylindrical coaxial design, especially for $g \ll 1$, but the first is an open framework whilst the second is a closed one.

Again, according to the conformal mapping theory [18], the equivalent electrostatic problem for the MSSCC in the plane (δ, θ) is just the problem of finding the potential distribution between two finite coordinate-plates like in the cylindrical case [18].

The potential $\Phi(\delta, \theta)$ for the TEM wave is now subject to the following boundary conditions:

$$\Phi(\delta_1, \theta) = 0 = C_m\delta_1 + D_m,$$

$$\Phi\left(\delta_1 + \frac{2n\pi g^2}{1+g^2}, \theta\right) = V_0 = C_m\left(\delta_1 + \frac{2n\pi g^2}{1+g^2}\right) + D_m, \quad (70)$$

$$\forall\theta \in \left[\theta_i, \theta_f\right], \quad \left|\theta_i - \theta_f\right| \le 2\pi.$$

Consequently the solution in (36) to Laplace's electrostatic equation (25) takes the form

$$\Phi(\delta, \theta) = V_0\frac{1+g^2}{2n\pi g^2}(\delta - \delta_1). \quad (71)$$

The electric and magnetic field for the MSSCC is simplified compared to the MDSCC; that is,

$$\vec{E}_\perp = E_\delta\hat{e}_\delta = -\nabla_\perp\Phi = \frac{e^{(-\delta/g+g\theta)}}{\sqrt{1+g^2}}\frac{gV_0\left(1+g^2\right)}{2n\pi g^2}\hat{e}_\delta,$$

$$E_\theta = 0,$$

$$\vec{H}_\perp = H_\theta\hat{e}_\theta = \frac{1}{\eta}\hat{e}_z \times E_\delta\hat{e}_\delta = -\frac{e^{(-\delta/g+g\theta)}}{\eta\sqrt{1+g^2}}\frac{V_0\left(1+g^2\right)}{2n\pi g}\hat{e}_\theta,$$

$$H_\delta = 0,$$

$$\forall\theta \in \left[\theta_{i_1}, \theta_{f_2}\right], \quad \delta \in \left[\delta_1, \delta_1 + \frac{2n\pi g^2}{1+g^2}\right]. \quad (72)$$

The total charge Q on the inner/outer conductors of MSSCC of length W is

$$Q = \int_{S_m}\sigma dS_{\theta z} = W\epsilon\frac{V_0\left(1+g^2\right)}{ng}. \quad (73)$$

TABLE 1: Values of capacitance for an average human myelinated nerve obtained with the SSCC and the cylindrical coax models.

Fibre diameter [D]	Axon diameter [d]	g_{mye}	ϵ_{mye}	Number of lamellae n_l	Core-conductor capacitance C_{mye} [34]	Single-coax capacitance C_{mye}	Cole's inductance L_{mye} [36]	Single-coax inductance L_{mye}
$\simeq 2\,\mu m$	$\simeq 1.4\,\mu m$	$\simeq 0.0009$	$\simeq 13$	$\simeq 16$	$\epsilon_0\epsilon_{\mathrm{mye}}\dfrac{2\pi}{\log(D/d)}$ $\simeq 4.6\dfrac{nF}{m}$	$\epsilon_0\epsilon_{\mathrm{mye}}\dfrac{1+g_{\mathrm{mye}}^2}{2n_l g_{\mathrm{mye}}}$ $\simeq 4\dfrac{nF}{m}$	$\dfrac{\mu_{\mathrm{mye}}}{2\pi}\log(D/d)$ $\simeq 30\dfrac{nHenry}{m}$	$\mu_{\mathrm{mye}} n_l \dfrac{g_{\mathrm{mye}}}{1+g_{\mathrm{mye}}^2}$ $\simeq 20\dfrac{nHenry}{m}$

Since the potential difference between the two conductors is $\Delta V = V_0$, the capacitance per unit length of the MSSCC with n turns between the two spiral conductors takes the following simplified form:

$$C' = \epsilon\frac{1+g^2}{ng}. \tag{74}$$

The myelin sheath in the "core-conductor" model is an electrically insulating phospholipid multilamellar spiral membrane surrounding the conducting axons of many neurons; it consists of units of double bilayers separated by 3 to 4 nm thick aqueous layers composed of 75–80% lipid and 20–25% protein. The two conductors in myelinated fibres coincide with the inner conducting axon and the outer conducting extracellular fluid (see Figure 6(b)).

The myelin sheath acts as an electrical insulator, forming a capacitor surrounding the axon, which allows for faster and more efficient conduction of nerve impulses than unmyelinated nerves.

In Table 1, a comparison between the SSCC and the core conductor models [34] of an average human myelinated nerve is proposed.

The diameter of the myelinated nerve fibre [35] grows according to the formula

$$D = d + 2 \times n_l \times k_l, \tag{75}$$

where n_l is the number of lamellae bilayers, k_l is their average width, d is the diameter of the axon, and D is the diameter of the fibre.

Now, using the formula of the spiral mapping we have

$$
\begin{aligned}
d &= 2e^{\delta_m/g_m - g_m\theta_{i_1}}, \\
D &= 2e^{\delta_m/g_m - g_m\theta_{f_2}},
\end{aligned} \tag{76}
$$

where θ_{i_1,f_2} are the initial and final angles of the myelin sheaths, and δ_m determine the lipid membrane spiral contour.

For $g_m \ll 1$ as in the case of the myelin, the thickness of the nth bilayer is nearly constant and the radius at which it occurs is $r_n = e^{\delta_m/g - 4n\pi g}$.

By taking as value of the thickness $k_l \simeq r_1 - r_0 = r_0(1 - e^{-4g_m\pi}) \simeq 0.018\,\mu m$ [35] we have

$$g_{\mathrm{mye}} \simeq \frac{1}{4\pi}\ln\left(\frac{d}{d-2k_l}\right). \tag{77}$$

According to the statistics [35], the nerve fiber diameter D is linearly related to the axon d diameter; that is, $D = C_0 + C_1 d$.

By taking $4\pi n_l = \theta_{i_1} - \theta_{f_2}$ (each lipid bilayer consists of two spiral turns $\theta_{i_1} \gg \theta_{f_2}$) and using (76), we have the following relation between the number of myelin lamellae n_l and the diameter d of the axon:

$$n_l(d) = \mathrm{Int}\left\{\frac{1}{4\pi g_m}\log\left[\frac{C_0 + C_1 d}{d}\right]\right\} \tag{78}$$

which is confirmed by the statistics [35].

In the case of the SCC we have

$$
\begin{aligned}
L' &= \mu n\frac{g}{1+g^2}, \\
Z_0 &= \eta n\frac{g}{1+g^2},
\end{aligned} \tag{79}
$$

where n represents the number of spiral turns between the outer spiral conductor and the inner one.

The transmitted power in SCC depends inversely on the impedance of the line Z_0 which is proportional to the g factor of the spiral and on the number of turns.

During 1960's Cole [36] presented a circuit model of the nerves including the inductive effects of the small membrane currents.

In Table 1 a comparison between the Cole and the SCC inductances is proposed.

The expressions R' and G' for the SCC, related to the power loss per unit length due to the finite conductivity of the two spiral conductor strips and to the time-average power dissipated per unit length in the dielectric, respectively, are reported in Table 2 in a comparison with various types of transmission lines.

The inductance $L' \simeq 0$ [37] for the core-conductor model is negligible; (59) is then rewritten in the form

$$
\begin{aligned}
V &= \lambda^2\frac{\partial^2 V}{\partial z^2} - \tau\frac{\partial V}{\partial t}, \\
\lambda &= \frac{1}{\sqrt{R'G'}}, \\
\tau &= \frac{C'}{G'}, \\
T &= \frac{\tau\ell^2}{\lambda^2} = R'C'\ell^2,
\end{aligned} \tag{80}
$$

where λ and τ are called the cable space and time constants, respectively, while T is called the time per internodal distance ℓ [37].

TABLE 2: Transmission parameters for the MDSCC, MSSCC, the cylindrical coax, and the parallel plate lines.

	Double spiral coax	Single spiral coax	Cylindrical coax	Parallel plate
L'	$\dfrac{\mu}{g}\left((\theta_2-\theta_1)/(\delta_1-\delta_2)+((\theta_2-\theta_1-(2\pi/(1+g^2)))(\delta_2-\delta_1+(2\pi g^2/(1+g^2))))/(\delta_2-\delta_1+(2\pi g^2/(1+g^2)))\right)$	$\mu\,\dfrac{ng}{1+g^2}$	$\dfrac{\mu}{2\pi}\ln\dfrac{b}{a}$	$\mu\dfrac{d}{D}$
C'	$\epsilon' gW\left(\dfrac{\theta_2-\theta_1}{\delta_1-\delta_2}+\dfrac{\theta_2-\theta_1-(2\pi/(1+g^2))}{\delta_2-\delta_1+(2\pi g^2/(1+g^2))}\right)$	$\epsilon'\dfrac{(1+g^2)}{ng}$	$\epsilon'\dfrac{2\pi}{\ln b/a}$	$\epsilon'\dfrac{D}{d}$
R'	$\dfrac{R_S\sqrt{1+g^2}(\theta_2-\theta_1-(\pi/(1+g^2)))^2\,((1/(\delta_1-\delta_2))+1/(\delta_2-\delta_1)+(2\pi g^2/(1+g^2)))^2}{16g}$ $\times\left[\dfrac{1}{a_{22}}\left(\dfrac{1}{(\delta_2-\delta_1)^2}+\dfrac{e^{-(2g\pi/(1+g^2))}}{(\delta_1-\delta_2+(2\pi g^2/(1+g^2)))^2}\right)-\dfrac{1}{a_{21}}\left(\dfrac{1}{(\delta_2-\delta_1)^2}+\dfrac{e^{-(4g\pi/(1+g^2))}}{(\delta_1-\delta_2+(2\pi g^2/(1+g^2)))^2}\right)+\dfrac{1}{a_{12}}\left(\dfrac{1}{(\delta_2-\delta_1)^2}+\dfrac{e^{-(2g\pi/(1+g^2))}}{(\delta_1-\delta_2+(2\pi g^2/(1+g^2)))^2}\right)-\dfrac{1}{a_{11}}\left(\dfrac{1}{(\delta_2-\delta_1)^2}+\dfrac{e^{-(2g\pi/(1+g^2))}}{(\delta_1-\delta_2+(2\pi g^2/(1+g^2)))^2}\right)\right]$ $a_{pq}=e^{(\delta_p/g)-g\theta_q},\quad p,q=1,2.$	$\dfrac{R_S\sqrt{1+g^2}}{8\pi^2}$ $\times\left[\dfrac{1}{a_{11}}-\dfrac{1}{a_{12}}+\dfrac{e^{-(2\pi ng^2/(1+g^2))}}{a_{21}}-\dfrac{e^{-(2\pi ng^2/(1+g^2))}}{a_{22}}\right]$ $a_{pq}=e^{((\delta_1/g)-2g(p-q)\pi-g\theta_{ip})}$, $p,q=1,2.$	$\dfrac{R_S}{2\pi}\left(\dfrac{1}{a}+\dfrac{1}{b}\right)$	$\dfrac{2R_S}{D}$
G'	$\omega\epsilon'' g\left(\dfrac{\theta_2-\theta_1}{\delta_2-\delta_1}+\dfrac{\theta_2-\theta_1-(2\pi/(1+g^2))}{\delta_1-\delta_2+(2\pi g^2/(1+g^2))}\right)$	$\dfrac{\omega\epsilon''}{g}(1+g^2)$	$\dfrac{2\pi\omega\epsilon''}{\ln b/a}$	$\dfrac{\omega\epsilon'' D}{d}$

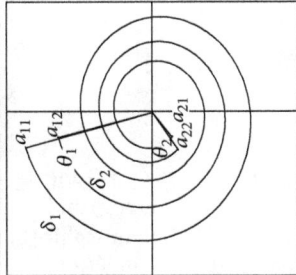

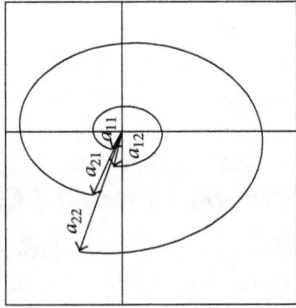

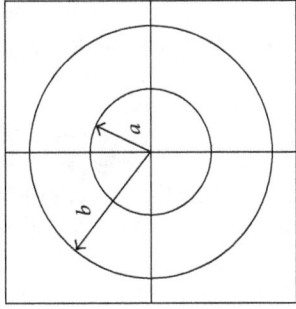

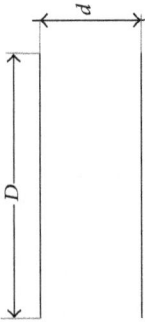

7. The Spiral Poynting Vector

On a matched spiral coaxial line the rms voltage V_0 is related to the total average power flow $P_z = (1/2) \int_{S_\perp} \vec{E} \times \vec{H}^* \cdot \hat{e}_z dS_\perp$ by

$$P_z$$

$$
= \begin{cases}
\dfrac{1}{2} \displaystyle\int_{\delta_1}^{\delta_2} \int_{\theta_1}^{\theta_2} \vec{E} \times \vec{H}^* \cdot \hat{e}_z dS_\perp \\
\quad + \displaystyle\int_{\delta_2}^{\delta_1 + 2\pi g^2/(1+g^2)} \int_{\theta_1 + 2\pi/(1+g^2)}^{\theta_2} \vec{E} \times \vec{H}^* \cdot \hat{e}_z dS_\perp \\
= \dfrac{1}{2}\sqrt{\dfrac{\epsilon}{\mu}} g V_0^2 \left(\dfrac{\theta_2 - \theta_1}{\delta_2 - \delta_1} + \dfrac{\theta_2 - \theta_1 - 2\pi/\left(1+g^2\right)}{\delta_1 - \delta_2 + 2\pi g^2/\left(1+g^2\right)} \right), \\
\hfill \text{double coax,} \\
\\
\dfrac{1}{2} \displaystyle\int_{\delta_1}^{\delta_1 + 2n\pi g^2/(1+g^2)} \int_{\theta_1}^{\theta_2} \vec{E} \times \vec{H}^* \cdot \hat{e}_z dS_\perp \\
= \dfrac{1}{\eta} \dfrac{1+g^2}{2gn} V_0^2, \hfill \text{single coax,}
\end{cases}
$$

$$(81)$$

where the infinitesimal cross section is $dS_\perp \equiv \|d\vec{S}_{\delta\theta}\|$ of (20).

As the g factor decreases, for example in the evolution of the Schwann's cell around the axon, progressively a higher number of spiral turns are required to yield the same value of transmitted power. Likewise, overcoming the power threshold in neural networks may provoke nerve inflammation and disorders or vice versa an amount of power below the natural required level could cause the neural signal to be blocked.

In order to change the transmitted power, the neural system can modify the number n of turns or the g factor.

Peters and Webster [27, 38, 39] showed that the angles subtended at the centre of the axon between the internal mesaxon and outer tongue of cytoplasm obey a precise statistic, that is, in about 75% of the mature myelin sheaths they examined the angle that lied within the same quadrant. This work refines the coaxial model for myelinated nerves introducing the spiral geometry and gives an explanation for the *Peters quadrant mystery* [38]. The surprising tendency for the start and finish of the myelin spiral to occur close together, according to this spiral coaxial model, comes out from the need of handling power throughout the nervous system.

In fact, the Poynting vector of (81) depends linearly on the Peters angle β_p which represents a finicky control of the power delivered along the myelinated nerves. A malformation of the Peters angle causes higher/lower power to be transmitted in the neural networks with respect to the required normal level.

8. Conclusions

In this paper two types of metal spiral coaxial cables have been proposed, the MSCC and the MDSCC.

A generalization of the *Schwarz-Christoffel* [40] conformal mapping was used to map the transverse section of the MSCC into a rectangle and to find the solution to its equivalent electrostatic *Laplace's equation*.

The fundamental TEM wave propagating along the MSCC has been determined together with the impedances of the line.

Comparisons of the MSCC with the classical cylindrical coax as well as with the hollow polar waveguide have been done.

The myelinated nerves, whose elm model is still based on the core-conductor theory, are analyzed by using the spiral coaxial model and their spiral geometrical factors are precisely related to the electrical impedances and propagating elm fields. The spiral model could be used to better analyze the neurodegenerative diseases, which are strictly connected to the geometrical malformations of the myelin bundles.

The MDSCC has many advantages compared to the cylindrical coaxial cable because it can be made multiturn, thus distributing the energy over a larger area and protecting the small signals from interference due to external electric fields.

The MSCC could have many interesting applications in the field of video and data transmission, as well as for sensing, instrumentation/control, communication equipment, and plasmonic nanostructure at optical wavelength.

Appendix

Spiral Generalization of the *Schwarz-Christoffel* Conformal Mapping

We define a spiral conformal coordinate system (u, v) as one as specified by a complex analytic function

$$w = u + iv, \quad w = f(z), \tag{A.1}$$

$$f(z) = A_0 \int_{z_0}^{z} \frac{1}{\zeta} d\zeta, \quad A_0 = 1 - ig, \ z_0 \neq 0, \tag{A.2}$$

where $g \in \mathbb{R}$ is a constant [40] and the function $f(z)$ is a generalization of the well-known holomorphic Schwarz-Christoffel [41] formula:

$$W(z) = A_0 \int_{z_0}^{z} \prod_{k=1}^{n} (\zeta - \zeta_k)^{-\alpha_k/\pi} d\zeta + B_0 \quad A_0, B_0 \in \mathbb{C}, \tag{A.3}$$

because for $\alpha_1 = \pi, \zeta_1 = 0$ and $\alpha_k = 0, \forall k > 1, \zeta_k = 0, \forall k \geq 1$, the two formulas of (A.2) and (A.3) are identical.

Since $f(z)$ is holomorphic the derivative $f'(z)$ exists and it is independent of direction.

For $g = 0$ or $A_0 \in \mathbb{R}$, the spiral conformal mapping of (A.1)-(A.2) coincides with the polar mapping (see [18] page 135); the elm propagation along the circular waveguide is then included in the theoretical treatment of this paper as a particular case.

In terms of cartesian (x, y) or polar (r, φ) coordinates

$$z = x + iy = re^{i\varphi}. \tag{A.4}$$

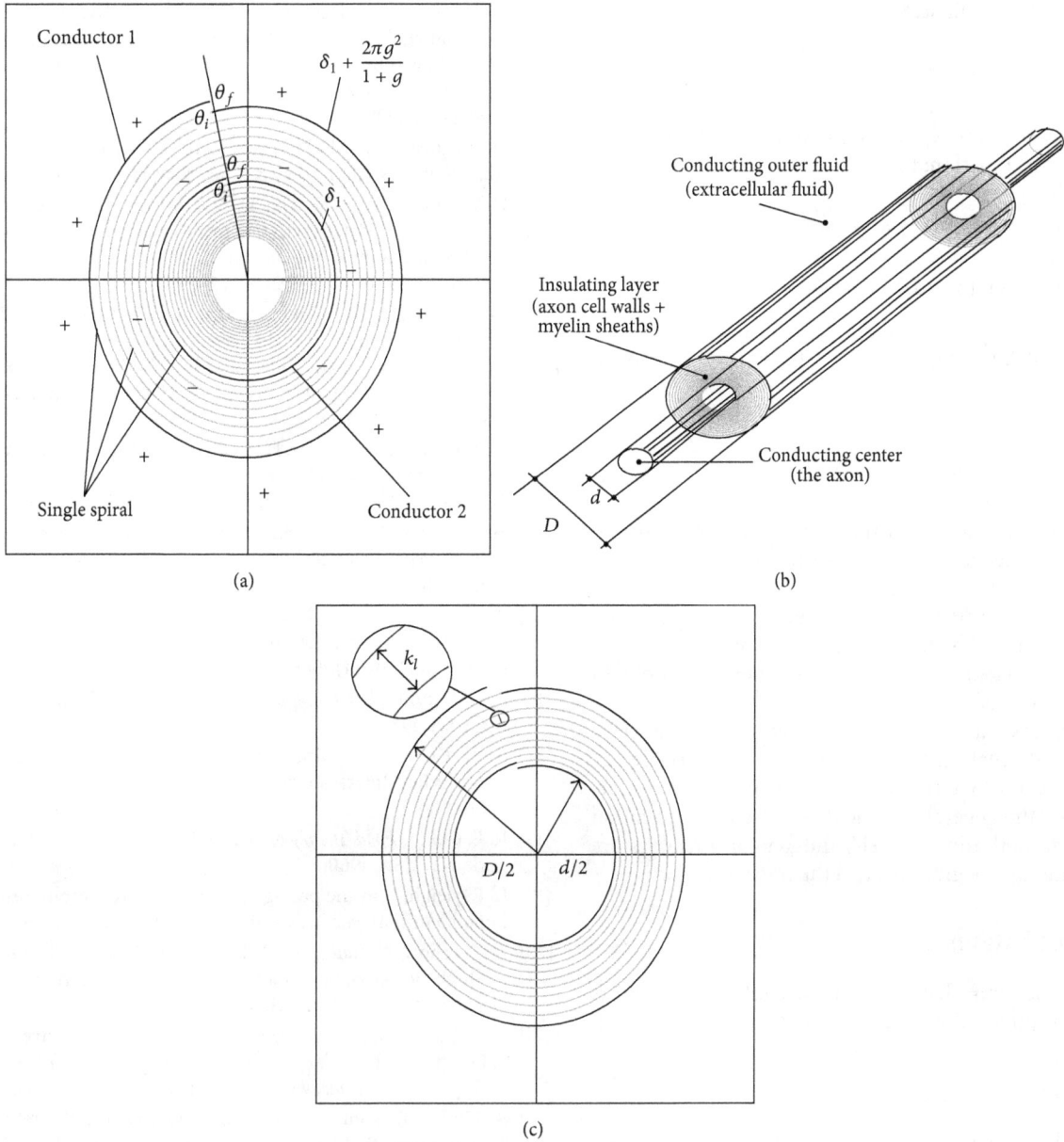

FIGURE 6: SSCC (a) transverse section, (b) longitudinal view, and (c) the myelin sheaths.

Substituting (A.2) into (A.1) we obtain

$$u + iv = (1 - ig)\log z + K = f(z). \qquad (A.5)$$

The value of the constant K represents the phase of the transformation and is related to $z_0 = e^{-K}$.

In order to study the spiral coaxial cable a further normalization of the angles u and v is introduced:

$$u + iv = \frac{1 + g^2}{g}\delta + i\left(1 + g^2\right)\theta. \qquad (A.6)$$

θ, δ are the two normalized variables. Using (A.1), (A.4), (A.6), and

$$w = (1 - ig)(\log r + i\varphi) + K \qquad (A.7)$$

we obtain the direct complex spiral coordinate transformation; that is,

$$z = e^{\delta/g - g\theta + i(\delta + \theta)}, \qquad (A.8)$$

where $K = 0$.

If $g = 0$ and $K = 0$ the two variables u, v coincide with the polar variables $\ln r$, φ (see [18] page 135).

The transverse arclength in cartesian or polar coordinates becomes

$$(d\ell)^2 = |dz|^2 = (dx)^2 + (dy)^2 = (dr)^2 + (rd\varphi)^2, \qquad (A.9)$$

where

$$|dz|^2 = |f'(z)|^{-2}|dw|^2, \qquad (A.10)$$

or in conformal coordinates

$$(d\ell)^2 = |s|^2 \left((du)^2 + (dv)^2\right), \quad |s| \equiv \frac{1}{|f'(z)|}, \quad (A.11)$$

where the scale factor is the inverse of the modulus of the derivative of the function; that is,

$$f'(z) = \frac{1 - ig}{z}. \quad (A.12)$$

Substituting (A.6) into (A.11) we have

$$(d\ell)^2 = |S|^2 \left(\left(\frac{d\delta}{g}\right)^2 + (d\theta)^2\right), \quad (A.13)$$

where

$$|S| = \left(1 + g^2\right)|s|. \quad (A.14)$$

Although the scale factors of the variables δ and θ are not equal, their normalized coordinate system is orthogonal and the potential satisfies the same differential equation that it does in the x, y coordinates [18]. By using the variables u and v of the original conformal mapping presented in [40], for which the scale factors are identical, it is possible to obtain exactly the same results of this paper.

The complex variable $z = x + iy$ here used to describe the spiral conformal mapping is not the same variable "z" that represents the longitudinal coordinate of the waveguide. Nevertheless, the general treatment of the elm propagation in waveguide [28] and Maxwell's differential operators are separated into the longitudinal and the transverse parts.

Conflict of Interests

The author declares that there is no conflict of interests regarding the publication of this paper.

References

[1] O. Heaviside, *Electromagnetic Theory*, vol. 1, Dover, New York, NY, USA, 1950.

[2] D. M. Pozar, *Microwave Engineering*, John Wiley & Sons, 4th edition, 2011.

[3] A. S. Khan, *Microwave Engineering: Concepts and Fundamentals*, CRC Press, New York, NY, USA, 2014.

[4] S. Ramo, J. R. Whinnery, and T. Van Duzer, *Fields and Waves in Communication Electronics*, John Wiley & Sons, 3rd edition, 1993.

[5] G. Lifante, *Integrated Photonics: Fundamentals*, John Wiley & Sons, Chichester, UK, 2003.

[6] C. H. Lee, *Microwave Photonics*, CRC Press, New York, NY, USA, 2006.

[7] R. de Waele, S. P. Burgos, A. Polman, and H. A. Atwater, "Plasmon dispersion in coaxial waveguides from single-cavity optical transmission measurements," *Nano Letters*, vol. 9, no. 8, pp. 2832–2837, 2009.

[8] M. S. Kushwaha and B. D. Rouhani, "Surface plasmons in coaxial metamaterial cables," *Modern Physics Letters B*, vol. 27, no. 17, Article ID 1330013, 2013.

[9] J.-C. Weeber, A. Dereux, C. Girard, J. R. Krenn, and J.-P. Goudonnet, "Plasmon polaritons of metallic nanowires for controlling submicron propagation of light," *Physical Review B: Condensed Matter and Materials Physics*, vol. 60, no. 12, pp. 9061–9068, 1999.

[10] H. Regneault, J. M. Lourtioz, and C. Delalande, *Levenson Nanophotonics*, John Wiley & Sons, New York, NY, USA, 2010.

[11] G. Veronis, Z. Yu, S. Kocaba, D. A. B. Miller, M. L. Brongersma, and S. Fan, "Metal-dielectric-metal plasmonic wave guide devices for manipulating light at the nanoscale," *Chinese Optics Letters*, vol. 7, no. 4, pp. 302–308, 2009.

[12] M. L. Brongersma, J. W. Hartman, and H. A. Atwater, "Electromagnetic energy transfer and switching in nanoparticle chain arrays below the diffraction limit," *Physical Review B—Condensed Matter and Materials Physics*, vol. 62, no. 24, pp. R16356–R16359, 2000.

[13] T. W. Ebbesen, H. J. Lezec, H. F. Ghaemi, T. Thio, and P. A. Wolff, "Extraordinary optical transmission through sub-wavelenght hole arrays," *Nature*, vol. 391, no. 6668, pp. 667–669, 1998.

[14] G. Boisde and A. Harmer, *Chemical and Biochemical Sensing with Optical Fibers and Waveguides*, Arthech House, Boston, Mass, USA, 1996.

[15] A. V. Zayats, I. I. Smolyaninov, and A. A. Maradudin, "Nano-optics of surface plasmon polaritons," *Physics Reports*, vol. 408, no. 3-4, pp. 131–314, 2005.

[16] J. A. Stratton, *Electromagnetic Theory*, McGraw-Hill, New York, NY, USA, 1941.

[17] R. E. Collin, *Foundations for Microwave Engineering*, IEEE Press, Wiley Interscience, New York, NY, USA, 2nd edition, 2001.

[18] R. E. Collin, *Field Theory of Guided Waves*, Mc-Graw Hill, New York, NY, USA, 1960.

[19] L. Rayleigh, "On the passage of electric waves through tubes," *Philosophical Magazine*, vol. 43, no. 261, pp. 125–132, 1897.

[20] I. M. Fabbri, A. Lauto, and A. Lucianetti, "A spiral index profile for high power optical fibers," *Journal of Optics A: Pure and Applied Optics*, vol. 9, no. 11, pp. 963–971, 2007.

[21] I. M. Fabbri, A. Lucianetti, and I. Krasikov, "On a Sturm Liouville periodic boundary values problem," *Integral Transforms and Special Functions*, vol. 20, no. 5-6, pp. 353–364, 2009.

[22] K. Guven, E. Saenz, R. Gonzalo, E. Ozbay, and S. Tretyakov, "Electromagnetic cloaking with canonical spiral inclusions," *New Journal of Physics*, vol. 10, Article ID 115037, 2008.

[23] W. T. Kelvin, "On the theory of the electric telegraph," *Proceedings of the Royal Society of London*, vol. 7, pp. 382–389, 1855.

[24] W. Rall, "Core conductor theory and cable properties of neurons," in *Handbook of Physiology, the Nervous System, Cellular Biology of Neurons*, John Wiley & Sons, New York, NY, USA, 2011.

[25] A. H. Buck, *Reference Handbook of the Medical Sciences*, vol. 3 of *edited by A. H. Buck*, Book on Demand, New York, NY, USA, 1901.

[26] A. L. Hodgkin and A. F. Huxley, "A quantitative description of membrane current and its application to conduction and excitation in nerve," *The Journal of Physiology*, vol. 117, no. 4, pp. 500–544, 1952.

[27] A. Peters, "Further observations on the structure of myelin sheaths in the central nervous system," *The Journal of Cell Biology*, vol. 20, pp. 281–296, 1964.

[28] N. Marcuvitz, *Waveguide Handbook*, Peter Peregrinus, New York, NY, USA, 1986.

[29] I. Boscolo and I. M. Fabbri, "A tunable bragg cavity for an efficient millimeter FEL driven by electrostatic accelerators," *Applied Physics B Photophysics and Laser Chemistry*, vol. 57, no. 3, pp. 217–225, 1993.

[30] J. D. Jackson, *Classical Electrodynamics*, John Wiley & Sons, New York, NY, USA, 1962.

[31] G. Mie, "Beiträge zur Optik trüber Medien, speziell kolloidaler Metallösungen," *Annalen der Physik*, vol. 330, no. 3, pp. 337–445, 1908, English translated by B. Crossland, Contributions to the optics of turbid media, particularly of colloidal metal solutions, Nasa Royal Aircraft Establishment no. 1873, 1976.

[32] M. Born and E. Wolf, *Principles of Optics: Electromagnetic Theory of Propagation*, Cambridge University Press, Cambridge,UK, 1999.

[33] V. M. Agranovich and D. L. Mills, Eds., *Surface Polaritons*, North-Holland, Amsterdam, The Netherlands, 1982.

[34] Y. Min, K. Kristiansen, J. M. Boggs, C. Husted, J. A. Zasadzinski, and J. Israelachvili, "Interaction forces and adhesion of supported myelin lipid bilayers modulated by myelin basic protein," *Proceedings of the National Academy of Sciences of the United States of America*, vol. 106, no. 9, pp. 3154–3159, 2009.

[35] C. H. Berthold, I. Nilsson, and M. Rydmark, "Axon diameter and myelin sheath thickness in nerve fibres of the ventral spinal root of the seventh lumbar nerve of the adult and developing cat," *Journal of Anatomy*, vol. 136, no. 3, pp. 483–508, 1983.

[36] K. Cole, *Membranes, Ions and Impulses: A Chapter of Classical Biophysics*, University of California Press, Los Angeles, Calif, USA, 1968.

[37] A. F. Huxley and R. Stampfli, "Evidence for saltatory conduction in peripheral myelinated nerve fibres," *The Journal of Physiology*, vol. 108, no. 3, pp. 315–339, 1949.

[38] R. R. Traill, *Strange Regularities in the Geometry of Myelin Nerve-Insulation—A Possible Single Cause*, Ondwelle Short Monograph no. 1, 2005.

[39] H. D. Webster, "The geometry of peripheral myelin sheaths during their formation and growth in rat sciatic nerves," *The Journal of Cell Biology*, vol. 48, no. 2, pp. 348–367, 1971.

[40] L. M. B. Campos and P. J. S. Gil, "On spiral coordinates with application to wave propagation," *Journal of Fluid Mechanics*, vol. 301, pp. 153–173, 1995.

[41] Z. Nehari, *Conformal Mapping*, Dover, New York, NY, USA, 1975.

14

Design of Miniaturized Multiband Filters Using Zero Order Resonators for WLAN Applications

Maryam Shafiee,[1] Mohammad Amin Chaychi Zadeh,[2] and Homayoon Oraizi[2]

[1]Department of Electrical Engineering, Arizona State University, Tempe, AZ 85287, USA
[2]Department of Electrical Engineering, Iran University of Science and Technology, Narmak, Tehran 16846 13114, Iran

Correspondence should be addressed to Maryam Shafiee; mshafiei@asu.edu

Academic Editor: Giancarlo Bartolucci

The objective of this paper is to design miniaturized narrow- and dual-band filters for WLAN application using zero order resonators by the method of least squares. The miniaturization of the narrow-band filter is up to 70% and that of the dual-band filter is up to 64% compared to the available models in the literature. Two prototype models of the narrow-band and dual-band filters are fabricated and measured, which verify the proposed structure for the filter and its design by the presented method, using an equivalent circuit model.

1. Introduction

Mobility of components and equipment is a requirement in mobile communication systems. Therefore, considerable effort has gone into the miniaturization of devices. Furthermore, various techniques of multibanding have been devised for various functions. Particularly, different techniques have been proposed to make components dual-band, which are considered conventional methods, such as series connection of two separate band pass filters, step impedance resonators, and defected ground structures. All of these methods have some limitations on the reduction of device dimensions, because they are based on half-wave resonators [1, 2].

Recently, metamaterials have found wide applications, such as miniaturization of microwave devices, due to their unique properties and nonlinear dispersion curves. Zero order resonance (ZOR) metamaterials have the distinct property of infinite wavelength, which makes the resonance frequency independent of their dimensions. Consequently, the ZOR component may be fabricated as small as specified to realize the desired circuit elements for the required resonance frequency [3]. This unique characteristic of ZORs has been used in microwave components, such as filters [4–6].

In this paper, we use the method of least mean square error (LMS) as a contribution to design and fabricate a narrow-band metamaterial filter and also a dual-band filter for application in WLAN systems. The main advantage and feature of the proposed filters are their compactness and miniaturization, which is up to 70% smaller than conventional components in the literature [7].

2. Design Procedure

The proposed zero order resonance (ZOR) cell is depicted in Figure 1. The two interdigital capacitors provide the left-handed capacitance and right-handed inductance. The meander inductance and T-junction are inserted in the circuit to control and adjust the right-handed section of the device. Its equivalent circuit is shown in Figure 2. The values of capacitors and inductors are extracted from the formulas in the literature [8–11].

The error function is explicitly expressed as the function of geometrical dimensions of various sections of the filter. In other words, the lumped elements (L and C) of the equivalent circuit of the filter in Figure 2 are expressed in terms of the geometrical dimensions of the filter structure, which are given in literature. The overall transmission matrix of the equivalent circuit is first derived, which is then converted to the scattering parameters. Thus, the scattering parameters of equivalent circuit might be obtained by the available relations of the T-matrix in terms of its physical dimensions [12].

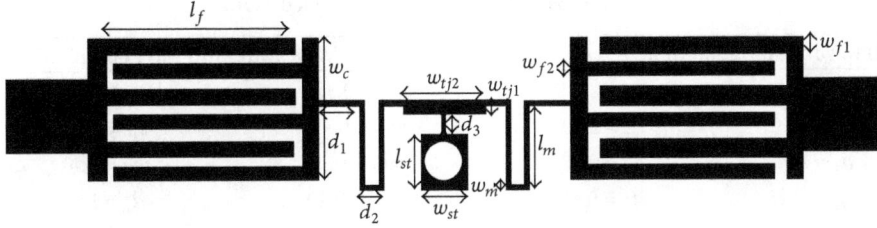

FIGURE 1: The microstrip layout of the metamaterial unit cell: $w_m = 0.1\,\text{mm}$, $l_f = 4.55\,\text{mm}$, $w_{f1} = 0.3\,\text{mm}$, $w_{f2} = 0.4\,\text{mm}$, $\text{gap}_1 = 0.3\,\text{mm}$, $\text{gap}_2 = 0.2\,\text{mm}$, $w_{st} = 1.14\,\text{mm}$, $l_{st} = 1.3\,\text{mm}$, $d_1 = 1\,\text{mm}$, $d_2 = 0.45\,\text{mm}$, $l_m = 2.0\,\text{mm}$, $w_{tj1} = 0.3\,\text{mm}$, and $w_{tj2} = 0.1\,\text{mm}$.

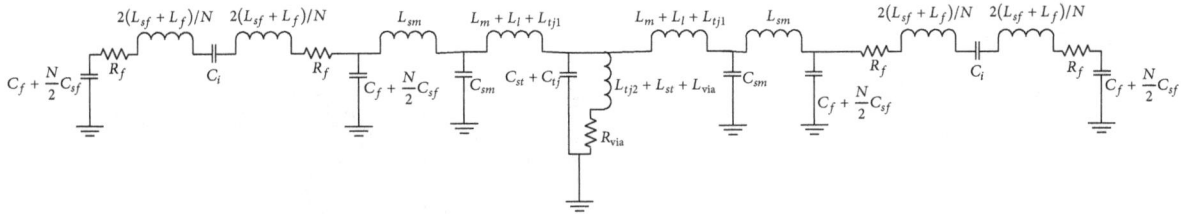

FIGURE 2: The proposed equivalent circuit of metamaterial unit cell.

We now refer to the specified frequency response of the bandpass filter, as shown in Figure 3. It is composed of the lower (f_L to f_f) and upper (f_l to f_U) stop bands, with attenuation g_s, the lower (f_f to $f_0 - \text{BW}/2$) and upper ($f_0 + \text{BW}/2$ to f_l) transition bands, with the attenuations $g_{tf} = (g_s/(f_f - (f_0 - \text{BW}/2)))(f_i - f_f)$ and $g_{ts} = (g_s/(f_f - (f_0 + \text{BW}/2)))(f_i - f_l)$, and the pass band with center frequency f_0 and bandwidth BW and insertion loss g_p. Each frequency interval is divided into N_i discrete frequencies. The accuracy of the error function increases for larger number of discrete frequencies, but the CPU time of computations also increases. Therefore, a trade-off should be considered between accuracy and computing time. We then construct an error function as

$$\varepsilon_s = \sum_{i \in \text{stop}}^{N_1} \left(\left| s_{21}^i \right|_{\text{dB}} - g_s \right)^2,$$

$$\varepsilon_p = \sum_{i \in \text{pass}}^{N_2} \left(\left| s_{21}^i \right|_{\text{dB}} - g_p \right)^2,$$

$$\varepsilon_{tf} = \sum_{i \in \text{first.transition}}^{N_3} \left(\left| s_{21}^i \right|_{\text{dB}} - g_{tf} \right)^2, \qquad (1)$$

$$\varepsilon_{ts} = \sum_{i \in \text{sec.transition}}^{N_4} \left(\left| s_{21}^i \right|_{\text{dB}} - g_{ts} \right)^2$$

$$\Longrightarrow \varepsilon_{\text{total}}$$

$$= w_p \times \varepsilon_p + w_t \times \left(\varepsilon_{ts} + \varepsilon_{tf} \right) + w_s \times \varepsilon_s,$$

where ε_s is the error due to the lower and upper stop bands, ε_p is the error related to the pass band, and ε_{tf} and ε_{ts} are the errors from the lower and upper transition bands, as depicted in the filter frequency response in Figure 3. The weighting

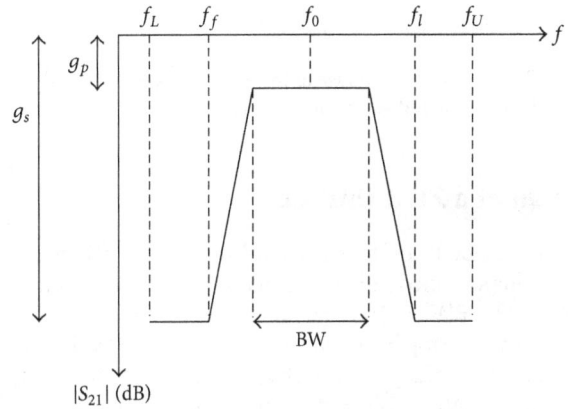

FIGURE 3: Design specifications of filter frequency response.

functions are w_s, w_p, and w_t in order to adjust the value of each error term in the function of total error.

The error function can be constructed for multiband filters. The minimization of error function gives the optimum physical dimensions of the filter. The genetic algorithm (GA) as a global minimum seeking algorithm does not need initial values for variable, but it is quite slow. However, the conjugate gradient (CG) algorithm is a local minimum seeker algorithm, which needs initial values but it is quite fast. Therefore, a MATLAB code is written which combines GA and CG. First, GA is run to reach the vicinity of the absolute minimum point, but it is aborted prematurely and GC is activated to locate the minimum point quite fast. Since the design procedure based on the transmission line circuit model suffers some approximation and does not account for full wave performance of the filter, the CST Microwave Studio computer simulator is used to adjust the filter performance and obtain its optimum design [13].

TABLE 1: Line widths and lengths for the designed ZOR.

Values **before full wave** optimization (mm)			Values **after full wave** optimization (mm)		
$w_f = 0.47$	$l_{st} = 0.90$	$w_{tj2} = 0.10$	$w_{f1} = 0.305$	$w_{st} = 1.14$	$w_{tj2} = 0.10$
gap $= 0.23$	$w_m = 0.10$	$l_m = 2.00$	$w_{f2} = 0.40$	$l_{st} = 1.30$	$l_m = 2.00$
$l_f = 4.95$	$d_1 = 1.14$	$d_2 = 0.67$	gap$_1 = 0.30$	$w_m = 0.10$	$d_2 = 0.45$
$w_{st} = 2.0$	$w_{tj1} = 0.12$	$d_3 = 0.91$	gap$_2 = 0.20$	$d_1 = 1.12$	$d_3 = 0.50$
			$l_f = 4.55$	$w_{tj1} = 0.30$	

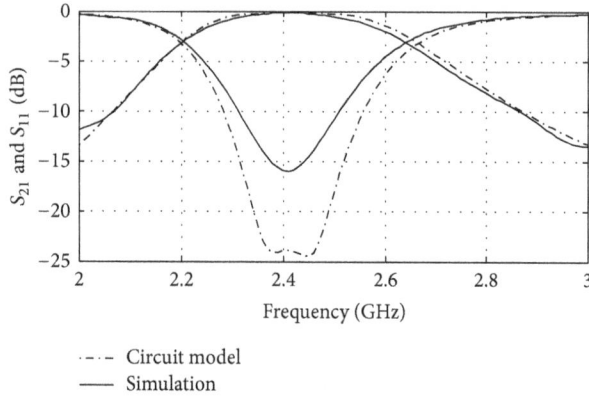

FIGURE 4: The frequency response of the ZOR unit cell obtained by the equivalent circuit and simulation software.

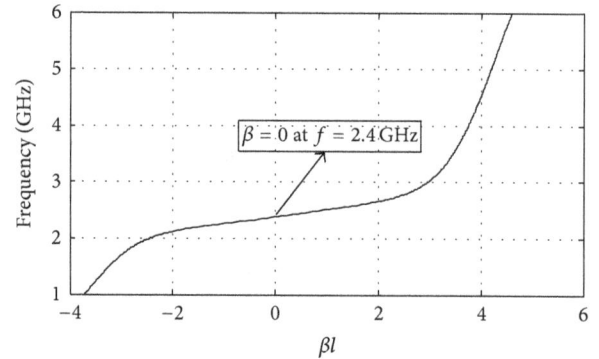

FIGURE 5: The dispersion diagram of the ZOR unit cell.

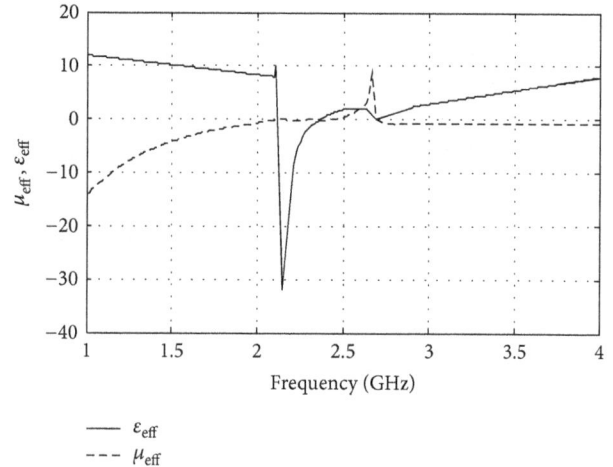

FIGURE 6: Constitutive parameters of the ZOR unit cell.

3. Design of a ZOR Unit Cell

We design a ZOR unit cell according to the following specifications specified in Figure 3, namely, centre frequency $f_0 = 2.4$ GHz, BW $= 10\%$, maximum insertion loss $g_p = 0.5$ dB, maximum stop band rejection $g_s = 20$ dB in the bands 1-2 GHz and 3-4 GHz, $f_f = 2.0$ GHz, $f_l = 3.0$ GHz, $N_1 = 51$, $N_2 = 212$, and $N_3 = N_4 = 45$. The minimization of error function determines the dimensions of the ZOR filter.

The frequency response of ZOR unit cell as obtained by the circuit model and full wave simulation using CST are shown in Figure 4. They are in a good agreement. In the circuit model of the unit cell, the interdigital sections of the filter are assumed identical, whereas in the full wave filter simulation they are assumed different by defining two separate variables for both gaps and widths to achieve the best possible result. Also, another optimization is performed on dimensions in full wave simulation as reported in Table 1.

The propagation constant of the structure could be computed from transmission parameters:

$$\beta = -\varphi^{\text{unwrapped}}(S_{21}) + \zeta. \tag{2}$$

The unwrapped phase of S_{21} is obtained by connecting the discontinuous sections of phase curves at $+\pi$ and $-\pi$, as shown in Figure 5. For the determination of the reference, where the electrical length ($\theta = \beta l$) is zero, it is necessary to shift the unwrapped phase curve by ζ to remove the phase ambiguity [3]. Its value is obtained from the location of zero crossing point at the phase response curve.

For the accurate characterization of metamaterial cell, its constitutive parameters are obtained by the procedure of effective media and are drawn in Figure 6 [14].

4. Design of a Narrow-Band Filter

In order to design a narrow-band filter with high rejection in the stop band, low insertion loss in the pass band, and sharp drop in the transition band, two ZOR unit cells are connected in series by an inductive coupling, as shown in Figure 7. The design procedure based on the LMS design method is similar to that of the unit cell. The scattering parameters are obtained

TABLE 2: Dimensions of narrow-band filter designed by the proposed method.

(a) Dimensions of first unit cell

Values **before full wave** optimization (mm)			Values **after full wave** optimization (mm)		
$w_f = 0.3$	$l_{st} = 1.00$	$w_{tj2} = 0.10$	$w_f = 0.43$	$l_{st} = 1.00$	$w_{tj2} = 0.10$
gap $= 0.588$	$w_m = 0.218$	$l_m = 2.65$	gap $= 0.56$	$w_m = 0.25$	$l_m = 2.70$
$l_f = 4.226$	$d_1 = 1.20$	$d_2 = 0.50$	$l_f = 3.44$	$d_1 = 1.2$	$d_2 = 0.80$
$w_{st} = 1.17$	$w_{tj1} = 0.20$	$d_3 = 0.00$	$w_{st} = 1.33$	$w_{tj1} = 0.20$	$d_3 = 0.00$

(b) Dimensions of second unit cell

Values **before full wave** optimization (mm)			Values **after full wave** optimization (mm)		
$w_f = 0.31$	$l_{st} = 1.05$	$w_{tj2} = 0.10$	$w_f = 0.52$	$l_{st} = 1.00$	$w_{tj2} = 0.10$
gap $= 0.69$	$w_m = 0.218$	$l_m = 2.65$	gap $= 0.7$	$w_m = 0.25$	$l_m = 2.70$
$l_f = 3.59$	$d_1 = 1.10$	$d_2 = 0.70$	$l_f = 3.47$	$d_1 = 0.92$	$d_2 = 0.70$
$w_{st} = 1.33$	$w_{tj1} = 0.20$	$d_3 = 0.00$	$w_{st} = 1.75$	$w_{tj1} = 0.20$	$d_3 = 0.00$

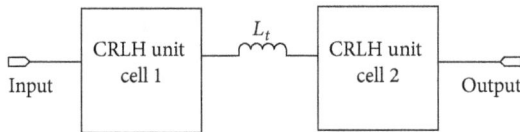

FIGURE 7: The equivalent circuit of the two metamaterial ZOR cells in the narrow-band and dual-band filters.

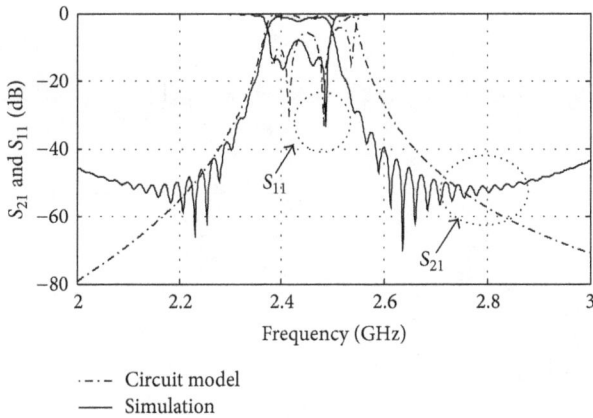

·—·— Circuit model
——— Simulation

FIGURE 8: The frequency response of the narrow-band filter as S_{21} obtained by simulations and circuit model.

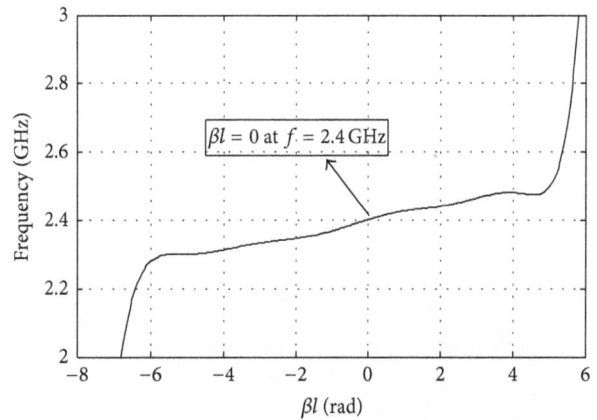

FIGURE 9: The dispersion response of the narrow-band filter.

from the transmission matrix, which are then used in the error function. The specifications of the narrow-band filter are $f_0 = 2.44\,\text{GHz}$, BW $= 100\,\text{MHz}$, $g_p = 0.5\,\text{dB}$, $g_s = 30\,\text{dB}$, $N_1 = 51$, $N_2 = 212$, and $N_3 = N_4 = 45$.

The physical dimensions of the narrow-band filter obtained by the LMS design procedure and CST are given in Table 2.

The frequency responses of filter as the amplitudes of S_{11} and S_{21} obtained by the circuit model and computer simulation using CST Microwave Studio are illustrated in Figure 8. The full wave result shows a bandwidth of 80 MHz from 2.36 GHz to 2.48 GHz. The dispersion diagram of the proposed filter is depicted in Figure 9. The appearance of

ripples in the response curve is due to the CST simulation software, which is based on the time-domain analysis using a truncated Gaussian excitation signal. Consequently, the resulting bounded frequency band of excitation and the resonance nature of filter structure generate the ripples in the simulated response. The magnitude of ripples may be reduced by increasing the accuracy of simulations, but the time of computation greatly increases.

The frequency response of the single-band filter is shown in Figure 10. The graph covers a frequency range up to 10 GHz to show that spurious response suppression is very good and is about 65 dB.

The main advantage of the designed filter is its compactness and miniaturization. It is almost 70% smaller than those reported in the literature [7].

5. Design of a Dual-Band Filter

At first, a ZOR unit cell filter is designed for dual-band applications. The lower pass band is due to its left-handed resonance and the upper pass band is due to the right-handed

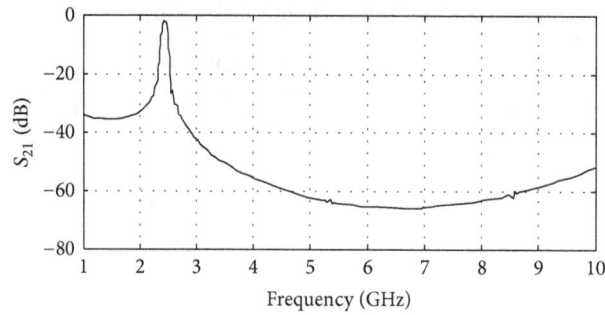

FIGURE 10: The spurious response of narrow-band filter as S_{21} in the stop band up to 10 GHz.

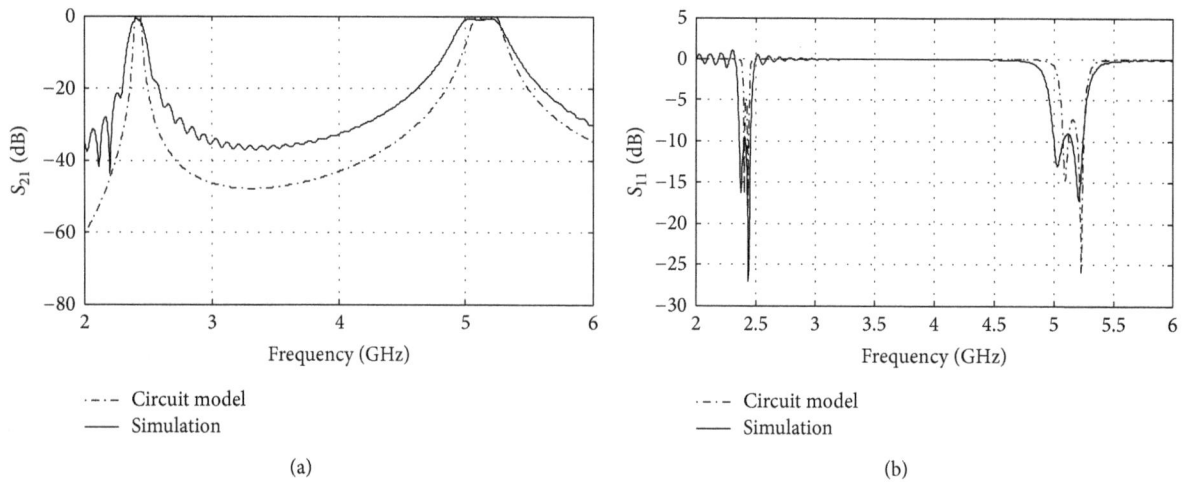

(a)

(b)

FIGURE 11: The frequency response of the dual-band filter as obtained by the circuit model and simulation. (a) S_{21}; (b) S_{11}.

resonance. Both upper and lower bands can be designed independently. The frequency response of the designed filter has achieved its specification. In order to improve the isolation between the two bands and also higher quality factor (Q) at each band, two ZOR unit cells are connected again using inductive coupling. Its design specifications are as follows:

first band: center frequency, f_0 = 2.44 GHz, bandwidth, BW = 80 MHz, and maximum insertion loss, IL = 0.5 dB;

second band: center frequency, f_0 = 5.2 GHz, bandwidth, BW = 150 MHz, and maximum insertion loss, IL = 0.5 dB;

isolation between the two pass bands = 30 dB.

For the construction of error function for the dual-band filter, the frequency intervals of each pass-band, each stop band, and each transition band are divided into 51, 212, and 45 discrete frequencies. The physical dimensions of each unit cell of the dual-band filter obtained for the optimum design are given in Table 3.

The frequency responses by circuit model and full wave computer simulation using CST Microwave Studio are illustrated in Figure 11. An isolation of 37 dB is observed between two pass bands in full wave simulation. The dispersion characteristic is extracted and shown in Figure 12.

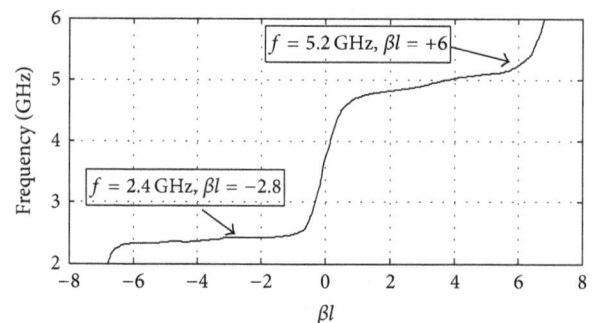

FIGURE 12: The dispersion curve of the dual-band filter.

The compactness of this filter is 64 percent better than the available designs reported in the literature [2, 15].

The difference between circuit model and full wave simulation results in Figures 8 and 11 is due to the fact that we used two different variables for gaps and widths of interdigital structures to provide more degree of freedom in full wave optimization.

To ensure maximum power transfer and reduction of return loss, a single open stub matching network is designed and added in parallel to the input of both filters to match source and load impedances.

TABLE 3: Dimensions of dual-band filter designed by the proposed method.

(a) Dimensions of first unit cell

Values **before full wave** optimization (mm)			Values **after full wave** optimization (mm)		
$w_f = 0.23$	$l_{st} = 2.00$	$w_{tj2} = 0.10$	$w_{f1,2} = 0.27 \ \& \ 0.20$	$l_{st} = 1.10$	$w_{tj2} = 0.10$
gap = 0.48	$w_m = 0.24$	$l_m = 2.00$	gap = 0.45	$w_m = 0.10$	$l_m = 2.00$
$l_{f1,2} = 1.27 \ \& \ 1.26$	$d_1 = 0.35$	$d_2 = 0.37$	$l_{f1,2} = 1.30 \ \& \ 1.70$	$d_1 = 0.40$	$d_2 = 0.40$
$w_{st} = 1.17$	$w_{tj1} = 0.20$	$d_3 = 4.03$	$w_{st} = 1.12$	$w_{tj1} = 0.20$	$d_3 = 4.95$

(b) Dimensions of second unit cell

Values **before full wave** optimization (mm)			Values **after full wave** optimization (mm)		
$w_f = 0.3$	$l_{st} = 1.90$	$w_{tj2} = 0.10$	$w_f = 0.25$	$l_{st} = 1.20$	$w_{tj2} = 0.10$
gap = 0.47	$w_m = 0.24$	$l_m = 2.00$	gap = 0.40	$w_m = 0.10$	$l_m = 2.00$
$l_{f1,2} = 1.40 \ \& \ 1.30$	$d_1 = 0.50$	$d_2 = 0.50$	$l_{f1,2} = 1.10 \ \& \ 1.30$	$d_1 = 0.50$	$d_2 = 0.50$
$w_{st} = 1.10$	$w_{tj1} = 0.20$	$d_3 = 4.30$	$w_{st} = 1.14$	$w_{tj1} = 0.20$	$d_3 = 4.95$

(a)   (b)

FIGURE 13: Photographs of the fabricated prototype models. (a) Narrow-band filter; (b) dual-band filter.

6. Measurement Result

The prototypes of the fabricated narrow-band and dual-band band pass filters are depicted in Figure 13. Substrate Rogers 4003 is used with dielectric constant 3.55, thickness 20 mil, and $\tan \delta = 0.0027$. For the miniaturization of filter profiles, the two unit cells of filters are located together back to back by a pin, so that the ground plane is placed between them. The dimensions of the pin are optimized in the full wave simulation. The frequency responses of the narrow-band and dual-band filters as obtained by full wave simulation and measurement are illustrated in Figures 14 and 15, for S_{21} versus frequency.

The agreement among simulation and measurement is fairly good. The difference between the simulation results and measurement data is due to the mediocre fabrication techniques and low precision of the equipment in our laboratory.

7. Conclusion

The zero order resonators derived from the theory of metamaterials are used as the building blocks to design narrow-band and dual-band filters. A circuit model is derived for

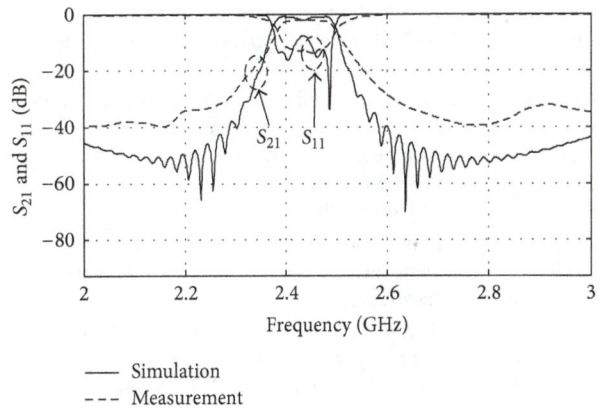

FIGURE 14: Results for simulation and measurement of the narrow-band filter (the dimensions of substrate are $1.4 \times 3.2 \ \text{mm}^2$).

the filters by the method of least squares. The filter designs are for single- and dual-band application in WLAN systems. The proposed designs have achieved 70% and 64% miniaturization for the narrow-band and dual-band filters compared to the available models in the literature. The measurement and computer simulation data of fabricated prototypes agree quite

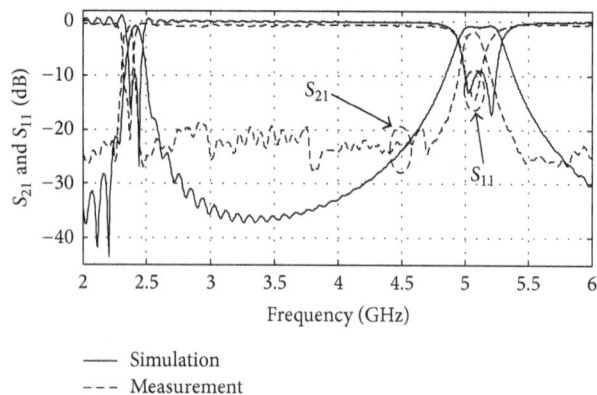

FIGURE 15: Results for simulation and measurement of S_{21} for the dual-band filter (the dimensions of substrate are $2 \times 1.4\,\mathrm{mm}^2$).

well with the results of the proposed equivalent circuits of the filter. Consequently, the proposed filter structures and design method are verified for application in microwave systems.

Conflict of Interests

The authors declare that there is no conflict of interests regarding the publication of this paper.

References

[1] J.-T. Kuo and H.-S. Cheng, "Design of quasi-elliptic function filters with a dual-passband response," *IEEE Microwave and Wireless Components Letters*, vol. 14, no. 10, pp. 472–474, 2004.

[2] S. Sun and L. Zhu, "Coupling dispersion of parallel-coupled microstrip lines for dual-band filters with controllable fractional pass bandwidths," in *Proceedings of the IEEE MTT-S International Microwave Symposium*, p. 4, IEEE, June 2005.

[3] C. Caloz and T. Itoh, *Electromagnetic Metamaterials: Transmission Line Theory and Microwave Applications*, John Wiley & Sons, 2005.

[4] S. Kahng, G. Jang, B. Lee, J. Ju, and S. Lee, "Compact UHF bandpass filter with the subwavelength metamaterial ZORs and transmission zeros for enhanced channel selectivity," in *Proceedings of the 41st European Microwave Conference (EuMC '11)*, pp. 567–570, IEEE, October 2011.

[5] G. Jang and S. Kahng, "Design of a dual-band metamaterial band-pass filter using zeroth order resonance," *Progress In Electromagnetics Research C*, vol. 12, pp. 149–162, 2010.

[6] C. H. Tseng and T. Itoh, "Dual-band bandpass and bandstop filters using composite right/left-handed metamaterial transmission lines," in *Proceedings of the IEEE MTT-S International Microwave Symposium Digest*, pp. 931–934, IEEE, June 2006.

[7] X. Chen, J. Li, and Y. Shen, "WLAN bandpass filter with wide stopband using spurlines and microstrip stubs," in *Proceedings of the 9th International Symposium on Antennas Propagation and EM Theory (ISAPE '10)*, pp. 1188–1191, December 2010.

[8] T. C. Edwards, *Foundations for Microstrip Circuit Design*, Wiley, Chichester, UK, 1981.

[9] G. D. Alley, "Interdigital capacitors and their application to lumped-element microwave integrated circuits," *IEEE Transactions on Microwave Theory and Techniques*, vol. 18, no. 12, pp. 1028–1033, 1970.

[10] R. Siragusa, H. V. Nguyen, P. Lemaître-Auger, S. Tedjini, and C. Caloz, "Modeling and synthesis of the interdigital/stub composite right/left-handed artificial transmission line," *International Journal of RF and Microwave Computer-Aided Engineering*, vol. 19, no. 5, pp. 549–560, 2009.

[11] B. C. Wadell, *Transmission Line Design Handbook*, Artech House, 1991.

[12] D. M. Pozar, *Microwave Engineering*, John Wiley & Sons, 2009.

[13] H. Oraizi and N. Azadi-Tinat, "Optimum design of novel UWB multilayer microstrip hairpin filters with harmonic suppression and impedance matching," *International Journal of Antennas and Propagation*, vol. 2012, Article ID 762790, 7 pages, 2012.

[14] S.-G. Mao, S.-L. Chen, and C.-W. Huang, "Effective electromagnetic parameters of novel distributed left-handed microstrip lines," *IEEE Transactions on Microwave Theory and Techniques*, vol. 53, no. 4, pp. 1515–1521, 2005.

[15] J. T. Kuo and H. S. Cheng, "Design of quasi-elliptic function filters with a dual-passband response," *IEEE Microwave and Wireless Components Letters*, vol. 14, no. 10, pp. 472–474, 2004.

15

A Multifrequency Radar System for Detecting Humans and Characterizing Human Activities for Short-Range Through-Wall and Long-Range Foliage Penetration Applications

Ram M. Narayanan, Sonny Smith, and Kyle A. Gallagher

The Pennsylvania State University, University Park, PA 16802, USA

Correspondence should be addressed to Ram M. Narayanan; ram@ee.psu.edu

Academic Editor: Xianming Qing

A multifrequency radar system for detecting humans and classifying their activities at short and long ranges is described. The short-range radar system operates within the S-Band frequency range for through-wall applications at distances of up to 3 m. It utilizes two separate waveforms which are selected via switching: a wide-band noise waveform or a continuous single tone. The long-range radar system operating in the W-Band millimeter-wave frequency range performs at distances of up to about 100 m in free space and up to about 30 m through light foliage. It employs a composite multimodal signal consisting of two waveforms, a wide-band noise waveform and an embedded single tone, which are summed and transmitted simultaneously. Matched filtering of the received and transmitted noise signals is performed to detect targets with high-range resolution, whereas the received single tone signal is used for the Doppler analysis. Doppler measurements are used to distinguish between different human movements and gestures using the characteristic micro-Doppler signals. Our measurements establish the ability of this system to detect and range humans and distinguish between different human movements at different ranges.

1. Introduction

The ability to detect human targets and identify their movements through building walls and behind light foliage is increasingly important in military and security applications. Expeditionary warfighters and law enforcement personnel are commonly faced with unknown enemy threats from behind different types of walls as well as those concealed behind shrubs and trees. Technology that can be used to unobtrusively detect and monitor the presence of human subjects from stand-off distances and through walls and foliage can be a powerful tool to meet such challenges. Although optical systems achieve excellent angular resolution, optical signals are unable to penetrate solid barriers and foliage cover and therefore are totally ineffective in detecting humans in defilade. However, signals in the microwave frequency range can penetrate barriers to an acceptable degree and are therefore the sensors of choice in detection of targets through optically opaque walls. In this case, the choice

of the frequency of operation depends on the application, specifically on the barrier type, target position behind the wall, stand-off requirement, and resolution requirements, all of which are somewhat interrelated. Furthermore, since signals in the millimeter-wave frequency range are able to penetrate light foliage cover to an acceptable degree and can be focused to isolate a single human being, they are emerging as the sensors of choice in detection of targets hidden in foliage. The choice of the frequency of operation depends on the application, specifically on the atmospheric attenuation, stand-off requirements, and resolution requirements, all of which are somewhat interrelated.

Low-frequency microwave signals, less than 5 GHz in frequency, can penetrate building walls made of concrete, brick, or cinder blocks, with reasonably low loss. A noteworthy point is that humans behind walls are located at much shorter range from the radar sensor (typically 6–10 feet); thus portable antennas with relatively wider beamwidths can easily isolate a single human. Millimeter-wave systems typically operate in one of the atmospheric "windows," which offer low propagation loss. These windows exist around 35, 95, 140, and 220 GHz frequencies. The W-Band of the microwave part of the electromagnetic spectrum ranges from 75 to 110 GHz, thus covering the 95 GHz window. The short wavelengths at these frequencies permit the use of small portable antennas to achieve the required angular resolution in order to isolate a single human.

The antenna beamwidth θ, that is, the "field of view" of the antenna beam, of a circular aperture antenna (in radians) with a typically used parabolic aperture taper is given by

$$\theta = \frac{1.27\lambda}{D}, \tag{1}$$

where λ is the wavelength and D is the antenna size [1]. In radar applications, the two-way beamwidth is needed, which takes into account the combined transmit/receive antenna pattern. The two-way beamwidth $\overline{\theta}$ is given by

$$\overline{\theta} = \frac{\theta}{\sqrt{2}} = \frac{1.27\lambda}{D\sqrt{2}}, \tag{2}$$

by reasonably assuming a Gaussian-shaped main beam antenna radiation pattern. The $\sqrt{2}$ term in the denominator appears due to the fact that we are considering the angle between the half-power points of the two-way, that is, transmit/receive, antenna pattern. As an example, assuming a 3 GHz transmit frequency in the S-Band frequency range (2–4 GHz) corresponding to a wavelength of 10 cm (4 in) and a manageable antenna size of 6 inches, the two-way beamwidth is computed as 0.6 radians or 34.2 degrees. At a target range R of 2 m (~6 ft), the azimuth or cross range resolution ΔR_{CR} for a real-aperture radar, given by

$$\Delta R_{CR} = R\overline{\theta}, \tag{3}$$

is computed as 1.1 m (~3.6 ft), which is considered adequate for isolating a single human. In addition, assuming a 95 GHz transmit frequency in the W-Band corresponding to a wavelength of 3.15 mm (~1/8 in) and a manageable antenna size of 15 cm (6 in), the two-way beamwidth is computed as 0.018 radians (~1 degree). At a target range R of 100 m (~300 ft), the azimuth or cross range resolution is computed as 1.7 m (~5.4 ft), which is considered reasonably adequate for isolating a single human. While higher millimeter-wave frequencies achieve narrower beamwidths, the W-Band frequency range covering the 95 GHz window is preferred due to the lower cost and more extensive availability of components in this range.

The down-range resolution ΔR_{DR} is solely determined by the transmit bandwidth B and is expressed as

$$\Delta R_{DR} = \frac{c}{2B}, \tag{4}$$

where c is the speed of light [2]. Therefore, a 500 MHz transmit bandwidth yields a down-range resolution of 30 cm or 1 ft, quite adequate for isolating a single human. Several frequency-modulated waveforms operating over the bandwidth required for achieving the desired down-range resolution can be employed for through-wall imaging applications [3].

This paper discusses the architecture of the multifrequency radar system and presents data showing that human detection and human activity characterization are possible through different types of barriers. Section 2 provides an overview of noise radar and the micro-Doppler signal analysis. Sections 3 and 4 provide details of the design of the S-Band and W-Band portions of the multifrequency radar system, respectively. Experimental results are shown in Section 5 and conclusions are presented in Section 6.

2. Principles of Noise Radar and the Micro-Doppler Analysis

2.1. Noise Radar. While adequate cross range resolution can be achieved using small size antennas for short-range wall penetration, a suitable modulation scheme must be used to obtain the wide transmit bandwidth of 500 MHz to achieve the desired down-range resolution. Random noise modulation is an ideal candidate for military applications since it possesses several desirable properties, such as covertness, low probability of detection (LPD), low probability of intercept (LPI), immunity from jamming, and resistance to interference, owing to its totally featureless characteristics [4]. Only the basic principles of random noise radar are presented here for the sake of completeness. In a random noise radar, target detection and ranging are accomplished by cross-correlating the target reflected signal with a time-delayed replica of the transmit waveform [5]. The round-trip return time τ_R for a target located at a range of R is given by

$$\tau_R = \frac{2R}{c}. \tag{5}$$

Let $n(t)$ represent the transmitted wideband noise waveform whose autocorrelation is given by

$$R_{nn}(\tau) = \delta(\tau). \tag{6}$$

From (6), we see that the autocorrelation shows a peaked response at zero lag and is zero elsewhere. The reflected signal from the target is delayed by τ_R and can therefore be represented as $n(t - \tau_R)$. If a portion of the transmitted noise waveform is captured within the radar and internally delayed by a duration τ_D, it can be represented as $n(t - \tau_D)$. The cross correlation of the two signals $n(t - \tau_R)$ and $n(t - \tau_D)$ shows a peaked response only when $\tau_R = \tau_D$ and is zero when $\tau_R \neq \tau_D$. By stepping through various internal delays and determining the particular value of $\tau_D = \tau_{D,MAX}$ at which a peak occurs in the cross correlation response, target range R can be determined using

$$R = \frac{c\tau_{D,MAX}}{2}. \tag{7}$$

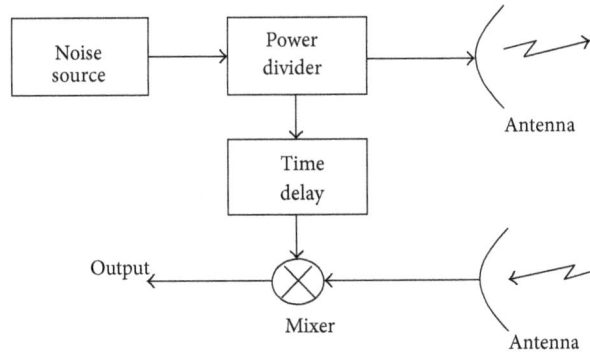

FIGURE 1: Simplified block diagram of a noise radar.

A simplified block diagram of a noise radar is shown in Figure 1, wherein the mixer acts as the cross correlator [6]. Current technological advances permit the implementation of fully digital radar architectures for noise signal generation and processing in the microwave frequency range and thereby achieve a great degree of flexibility [7, 8]. Signals in the millimeter-wave range of frequencies can be generated and processed using up- and downconversion of the digitally generated lower frequency microwave signals.

2.2. The Micro-Doppler Analysis. When a moving target is illuminated with a single tone frequency of f_0 corresponding to a wavelength of λ_0, it induces a Doppler frequency shift f_d in the reflected signal, which is given by

$$f_d = \frac{2v_r}{\lambda_0} = \frac{2v_r f_0}{c}, \qquad (8)$$

where v_r is the target's radial velocity with respect to the radar antenna. Mechanical vibration or rotation of structures in a target may induce frequency modulation on the target reflected signals and generate sidebands about the center frequency of the target's body Doppler frequency [9]. These modulations, which are usually at very low frequencies relative to the body Doppler frequency, are known as the micro-Doppler signatures. A stationary target of course produces no Doppler shift. However, if a stationary target vibrates, rotates, or maneuvers, its structural parts are in motion, and these induce the micro-Doppler modulations. If there are $i = 1, 2, \ldots, N$ structures with $v_{r,i}$ being the radial velocity of the ith structure, the composite micro-Doppler signal has frequency components at $2v_{r,i}f_0/c \; \forall i = 1, 2, \ldots, N$, which are unique to the specific motional characteristics of the target. Analysis of the micro-Doppler signatures in the joint time-frequency domain can provide useful information for target detection, classification, and recognition.

The micro-Doppler signals are also present in human activity, such as breathing and swinging arms, since each activity involves different types of motions of the chest, torso, and limbs. Figure 2 shows the micro-Doppler signatures of humans behind a wooden wall performing several distinctive activities. The radar stand-off distance was 9 m (30 ft) [10]. There are significant differences in the signatures leading us

to believe that not only can we detect concealed humans using radar but we can also identify what they are doing by analyzing their micro-Doppler signatures, which may help us to infer intent. At millimeter-wave frequencies, the Doppler signals occur at much higher frequencies due to the shorter wavelengths. In addition, smaller scale movements can be more easily recorded at these shorter wavelengths.

3. S-Band Radar System Description

A brief summary description of the S-Band through-wall radar system is provided below. A more complete description can be obtained from [11].

3.1. Baseband Dual-Mode Waveform Generation. In order to both detect humans and characterize their micro-Doppler signatures, a composite waveform is used, consisting of a wideband noise waveform for ranging and a single tone continuous wave signal for micro-Doppler detection. These waveforms are generated at lower frequencies, called baseband, and then upconverted to the desired frequency range of operation. The noise waveform of 500 MHz bandwidth is generated over the frequency range 100 Hz to 500 MHz, while the single tone is located at 300 MHz. An RF switch is used to select either waveform; therefore, the system operates in either the ranging mode (using the noise waveform) or in the Doppler mode (using the single tone). Each waveform is split and one-half is upconverted to the desired frequency range of operation, while the other half of the signal is routed to the receiver for performing the ranging or the micro-Doppler processing with the received and downconverted signal.

3.2. S-Band Radar System Overview Description. A simplified block diagram of the system is shown in Figure 3. The noise source produces a noise waveform with over the 100 Hz to 500 MHz range at an output power of 10 mW, which is filtered to achieve a good spectral shape. After filtering, the waveform passes through a power splitter where one output serves as a reference signal and the other output goes to a switch (SPDT RF Switch) for upconversion. The single tone is generated by a voltage controlled oscillator (VCO), which produces an output of about 4 mW at the desired frequency. This signal is filtered and then split in a similar fashion, where one

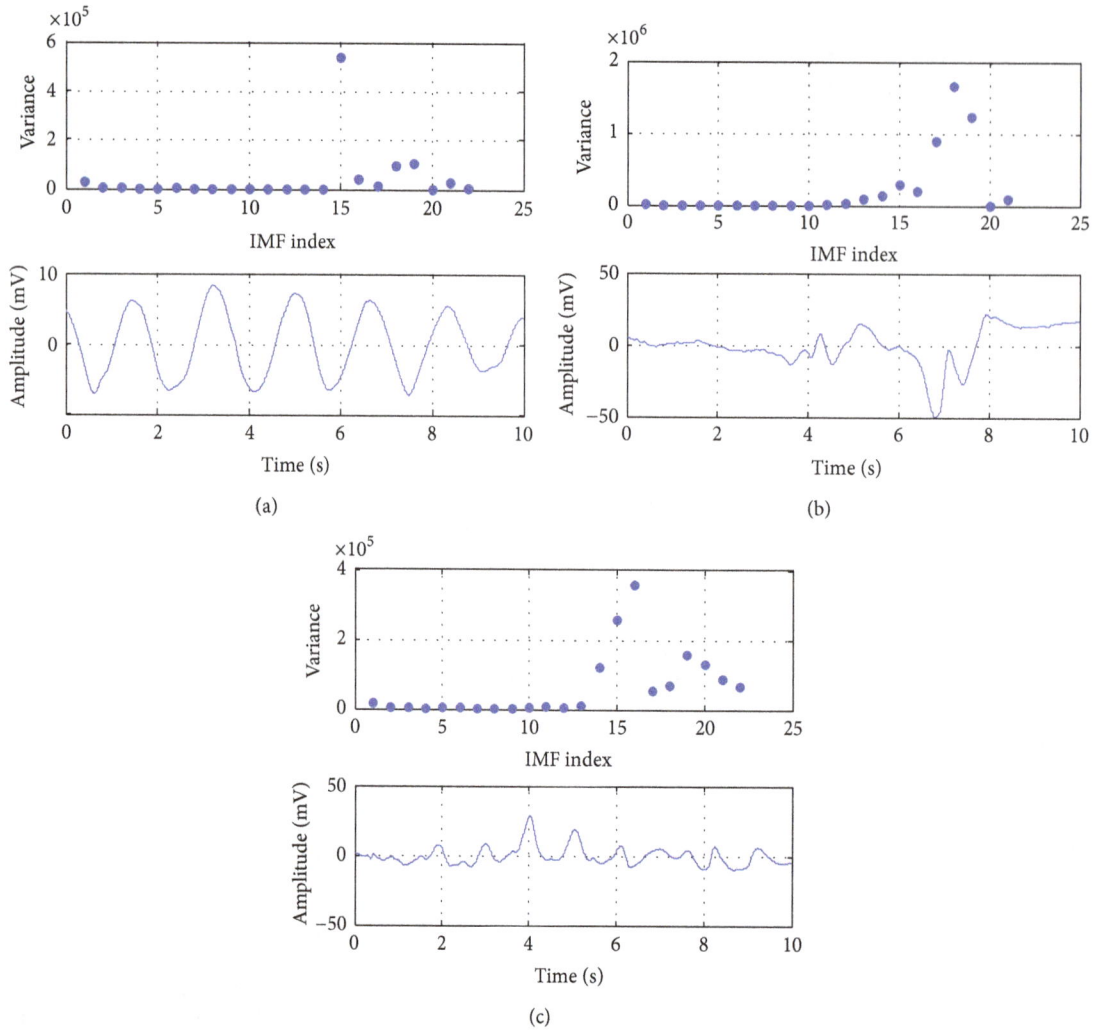

FIGURE 2: The micro-Doppler signatures of concealed human activities at 9 meters stand-off distance in front of a wooden shed: (a) breathing, (b) lifting a large object from the ground, and (c) moving arms up and down rapidly.

output leads to the switch awaiting upconversion while the other output terminal goes to the receiver. Depending on user preference, either the noise signal or single tone is selected by the RF switch and sent to an upper sideband upconverter, which is pumped by a high-frequency single tone S-Band local oscillator signal at frequency f_{LO} (in MHz). Thus, the upconverter selects the upper sideband of the mixing process. For the ranging mode, the upconverter output exists over a frequency range between f_{LO} and $f_{LO} + 500$ MHz, while, in the Doppler mode, the output frequency is equal to $f_{LO} + 300$ MHz. The upconverter output is amplified, filtered (again), and transmitted via a transmit antenna.

A two-stage downconversion receiver processor is used in the system. The time-delayed received signal is collected by an identical receive antenna, amplified, and filtered to remove out of band interference and noise. Then, it is downconverted using the same single tone S-Band signal at frequency f_{LO} as the local oscillator, which yields the reflected noise signal over 0 to 500 MHz range in the ranging mode or a micro-Doppler modulated single tone around 300 MHz in the Doppler

mode. These signals are separated into different paths via appropriate filters and sent to a data acquisition system and the digitizer. The digitizer also receives the transmitted samples in both the ranging and the Doppler modes. The reference and the received noise signals are digitally cross-correlated to obtain range to target, while the reference single tone and the received micro-Doppler modulated single tone are mixed together after being low-pass filtered and processed to extract the micro-Doppler modulation. The sampling frequency for each channel is 2 GS/s for ranging and 1.25 GS/s for the micro-Doppler signature, more than adequate to satisfy the Nyquist sampling criterion [12, 13].

The component layout of the S-Band radar is shown in Figure 4, while the fully packaged system (minus antennas) is shown in Figure 5.

3.3. Antennas. RF antennas are usually linearly polarized. However, most reinforced building walls contain a lattice of reinforcing bars, or rebars which may be either vertically oriented, or horizontally oriented, or both. Such a structure will

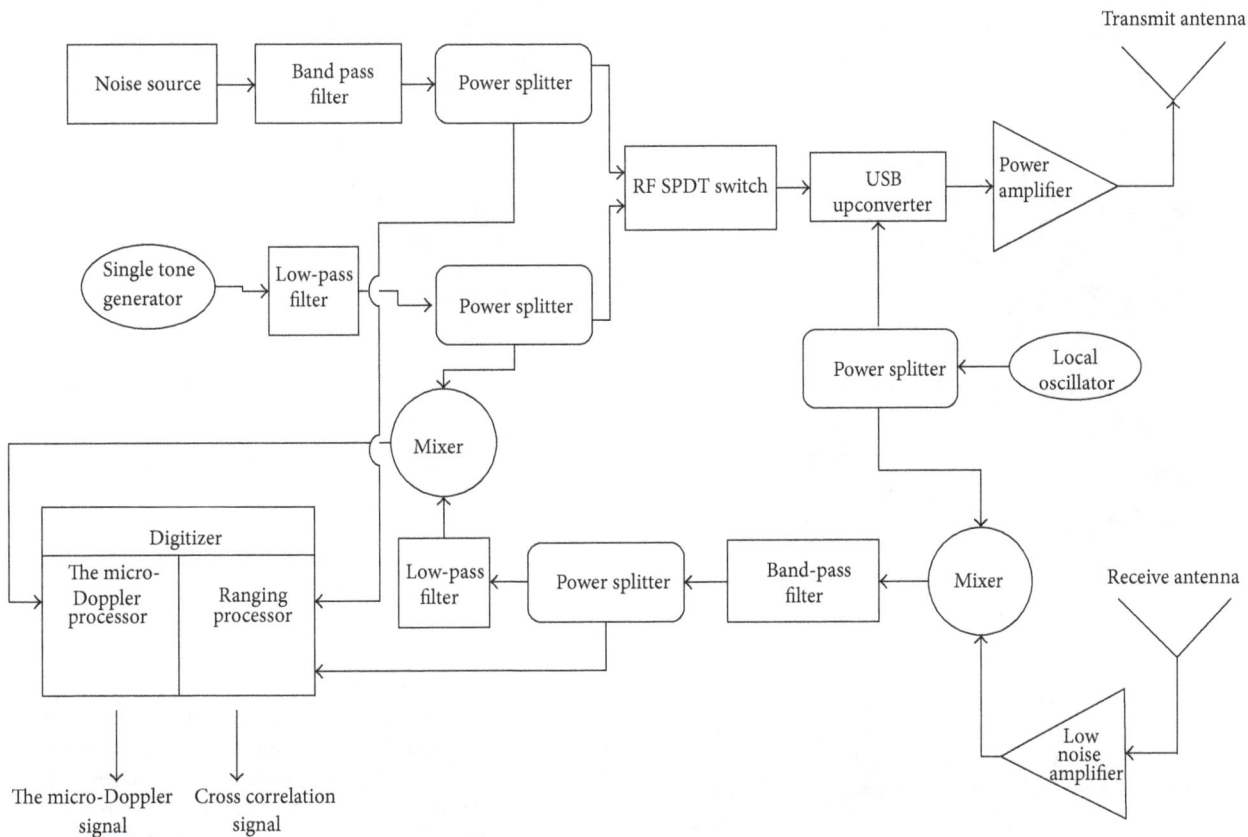

FIGURE 3: Simplified block diagram of the S-Band radar system.

FIGURE 4: Component layout of S-Band radar system.

FIGURE 5: Fully packaged S-Band radar system (minus antennas).

affect the propagation of EM waves through it, especially if the rebars are oriented in the direction of the wave polarization. A method to overcome this limitation is to employ a circularly polarized wave, wherein the instantaneous polarization of the wave moves around a circle, thereby allowing most of the wave to pass through with very little loss due to the choice of the wrong polarization. In our system, therefore, we employed helical antennas which are able to transmit and receive circularly polarized signals [14]. They consist of a conducting element wound in the geometrical shape of a helix. The conductors are supported by a central buttress frame, and together they are mounted on a ground plane.

To enhance the gain of the antenna and thereby reduce its beamwidth as well as to reduce the beam sidelobes and back lobe, the ground plane can be modified in the shape of a "salad bowl" curved towards the helix, as suggested in [15].

It is known that targets reflect the oppositely handed polarization when illuminated by a circularly polarized wave. The helices used for the transmit and the receive antennas are oppositely wound so that the transmitted wave is right-hand circularly polarized whereas the receive antenna is left-hand circularly polarized. The helical antennas were operated in the axial mode; that is, the antenna dimensions were comparable to the wavelength, wherein a directional endfire pattern,

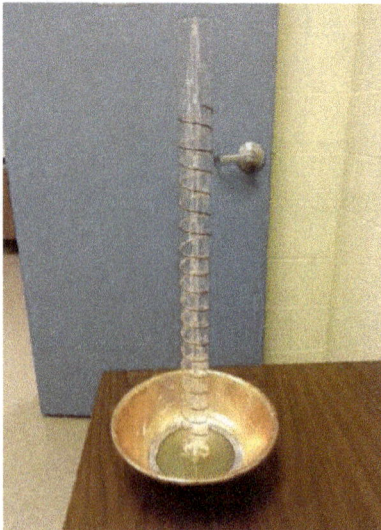

FIGURE 6: View of the helical antenna showing the construction details.

FIGURE 7: 20.32 cm (8 in) thick cinder block wall in the wall support frame.

along the axis of the helix, is achieved. The helical antennas designed for this application had the following dimensions: (a) outside rim diameter of the salad bowl shaped ground plane = 18.8 cm (7.4 in), (b) bottom diameter of the ground plane = 8.89 cm (3.5 in), and (c) overall axial length = 35.3 cm (13.9 in). The designed antenna is shown in Figure 6.

3.4. Wall Construction. A wall support frame was constructed to house different masonry materials in a dry-stack fashion. The frame was designed to support a wall (e.g., brick or cinder block) that was 2.44 m (8 ft) tall × 2.44 m (8 ft) wide. In addition, the frame had an adjustable width for wall thicknesses of 10.2, 20.3, or 30.5 cm (4, 8, or 12 in, resp.). The structure stood a foot above the ground (adding additional height to the wall) on castor wheels enabling the wall to be mobile. Figure 7 shows the constructed wall with 20.3 cm (8 in) thick cinder blocks.

To collect radar data, the antennas were mounted on a wooden stand that positioned the antennas approximately 1.37 m (54 in) above the ground and about 1.83 m (6 ft) from the front of the wall. Care was taken to align the antennas properly since poor alignment could negatively influence the results. Coaxial cables of adequate length were connected to the antennas from the radar system (allowing some separation). Targets, such as metallic trihedral corner reflectors and humans were located behind the wall at various distances.

4. W-Band Radar System Description

A brief summary description of the W-Band foliage penetration radar system is provided below. A more complete description is provided in reference [16].

4.1. Baseband Multimodal Waveform Generation. In order to both detect humans and characterize their micro-Doppler

signatures, a composite waveform is required, consisting of a wideband noise waveform for ranging and a single tone continuous wave signal for the micro-Doppler detection. These waveforms are generated at lower frequencies, called baseband, and then upconverted to the desired frequency range of operation. The noise waveform of 500 MHz bandwidth is generated over the frequency range of 1.1–1.6 GHz in the L-Band frequency range, while the embedded single tone is located at 1.1 GHz. Both signals are summed together, upconverted to the desired frequency range at W-Band, and transmitted as a composite multimodal waveform. Thus, the system operates simultaneously in both the ranging mode (exploiting the noise waveform component) and the Doppler mode (exploiting the single tone component). Just prior to waveform summation in the transmit chain, one-half of each signal is split and routed to the receiver for performing the cross correlation operation with the received and downconverted signal. Our system was designed in two main sections, a low-frequency L-Band section and a high-frequency mm wave section. Both the L-Band and mm wave sections can then be further subdivided into transmit and receive chains.

4.2. W-Band Radar System Overview Description. A simplified block diagram of the system is shown in Figure 8. The L-Band and the W-Band transmit and receive chains are clearly demarcated. The low-frequency noise source produces a noise waveform over the 1.1 to 1.6 GHz range at an output power of 6.25 mW, which is filtered to achieve a good spectral shape. The single tone output power at 1.1 GHz is 5 mW. Thus, the total power of the composite multimodal waveform is 11.25 mW. The W-Band transmit chain accepts the signal from the L-Band transmit chain and upconverts it to W-band via a mixer followed by a high-pass filter (HPF), which discards the lower sideband. The mixer affords good suppression of the W-Band local oscillator source leakage. The SSB upconversion is possible because the L-Band signal is offset in frequency

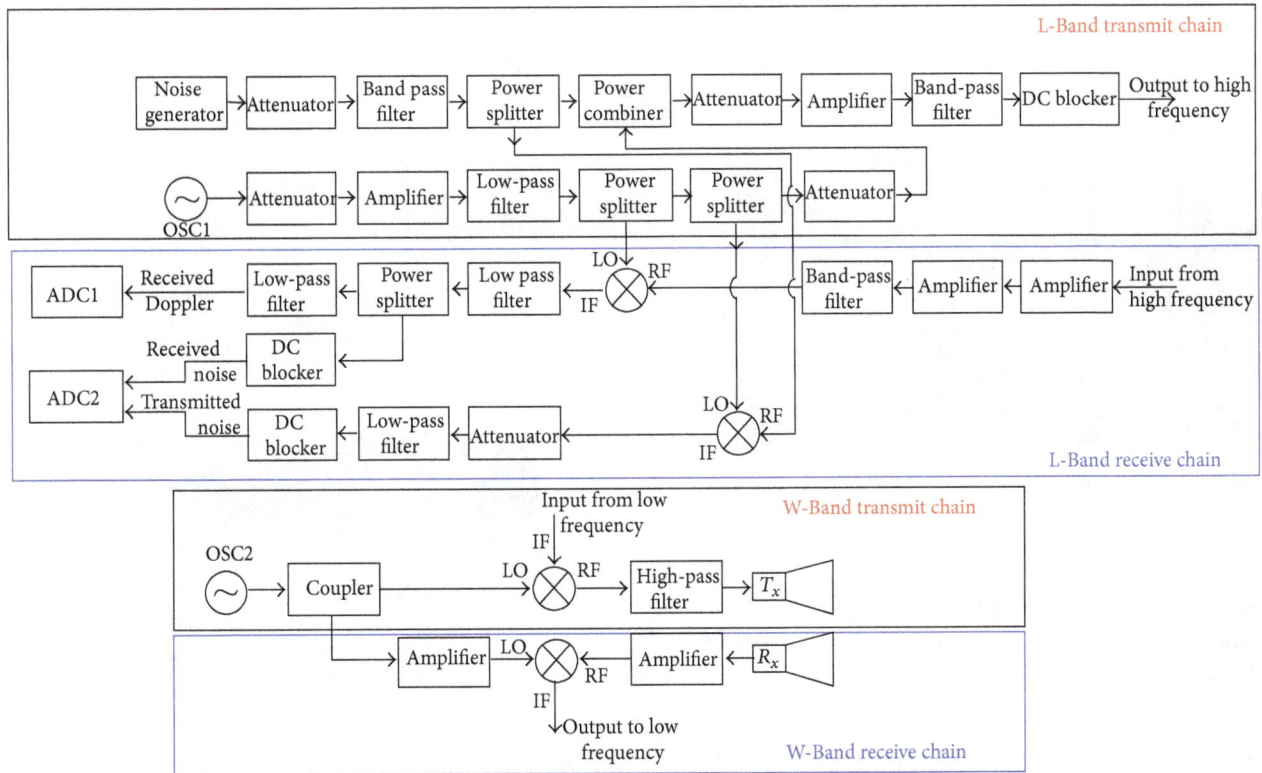

FIGURE 8: Block diagram of the W-Band radar system.

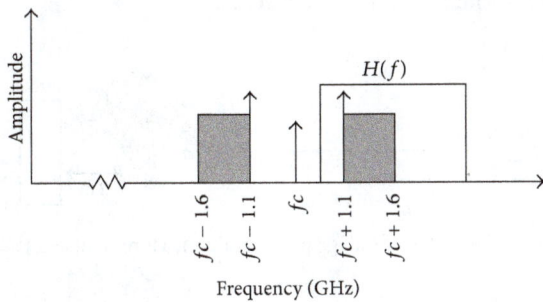

FIGURE 9: Upconverted spectrum at W-Band.

from baseband; that is, it is created from 1.1 to 1.6 GHz. This scheme avoids the necessity to use an expensive in-phase and quadrature (I/Q) SSB upconverter. Figure 9 illustrates the frequency spectrum of the signal after upconversion and filtering. After upconversion, the composite signal power is reduced to 4 mW due to conversion loss of the mixer. The multimodal signal at W-Band is then transmitted using a transmit antenna.

Once the signal reflects off of an object, the W-Band receive chain captures the backscattered signal through the receive antenna. The received signal is amplified using a low noise amplifier and downconverted back to the 1.1–1.6 GHz frequency range. The signal is then sent to the L-Band receive chain for further processing. The L-Band receive chain takes this signal and prepares it for the final downconverting stage. This process consists of amplifying and filtering the

signal before downconverting to baseband. Down converting both the Doppler and noise waveforms to baseband simultaneously is possible because the single tone is placed at the beginning edge of the band. After downconversion to baseband, the Doppler and noise waveforms are separated by splitting the signal and filtering appropriately. Since the Doppler signal is located in the range of DC to a few kHz, a low-pass filter (LPF) is used to band-limit the signal and to avoid aliasing unwanted signals components when the signal is digitized. The noise waveform contains a DC offset created by the single tone mixing with itself so a DC blocker in addition to a LPF is used to prepare the noise waveform for digitizing. A copy of the transmitted signal is needed as a reference to the matched filter. A copy of the noise waveform is sampled from the L-Band transmit chain and downconverted using an identical mixer and local oscillator as is used for the received signal. The signal is then filtered and attenuated before being digitized. Once the Doppler and noise waveforms are available at baseband, they are digitized using two separate digitizers. The Doppler signal is digitized using a low sample rate digitizer since these frequencies are quite low. Both the received and reference noise waveforms are digitized using high-speed digitizer with the sample rate set to 1 Gs/sec to satisfy the Nyquist sampling criterion [12, 13]. Both digitizers are connected to a laptop and interfaced with LabView for processing and data saving.

The component layout of the W-Band radar is shown in Figure 10. The W-Band component tray fits atop the L-Band tray in the fully packaged system. The antennas shown

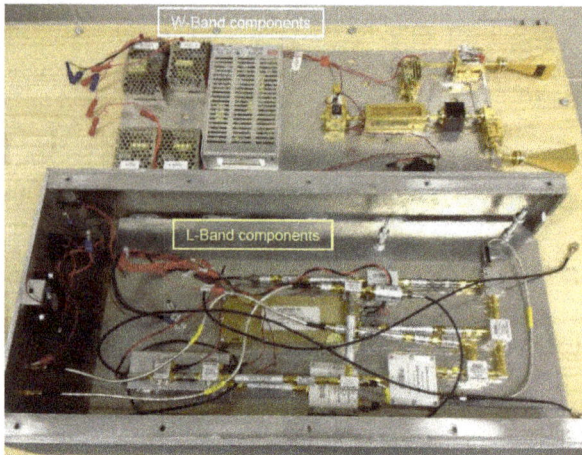

FIGURE 10: Component layout of W-Band radar system.

in Figure 11 are smaller pyramidal horn antennas for closer range measurements, while long-range field measurements used a larger circular dielectric horn lens antenna.

4.3. Antenna. Two different antennas were used, depending upon the target range considerations. For close range measurements, pyramidal horn antennas of aperture size 1.0625 inches × 0.875 inches, shown in Figure 11(a), were used. The one-way beamwidths in the principal planes for this antenna are computed as 8.4 degrees × 10.2 degrees. In this antenna, the feed waveguide aperture is flared in both dimensions to achieve a higher gain over the operating waveguide bandwidth [17]. These antennas achieve a cross range resolution of 1.7 m (5.4 ft) at a target range of approximately 14.7 m (48 ft). Beyond this range, it is not possible to isolate a single human; therefore, larger size antennas were used for longer range applications. For longer range measurements, circular dielectric horn lens antennas of 15.24 cm (6 in) aperture diameter, shown in Figure 11(b), were used. In this antenna, a dielectric lens is integrated into a dielectric loaded horn antenna to improve the antenna efficiency [18]. The principle involved in geometrical optics lens design is to collimate the rays from the primary source by refraction at the surface of the lens. These antennas had a one-way beamwidth of approximately 1.5 degrees, achieving a cross range resolution of 1.7 m (5.4 ft) at a target range of 91.4 m (300 ft).

4.4. Foliage Cover Description. In addition to unobstructed long-range measurements, data were also collected to investigate the radar system's ability to detect targets through light foliage. To do this, we aimed the radar at a Border Forsythia (*Forsythia × intermedia*) shrub of approximate dimensions 2 m × 2 m × 2 m and placed different targets behind it. This shrub has an upright habit with arching branches and grows to 3 to 4 m high. Data were collected from targets behind the bush when it had leaves and also after the leaves fell off to see the effect the leaves have on the system's ability to identify targets behind it. A diagram of experimental

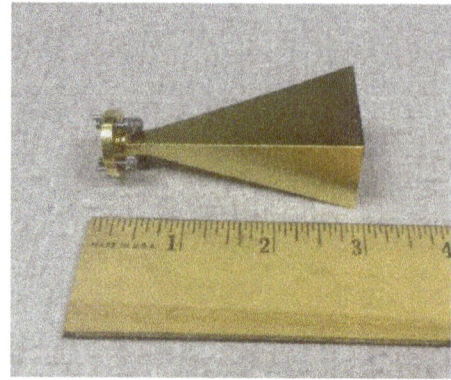

(a)

(b)

FIGURE 11: (a) Pyramidal horn antenna; (b) circular dielectric horn lens antenna with a sighting scope attached.

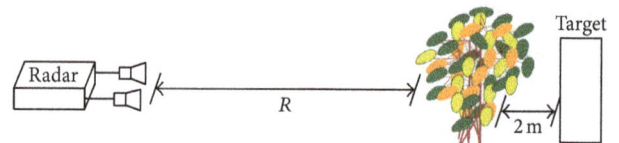

FIGURE 12: Geometry of foliage penetration measurement setup.

setup is shown in Figure 12 and photographs of the foliage penetration measurement setup are shown in Figure 13.

To collect radar data, the antennas were mounted on a wooden stand that positioned the antennas approximately 1.37 m (54 in) above the ground. Care was taken to align the antennas properly using sighting scopes since poor alignment could negatively influence the results. Targets, such as metallic trihedral corner reflectors and humans, were located at various distances.

5. Experimental Results and Data Analysis

5.1. Background Subtraction. A major problem in through-wall radar is the existence of large peaks in the reflected response due to direct antenna coupling as well as the reflection from the wall itself. These signals can obscure the target reflections, as can be seen in Figure 14. In order to overcome this limitation, background subtraction was used

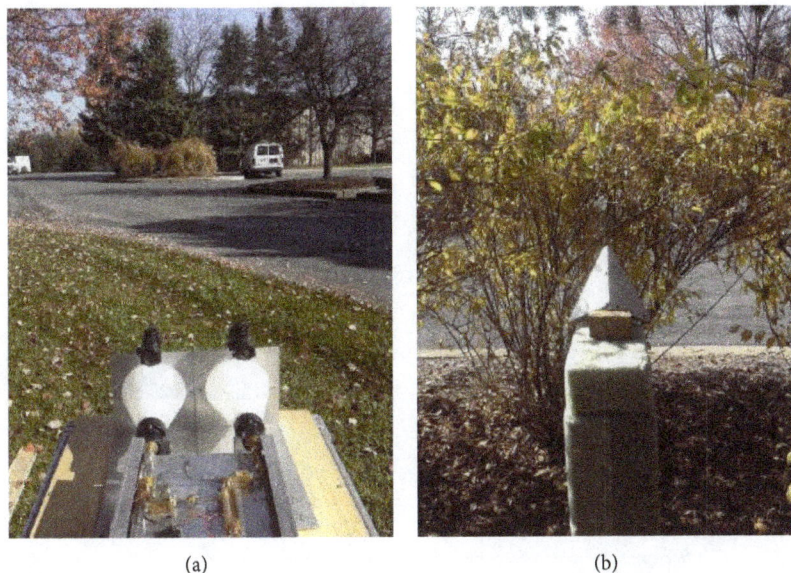

FIGURE 13: Photographs of the foliage penetration measurements. (a) shows the radar aimed at the bush, while (b) shows a corner reflector target placed behind the bush.

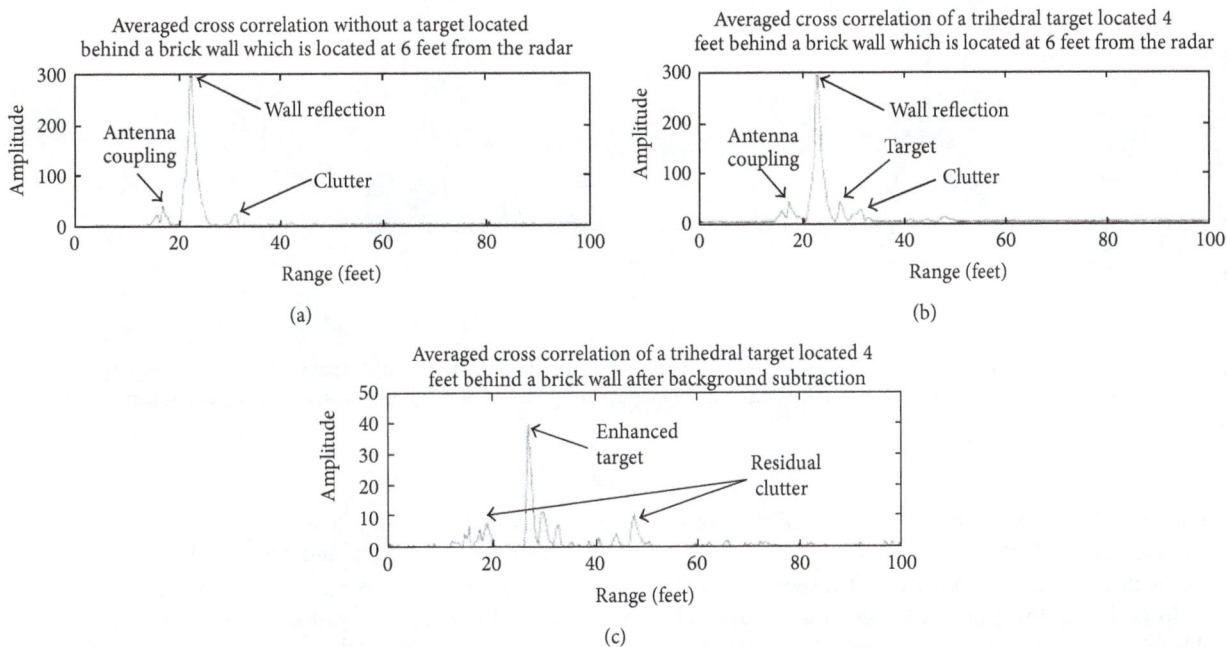

FIGURE 14: Detection of a small trihedral target placed 1.22 m (4 ft) behind a 10.2 cm (4 in) thick brick wall. Both no target and target present cases are shown, as well as the implementation of the background subtraction algorithm which suppresses non-target responses and enhances target response.

[19] for the S-Band radar, which resulted in the excellent suppression of constant nontarget induced responses and significant enhancement of the target-induced response, which is also shown in Figure 14. The residual clutter, which is not completely suppressed, is due to mutual interactions between different reflectors.

5.2. *Distance Correction.* Direct antenna coupling also obscured the low target reflections from longer ranges for the W-Band radar, as can be seen from the correlation plot in Figure 15(a) for the data acquired from a human target under unobstructed conditions at a range of 700 feet (213 m). In order to overcome this limitation, background subtraction

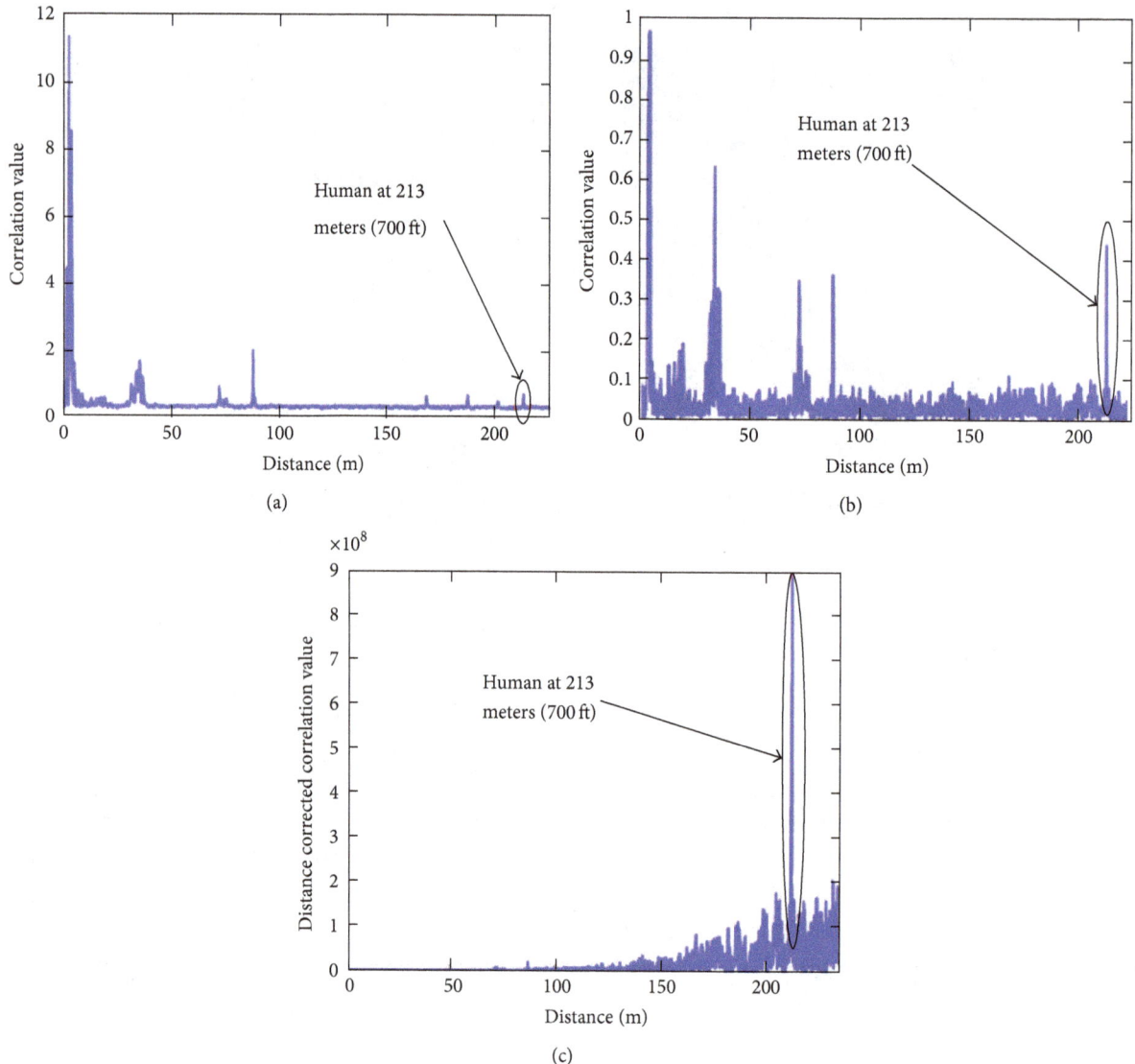

FIGURE 15: Detection of a human at a range of 213 m (700 ft). (a) No background subtraction; (b) background subtraction implemented; (c) background subtraction and distance correction implemented. Note the suppression of non-target responses and enhancement of target response at longer range.

was used [19] and the results are shown in the correlation plot in Figure 15(b). The residual clutter, which is not completely suppressed, is due to mutual interactions between different reflectors. In addition, there are still background peaks of comparable amplitudes to that of the human. These peaks are much closer to the radar. However, the radar range equation tells us that the received power from a target falls off as $1/R^4$, which is called the spreading loss. The spreading loss effect is more pronounced in the case of the long-range radars, where the maximum target distance is several orders of magnitude compared to the minimum target distance. Thus, if the same target is placed at twice the distance from the radar antenna, the power actually received at the radar is reduced by a factor of $2^4 = 16$. A straightforward method is to weight the data in order to compensate for the increasing signal attenuation as a function of range caused by the material

attenuation and spreading loss [20]. Therefore, in order to compensate for this roll-off and equalize the target response at different ranges, an inverse distance correction of R^4 was implemented to adaptively enhance the return from distant targets. Figure 15(c) shows the distance-corrected correlation plot, wherein the peaks from the human target can be more clearly detected and identified.

5.3. *Human Detection and Human Movement Tracking through Wall.* For the S-Band though-wall radar, a human was located at a distance of 1.22 m (4 ft) behind a 10.2 cm (4 in) thick brick wall with the antennas located at a distance of 1.83 m (6 ft) in front of the wall. Thus, the distance between the antennas and the human was about 3.05 m (10 ft). We note from Figure 16 that the background subtraction technique is able to detect the human quite clearly.

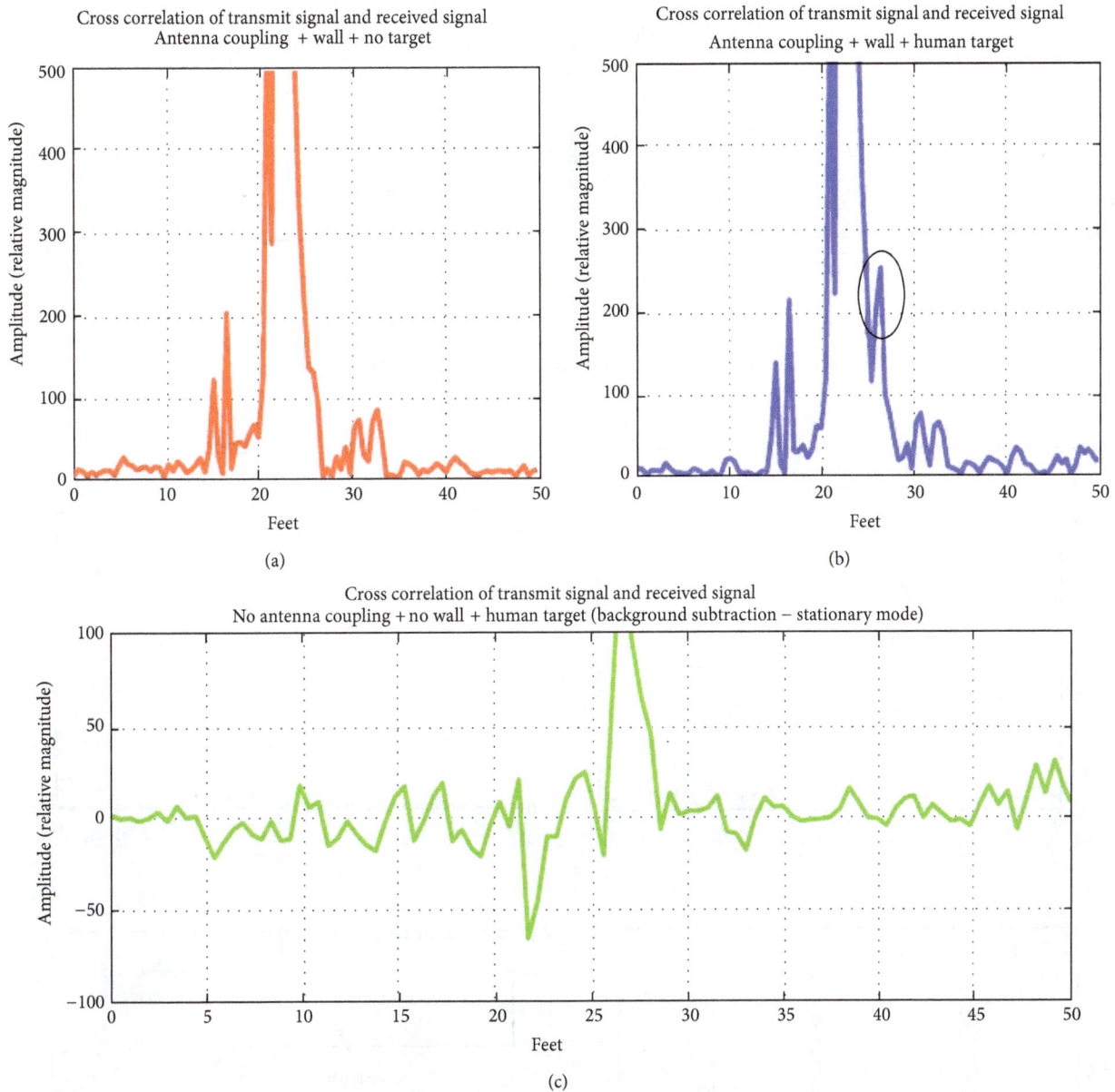

FIGURE 16: Detection of a human target placed 4 feet behind a 10.4 inch thick brick wall. Both no target and target present cases are shown, as well as the implementation of the background subtraction algorithm which suppresses non-target responses and enhances target response.

In addition, background subtraction can also be used for detecting moving targets. Subtraction of successive frames of the cross correlation signals between each received element signal and the transmitted signal has been shown to be able to isolate moving targets in heavy clutter [21]. To accomplish this, a total of 10 previous correlation scenes which were aggregated for averaging were also subsequently used to perform frame by frame subtraction. The plots shown in Figure 17 clearly demonstrate the ability of the algorithm to track a moving human.

5.4. *Target Detection through Foliage.* For the W-Band radar foliage penetration experiments, we used two targets, a corner reflector and a human. The radar was located at a stand-off distance of 30 m (98.4 ft) from the bush and each target was placed 2 m (6.6 ft) behind the bush. Correlation data were averaged over 100 looks to reduce the effects of noise. Figures 18(a) and 18(b) show baseline data for the bush with no leaves and corner reflector, respectively. The average correlation value, which is proportional to the received power, of the bush with no leaves is 0.09, while it is 5.0 for the unobstructed corner reflector. When the corner reflector was placed behind the bush, its average correlation value dropped to 0.92, as seen in Figure 18(c). This corresponds to a two-way RF signal loss of 7.4 dB for the bush with no leaves. We note that the reflection from the bush in this scene is consistent with the baseline bush data. Data were also taken when the bush had leaves on it and the corner reflector was placed behind

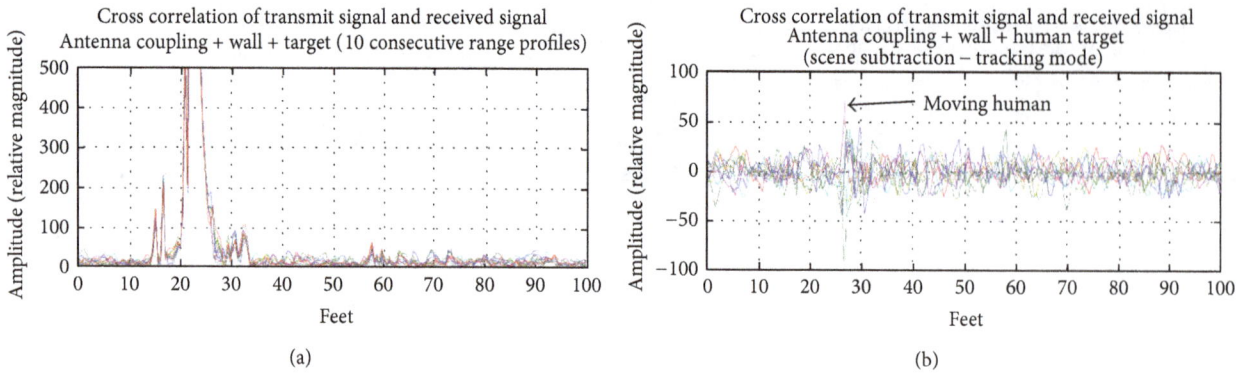

FIGURE 17: Tracking of a moving human using successive scene subtraction.

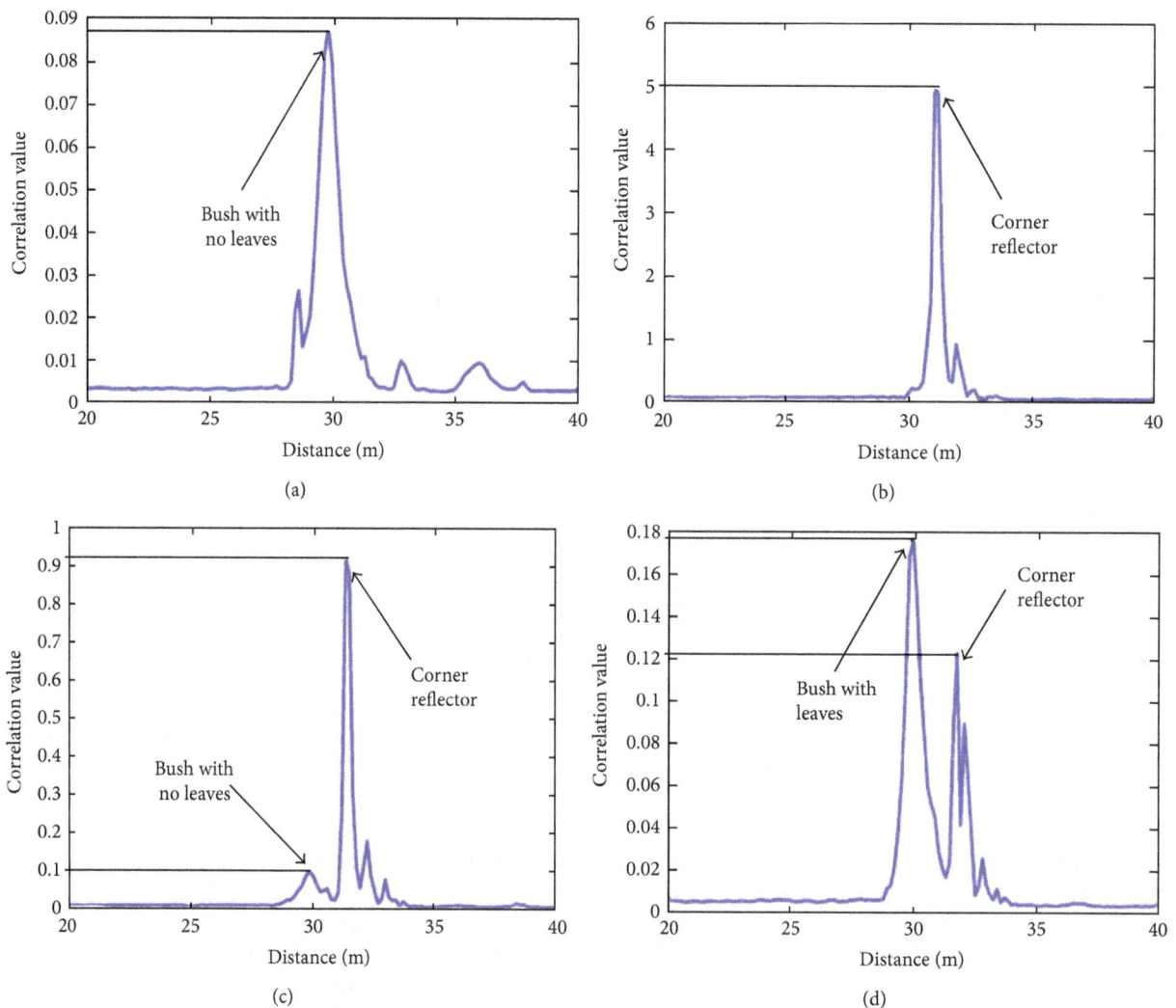

FIGURE 18: Detection of a corner reflector behind a bush.

the bush. The correlation plot in Figure 18(d) shows that the reflection from the bush has doubled to 0.18 (due to the leaves), while the response from the corner reflector is reduced to 0.12. This corresponds to a two-way loss through the bush with leaves of 16.2 dB.

In addition, data were also collected for a human target. Correlation data taken for the human behind the bush with no leaves is shown in Figure 19. This plot shows that the response of the human behind the bush with no leaves is about 0.04. This is 23 times less (13.6 dB less) than the

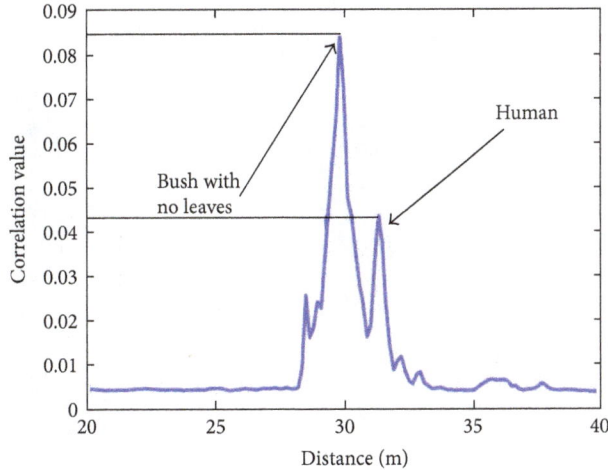

FIGURE 19: Detection of a human behind a bush.

response of the corner reflector for the same setup. When the loss through the bush with leaves is taken into account, we infer that detection of human is not possible under these conditions. However, we will show later on that the human can be detected behind a fully foliated bush by using the micro-Doppler characteristics of the human movement activity.

5.5. Human Activity Recognition Approach. Since different human activities result in different micro-Doppler signatures, a technique was developed and implemented for automatic classification of specific human activities, more fully described in [22], and hence not repeated. Since the radar micro-Doppler signals are generally nonlinear and nonstationary, conventional Fourier-based approaches are not optimal for their analysis. The basis of our approach is to decompose the micro-Doppler signals using the empirical mode decomposition (EMD) into their intrinsic oscillatory modes called the intrinsic mode functions (IMFs) [23]. The faster oscillations in the signal are present in the lower-indexed IMFs while the slower oscillations reside in the higher-indexed IMFs. These IMFs are components of the original signal and each IMF is orthogonal to all of the other IMFs. Each IMF comprises signal components that belong to a specific oscillatory time scale. The energy as a function of the IMF index provides us with a unique feature vector for human activity classification.

Classification of signals, such as the micro-Doppler signatures, requires a unique feature vector for each signal. EMD readily provides a feature vector by the calculation of the energy of each IMF component or the inner product of the signal with itself. When the EMD process is conducted on the micro-Doppler signals, the collection of IMF energies provides us with a vector that is unique to the movement that caused the Doppler frequency shift.

Support vector machines (SVMs) have proven to be an effective alternative to traditional classification techniques, such as the Bayesian classifiers and artificial neural networks (ANNs) [24]. The primary advantages of SVMs over other

methods are their ability to generalize and relative ease of implementation. The classifier is optimized to produce a model that is based on the training set feature vectors and their associated known class label. Using this model, the test set can be accurately classified using only their feature vectors, without knowledge of the class label.

The classification is performed using an SVM with a Gaussian kernel. The constrained optimization problem is formulated as

$$\max_{w,b,\xi} \quad \frac{1}{2}\mathbf{w}^T\mathbf{w} + C\sum_{i=1}^{l}\xi_i$$
$$\text{subject to}: \quad y_i\left(\mathbf{w}^T\phi(\mathbf{x}_i)+b\right) \geq 1-\xi_i$$
$$\xi_i \geq 0, \quad i=1,\dots,l,$$

(9)

where $\mathbf{w}$ is the weight vector that defines a linear hyperplane separating the two classes of data, b is the constant offset of the hyperplane, ξ_i is a measure of the error of any misclassification for the ith class, and C is a penalty parameter that allows the classifier to tolerate some errors. The vector function ϕ maps the feature vectors into an N-dimensional space. The parameters $\mathbf{x}_i$ and y_i are the feature vectors and their associated class label (±1), respectively. For the specific problem of classifying the micro-Doppler signals that arise from human motion, the $\mathbf{x}_i$ values are the energy feature vectors extracted using EMD.

SVMs were originally developed to solve the binary classification problem; therefore, modifications must be made in order to extend the binary problem to a multiclass problem. Multiple methods have been proposed to tackle this problem. Because of its intuitiveness and its ability to be easily adapted for additional classes, the one-against-all (1-a-a) method was chosen for the experiments [25]. The five human activity classes tested are as follows: (1) noise, that is, no human present, (2) breathing, (3) swinging arms, (4) picking up an object on the ground from a standing position, and (5) transitioning from crouching to standing.

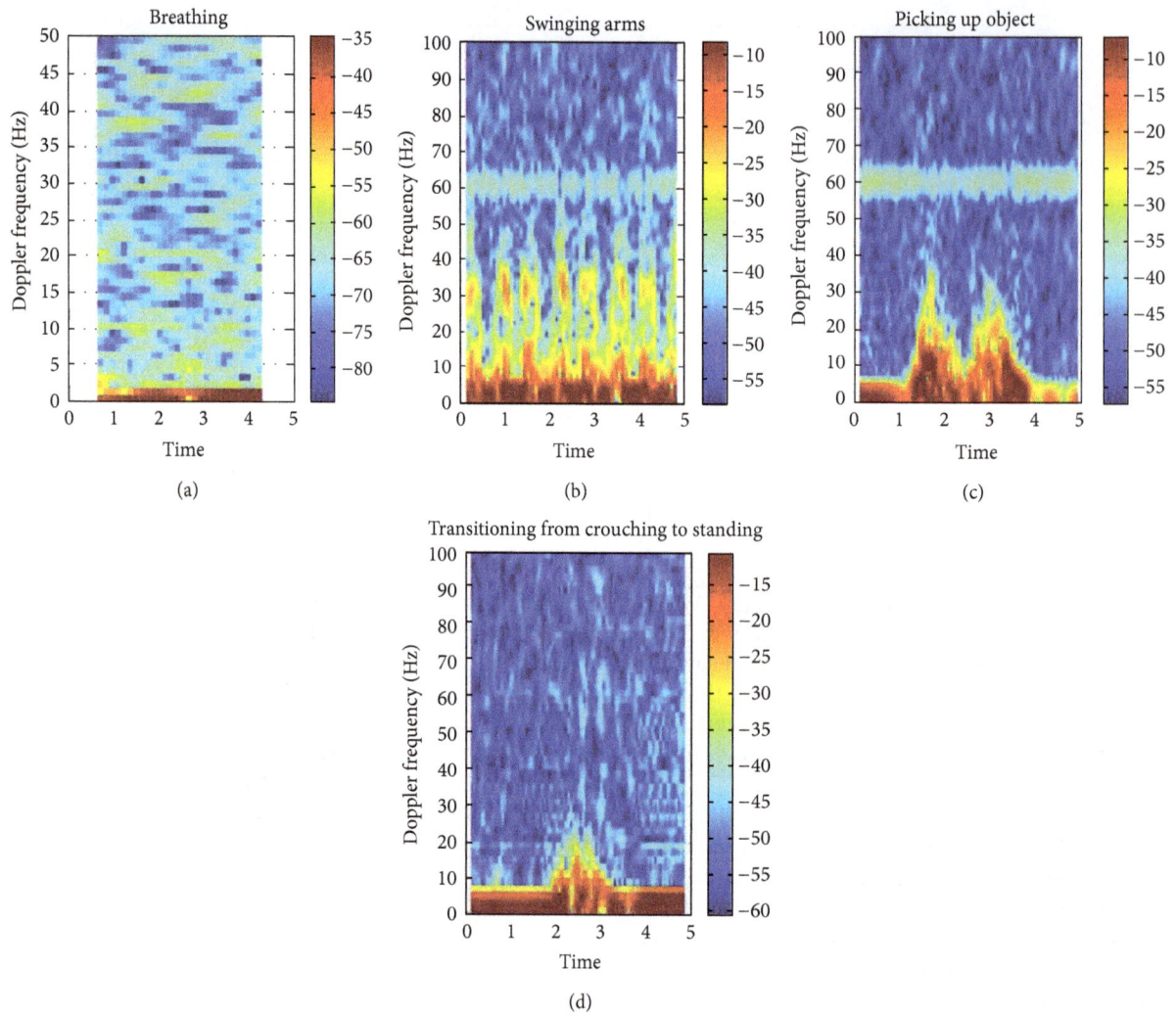

FIGURE 20: Time-frequency plots of different human activities as measured by the S-Band through-wall radar.

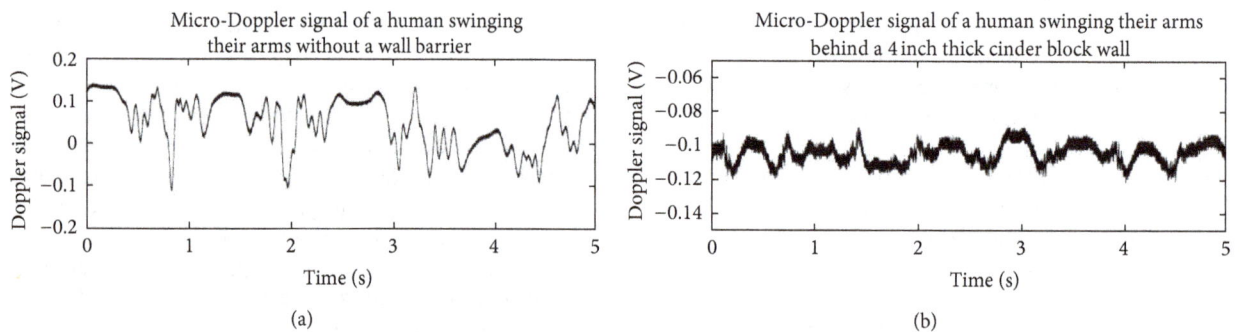

FIGURE 21: Comparison of the micro-Doppler signal of humans swinging their arms without a wall barrier and with a wall barrier using the S-Band through-wall radar.

5.6. Human Activity Classification Results for S-Band Through-Wall Radar. Experimentally observed time-frequency plots of unobstructed (without wall) human activities are shown in Figure 20 for the S-Band through-wall radar, wherein clear differences are observed. With the presence of a wall between the radar and the human, it was noted that the signal-to-noise ratio (SNR) was degraded and the micro-Doppler features

were distorted, as shown in Figure 21. This indicates that the feature vectors for through-wall recognition of human activities are expected to be different from those without a wall.

Three cases were examined for activity classification for the S-Band through-wall radar: (1) direct transmission without a wall barrier, (2) transmission through a 10.2 cm

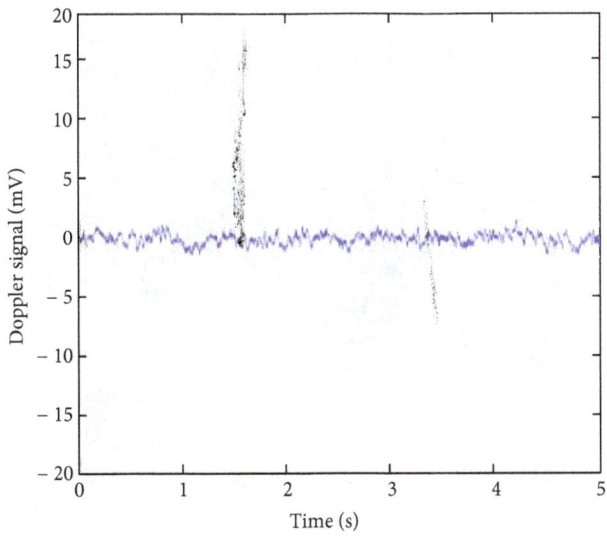
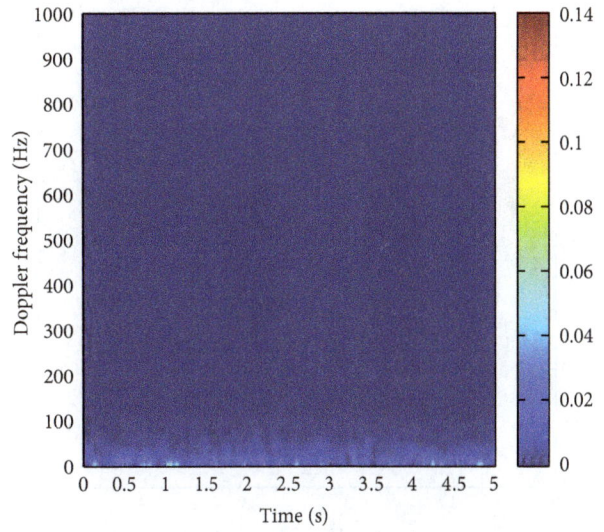

(a)

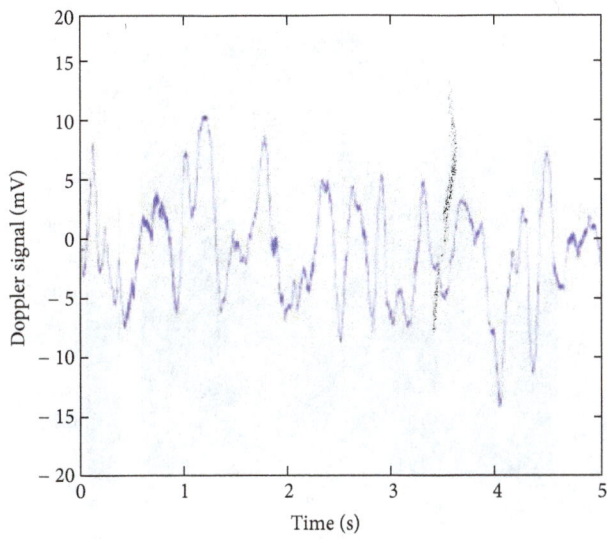
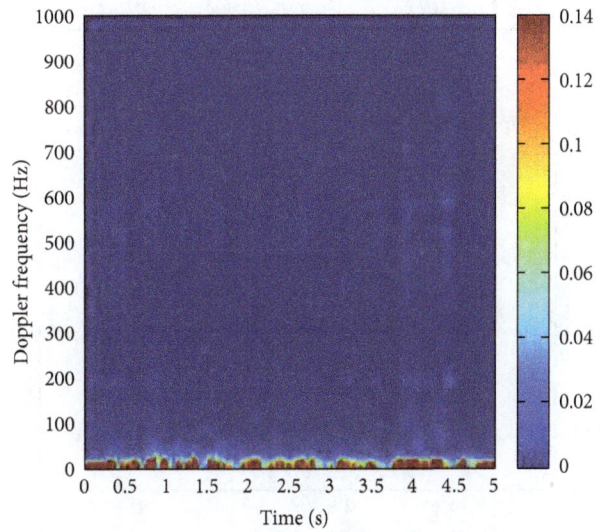

(b)

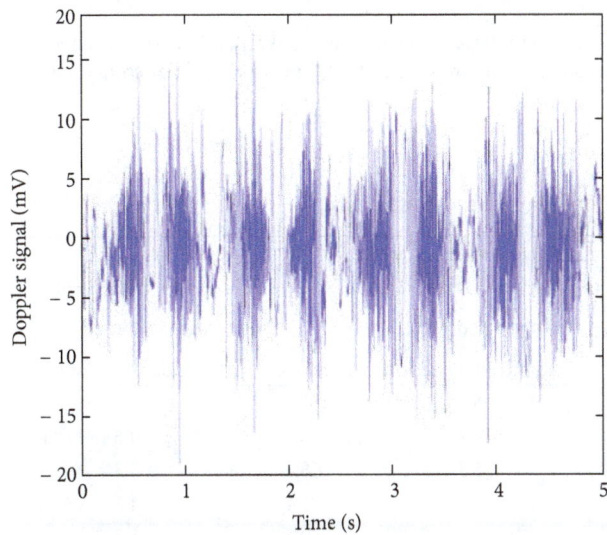
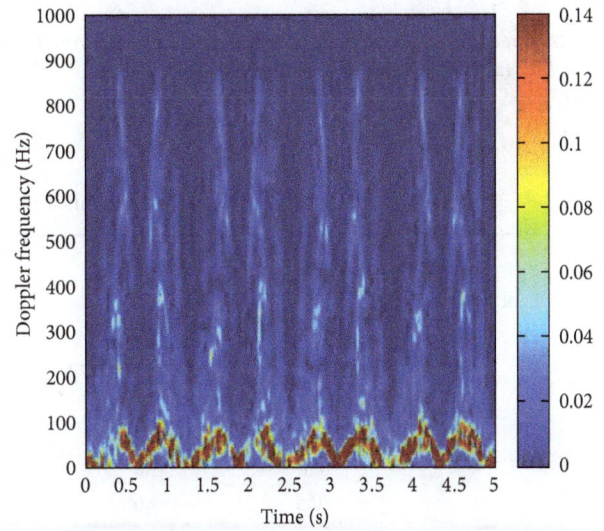

(c)

FIGURE 22: Continued.

(d)

(e)

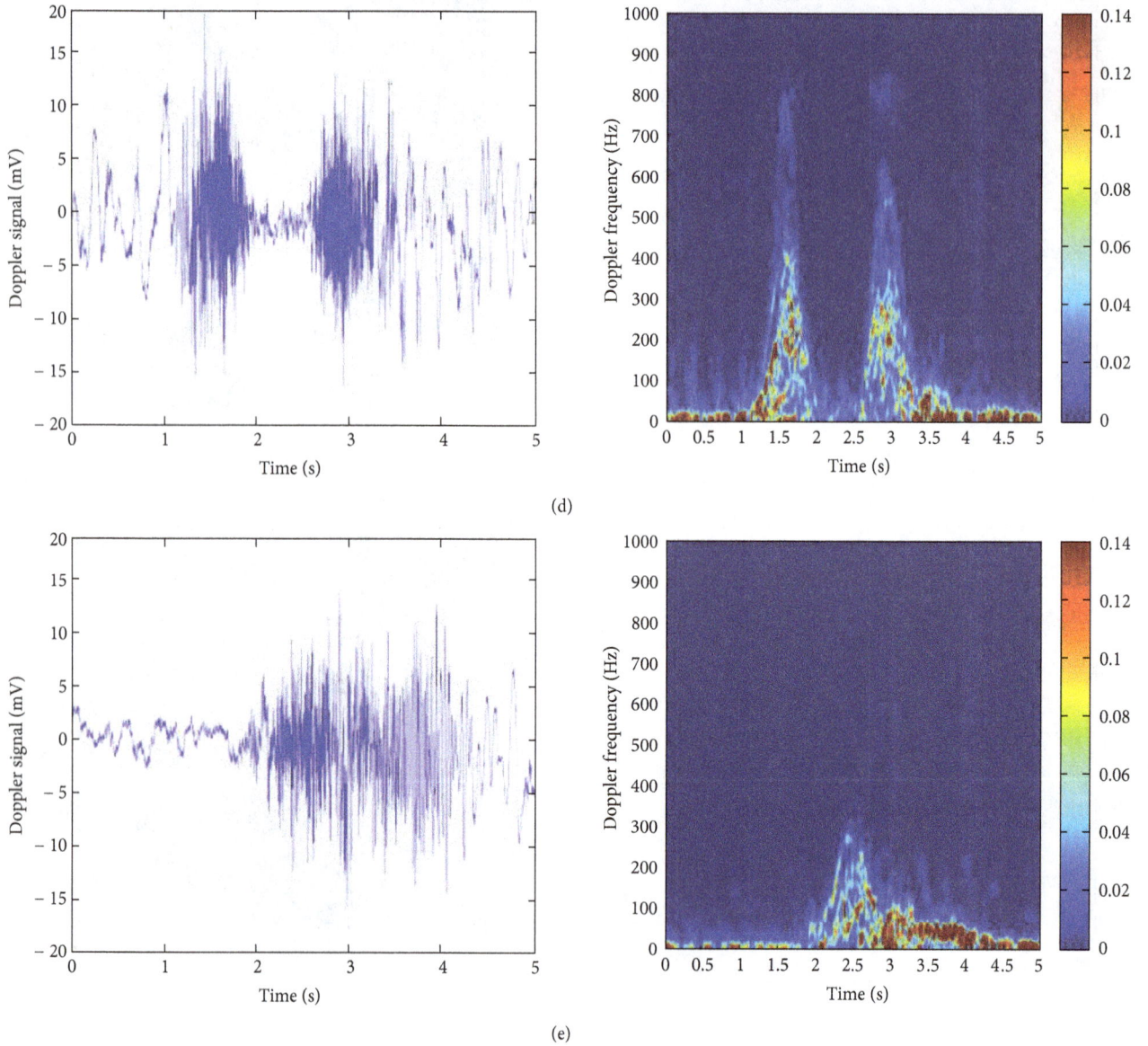

FIGURE 22: Time series and time-frequency plots of different unobstructed human activities as measured by the W-Band radar at a range of 30 m (98.4 ft). (a) No activity, (b) breathing, (c) swinging arms, (d) picking up object from ground, and (e) transitioning from crouching to standing.

TABLE 1: Human activity classification results for S-Band through-wall radar.

| Human test subject number | Average classification accuracy (%) | | | | | |
| | No wall (11 ft range to target) | | 4 in thick brick wall ($11\frac{1}{3}$ ft range to target) | | 8 in thick cinder block wall ($5\frac{2}{3}$ ft range to target) | |
	Mean	Standard deviation	Mean	Standard deviation	Mean	Standard deviation
1	76.0	3.3	61.4	5.1	57.2	5.0
2	52.8	6.9	48.6	9.8	66.0	4.5
3	61.0	3.8	44.2	4.3	68.4	5.3
4	70.4	3.6	48.0	4.5	71.4	4.8
5	66.8	3.6	46.6	5.0	59.6	5.8
6	56.2	3.3	49.8	2.7	61.8	6.0

TABLE 2: Human activity classification results for W-Band radar under free space conditions.

Human test subject number	Average classification accuracy (%)					
	100 ft range to target		200 ft range to target		300 ft range to target	
	Mean	Standard deviation	Mean	Standard deviation	Mean	Standard deviation
1	88.6	1.3	86.0	6.0	94.8	3.7
2	87.0	1.9	80.4	1.3	95.2	3.2
3	85.4	1.9	86.6	1.6	94.0	3.4
4	80.6	3.1	84.4	4.6	95.2	1.9
5	80.2	5.8	71.2	3.9	93.4	4.1
6	53.4	9.3	69.2	7.1	93.4	5.0

(4 in) thick brick wall, and (3) transmission through a 20.3 cm (8 in) thick cinder block wall. For the direct transmission case, the person was located 3.35 m (11 ft) away from the radar antennas. For the brick wall case, the person was located 1.52 m (5 ft) behind the wall with the wall 1.83 m (6 ft) away from the radar antennas. For the cinder block case, the signals were very weak, so the distances were shortened. The person was located 91 cm (3 ft) behind the wall and the wall was located 61 cm (2 ft) away from the radar antennas. The five different motions listed above were considered for classification. The training set consisted of data from five of the six test subjects and was used both for training the classifier and for the cross-validation. The cross-validation set also consists of data from five of the six test subjects, but these data were used only for cross-validation and not used for training the classifier. The test set consists of data from one of the six test subjects and these data were used neither for training the classifier nor for cross-validation. Classification results are shown in Table 1. The average accuracy when combining the results of all six of the test subjects is 63.9%, 49.8%, and 64.1% for no barrier, brick wall, and cinder block wall, respectively. Although the classification accuracies appear to be low, it must be borne in mind that different individuals perform physical activities quite differently based on body shape and cultural factors.

5.7. Human Activity Classification Results for W-Band Foliage Penetration Radar. Five different motions were investigated for the W-Band foliage penetration radar. These include (a) no activity (for reference), (b) breathing, (c) swinging arms, (d) picking up object from ground, and (e) transitioning from crouching to standing. Experimentally observed time series and time-frequency plots of unobstructed (without foliage cover) human activities at a range of 30 m (98.4 ft) are shown in Figure 22, wherein clear differences between different activities are observed. The time series plots are just the raw data versus time recorded by the system when the specified activity takes place, while the time-frequency plots show the Doppler frequency as a function of time, with the higher amplitudes signifying faster movement. From Figure 22(b), it is evident that breathing produces the Doppler frequency shifts on the order of a few tens of Hz. Figure 22(c) shows that the arm swinging motion is periodic every 1/2 second

or so. The time-frequency plot also shows that there are multiple components of the body moving while the arms are swinging. The Doppler frequencies lower than 100 Hz correspond to the larger, slower torso swaying while the smaller, faster Doppler response from the arms swinging goes up to 900 Hz. In the picking up object motion in Figure 22(d), there are two distinct high-frequency pulses that correspond to the person first bending over and then standing back up. These pulses last about 1 second, and within each pulse there are multiple components in the Doppler signal relating to different parts of the body. The higher frequency lower amplitude component corresponds to the motion of the head and reaches frequencies up to 800 Hz. This is because the pivot point is at the waist, the head is the fast moving object when bending over and standing back up, and the head is smaller in physical size than the torso and shoulders. The lower frequency higher amplitude components correspond to the high RCS torso, arms, and shoulders moving slower because they are close to the pivot point. These motions produce the Doppler frequencies up to 400 Hz. For the crouching to standing motion in Figure 22(e), there is one high-frequency pulse that lasts about 1 second and tapers off into lower frequency swaying. The lower frequency swaying is caused by the human being slightly unstable from the process of standing from the crouch position. This motion does not have a pivot point so all body parts move at the same time with similar speeds, making it difficult to identify different body parts based on speed. The highest Doppler frequencies reached in this motion are about 300 Hz and the low speed swaying causes the Doppler frequencies of <100 Hz. Figure 23 shows just the time-frequency plots of the same activities at a range of 90 m (295 ft), which are very similar to the 30 m measurements, except for a reduction in amplitude. Note that the Doppler frequencies and the associated time durations are also nearly identical. The radar's ability to detect the human Doppler signal through light foliage was also tested. The test setup was the same as Section 4.4 and is as shown in Figures 12 and 13. In this scenario, the received signal contains Doppler signatures of both the wind-influenced foliage and the human activity behind it. For this case, the background Doppler data are very important. This is because, unlike the ranging data, the Doppler background cannot be averaged or subtracted out. In this case, filtering out the dominant frequencies produced

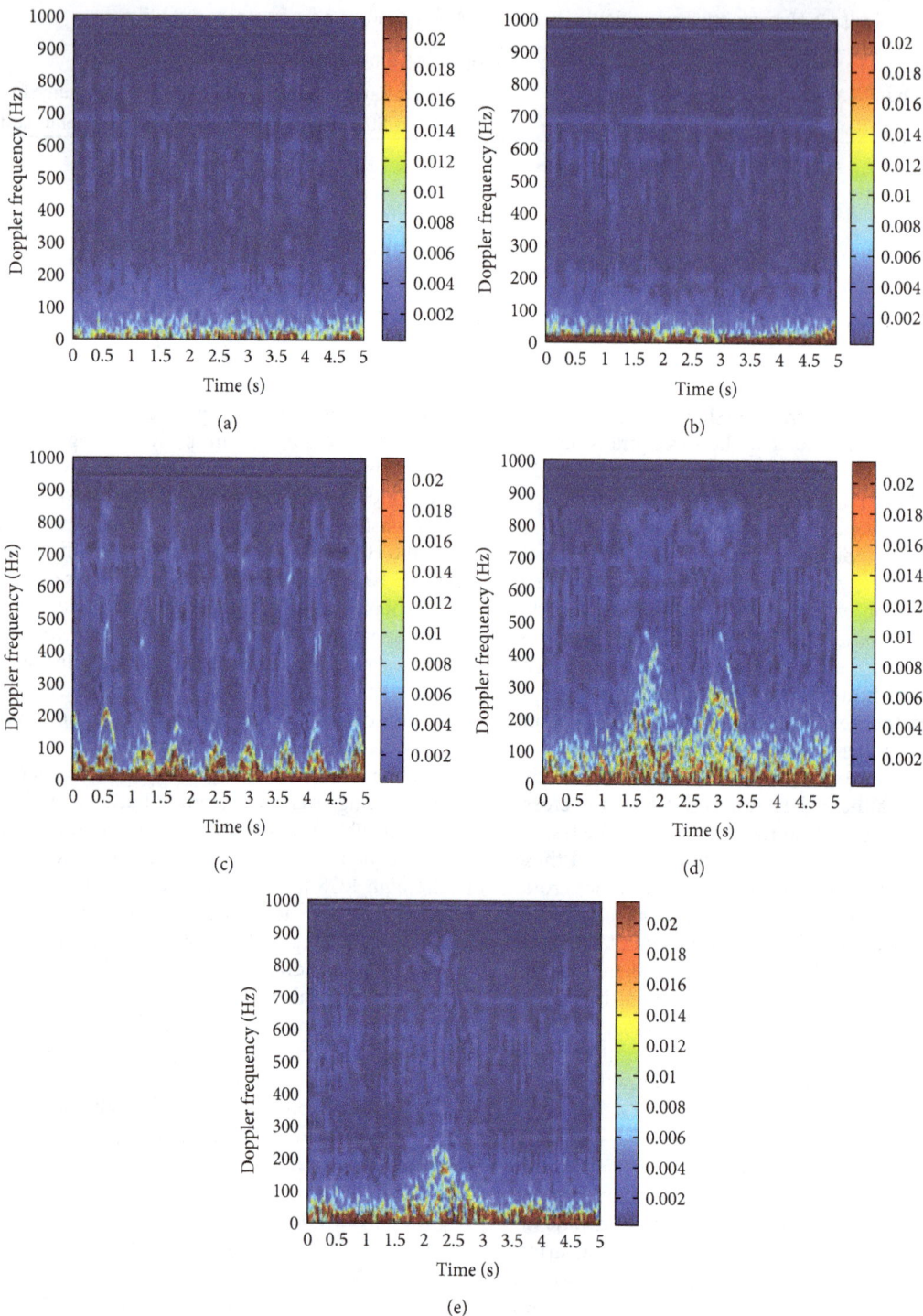

FIGURE 23: Time-frequency plots of different unobstructed human activities as measured by the W-Band radar at a range of 90 m (295 ft). (a) No activity, (b) breathing, (c) swinging arms, (d) picking up object from ground, and (e) transitioning from crouching to standing.

by the foliage is needed. When filtering is performed, it consequently filters out some of our desired signals, but in many cases there is enough information remaining to identify human motion. Figure 24(a) shows the time series and time-frequency plots of the background Doppler data from the foliage. As seen in time-frequency plot, the majority of the

background Doppler is present in the frequencies less than 50 Hz, which is much lower than the Doppler from the human activity (except for the breathing). For this reason, a high-pass filter with a cutoff frequency of 50 Hz was applied to reduce the effect of the foliage. Figure 24(b) shows the corresponding background Doppler data plots after the filter

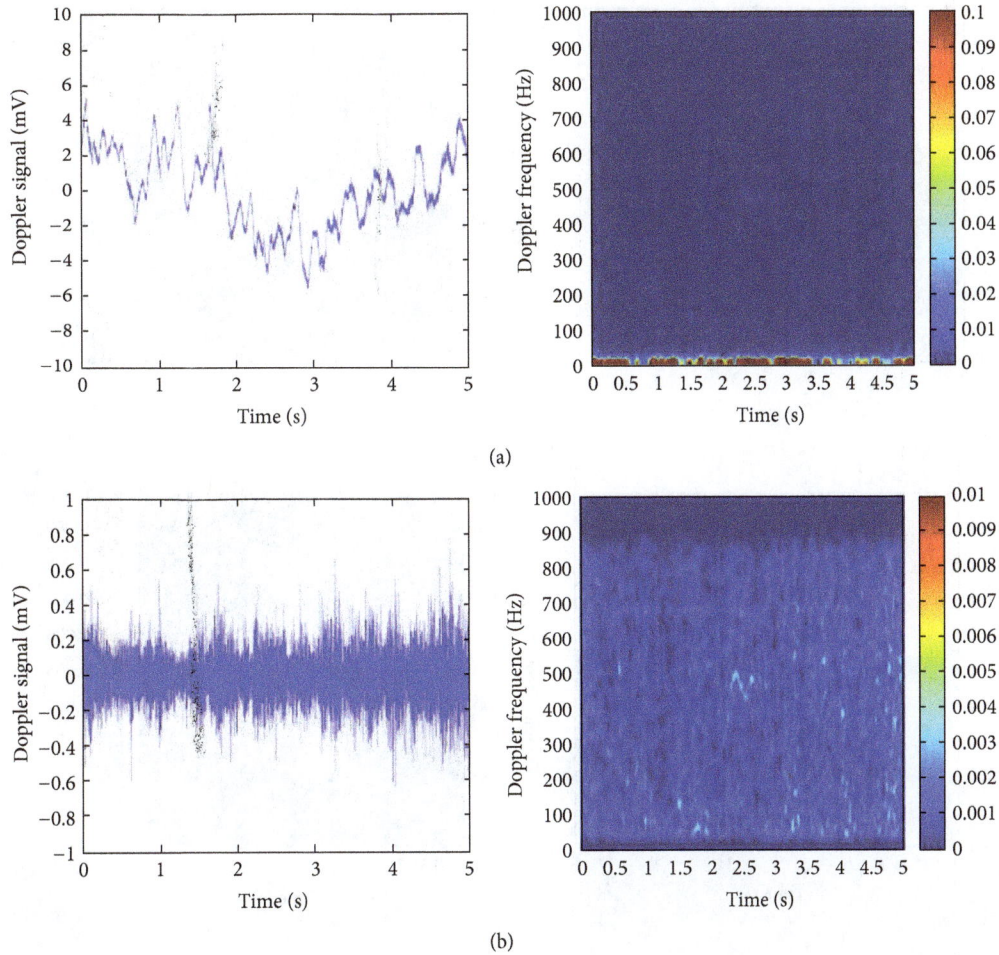

FIGURE 24: Time series and time-frequency plots from the fully-foliated bush with leaves at 30 m (98.4 ft). (a) Unfiltered, and (b) filtered.

is applied, from which we note that the response from the foliage is reduced by a factor of 10–20 dB. An example of how effectively the filtering works is shown in Figure 25 for a human picking up an object behind the bush. The filtered time series and time-frequency plots in Figure 25(b) are almost identical to the free-space measurement shown in Figure 23(d). Similar results were obtained for swinging arms and transitioning from crouching to standing. Unfortunately, since the breathing Doppler signals are of the same order as those caused by the swaying of the bush, it is not possible to detect human breathing while standing still behind a bush.

All five different motions listed above were considered for classification. The training set consisted of data from five of the six test subjects and was used for both training the classifier and for the cross-validation. The cross-validation set also consists of data from five of the six test subjects, but these data were used only for cross-validation and not used for training the classifier. The test set consists of data from one of the six test subjects and these data were used neither for training the classifier nor for cross-validation. Classification results are shown in Table 2. The average accuracy when combining the results of all six of the test subjects is 79.2%, 79.6%, and 94.3% for human targets at ranges of 30.5, 61, and

91.4 m (100, 200, and 300 ft, resp.) from the radar antennas, respectively. These are good classification accuracies. The reasoning behind the high accuracy at 91.4 m (300 ft) is because, at the shorter distances, the human target will move outside of the antenna beam for portions of the motions, especially for picking up an object and transitioning from crouching to standing. At the longer distances, the entire body is within the antenna beam, but the illuminated area is still small enough to isolate the human target. As stated earlier, different individuals perform physical activities quite differently based on body shape and cultural factors.

6. Conclusions

To the best of our knowledge, this is the first reporting of the ability to classify different types of human activity behind opaque walls and foliage cover. While the results obtained thus far are quite encouraging, more research and system development are needed to improve the classification accuracies in the presence of barriers and to include additional movements and gestures. We are currently working on expanding the range of human activities as well as the variety of humans for additional data collection.

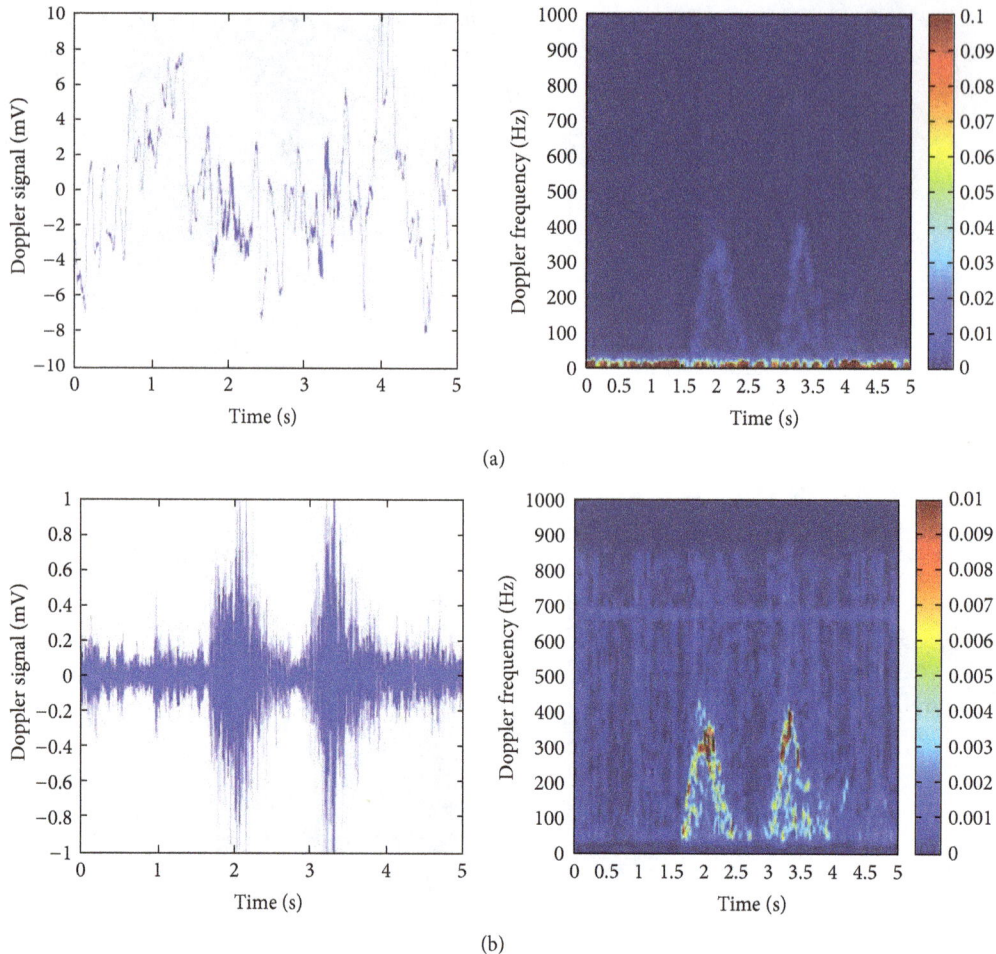

FIGURE 25: Time series and time-frequency plots from arm swinging behind fully foliated bush at 30 m (98.4 ft). (a) Unfiltered and (b) filtered.

Conflict of Interests

The authors declare that there is no conflict of interests regarding the publication of this paper.

Acknowledgments

This work was supported by the U.S. Army ARDEC Joint Service Small Arms Program (JSSAP) under Contract no. W15QKN-09-C-0116. The authors appreciate helpful comments provided by E. Beckel, W. Luk, and G. Gaeta of ARDEC.

References

[1] W. L. Stutzman and G. A. Thiele, *Antenna Theory and Design*, John Wiley & Sons, Hoboken, NJ, USA, 3rd edition, 2013.

[2] D. N. Keep, "Frequency-modulation radar for use in the mercantile marine," *Proceedings of the Institution of Electrical Engineers B: Radio and Electronic Engineering*, vol. 103, no. 10, pp. 519–523, 1956.

[3] F. Ahmad and R. M. Narayanan, "Conventional and emerging waveforms for detection and imaging of targets behind walls," in *Through-the-Wall Radar Imaging*, M. G. Amin, Ed., pp. 157–184, CRC Press, Boca Raton, Fla, USA, 2010.

[4] L. Turner, "The evolution of featureless waveforms for LPI communications," in *Proceedings of the IEEE National Aerospace and Electronics Conference (NAECON '91)*, pp. 1325–1331, Dayton, Ohio, USA, May 1991.

[5] B. M. Horton, "Noise-modulated distance measuring systems," *Proceedings of the IRE*, vol. 47, no. 5, pp. 821–828, 1959.

[6] R. M. Narayanan, "Through-wall radar imaging using UWB noise waveforms," *Journal of the Franklin Institute*, vol. 345, no. 6, pp. 659–678, 2008.

[7] B. Ferguson, S. Mosel, W. Brodie-Tyrrell, M. Trinke, and D. Gray, "Characterisation of an L-band digital noise radar," in *Proceedings of the IET International Conference on Radar Systems (RADAR '07)*, pp. 1–5, Edinburgh, UK, October 2007.

[8] E. K. Walton, "Digital noise radar prototype development," in *Proceedings of the 30th Antenna Measurement Techniques Association Annual Symposium*, pp. 23–26, Boston, Mass, USA, November 2008.

[9] V. C. Chen, "Analysis of radar micro-doppler signature with time-frequency transform," in *Proceedings of the IEEE Signal Processing Workshop on Statistical Signal and Array Processing (SSAP '00)*, pp. 463–466, Pocono Manor, Pa, USA, August 2000.

[10] P.-H. Chen, M. C. Shastry, C.-P. Lai, and R. M. Narayanan, "A portable real-time digital noise radar system for through-the-wall imaging," *IEEE Transactions on Geoscience and Remote Sensing*, vol. 50, no. 10, pp. 4123–4134, 2012.

[11] S. Smith and R. M. Narayanan, "Ranging and target detection performance through lossy media using an ultrawideband S-band through-wall sensing noise radar," in *Radar Sensor Technology XVII*, vol. 8714 of *Proceedings of SPIE*, pp. 1–12, Baltimore, Md, USA, April 2013, 871408.

[12] H. Nyquist, "Certain topics in telegraph transmission theory," *Transactions of the American Institute of Electrical Engineers*, vol. 47, no. 2, pp. 617–644, 1928.

[13] C. E. Shannon, "Communication in the presence of noise," *Proceedings of the IRE*, vol. 37, no. 1, pp. 10–21, 1949.

[14] J. D. Kraus, "The helical antenna," *Proceedings of the IEEE*, vol. 37, no. 3, pp. 263–272, 1949.

[15] A. R. Djordjević, A. G. Zajić, and M. M. Ilić, "Enhancing the gain of helical antennas by shaping the ground conductor," *IEEE Antennas and Wireless Propagation Letters*, vol. 5, pp. 138–140, 2006.

[16] K. A. Gallagher and R. M. Narayanan, "Human detection and ranging at long range and through light foliage using a W-band noise radar with an embedded tone," in *Radar Sensor Technology XVII*, vol. 8714 of *Proceedings of SPIE*, pp. 1–12, Baltimore, Md, USA, April 2013, 871402.

[17] M. J. Maybell and P. S. Simon, "Pyramidal horn gain calculation with improved accuracy," *IEEE Transactions on Antennas and Propagation*, vol. 41, no. 7, pp. 884–889, 1993.

[18] A. D. Olver and B. Philips, "Integrated lens with dielectric horn antenna," *Electronics Letters*, vol. 29, no. 13, pp. 1150–1152, 1993.

[19] M. Piccardi, "Background subtraction techniques: a review," in *Proceedings of the IEEE International Conference on Systems, Man and Cybernetics (SMC '04)*, pp. 3099–3104, The Hague, The Netherlands, October 2004.

[20] D. J. Daniels, "Radar for non destructive testing of materials," in *Proceedings of the IEE Colloquium on Measurements, Modelling and Imaging for Non-Destructive Testing*, pp. 9-1–9-3, London, UK, March 1991.

[21] H. Wang, R. M. Narayanan, and Z. O. Zhou, "Through-wall imaging of moving targets using UWB random noise radar," *IEEE Antennas and Wireless Propagation Letters*, vol. 8, pp. 802–805, 2009.

[22] D. P. Fairchild and R. M. Narayanan, "Micro-doppler radar classification of human motions under various training scenarios," in *Active and Passive Signatures IV*, vol. 8734 of *Proceedings of SPIE*, pp. 1–11, Baltimore, Md, USA, April 2013, 873407.

[23] N. E. Huang, Z. Shen, S. R. Long et al., "The empirical mode decomposition and the Hubert spectrum for nonlinear and non-stationary time series analysis," *Proceedings of the Royal Society A: Mathematical, Physical and Engineering Sciences*, vol. 454, no. 1971, pp. 903–995, 1998.

[24] C. Cortes and V. Vapnik, "Support-vector networks," *Machine Learning*, vol. 20, no. 3, pp. 273–297, 1995.

[25] R. Rifkin and A. Klautau, "In defense of one-vs-all classification," *The Journal of Machine Learning Research*, vol. 5, pp. 101–141, 2004.

16

Ultrawide Bandwidth 180°-Hybrid-Coupler in Planar Technology

Steffen Scherr, Serdal Ayhan, Grzegorz Adamiuk, Philipp Pahl, and Thomas Zwick

Institut für Hochfrequenztechnik und Elektronik, Karlsruhe Institute of Technology (KIT), Kaiserstraße 12, 76131 Karlsruhe, Germany

Correspondence should be addressed to Steffen Scherr; steffen.scherr@kit.edu

Academic Editor: Yo Shen Lin

A new concept of an ultrawide bandwidth 180°-hybrid-coupler is presented. The ultrawideband design approach is based on the excitation of a coplanar waveguide (CPW) mode and a coupled slot line (CSL) mode in the same double slotted planar waveguide. The coupler is suitable for realization in planar printed circuit board technology. For verification of the new concept a prototype was designed for the frequency range from 3 GHz to 11 GHz, built, and measured. The measurement results presented in this paper show a good agreement between simulation and measurement and demonstrate the very broadband performance of the new device. The demonstrated coupler with a size of 40 mm × 55 mm exhibits a fractional bandwidth of 114% centered at 7 GHz with a maximum amplitude imbalance of 0.8 dB and a maximum phase imbalance of 5°.

1. Introduction

The main task of a 180°-hybrid-coupler is the division of an input signal into two autonomous signals, which possess the same amplitude and are in-phase or out-of-phase. The phase relation between the output signals depends on the feeding port, referred to as Σ-port or Δ-port, respectively. A 180°-hybrid-coupler consists all in all of four ports, two inputs and two outputs. The division into in-phase signals introduces a principle of a basic power divider, which can be easily realized even for very large bandwidths. The creation of differential signals can be achieved by a differential power divider. Bialkowski and Abbosh [1] describe such a differential power divider for UWB technology. A 180°-hybrid-coupler combines the two aforementioned power dividing principles in one single device.

180°-hybrid-couplers are used in many microwave circuits such as push-pull amplifiers [2], balanced mixers [3], and pattern diversity antennas [4]. A further application of such couplers is in the monopulse radar technique, where sum and difference beams are created for an accurate angular tracking of the target [5]. The possibility of the creation of a sum and difference beam over UWB bandwidth is verified by the authors in [6].

The general advantage of the 180°-hybrid-coupler over conventional or differential power dividers is the possibility to process sum and differential signals in the same device. In narrowband systems, those kinds of couplers are well known as rat-race couplers or magic-tees. However, there is a demand for systems combining the new broadband possibilities (e.g., UWB technology) with traditional narrowband concepts (e.g., monopulse radar) [7]. Hence, hybrid-couplers that cover the UWB frequency band, for example, from 3 GHz to 11 GHz are of interest [8]. As traditional hybrid-coupler designs need wavelength-dependent structures, it is necessary to find wideband design approaches.

The presented coupler can be fabricated in planar technology or low temperature cofired ceramics (LTCC) to cover the aforementioned frequency band. Even on-chip fabrication in standard silicon technologies at mm-wave is possible with the new concept.

The coupler should possess a satisfactory impedance matching at the Σ- and Δ-port, low coupling between the Σ- and Δ-port, and high transition from the Σ- and Δ-port to the outputs. Under ideal conditions the amplitudes of the signals at the output ports are equal and the phase difference is 0° when fed at the Σ-port and 180° when fed at the Δ-port. All port impedances are optimized to 50 Ω in the presented approach.

The following sections present the principle of the new coupler concept and the measurements which validate the functionality of the coupler.

2. Principle of the Coupler

Figure 1 shows the structure of the coupler. The dark gray and the light gray areas mark the metallization at the top of the coupler and at the bottom, respectively. For the explanation of the coupler's functioning port 1 (Σ-port) and port 2 (Δ-port) are considered as input ports, whereas port 3 and port 4 are output ports if one of the input ports is excited (Figure 1). The principle of the coupler can best be explained if the behavior of signals separately excited at port 1 and port 2 on their way to port 3 and port 4 is described.

At first the behavior of signals coupled into port 1 (Σ-port) is investigated. An excitation of port 1 leads to signals having the same amplitude and phase at the outputs. The corresponding simplified electric field distribution is depicted in Figure 1(a), showing the signal propagation from port 1 to port 3 and to port 4. The CPW mode [4] is excited in the two adjacent slots and the signal is further transferred through the set of vias. The vias realize a short circuit of the outer coplanar waveguide metallizations by connecting a metallized patch, placed on the opposite side of the substrate. As a CPW line is formed by a conductor with a separated pair of ground planes, there is no change in the potential introduced by the vias. The CPW mode can pass the vias without interruption. Behind the vias the slot lines cross the microstrip line, which is placed on the top side of the coupler. The electric field is concentrated mainly in the slots, that is, at the top of the coupler; hence, the influence of the microstrip line on the transmission of the CPW mode can be neglected. The power divider is realized by separating the adjacent slots of the coplanar waveguide into separate slot lines. To avoid an abrupt impedance change, the coplanar waveguide before the power divider is tapered. At the end of each of the slot lines an aperture coupling transfers the signals to the microstrip lines [14]. As can be seen from Figure 1(a), the direction of the electric field components at the outputs is the same. Thus, the output signals are in-phase. Due to the design of the hybrid-coupler, the output signals possess the same amplitude if a CPW mode is excited at the Σ-port, for example, by an SMA connector.

If a signal is coupled into port 2 (Δ-port), the signals at port 3 and port 4 are differential to each other. The signal travels from the Δ-port to the aperture coupling at the end of the microstrip line (Figure 1(b)). The aperture coupling of the signal causes an excitation of the CSL mode [4] in the adjacent slot lines. This form of the coupling is similar to the traditional one [14]; however, in the proposed structure two adjacent slots are used instead of a single one. By proper adjustment of the aperture coupling both slots get excited with electric fields that are oriented in the same direction (Figure 1(b)). A hybrid-coupler isolates the Σ-port and the

Δ-port from each other. This isolation is realized by the implementation of the vias. In contrast to the CPW mode, the CSL mode sees a different potential at the ground planes. Hence, the short circuit caused by the vias blocks the CSL mode. In succession the signal propagated in this CSL mode is reflected. This effect is similar to the reflection at the slot widening in the traditional wideband transition from microstrip line to slot line [14]. As the set of vias allows the transition of the CPW mode and reflects the CSL mode, the vias serve as a mode filter. In order to achieve a higher reflection of the CSL mode, three sets of vias are used. The CSL mode is transmitted through the slot separation to the aperture couplings and finally to port 3 and port 4. The signals at these ports are out-of-phase. In order to guarantee the equality of the amplitudes at port 3 and port 4 both slots have to be excited with the same amplitude. This can be achieved by a proper optimization of the slots' relative orientation with respect to the coupling microstrip line. Furthermore, the distance between the slots should be kept small; otherwise, it is impossible to excite the slots with equal signals.

3. Proposed UWB 180°-Hybrid-Coupler

The coupler is built on Duroid 4003 with a thickness of $h_{\mathrm{sub}} = 0.79$ mm and a dielectric constant of $\epsilon_r = 3.55$. The dimensions of the prototype (Figures 2 and 3) are 40 mm × 55 mm. As the microstrip lines and the coplanar lines do not limit the bandwidth of the coupler [15], the only components with a frequency dependent and hence bandwidth limiting geometry are the aperture couplings. Therefore, the main objective of the optimization for this coupler is to optimize the wideband microstrip line to coplanar line transition. Simulations (CST Microwave Studio [16]) show that the best results can be achieved, if the width of the coplanar line at this transition is minimized. In this case the transition in combination with the vias behaves like a traditional aperture coupling from microstrip line to single slot line.

A width of 60 μm for the slot of the coplanar line is chosen at the microstrip line to coplanar line transition. Since this part is very sensitive to etching tolerances, a smaller slot width could lead to strong deviations between simulation and measurement. A slot width of less than 60 μm would be advantageous, but it results in very low yield for the applied etching process. However, a further miniaturization of this part of the coupler and the possibility to use a substrate with a higher dielectric constant (e.g., LTCC fabrication) would lead to a better overall performance. After the slot size and the impedance of the coplanar line are determined, the width can be calculated to 0.35 mm [15]. All port impedances are optimized for 50 Ω and the size of the microstrip line and the coplanar line is limited by the size of the connectors. Hence, tapers are used to achieve wideband behavior while providing a solderable geometry at the connectors. The width of the slot line at the end of the coplanar line determines the matching of the CSL and the CPW mode. A larger width of the slot line improves the matching of the CSL mode and a smaller width of the slot line improves the matching of the CPW mode. As a trade-off, a width of 0.26 mm is selected here. The aperture

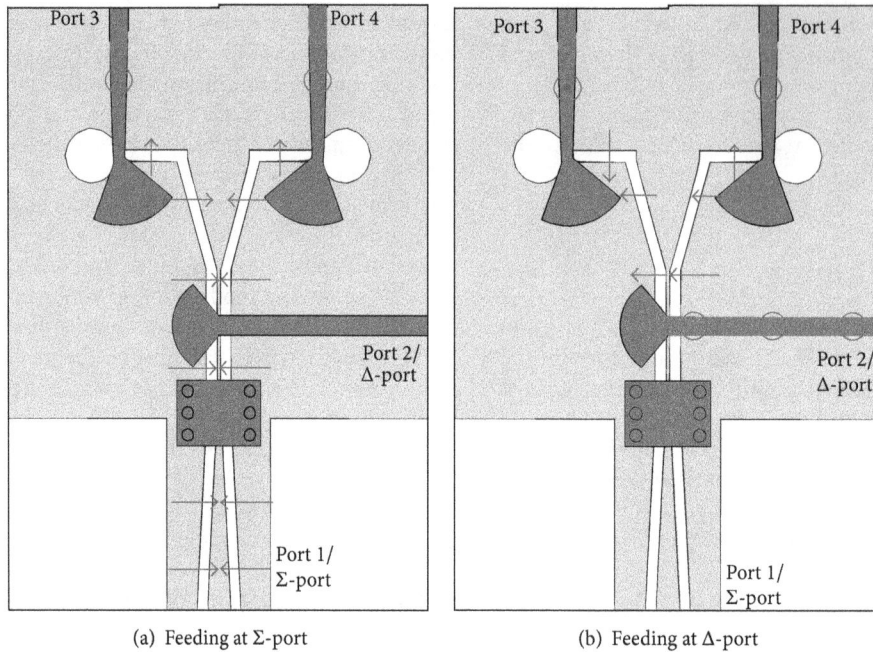

(a) Feeding at Σ-port   (b) Feeding at Δ-port

FIGURE 1: Schematical electric field distribution in the coupler. The slot lines are enlarged to get a better understanding of the coupler's principle.

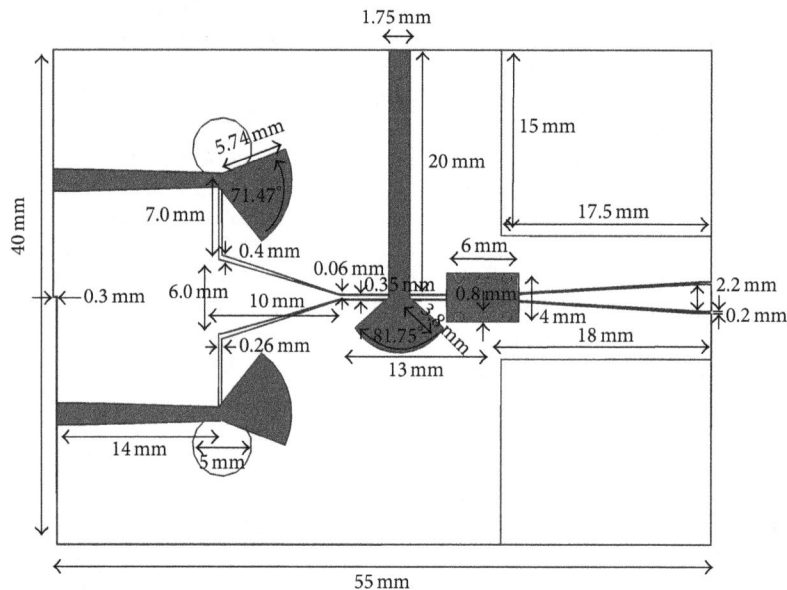

FIGURE 2: Size of the hybrid coupler.

couplings from microstrip line to slot line are realized by $\lambda/4$-stubs (5.3 mm at 7.5 GHz, Figure 2), which are extended to circles and segments of circles to increase the bandwidth [6, 14, 15].

4. Simulation and Measurement Results

The three most important parameters of a hybrid-coupler are the S-parameters, the amplitude imbalance, and the phase difference, which will be considered in this section.

Figure 4 shows the simulated and measured amplitudes of the S-parameters of the coupler. The coupler exhibits an ultrabroadband behavior in the frequency range from 3 GHz to 11 GHz.

The input impedance matching at port 1 and port 2 as well as the decoupling between the ports is sufficient in the desired frequency range. The transmission factor from port 1 (Σ-port) to port 3 is close to the optimum value of -3 dB in the lower frequency range and decreases slightly with increasing frequency. The transmission from port 2 (Δ-port) to port 3

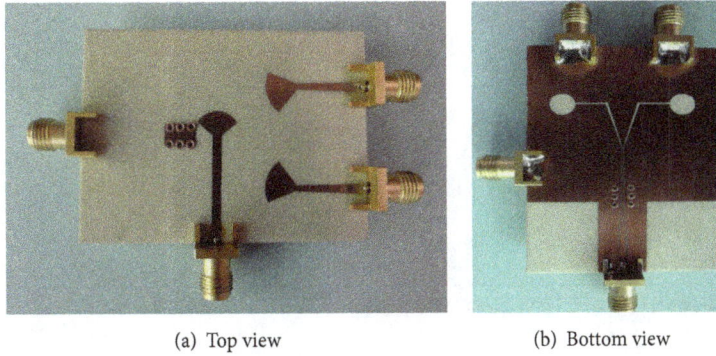

(a) Top view   (b) Bottom view

FIGURE 3: Photograph of the prototype.

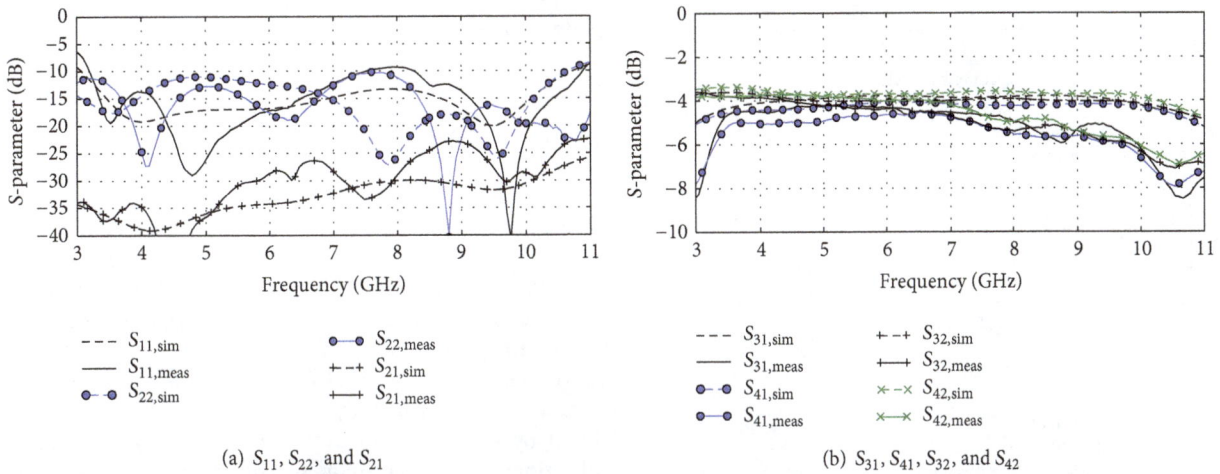

$S_{11,sim}$ $S_{22,meas}$
$S_{11,meas}$ $S_{21,sim}$
$S_{22,sim}$ $S_{21,meas}$

(a) S_{11}, S_{22}, and S_{21}

$S_{31,sim}$ $S_{32,sim}$
$S_{31,meas}$ $S_{32,meas}$
$S_{41,sim}$ $S_{42,sim}$
$S_{41,meas}$ $S_{42,meas}$

(b) S_{31}, S_{41}, S_{32}, and S_{42}

FIGURE 4: Simulated and measured amplitudes of the S-parameters of the presented coupler.

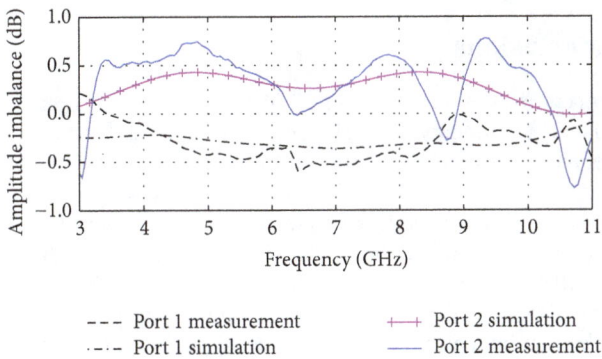

Port 1 measurement   Port 2 simulation
Port 1 simulation   Port 2 measurement

FIGURE 5: Amplitude imbalance of the output signals for feeding at port 1 and port 2.

encounters similar losses in the high frequency range. The simulations show that these losses are mainly caused by radiation. They can be decreased by using a substrate with lower thickness h_{sub} and higher dielectric constant ϵ_r or by implementing the coupler in LTCC technology. Due to the lossless materials assumed in the simulation, the measured transmission losses are slightly different compared to the simulated losses. Above 7 GHz increased losses have been observed due to tolerances in the manufacturing process at the microstrip line to coplanar line transition.

The measured amplitude imbalance curves are shown in Figure 5. The imbalance is calculated between the output signals by exciting separately port 1 (Σ-port) and port 2 (Δ-port) and calculating the relation of the S-parameters at the output ports, respectively. For the feeding at port 1, smaller differences in the amplitudes (max. 0.6 dB) are observed than for port 2 (max. 0.8 dB). This is mainly due to the less complicated excitation procedure of the CPW mode. Based on the presented curves, it can be concluded that the coupler possesses a good amplitude behavior over the UWB frequency range.

In 180°-hybrid-couplers the phase behavior of the transmission factors is crucial. Figure 6 shows the simulated and measured phase differences between the outputs of the coupler for separate feeding at port 1 and port 2. The values of the phase difference, while port 1 (Σ-port) is excited, are relatively small and do not exceed 3.5°. During the excitation of the Δ-port, the phase imbalance has shown a linearly increasing error over frequency. This has been compensated

TABLE 1: Comparison between planar 180°-hybrid-couplers.

Reference	Relative bandwidth B%	Center frequency f_c	Maximal amplitude imbalance	Maximal phase imbalance	Size [mm]	Layers
[9]	60%	2 GHz	0.2 dB	2°	—	2
[10]	69%	10.1 GHz	0.6 dB	1.0°	20 × 30	2
[11]	100%	4 GHz	0.4 dB	2.5°	—	1
[12]	100%	4 GHz	0.4 dB	4°	—	1
[13]	100%	6 GHz	—	—	24 × 35	2
This work	114%	7 GHz	0.8 dB	5°	40 × 50	2

Note: the values are partly estimated from figures if not explicitly given in the paper.

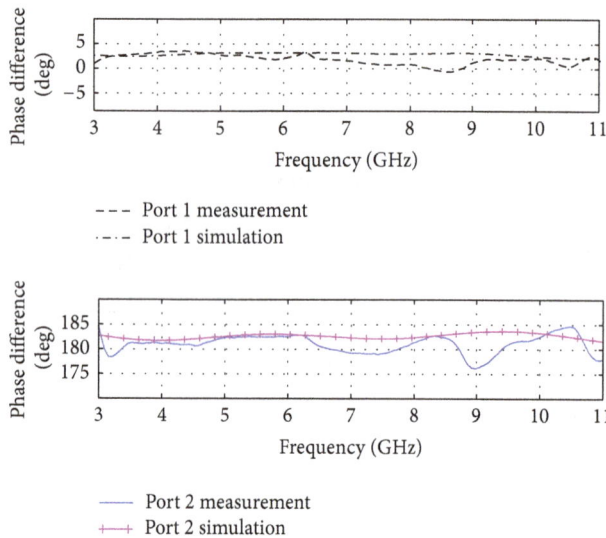

--- Port 1 measurement
·--· Port 1 simulation

—— Port 2 measurement
+—+ Port 2 simulation

FIGURE 6: Phase difference of the output signals for feeding at port 1 and port 2.

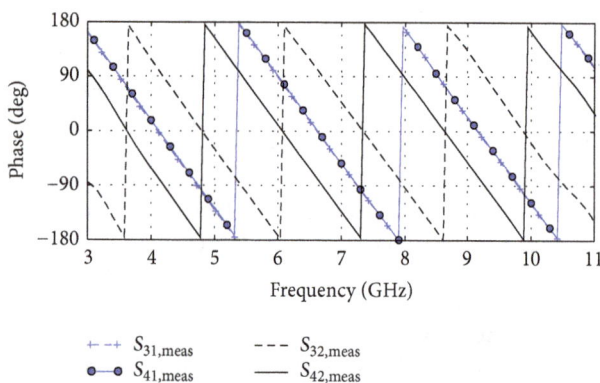

+–-+ $S_{31,meas}$ - - - $S_{32,meas}$
●—● $S_{41,meas}$ —— $S_{42,meas}$

FIGURE 7: Measured phase response of the hybrid coupler.

by an adjustment of the length of the microstrip line at port 4 (Figures 1 and 2). The linear dependency of the phase imbalance on the frequency is due to the different arrangements of both slot lines with respect to the microstrip line coming from port 2. The phase balance when this port is excited can be improved, for example, by a closer arrangement of the adjacent slot lines; however, the compensation of the nonideality is still necessary in the presented design. Due

to the extension of the microstrip line at port 4, the phase deviations do not exceed 5° for feeding at the Δ-port.

Table 1 compares the performance of different planar 180°-hybrid-couplers. The relative bandwidth is defined as

$$B\% = 2 \cdot \frac{f_h - f_l}{f_h + f_l} \cdot 100 = \frac{f_h - f_l}{f_c} \cdot 100, \quad (1)$$

where f_h and f_l are the highest and lowest frequency, which cover a frequency band in which the coupler possesses a return loss of mostly less than −10 dB and an amplitude imbalance of less than 1 dB. It can be found that the proposed 180°-hybrid-coupler is better than the referenced ones in terms of relative bandwidth. Compared to the other couplers, the new coupler covers the whole UWB frequency band defined by the FCC and can be used for high resolution applications. All subcomponents of the coupler (aperture couplings, line separation, vias, etc.) maintain a nearly linear phase response (Figure 7). Hence, the coupler can also be applied to pulse-based UWB systems (IR-UWB), where a small distortion of the pulse is desired [1, 6, 17].

5. Conclusions

A concept for the realization of a UWB 180°-hybrid-coupler in planar technology requiring only two metal layers is presented. This concept is verified by measurements of the prototype, which is optimized for the frequency range from 3 GHz to 11 GHz. Across this frequency range, the return loss for the Δ-port is mostly less than −10 dB and always better than −10 dB at the Σ-port. The maximal phase imbalance for the Σ-port and the Δ-port is 3.5° and 5°, respectively, and the maximal amplitude imbalance is 0.6 dB and 0.8 dB, respectively. The linear phase response of the coupler allows the cost-effective usage for pulse-based UWB systems. The presented prototype can be applied to many applications where ultrawide bandwidth is desired in the creation of in-phase and differential signals, for example, the monopulse radar principle in a combination with the UWB technique.

Conflict of Interests

The authors declare that there is no conflict of interests regarding the publication of this paper.

References

[1] M. E. Bialkowski and A. M. Abbosh, "Design of a compact UWB out-of-phase power divider," *IEEE Microwave and Wireless Components Letters*, vol. 17, no. 4, pp. 289–291, 2007.

[2] S. Toyoda, "Broad-band push-pull power amplifier," in *Proceedings of the IEEE MTT-S International Microwave Symposium Digest*, vol. 1, pp. 507–510, May 1990.

[3] R. Blight, "Microstrip hybrid couplers and thier integration into balanced mixers at X and K-bands," in *Proceedings of the G-MTT International Microwave Symposium Digest*, pp. 136–138, Boston, Mass, USA, May 1967.

[4] E. Gschwendtner and W. Wiesbeck, "Ultra-broadband car antennas for communications and navigation applications," *IEEE Transactions on Antennas and Propagation*, vol. 51, no. 8, pp. 2020–2027, 2003.

[5] M. Skolnik, *Introduction to Radar Systems*, McGraw-Hill, New York, NY, USA, 1962.

[6] G. Adamiuk, W. Wiesbeck, and T. Zwick, "Multi-mode antenna feed for ultra wideband technology," in *Proceedings of the IEEE Radio and Wireless Symposium (RWS '09)*, pp. 578–581, San Diego, Calif, USA, January 2008.

[7] G. Adamiuk, C. Heine, W. Wiesbeck, and T. Zwick, "Antenna array system for UWB-monopulse-radar," in *Proceedings of the International Workshop on Antenna Technology (iWAT '10)*, pp. 1–4, Lisbon, Portugal, March 2010.

[8] Federal Communications Commission (FCC), "Revision of part 15 of the commissions rules regarding ultrawideband transmission systems," First Report and Order, ET Docket 98-153, FCC 02-48, 2002.

[9] J.-P. Kim and W. S. Park, "Novel configurations of planar multilayer magic-T using microstrip-slotline transitions," *IEEE Transactions on Microwave Theory and Techniques*, vol. 50, no. 7, pp. 1683–1688, 2002.

[10] K. U-yen, E. J. Wollack, J. Papapolymerou, and J. Laskar, "A broadband planar magic-T using microstrip-slotline transitions," *IEEE Transactions on Microwave Theory and Techniques*, vol. 56, no. 1, pp. 172–177, 2008.

[11] L. Fan, C.-H. Ho, S. Kanamaluru, and K. Chang, "Wide-band reduced-size uniplanar magic-T, hybrid-ring, and de Ronde's CPW-slot couplers," *IEEE Transactions on Microwave Theory and Techniques*, vol. 43, no. 12, pp. 2749–2758, 1995.

[12] B. R. Heimer, L. Fan, and K. Chang, "Uniplanar hybrid couplers using asymmetrical coplanar striplines," *IEEE Transactions on Microwave Theory and Techniques*, vol. 45, no. 12, pp. 2234–2240, 1997.

[13] M. E. Bialkowski and Y. Wang, "Wideband microstrip 180° hybrid utilizing ground slots," *IEEE Microwave and Wireless Components Letters*, vol. 20, no. 9, pp. 495–497, 2010.

[14] M. M. Zinieris, R. Sloan, and L. E. Davis, "A broadband microstrip-to-slot-line transition," *Microwave and Optical Technology Letters*, vol. 18, no. 5, pp. 339–342, 1998.

[15] K. Gupta, *Microstrip Lines and Slotlines*, Artech House, Norwood, Mass, USA, 2nd edition, 1996.

[16] CST Microwave Studio, http://www.cst.com.

[17] W. Sörgel and W. Wiesbeck, "Influence of the antennas on the ultra-wideband transmission," *EURASIP Journal on Applied Signal Processing*, vol. 2005, no. 3, Article ID 843268, 2005.

17

An Iterative Approach to Improve Images of Multiple Targets and Targets with Layered or Continuous Profile

Yu-Hsin Kuo and Jean-Fu Kiang

Department of Electrical Engineering and the Graduate Institute of Communication Engineering, National Taiwan University, Taipei 106, Taiwan

Correspondence should be addressed to Jean-Fu Kiang; jfkiang@ntu.edu.tw

Academic Editor: Ramon Gonzalo

An iterative approach, based on the linear sampling method (LSM) and the contrast source inversion (CSI) method, is proposed to improve the recovered images of multiple targets and targets with layered or continuous profile, including shape and distribution of electric properties. The difficulties in dealing with large targets or high contrast are partly overcome with this approach. Typical targets studied in the literatures are chosen for simulations and comparison.

1. Introduction

Electromagnetic inverse techniques have been widely explored in geophysical survey, target detection, nondestructive testing, medical imaging, and so forth to retrieve the electric properties of possible targets. The linear sampling method (LSM) has been proposed to estimate the target shape [1, 2], implemented with the techniques of singular value decomposition (SVD) and Tikhonov regularization [3]. A level set process (LSP) has also been developed to highlight the target shape more accurately, especially when it has a high conductivity [4, 5].

When the target shape is properly acquired, the contrast source inversion (CSI) method can be applied to retrieve the permittivity and conductivity in both the target and the background medium. An IE-CSI (integral equation CSI) method was developed to recover targets immersed in homogeneous or layered background media; and an FD-CSI (finite difference CSI) method was claimed to be suitable for recovering inhomogeneous targets embedded in an inhomogeneous background medium [6–8]. A typical CSI algorithm, facilitated with the SVD technique, usually takes many iterations to converge [1].

The LSM and the CSI method have also been applied to recover multiple targets. For example, nine square cylinders, of width 0.3λ, $\epsilon_r = 2$, and $\sigma = 10\,\text{mS/m}$, were placed in free space and probed at 1 GHz [9]. The corners of each cylinder look blurred, and the recovered electric properties appear uneven. It was observed that the permittivity of dielectric targets could be underestimated [10].

In [11], four different inverse methods are compared on the Fresnel dataset [12]. When applying the CSI method to two separated cubes, the dielectric constant in the gap between the cubes is overestimated, especially at lower frequencies. In the case of two jointed spheres, the dielectric constant near the joint is strongly overestimated.

Similarly, in other cases of multiple lossy dielectric cylinders, the electric properties in between cylinders appear to be overestimated, and those near the center of cylinders are underestimated [13, 14]. When the radius of a cylinder is increased beyond half a wavelength, the dielectric constant near the center of the cylinder is severely underestimated [14].

In [15], a layered target was considered, which is composed of a cylinder of radius 0.15λ, $\epsilon_r = 2$, and $\sigma = 0\,\text{mS/m}$, enclosed within a shell of inner radius 0.4λ, outer radius 0.6λ, $\epsilon_r = 1.6$, and $\sigma = 0\,\text{mS/m}$. In [16, 17], the same geometry was studied, with larger permittivity of both the cylinder and the shell. In all these cases, the permittivity of the shell appears to be underestimated. In [18], two lossy dielectric cylinders were placed within a shell. It was found that the recovered shell image appears to shift inwards, with its dielectric constant being underestimated. On the other hand, the dielectric

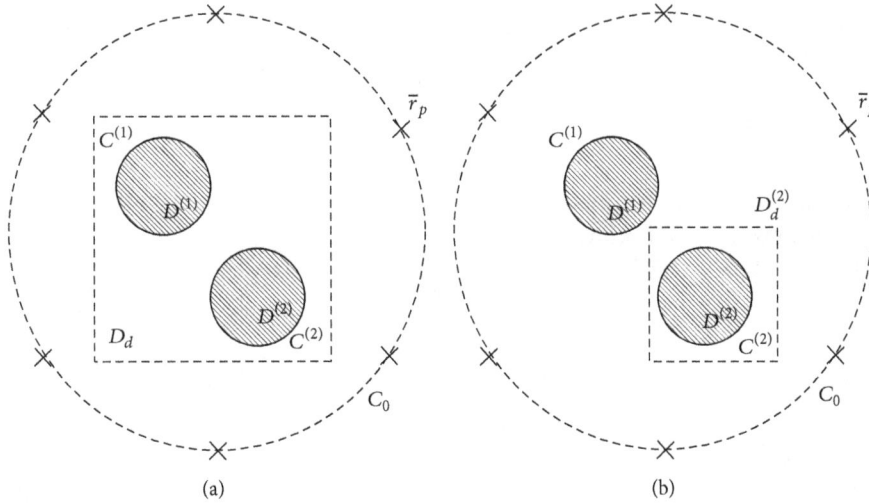

FIGURE 1: Detection domain (in dashed square) of a two-dimensional inverse problem: (a) the first stage and (b) the second stage.

constant of the two cylinders is overestimated, which may be attributed to the blockage by the shell.

In [19], a coated cube of inner width 0.6λ and outer width 1.6λ, with $\epsilon_r = 1.6$ and 1.3, respectively, was considered. The recovered permittivity near the center of the inner cube appears to be underestimated. Similar types of geometry reported in the literatures indicate that the permittivity within a target tends to be underestimated, especially when its electric size is large. With a larger permittivity difference between the target and the background, or between adjacent targets, the recovered permittivity profile becomes less accurate.

In this work, we propose an iterative approach, based on the LSM and the CSI method, to improve the recovered shape and the profile of electric properties of multiple targets, layered targets, and targets with a continuous profile. This paper is organized as follows. The iterative approach is described in Section 2 and simulations results of multiple targets, layered targets, and targets with a continuous profile are presented and discussed in Sections 3, 4, and 5, respectively. Finally, some conclusions are drawn in Section 6.

2. Description of Iterative Approach

The inverse process is executed in two stages. In the first stage, apply the linear sampling method (LSM) to estimate the geometrical shape of the target embedded in the detection domain, D_d, as shown in Figure 1(a). The probes are deployed on the perimeter C_0, which is outside the detection domain and encloses all the targets. Then, apply the contrast source inversion (CSI) method to estimate the electric parameters in the target domains, $D^{(1)}$ and $D^{(2)}$.

In the second stage, as shown in Figure 1(b), select part of the target domain, $D^{(1)}$, for example, in the first stage and treat it as part of the background. Then, the detection domain is reduced from D_d to $D_d^{(2)}$; the background field and the scattered field are updated accordingly. Next, apply the LSM

to the new detection domain $D_d^{(2)}$ to estimate the geometrical shape of the remaining targets; then apply the CSI method to estimate the electric parameters in the remaining target domain.

The LSM and the CSI algorithms are the same as those in [1], which will be briefly described in Sections 2.1 and 2.2. The iterative strategy proposed to improve the accuracy of the target shape and to reduce the rippling artifacts in the inverse results will be presented in Section 2.3.

2.1. LSM in Stage 1. The scattered field is used to estimate the target shape with the LSM [1]. Given an excitation source at $\bar{r}''$, the scattered field can be expressed as

$$\bar{E}_s\left(\bar{r},\bar{r}''\right) = k_b^2 \iint_{D_d} G\left(\bar{r},\bar{r}'\right) \chi\left(\bar{r}'\right) \bar{E}_t\left(\bar{r}',\bar{r}''\right) d\bar{r}', \quad (1)$$

where $\bar{E}_s(\bar{r},\bar{r}'')$ is the scattered field at $\bar{r}$, $\bar{E}_t(\bar{r}',\bar{r}'')$ is the total field at $\bar{r}'$, $\chi(\bar{r}')$ is the contrast function of the medium, and $G(\bar{r},\bar{r}')$ is the two-dimensional Green's function, which satisfies the wave equation

$$\left(\nabla^2 + k_b^2\right) G\left(\bar{r},\bar{r}'\right) = -\delta\left(\bar{r} - \bar{r}'\right). \quad (2)$$

It has the explicit form $G(\bar{r},\bar{r}') = -(j/4)H_0^{(2)}(k_b|\bar{r}-\bar{r}'|)$, where $H_0^{(2)}$ is the zeroth-order Hankel's function of the second kind and k_b is the wavenumber of the background medium. The contrast function is defined as

$$\chi\left(\bar{r}\right) = \frac{\epsilon\left(\bar{r}\right) - \epsilon_b}{\epsilon_b}, \quad (3)$$

where $\epsilon(r)$ and ϵ_b are the complex permittivity of the target and the background medium, respectively. The complex permittivity ϵ can be expressed as $\epsilon = \epsilon' - j\epsilon''$, where ϵ' is the real dielectric constant and ϵ'' is related to the conductivity σ as $\epsilon'' = \sigma/\omega$.

Define an adjoint field $\xi(\bar{r}, \bar{r}'')$, which satisfies the adjoint equation

$$\iint_{D_d'} \xi\left(\bar{r}, \bar{r}''\right) E_s\left(\bar{r}', \bar{r}''\right) d\bar{r}'' = G\left(\bar{r}, \bar{r}'\right), \qquad (4)$$

where all possible sources are located in D_d', which is practically outside of D_d. To solve (4) for ξ, consider M excitation probes placed at $\bar{r}_{pn}$'s, with $1 \le n \le M$ [1]. Hence, (4) can be discretized into

$$\sum_{n=1}^{M} \zeta_n \xi\left(\bar{r}, \bar{r}_{pn}\right) E_s\left(\bar{r}', \bar{r}_{pn}\right) = G\left(\bar{r}, \bar{r}'\right), \qquad (5)$$

where ζ_n is a weighting factor associated with the nth excitation probe.

The detection domain D_d is divided into N_d cells, with the center of the ℓth cell being at $\bar{r}_\ell$. For a specific $\bar{r}_\ell$, (5) can be discretized into a matrix form as

$$\overline{\overline{A}} \cdot \overline{f} = \overline{g}, \qquad (6)$$

where $A_{mn} = \zeta_n E_s(\bar{r}_{pm}, \bar{r}_{pn})$, $f_n = \xi(\bar{r}_\ell, \bar{r}_{pn})$, and $g_m = G(\bar{r}_\ell, \bar{r}_{pm})$, with $1 \le m, n \le M$. Then, apply the singular value decomposition (SVD) and the Tikhonov regularization techniques to solve (6) for $\overline{f}$.

An LSM indicator for cell centered at $\bar{r}_\ell$ is calculated as

$$I_\xi\left(\bar{r}_\ell\right) = \iint_{D_d'} \left|\xi\left(\bar{r}_\ell, \bar{r}''\right)\right|^2 d\bar{r}'' = \sum_{n=1}^{M} \zeta_n \left|\xi\left(\bar{r}_\ell, \bar{r}_{pn}\right)\right|^2. \qquad (7)$$

If $I_\xi(\bar{r}_\ell)$ is smaller than a threshold value, the cell centered $\bar{r}_\ell$ is categorized into part of the target.

2.2. CSI in Stage 1. The right-hand side of (4) can be viewed as an adjoint scattered field, $\Psi_s(\bar{r}, \bar{r}') = G(\bar{r}, \bar{r}')$, which is the scattered field $\overline{E}_s$ operated by $\overline{\overline{\xi}}$. Similarly, define an adjoint incident field Ψ_i and an adjoint total field Ψ_t as

$$\Psi_i\left(\bar{r}, \bar{r}'\right) = \iint_{D_d'} \xi\left(\bar{r}, \bar{r}''\right) E_i\left(\bar{r}', \bar{r}''\right) d\bar{r}'', \qquad (8)$$

$$\Psi_t\left(\bar{r}, \bar{r}'\right) = \iint_{D_d'} \xi\left(\bar{r}, \bar{r}''\right) E_t\left(\bar{r}', \bar{r}''\right) d\bar{r}''$$

$$= \Psi_i\left(\bar{r}, \bar{r}'\right) + \Psi_s\left(\bar{r}, \bar{r}'\right). \qquad (9)$$

The last equality holds because $E_t = E_i + E_s$. By substituting (1) into (4) and using (9), we have

$$G\left(\bar{r}, \bar{r}'\right) = k_b^2 \iint_{D_d} G\left(\bar{r}, \bar{r}''\right) \Psi_t\left(\bar{r}', \bar{r}''\right) \chi\left(\bar{r}''\right) d\bar{r}''. \qquad (10)$$

To solve for $\chi(\bar{r}'')$ in the detection domain D_d, (10) is transformed to a matrix form

$$\overline{\overline{L}} \cdot \overline{\chi} = \overline{g}, \qquad (11)$$

where $\overline{\chi} = [\chi(\bar{r}_1''), \chi(\bar{r}_2''), \dots, \chi(\bar{r}_{N_d}'')]^t$, $\overline{g} = [G(\bar{r}_{p1}, \bar{r}_1'),$ $G(\bar{r}_{p2}, \bar{r}_1'), \dots, G(\bar{r}_{pM}, \bar{r}_1'), G(\bar{r}_{p1}, \bar{r}_2'), G(\bar{r}_{p2}, \bar{r}_2'), \dots, G(\bar{r}_{pM}, \bar{r}_2')$ $\dots, G(\bar{r}_{p1}, \bar{r}_{N_s}'), G(\bar{r}_{p2}, \bar{r}_{N_s}'), \dots, G(\bar{r}_{pM}, \bar{r}_{N_s}')]^t$, and

$$\overline{\overline{L}} = \begin{bmatrix} L_{11,1} & \cdots & L_{11,N_d} \\ L_{12,1} & \cdots & L_{12,N_d} \\ \vdots & \vdots & \vdots \\ L_{1M,1} & \cdots & L_{1M,N_d} \\ \vdots & \vdots & \vdots \\ L_{N_sM,1} & \cdots & L_{N_sM,N_d} \end{bmatrix}. \qquad (12)$$

The explicit form of $L_{nm,\ell}$ is

$$L_{nm,\ell} = k_b^2 \iint_{\Delta D_{d\ell}} G\left(\bar{r}_{pm}, \bar{r}_\ell''\right) \Psi_t\left(\bar{r}_n', \bar{r}_\ell''\right) d\bar{r}_\ell'', \qquad (13)$$

with $1 \le n \le N_s$, $1 \le \ell \le N_d$, $1 \le m \le M$. Similar to the discretization of (4) to derive (5), $\Psi_t(\bar{r}, \bar{r}')$ can be calculated by discretizing (9) as

$$\Psi_t\left(\bar{r}, \bar{r}'\right) = \sum_{n=1}^{M} \zeta_n \xi\left(\bar{r}, \bar{r}_{pn}\right) E_t\left(\bar{r}', \bar{r}_{pn}\right). \qquad (14)$$

Note that N_s is the number of cells in the target domain, which is smaller than N_d. The SVD and the Tikhonov regularization techniques can then be applied to solve (11) for $\overline{\chi}$.

2.3. LSM and CSI in Stage 2. The definition of scattered field is extended to

$$\overline{E}_s\left(\bar{r}, \bar{r}''\right) = \overline{E}_t\left(\bar{r}, \bar{r}''\right) - \overline{E}_b\left(\bar{r}, \bar{r}''\right), \qquad (15)$$

where $\overline{E}_t(\bar{r}, \bar{r}'')$ is total field and $\overline{E}_b(\bar{r}, \bar{r}'')$ is the background field, which is the total field in a given background medium. The background field will reduce to the incident field if the background medium is free space.

A portion of the target area can be selected and merged into the background medium. For example, the shape of $D^{(1)}$, as shown in Figure 1, and the electric parameters are estimated in stage 1 and merged as part of the background. The background field is then numerically calculated and stored as $\overline{E}_b^{(1)}(\bar{r}, \bar{r}'')$.

Then, the scattered field $\overline{E}_s^{(2)}(\bar{r}, \bar{r}'')$ is updated as

$$\overline{E}_s^{(2)}\left(\bar{r}, \bar{r}''\right) = \overline{E}_t\left(\bar{r}, \bar{r}''\right) - \overline{E}_b^{(1)}\left(\bar{r}, \bar{r}''\right). \qquad (16)$$

Based on $\overline{E}_s^{(2)}(\bar{r}, \bar{r}'')$, an adjoint field $\xi^{(2)}(\bar{r})$ is defined, which satisfies the adjoint equation

$$\iint_{D_d'} \xi^{(2)}\left(\bar{r}, \bar{r}''\right) E_s^{(2)}\left(\bar{r}', \bar{r}''\right) d\bar{r}'' = G\left(\bar{r}, \bar{r}'\right). \qquad (17)$$

The LSM as used in stage 1 is applied to solve (17) for $\xi^{(2)}$, which is then used to estimate the target shape.

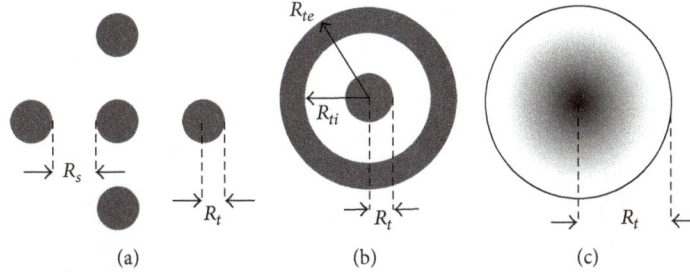

FIGURE 2: Configurations of (a) five cylinders, (b) a cylinder enclosed by a shell, and (c) a cylinder with continuous permittivity profile.

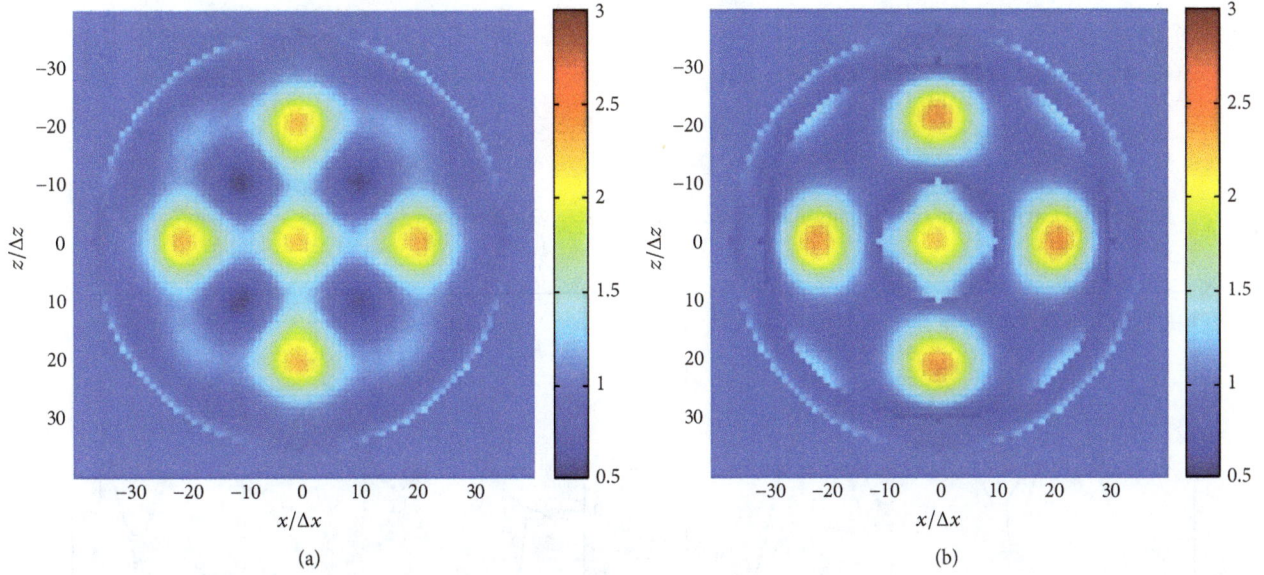

FIGURE 3: Distribution of relative dielectric constant after (a) stage 1 and (b) stage 2; $R_t = 0.125\,\text{m}$, $R_s = 0.25\,\text{m}$, $\epsilon_r = 2.5$, $\sigma = 5\,\text{mS/m}$, $M = 48$, and $R_d = 0.875\,\text{m}$.

Similar to the CSI method in stage 1, the right-hand side of (17) can be viewed as an adjoint scattered field, $\Psi_s^{(2)}(\bar{r}, \bar{r}')$. The corresponding adjoint incident field $\Psi_i^{(2)}$ and adjoint total field $\Psi_t^{(2)}$ can be defined as

$$\Psi_i^{(2)}\left(\bar{r}, \bar{r}'\right) = \iint_{D_d'} \xi^{(2)}\left(\bar{r}, \bar{r}''\right) E_b^{(1)}\left(\bar{r}', \bar{r}''\right) d\bar{r}'',$$

$$\Psi_t^{(2)}\left(\bar{r}, \bar{r}'\right) = \iint_{D_d'} \xi^{(2)}\left(\bar{r}, \bar{r}''\right) E_t\left(\bar{r}', \bar{r}''\right) d\bar{r}'' \qquad (18)$$

$$= \Psi_i^{(2)}\left(\bar{r}, \bar{r}'\right) + \Psi_s^{(2)}\left(\bar{r}, \bar{r}'\right).$$

Next, substitute (16) into (17) to have

$$G\left(\bar{r}, \bar{r}'\right)$$

$$= k_b^2 \iint_{D_d^{(2)}} G\left(\bar{r}, \bar{r}''\right) \Psi_t^{(2)}\left(\bar{r}', \bar{r}''\right) \chi^{(2)}\left(\bar{r}''\right) d\bar{r}'' \qquad (19)$$

from which $\chi^{(2)}(\bar{r}'')$ in domain $D_d^{(2)}$ can be solved by applying the SVD and Tikhonov regularization techniques as used in stage 1, where $\Psi_t^{(2)}(\bar{r}, \bar{r}')$ is calculated as

$$\Psi_t^{(2)}\left(\bar{r}, \bar{r}'\right) = \sum_{n=1}^{M} \zeta_n \xi^{(2)}\left(\bar{r}, \bar{r}_{pn}\right) E_t\left(\bar{r}', \bar{r}_{pn}\right). \qquad (20)$$

3. Multiple Targets

Figure 2 shows three types of targets that have been commonly tested in the literatures. The efficacy of the proposed strategy will be studied by simulations on these types in the following three sections, respectively.

Figure 2(a) shows five cylindrical targets placed in free space. The radius of the detection domain is $R_d = 0.875\,\text{m}$ and $M = 48$ probes are used. At the operating frequency of 300 MHz, the separation between two adjacent probes is about 0.2λ. The cell size is $\Delta x = \Delta z = \lambda/40$, and $N_s/N_d = 0.0210$.

The recovered distributions of permittivity and conductivity after stage 1 are shown in Figures 3(a) and 4(a), respectively. In stage 2, the domain slightly larger than the center cylinder is selected to be part of the background. The results after stage 2 are shown in Figures 3(b) and 4(b), respectively; and the distributions at $z = 0$ are shown in Figure 5. The recovered images after stage 1 and stage 2 look similar. The spacing between the center cylinder and the other four seems to be large enough to allow sufficient probing signals to reach all the five cylinders.

Next, the separation between adjacent cylinders is reduced from $R_s = 0.25\,\text{m}$ to $R_s = 0.125\,\text{m}$. The recovered

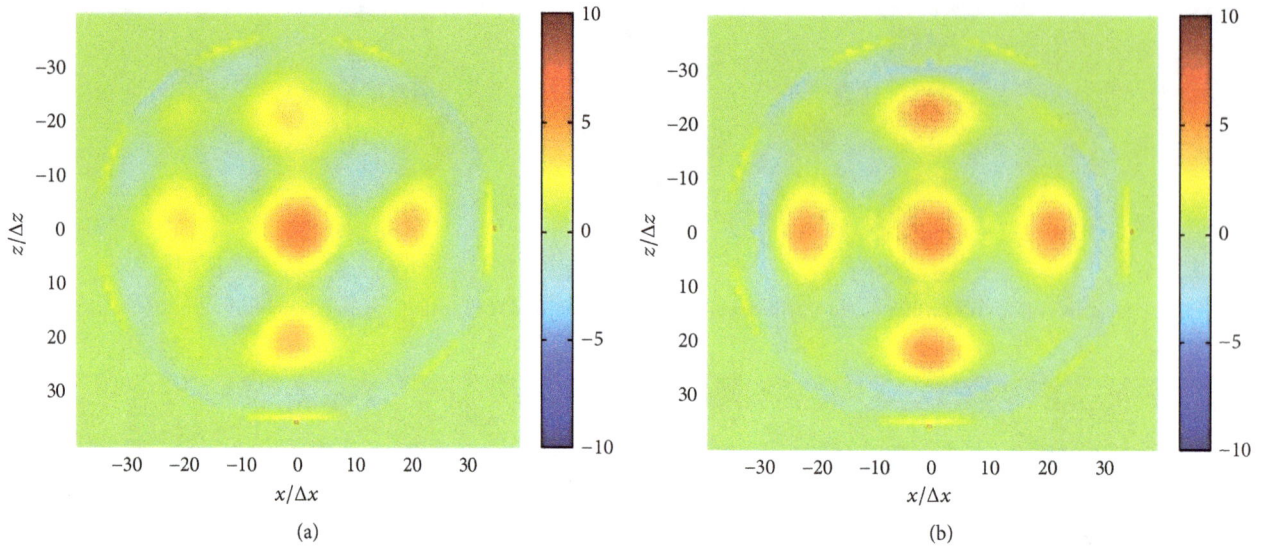

FIGURE 4: Distribution of conductivity after (a) stage 1 and (b) stage 2; parameters are the same as in Figure 3.

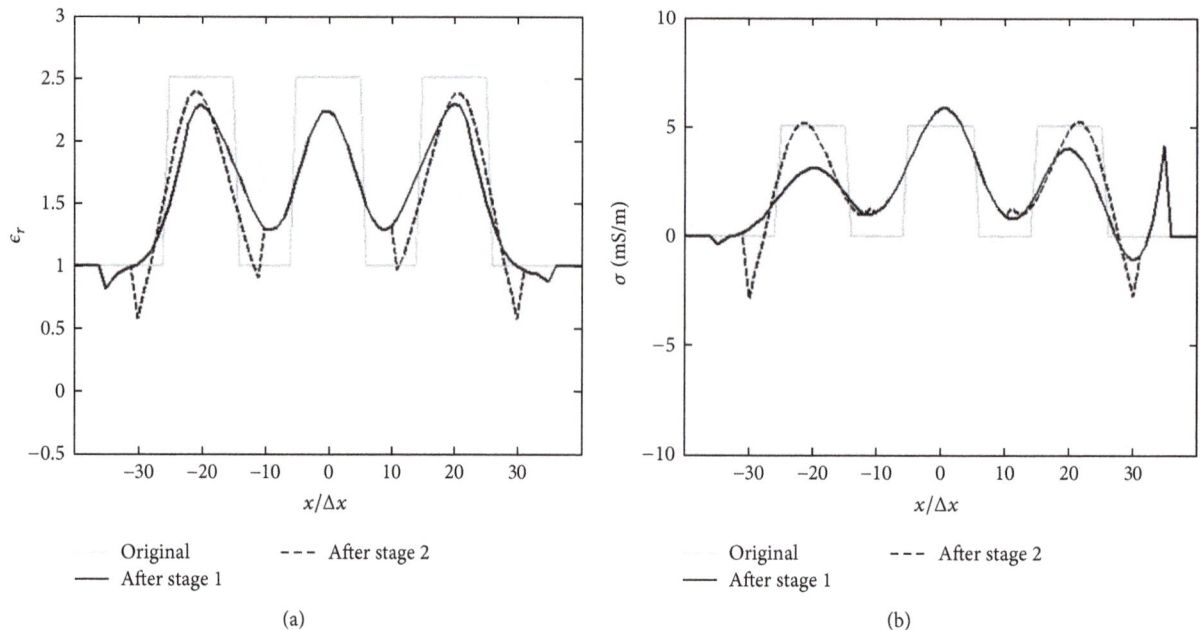

FIGURE 5: Distribution of (a) relative dielectric constant and (b) conductivity of multiple targets at $z = 0$; parameters are the same as in Figure 3.

permittivity and conductivity distributions after stage 1 are shown in Figures 6(a) and 7(a), respectively. The permittivity in the center cylinder is obviously underestimated.

In stage 2, an annular domain, enclosing the four outer cylinders but excluding the center one, is selected to be part of the background. The results after stage 2 are shown in Figures 6(b) and 7(b), respectively; and the distributions at $z = 0$ are shown in Figure 8.

In this case, the separation between the center cylinder and the other four seems to be too small. Some of the probing signals are blocked by the outer four cylinders from reaching the center one. Hence, the permittivity of the center cylinder

is underestimated, and that over the gap between the center cylinder and the other four is overestimated.

In summary, when $R_s = 0.25$ m, the results shown in Figures 4 and 5 indicate that both the conventional method and the proposed approach give similar results. When $R_s = 0.125$ m, the results shown in Figures 7 and 8 indicate that the conventional method underestimates the permittivity of the center cylinder because the outer four cylinders block some of the probing waves. After merging the four blocking cylinders to the background, the center cylinder is better observed with the probing waves.

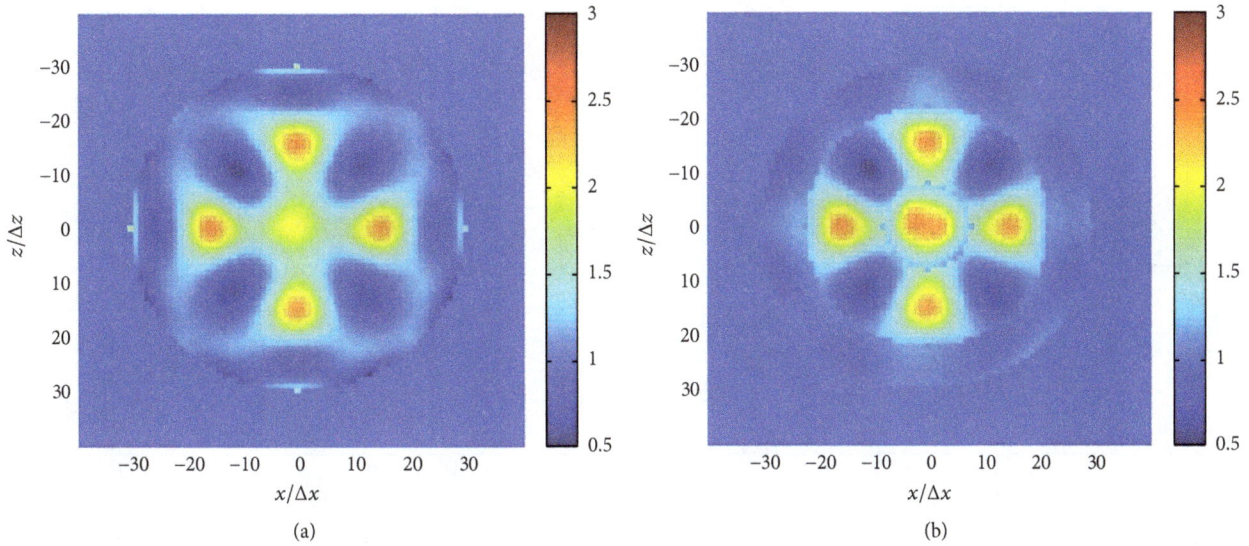

FIGURE 6: Distribution of relative dielectric constant after (a) stage 1 and (b) stage 2; $R_t = 0.125$ m, $R_s = 0.125$ m, $\epsilon_r = 2.5$, $\sigma = 5$ mS/m, $M = 48$, and $R_d = 0.875$ m.

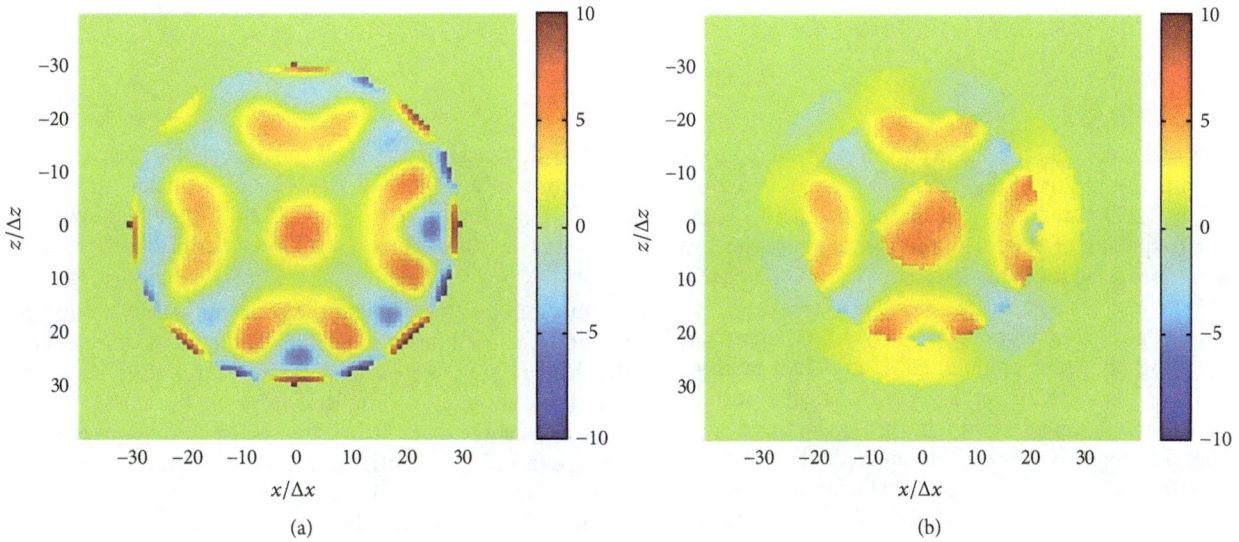

FIGURE 7: Distribution of conductivity after (a) stage 1 and (b) stage 2; parameters are the same as in Figure 6.

To quantify the improvement of accuracy by using the proposed strategy, we define the error indices on target shape, permittivity, and conductivity as

$$\varepsilon_s = 100 \times \frac{N_m}{N_t}\%,$$

$$\varepsilon_{\epsilon t} = 100 \times \sqrt{\frac{\sum_{n=1}^{N_t} \left| \epsilon_{tn}^e - \epsilon_{tn}^a \right|}{\sum_{n=1}^{N_t} \left| \epsilon_{tn}^a \right|}}\%, \qquad (21)$$

$$\varepsilon_{\sigma t} = 100 \times \sqrt{\frac{\sum_{n=1}^{N_t} \left| \sigma_{tn}^e - \sigma_{tn}^a \right|}{\sum_{n=1}^{N_t} \left| \sigma_{tn}^a \right|}}\%,$$

where ε_s is the shape-error index and $\varepsilon_{\epsilon t}$ and $\varepsilon_{\sigma t}$ are the error indices of permittivity and conductivity, respectively, within the target; N_t is the number of cells in the target, N_m is the number of cells in the target which are misrecognized as part of the background, and the superscripts e and a indicate the estimated value and the actual value, respectively.

By comparing the results after stages 1 and 2 as shown in Figures 6 and 7, it is observed that ε_s is reduced from 35% to 30%, $\varepsilon_{\epsilon t}$ is reduced from 25% to 23%, and $\varepsilon_{\sigma t}$ is reduced from 48% to 46%.

4. Layered Targets

Figure 2(b) shows a cylinder target enclosed by a cylindrical shell. The cylinder has a radius of $R_t = 0.1$ m, and the shell

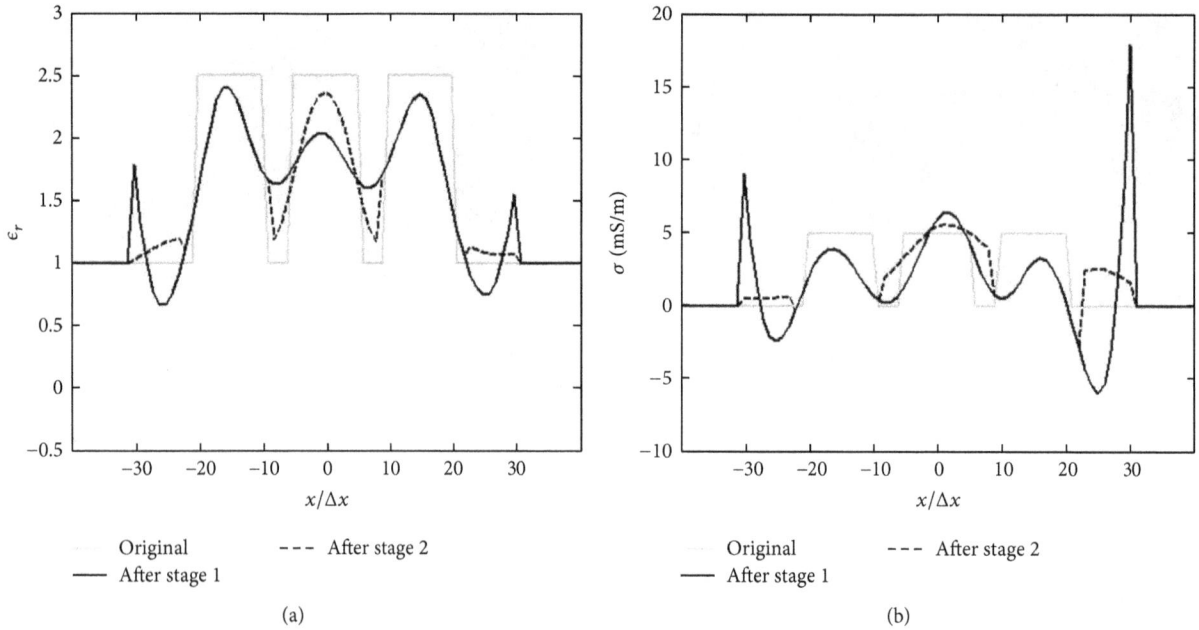

Original   - - - After stage 2
—— After stage 1

(a)

Original   - - - After stage 2
—— After stage 1

(b)

FIGURE 8: Distribution of (a) relative dielectric constant and (b) conductivity of multiple targets at $z = 0$; parameters are the same as in Figure 6.

has an internal radius of $R_{ti} = 0.3$ m and an external radius of $R_{te} = 0.4$ m. The radius of the detection domain is chosen to be $R_d = 1.25$ m, and $M = 48$ probes are used. At the operating frequency of 300 MHz, the spacing between two adjacent probes is about 0.2λ. The cell size is $\Delta x = \Delta z = \lambda/40$.

The solid curves in Figure 9 are the results after stage 1, by applying the conventional LSM and CSI method. Three different permittivities of target are simulated, and the position of the shell appears to shift inwards in all three cases.

When the iterative approach is applied, the electrical parameters of and around the center cylinder estimated in stage 1 are treated as part of the background medium in stage 2. The recovered distributions at $z = 0$ after stage 2, with $\epsilon_r = 2.0$ and 2.5, become closer to the original, as compared to the conventional method. For the case with $\epsilon_r = 3.0$, the permittivity in the shell is underestimated, but the shell position is also closer to the original than that predicted with the conventional method. By comparing the results after stages 1 and 2, it is found that ε_s is reduced from 120% to 31%, and ε_{et} is reduced from 42% to 29%.

As the shell is placed too close to the cylinder, the recovered position of the former may be shifted when using the conventional method. Next, we compare the effects of shell-cylinder separation by simulating cases with $(R_{ti}, R_{te}) = (0.4, 0.5)$ m and $(0.5, 0.6)$ m, with the distributions at $z = 0$ shown in Figures 10 and 11, respectively. In both cases, the position shift of shell becomes less severe than the previous case with $(R_{ti}, R_{te}) = (0.3, 0.4)$ m. The iterative approach not only improves the shell position as in the previous case but also obtains a better estimation of permittivity in the shell.

The thickness of the shell may affect the estimation of shell position and permittivity. Hence, a thicker shell with $(R_{ti}, R_{te}) = (0.3, 0.5)$ m is simulated. Figure 12 shows the recovered distributions at $z = 0$, where position shift is barely observable.

In summary, by applying the conventional LSM and CSI method, the shell position will be shifted if the cylinder and the shell are put too close or if the shell is too thin. Using the LSM indicator cannot completely separate the shell from the cylinder in some cases. When the shell is thin, more probing waves can reach the internal cylinder and the permittivity of the latter can be well recovered. However, the position of the shell is shifted and its permittivity is underestimated. On the other hand, if the shell is thick, less probing waves can reach the internal cylinder, leading to underestimation of permittivity of the cylinder. The iterative approach seems capable of overcoming this problem by merging the internal cylinder into the background to improve the image of the external shell.

5. Targets with Continuous Profile

Figure 2(c) shows a cylinder with a continuous permittivity profile. The radius of the cylinder is $R_t = 0.375$ m, the radius of the detection domain is $R_d = 1.25$ m, and $M = 48$ probes are used. At the operating frequency of 300 MHz, the spacing between two adjacent probes is about 0.2λ. The cell size is $\Delta x = \Delta z = \lambda/40$. Figures 13 and 14 show the recovered distributions of relative dielectric constant and conductivity at $z = 0$. The relative dielectric constant has a linear profile, with the maximum ϵ_r of 2.4 and 3.0, respectively. Each profile is approximated as a piecewise-constant function of 4, 8, and

(a)

(b)

(c)

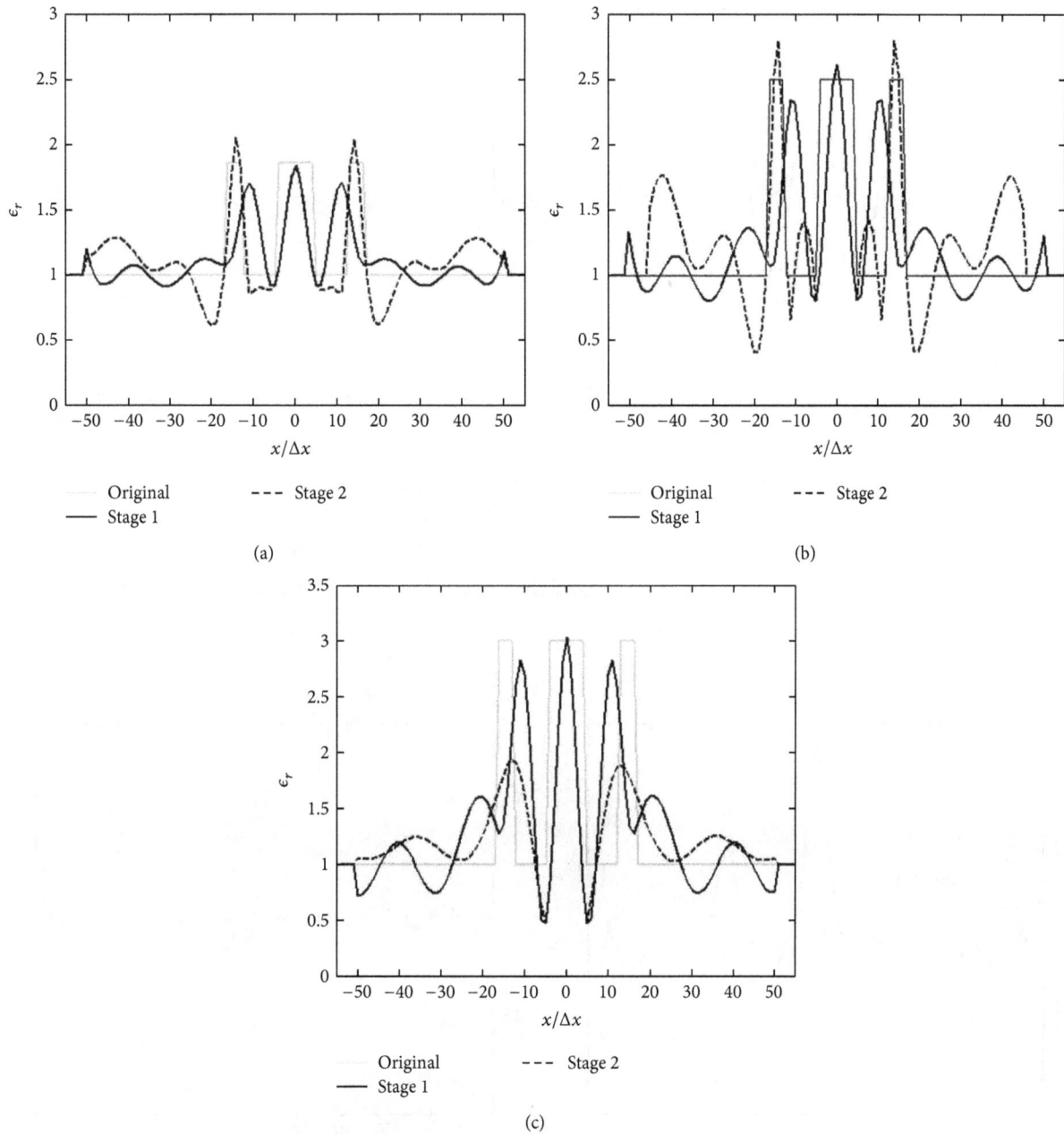

FIGURE 9: Distribution of relative dielectric constant of layered targets at $z = 0$; (a) $\epsilon_r = 2.0$, (b) $\epsilon_r = 2.5$, and (c) $\epsilon_r = 3.0$; $R_t = 0.1$ m, $R_{ti} = 0.3$ m, $R_{te} = 0.4$ m, $\sigma = 0$, $M = 48$, and $R_d = 1.25$ m.

15 stairs, respectively. The recovered results of the case with $\epsilon_{r,\max} = 2.4$ appear closer to the original profile than those with $\epsilon_{r,\max} = 3$. As shown in Figure 14, the permittivity in the internal portion of the 4-stair approximation is underestimated.

Figure 15 shows the recovered distributions with $R_t = 0.375, 0.4, 0.45$, and 0.5 m, respectively. A linear permittivity profile is assumed, with $\epsilon_{r,\max} = 2.4$. With $R_t = 0.45$ or 0.5 m, the permittivity in the internal portion is seriously underestimated.

Similar recovered images were observed in [20], where two concentric square cylinders were immersed in a background medium with $\epsilon_r = 1.2$ and $\sigma = 5$ mS/m at the operating frequency of 400 MHz. The external square cylinder has the width of 0.5 m, $\epsilon_r = 3.6$, and $\sigma = 50$ mS/m. The internal square cylinder has the width of 0.25 m, $\epsilon_r = 6$, and $\sigma = 80$ mS/m. The results using the contrast source-extended Born (CS-EB) approach are consistent with the original, but the results using the contrast source inversion approach of [8] are inconsistent. The permittivity in the internal portion is

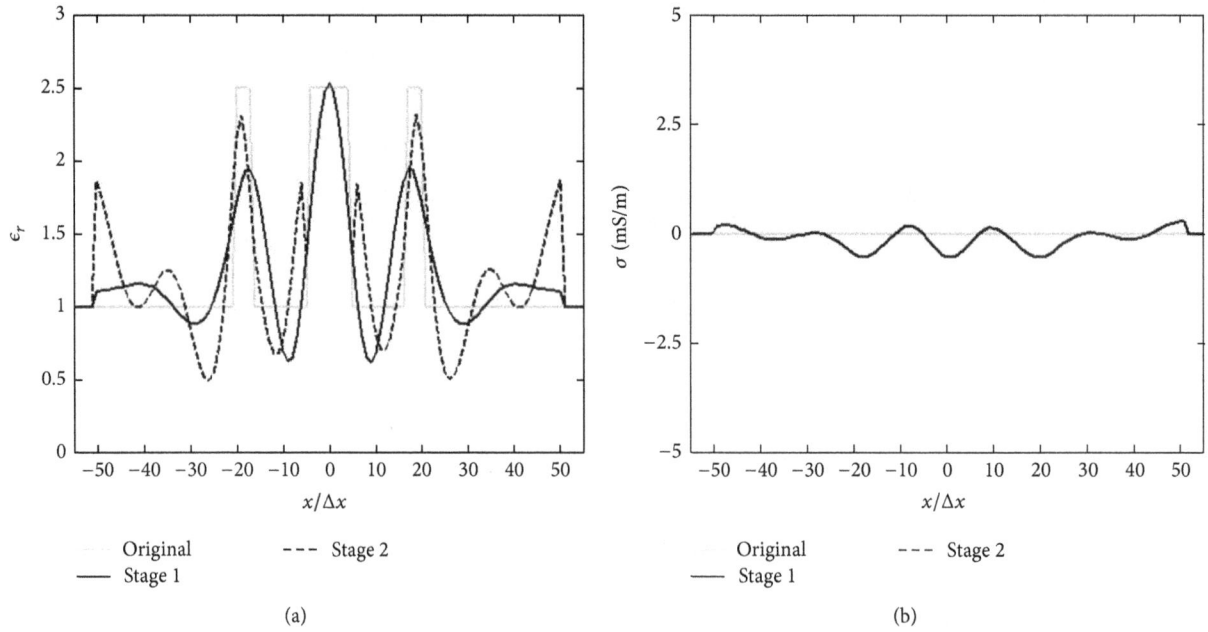

FIGURE 10: Distribution of (a) relative dielectric constant and (b) conductivity of layered targets at $z = 0$; $R_t = 0.1$ m, $R_{ti} = 0.4$ m, $R_{te} = 0.5$ m, $\epsilon_r = 2.5$, $\sigma = 0$, $M = 48$, and $R_d = 1.25$ m.

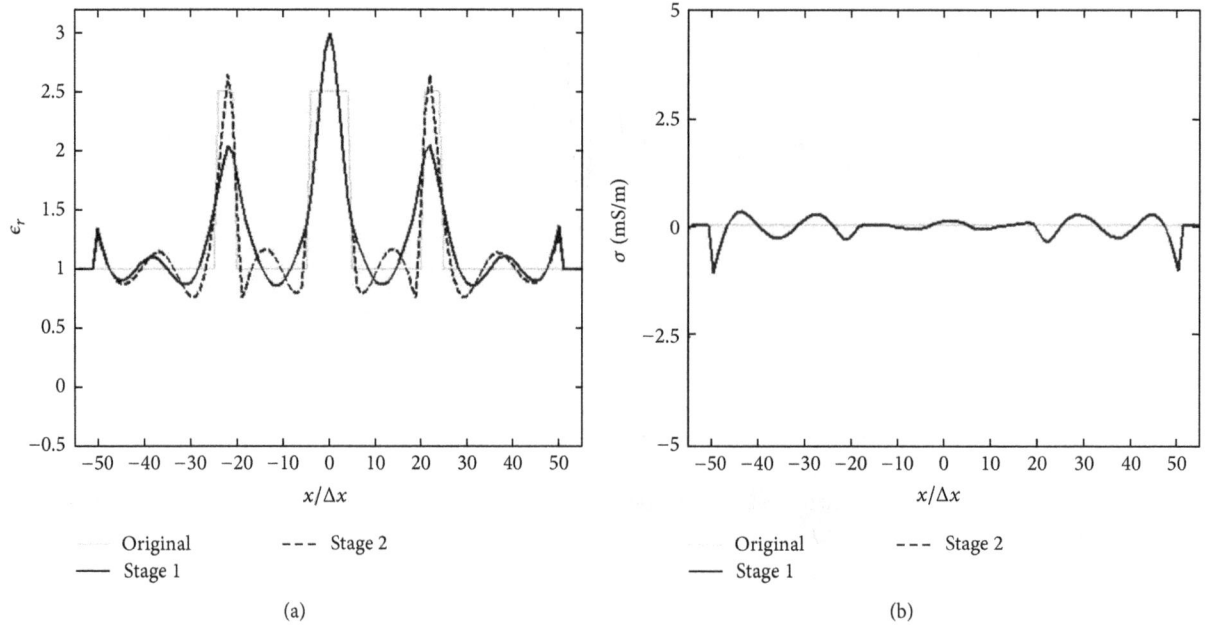

FIGURE 11: Distribution of (a) relative dielectric constant and (b) conductivity of layered targets at $z = 0$; $R_t = 0.1$ m, $R_{ti} = 0.5$ m, $R_{te} = 0.6$ m, $\epsilon_r = 2.5$ and $\sigma = 0$, $M = 48$, and $R_d = 1.25$ m.

seriously underestimated, similar to those shown in Figures 14 and 15.

The incident waves are partially reflected at the interfaces between adjacent stairs, and smaller discontinuity of ϵ_r over the interfaces leads to less reflection. Hence, more incident waves can reach the internal portion of the target domain

and get more information to estimate the internal permittivity.

Figure 16 shows the recovered distribution of relative dielectric constant at $z = 0$, using the iterative approach. A linear profile with $\epsilon_{r,\max} = 2.4$ is assumed, and $R_t = 0.5$ m. By merging the first few external layers to the background

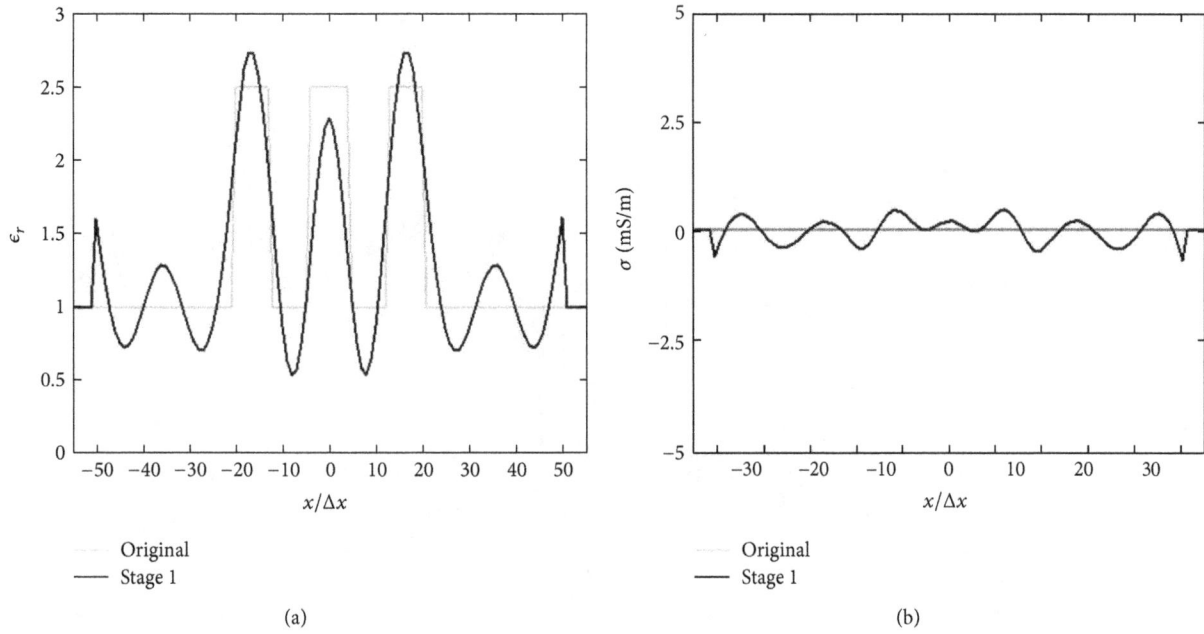

FIGURE 12: Distribution of (a) relative dielectric constant and (b) conductivity of layered targets at $z = 0$; $R_t = 0.1$ m, $R_{ti} = 0.3$ m, $R_{te} = 0.5$ m, $\epsilon_r = 2.5$, $\sigma = 0$, $M = 48$, and $R_d = 1.25$ m.

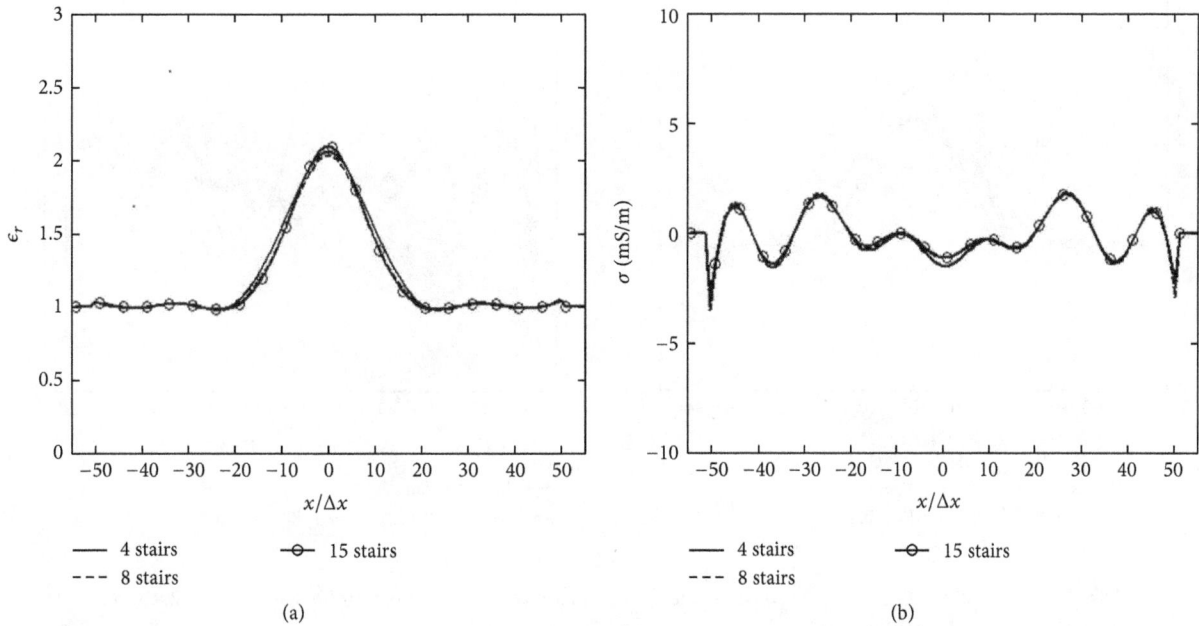

FIGURE 13: Distribution of (a) relative dielectric constant and (b) conductivity at $z = 0$; $R_t = 0.375$ m, linear profile with $\epsilon_{r,\max} = 2.4$, $\sigma = 0$, $M = 48$, and $R_d = 1.25$ m.

medium in stage 2, the internal portion is recovered more accurately. The recovered image with 0.35 m < R_t < 0.55 m merged as the background medium matches the most with the original one. However, the recovered distribution at the interface between the background medium and the target

becomes less accurate because the contrast function χ jumps from zero to a finite number at the interface.

In order to avoid the discontinuity of χ, the permittivity at the interface estimated in stage 1 is used as the background permittivity inside the interface in stage 2. As shown in

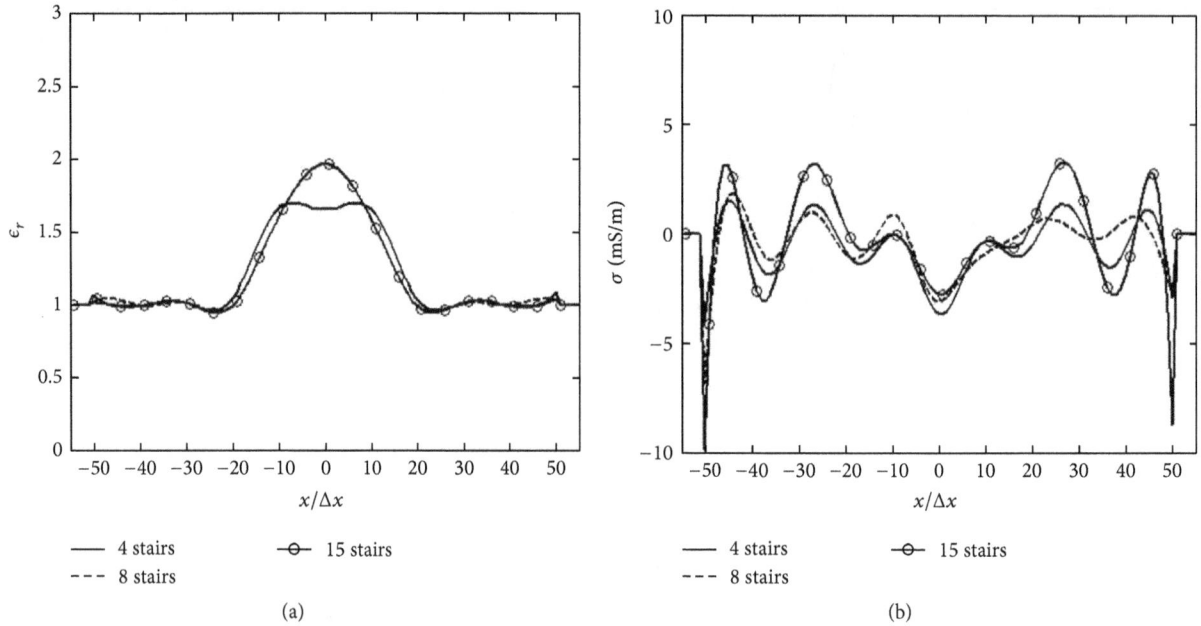

FIGURE 14: Distribution of (a) relative dielectric constant and (b) conductivity at $z = 0$; $R_t = 0.375$ m, linear profile with $\epsilon_{r,\max} = 3$, $\sigma = 0$, $M = 48$, and $R_d = 1.25$ m.

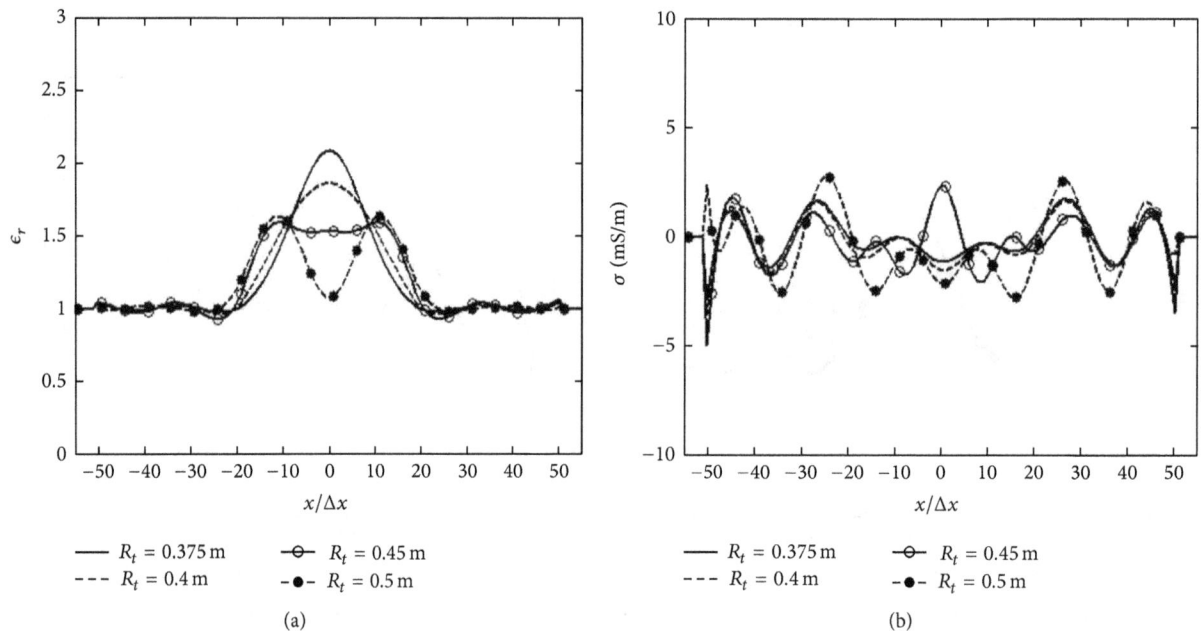

FIGURE 15: Distribution of (a) relative dielectric constant and (b) conductivity at $z = 0$; linear profile with $\epsilon_{r,\max} = 2.4$ and $\sigma = 0$.

Figure 17, the discontinuity problem is reduced, if not completely removed. When 0.35 m $< R_t < 0.55$ m is selected as the background, ε_s is reduced from 16% to 15% and ε_{et} is reduced from 34% to 5% by comparing the results after stages 1 and 2.

6. Conclusion

An iterative approach, based on LSM and CSI method, is proposed to improve the accuracy of recovered images for multiple targets, layered targets, and targets with a continuous permittivity profile. For multiple targets, when a target is partially blocked by other targets, its permittivity tends to be underestimated, and that of the gap between targets tends to be overestimated. For layered targets, the external layers tend to be shifted inwards, especially when the gap between layers is small or the external layer is thin. For a cylinder with continuous permittivity profile, when the radius or the spatial change rate of permittivity is large, the permittivity in the internal portion tends to be underestimated. All

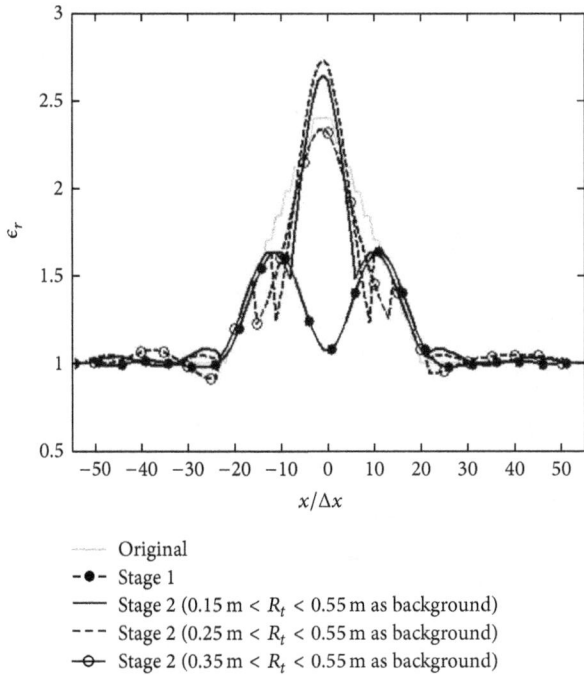

FIGURE 16: Distribution of relative dielectric constant at $z = 0$; linear profile with $\epsilon_{r,max} = 2.4$, $\sigma = 0$, and $R_t = 0.5$ m.

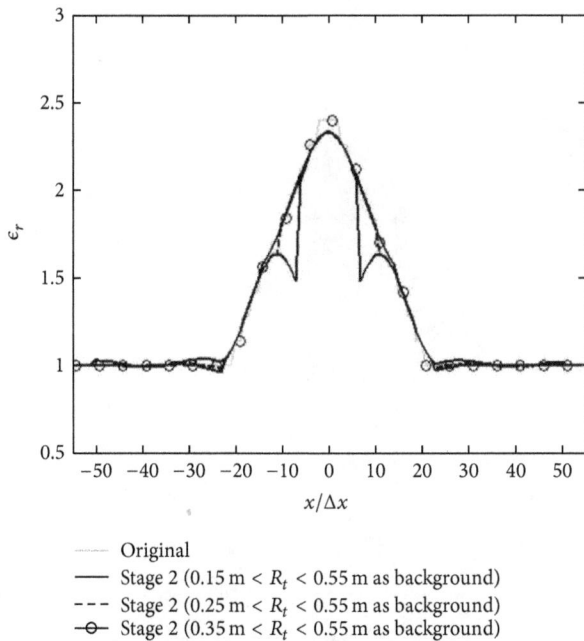

FIGURE 17: Distribution of relative dielectric constant at $z = 0$; linear profile with $\epsilon_{r,max} = 2.4$, $\sigma = 0$, and $R_t = 0.5$ m.

these symptoms can be relieved with the proposed iterative approach, which are validated by simulations.

Conflict of Interests

The authors declare that there is no conflict of interests regarding the publication of this paper.

Acknowledgments

This work was sponsored by the Ministry of Science and Technology, Taiwan, under Contract NSC 101-2221-E-002-129 and the Ministry of Education, Taiwan, under Aim for Top University Project 103R3401-1.

References

[1] L. Crocco, I. Catapano, L. D. Donato, and T. Isernia, "The linear sampling method as a way to quantitative inverse scattering," *IEEE Transactions on Antennas and Propagation*, vol. 60, no. 4, pp. 1844–1853, 2012.

[2] I. Catapano, L. Crocco, and T. Isernia, "Improved sampling methods for shape reconstruction of 3-D buried targets," *IEEE Transactions on Geoscience and Remote Sensing*, vol. 46, no. 10, pp. 3265–3273, 2008.

[3] D. Colton, H. Haddar, and M. Piana, "The linear sampling method in inverse electromagnetic scattering theory," *Inverse Problems*, vol. 19, no. 6, pp. S105–S137, 2003.

[4] A. M. Hassan, M. R. Hajihashemi, and M. El-Shenawee, "Inverse scattering shape reconstruction of 3D bacteria using the level set algorithm," *Progress in Electromagnetics Research B*, vol. 39, pp. 39–53, 2012.

[5] M. R. Hajihashemi and M. El-Shenawee, "Level set algorithm for shape reconstruction of non-overlapping three-dimensional penetrable targets," *IEEE Transactions on Geoscience and Remote Sensing*, vol. 50, no. 1, pp. 75–86, 2012.

[6] C. Gilmore, A. Abubakar, W. Hu, T. M. Habashy, and P. M. van den Berg, "Microwave biomedical data inversion using the finite-difference contrast source inversion method," *IEEE Transactions on Antennas and Propagation*, vol. 57, no. 5, pp. 1528–1538, 2009.

[7] A. Abubakar, W. Hu, P. M. van den Berg, and T. M. Habashy, "A finite-difference contrast source inversion method," *Inverse Problems*, vol. 24, no. 6, Article ID 065004, 2008.

[8] P. M. van den Berg and R. E. Kleinman, "A contrast source inversion method," *Inverse Problems*, vol. 13, no. 6, pp. 1607–1620, 1997.

[9] L. Di Donato, M. Bevacqua, T. Isernia, I. Catapano, and L. Crocco, "Improved quantitative microwave tomography by exploiting the physical meaning of the Linear Sampling Method," in *Proceedings of the 5th European Conference on Antennas and Propagation (EUCAP '11)*, pp. 3828–3831, Rome, Italy, April 2011.

[10] X. Ye, X. Chen, Y. Zhong, and R. Song, "Simultaneous reconstruction of dielectric and perfectly conducting scatterers via T matrix method," *IEEE Transactions on Antennas and Propagation*, vol. 61, no. 7, pp. 3774–3781, 2013.

[11] E. Mudry, P. C. Chaumet, K. Belkebir, and A. Sentenac, "Electromagnetic wave imaging of three-dimensional targets using a hybrid iterative inversion method," *Inverse Problems*, vol. 28, no. 6, 2012.

[12] J. M. Geffrin and P. Sabouroux, "Testing inversion algorithms against experimental data," *Inverse Problems*, vol. 17, no. 6, pp. 1565–1571, 2001.

[13] M. Ostadrahimi, A. Zakaria, J. Lovetri, and L. Shafai, "A near-field dual polarized (TE-TM) microwave imaging system," *IEEE Transactions on Microwave Theory and Techniques*, vol. 61, no. 3, pp. 1376–1384, 2013.

[14] A. Zakaria, C. Gilmore, and J. LoVetri, "Finite-element contrast source inversion method for microwave imaging," *Inverse Problems*, vol. 26, no. 11, 2010.

[15] L. Pan, Y. Zhong, X. Chen, and S. P. Yeo, "Subspace-based optimization method for inverse scattering problems utilizing phaseless data," *IEEE Transactions on Geoscience and Remote Sensing*, vol. 49, no. 3, pp. 981–987, 2011.

[16] L. Pan, X. Chen, and S. P. Yeo, "Nondestructive evaluation of nanoscale structures: inverse scattering approach," *Applied Physics A: Materials Science and Processing*, vol. 101, no. 1, pp. 143–146, 2010.

[17] L. Pan, X. Chen, Y. Zhong, and S. P. Yeo, "Comparison among the variants of subspace-based optimization method for addressing inverse scattering problems transverse electric case," *Journal of the Optical Society of America A*, vol. 27, no. 10, pp. 2208–2215, 2010.

[18] C. Gilmore, P. Mojabi, A. Zakaria, S. Pistorius, and J. Lovetri, "On super-resolution with an experimental microwave tomography system," *IEEE Antennas and Wireless Propagation Letters*, vol. 9, pp. 393–396, 2010.

[19] Y. Zhong, X. Chen, and K. Agarwal, "An improved subspace-based optimization method and its implementation in solving three-dimensional inverse problems," *IEEE Transactions on Geoscience and Remote Sensing*, vol. 48, no. 10, pp. 3763–3768, 2010.

[20] M. D'Urso, T. Isernia, and A. F. Morabito, "On the solution of 2-d inverse scattering problems via source-type integral equations," *IEEE Transactions on Geoscience and Remote Sensing*, vol. 48, no. 3, pp. 1186–1198, 2010.

18

Negative Group Delay Circuit Based on Microwave Recursive Filters

Mohammad Ashraf Ali and Chung-Tse Michael Wu

Department of Electrical and Computer Engineering, Wayne State University, 5050 Anthony Wayne Drive, Detroit, MI 48202, USA

Correspondence should be addressed to Chung-Tse Michael Wu; ctmwu@wayne.edu

Academic Editor: Tanmay Basak

This work presents a novel approach to design a maximally flat negative group delay (NGD) circuit based on microwave recursive filters. The proposed NGD circuit is realized by cascading N stages of quarter-wavelength stepped-impedance transformer. It is shown that the given circuit can be designed to have any prescribed group delay by changing the characteristic impedance of the quarter-wave transformers (QWTs) cascaded with each other. The proposed approach provides a systematic method to synthesize NGD of arbitrary amount without including any discrete lumped component. For various prescribed NGD, the characteristic impedance of QWT has been tabulated for two and three stages of the circuit. The widths and lengths of microstrip transmission lines can be obtained from characteristic impedance and the frequency of operation of the transmission line. The results are verified in both simulation and measurement, showing a good agreement.

1. Introduction

Following the classical paper published by Brillouin and Sommerfeld in 1960 [1], Garrett and McCumber were the first to analytically prove that group velocity of wave can be greater than the speed of light [2]. However, its first practical demonstration was performed after a decade by Chu and Wong [3]. At first, the concept of a wave having velocity greater than the speed of light (superluminal velocity [1]) seems to defy the causality. Nevertheless, there exist enough practical experiments showing that the concept of superluminal velocity follows the causal system definition [3–5]. Since the development of this concept, there has been a lot of work done by scientists to utilize this property and apply it to practical applications. This concept has been used to enhance the efficiency of the feed-forward amplifier, broadband and constant phase shifter, and shortening of delay lines [6–8]. By using NGD circuits, positive group delays introduced by the circuit components and electrical connections in electronic systems can be compensated. In NGD circuits, electromagnetic waves can have superluminal velocity in the region of anomalous dispersion such that the phase of the higher frequency component of the wave will move in advance with respect to the lower frequency components.

Although there have been a lot of efforts put towards generating NGD using both active and passive techniques, little work has been done towards generating the prescribed NGD at the desired frequency. Recently, a maximally flat NGD circuit based on transversal filter concept has been proposed to synthesize the desired group delay [9]. In this paper, we will demonstrate that a microwave recursive filter can also be used to generate NGD of predetermined values in the desired frequency band by using only distributed components [10]. It is noted that very recently distributed components-based NGD circuits have also been proposed; however, few systematic methods were provided to synthesize the desired NGD [11]. The contribution in this work expands the theory derived in [9] and provides a full analysis of a microwave recursive-filter based NGD circuitry.

To illustrate, multistage QWTs are used to generate NGD of predetermined values as shown in Figure 1. Essentially, it behaves as a recursive filter since the reflected waves will bounce back and forth among the impedance interfaces and form an infinite impulse response (IIR). Following this concept, we will show a comprehensive methodology to synthesize the desired NGD by treating the multistage QWT as an IIR filter. In addition, it is worth mentioning that there are no lumped components present, which makes this design easy to fabricate and to be scaled up to higher frequency. We will also demonstrate the synthesis procedure of a two-stage QWT with −0.5 ns, −1 ns, and −2 ns group delays. In addition, a branch-line coupler is used to transfer NGD from a one-port circuit to a two-port one, making the NGD circuit more applicable in practices.

2. Theory and Formulae

A microstrip-line based transmission line structure is used to fabricate QWTs. As depicted in Figure 1, the QWTs have the characteristic impedance of $Z_1, Z_2, Z_3 \cdots Z_N$ with an identical electrical length of θ. The impedance of all QWTs has been normalized to the termination impedance of 50 ohms. The design flow is as follows. First, we need to obtain the characteristic impedance of each of the microstrip transmission line QWTs that can generate the desired group delay. Once we obtain the characteristic impedance, we can calculate the width and length of the transmission line QWT according to the synthesis procedure shown in [12]. In order to solve for the desired characteristic impedance, we first relate it to the transmission matrix or ABCD matrix parameters. The relation between the characteristic impedance of the transmission line and the transmission matrix parameters for the ith stage of QWT is as follows:

$$\begin{bmatrix} \cos\theta & jZ_i \sin\theta \\ \left(\dfrac{j}{Z_i}\right)\sin\theta & \cos\theta \end{bmatrix} = \begin{bmatrix} A_i & jB_i \\ jC_i & D_i \end{bmatrix}. \quad (1)$$

Furthermore, when cascading N stages of QWT we will obtain

$$\prod_{i=1}^{N} \begin{bmatrix} \cos\theta & jZ_i \sin\theta \\ \left(\dfrac{j}{Z_i}\right)\sin\theta & \cos\theta \end{bmatrix} = \prod_{i=1}^{N} \begin{bmatrix} A_i & jB_i \\ jC_i & D_i \end{bmatrix}, \quad (2)$$

where N is the number of stages in the circuit, Z_i corresponds to the impedance of the ith QWT, and θ is the electrical length of each stage of the transmission line which is equal to $\pi/2$ at the center frequency ($\theta = \beta l$, where $l = \lambda/4$ and $\beta = 2\pi/\lambda$). We can then represent the product of the above transmission matrix parameter for N stage as follows:

$$\prod_{i=1}^{N} \begin{bmatrix} A_i & jB_i \\ jC_i & D_i \end{bmatrix} = \begin{bmatrix} A_n & jB_n \\ jC_n & D_n \end{bmatrix}, \quad (3)$$

in which A_n, B_n, C_n, and D_n represent the transmission matrix parameters for the entire N-stage quarter-wave transformer circuit. The reflection coefficient (Γ) for the N-stage circuit can be written as

$$\Gamma = \frac{V_{\text{in}}^-}{V_{\text{in}}^+} = \frac{(A_n - D_n) + j(B_n - C_n)}{(A_n + D_n) + j(B_n + C_n)}. \quad (4)$$

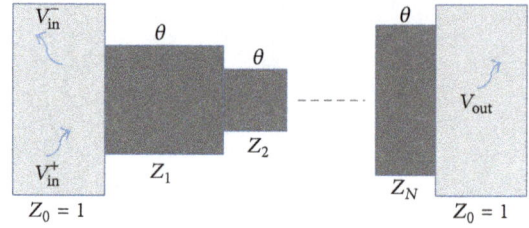

FIGURE 1: An N-stage quarter-wave transformer.

The phase of the reflection coefficient can be expressed as

$$\angle\Gamma = \tan^{-1}\frac{(B_n - C_n)}{(A_n - D_n)} - \tan^{-1}\frac{(B_n + C_n)}{(A_n + D_n)}. \quad (5)$$

In order to obtain the prescribed maximally flat group delay, we need to obtain the characteristic impedance of each of the transmission lines $[Z_1, Z_2, Z_3 \cdots Z_N]$ by solving the following equations:

$$\left|\Gamma\left(\theta = \frac{\pi}{2}\right)\right| = \Gamma_p, \quad (6)$$

$$-\frac{\partial\angle\Gamma(\theta)}{\partial\theta}\bigg|_{\theta=\pi/2} = \tau_{gp}, \quad (7)$$

$$-\frac{\partial^{2n-3}\angle\Gamma(\theta)}{\partial\theta}\bigg|_{\theta=\pi/2} = 0, \quad (8)$$

$$\text{where } n = 3, 4, \ldots, (N-1).$$

Here Γ_p is the prescribed magnitude, and τ_{gp} is the prescribed NGD with a unit of the sampling period T. In order to have the maximally flat response, we set all the higher order derivatives shown in (8) to zero. To give a quantitative example, let us take the two-stage QWTs case ($N = 2$) and assume τ_{gp} to be −4 and Γ_p to be 0.04 at the center frequency 1 GHz. Since the sampling period T of a 90-degree ($\pi/2$) delay line at 1 GHz is 0.25 ns, we can generate group delay of $-4 \times 0.25 = -1$ ns. Similarly, by changing τ_{gp} to −8 we can obtain group delay of −2 ns and so on. It is noted that we terminate our device with standard 50 ohms load, and the characteristic impedance of the transmission lines obtained here is normalized to the load impedance. Table 1 shows the different values of normalized characteristic impedance for various group delays by setting Γ_p equal to 0.04 derived from (6)–(8). After obtaining the characteristic impedance, we can easily calculate the dimension of QWT [12]. The width and length of the QWT corresponding to the characteristic impedance which we calculate from (6)–(8) for negative group delay of −0.5 ns, −1 ns, and −2 ns are tabulated in Table 2. The pictorial representation of width and length of the transmission line is shown in Figure 2.

3. Simulation and Measurement

To validate our concept of generating the prescribed NGD using transversal filter methodology, we designed a NGD circuit consisting of two-stage quarter-wave transformer

TABLE 1: Normalized characteristic impedance for two and three stages ($\Gamma_p = 0.04$).

$\tau_{gp}(T)$	N = 2		N = 3		
	Z_1	Z_2	Z_1	Z_2	Z_3
−1	1.0833	1.0408	1.10527	1.0834	1.02024
−2	1.1052	1.0618	1.15044	1.15051	1.0409
−3	1.1275	1.0833	1.20535	1.23801	1.06904
−4	1.1503	1.1052	1.27094	1.34942	1.1051
−5	1.1735	1.1274	1.34819	1.489	1.14954
−6	1.1971	1.1501	1.43799	1.66171	1.20276
−7	1.2211	1.1732	1.54106	1.87306	1.26507
−8	1.2455	1.1966	1.65783	2.12887	1.33657
−9	1.2703	1.2205	1.78835	2.43493	1.41715
−10	1.2956	1.2448	1.9323	2.79674	1.50646

TABLE 2: Dimension of transmission line for different group delay (substrates: Rogers RO3010 with dielectric constant of 10.2 and thickness of 25 mils).

Group delay (ns)	Dimensions	
	Width (mm)	Length (mm)
−0.5 ns		
$\quad$ TL$_1$(Z_1)	0.459072	29.3245
$\quad$ TL$_2$(Z_2)	0.503547	29.2085
−1.0 ns		
$\quad$ TL$_1$(Z_1)	0.417143	29.4401
$\quad$ TL$_2$(Z_2)	0.459072	29.3245
−2.0 ns		
$\quad$ TL$_1$(Z_1)	0.340949	29.6696
$\quad$ TL$_2$(Z_2)	0.376	29.5539

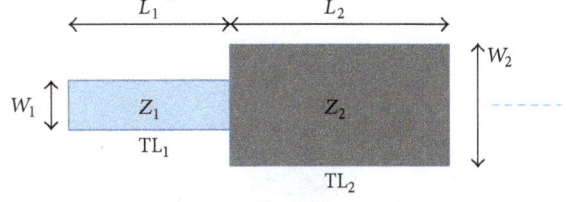

FIGURE 2: Dimensions of the quarter-wave transformers.

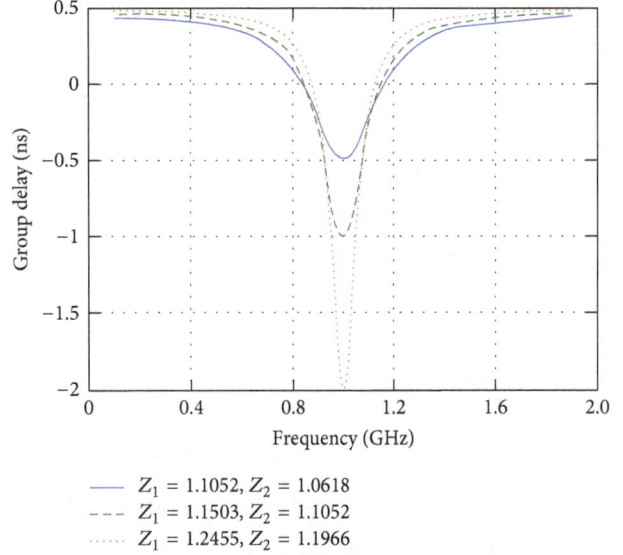

$\quad$ —— $Z_1 = 1.1052, Z_2 = 1.0618$
$\quad$ − − − $Z_1 = 1.1503, Z_2 = 1.1052$
$\quad$ ⋯⋯ $Z_1 = 1.2455, Z_2 = 1.1966$

FIGURE 3: Simulated results for prescribed group delay of −0.5 ns, −1 ns, and −2 ns for a two-stage quarter ($N = 2$) wave transformer.

using microstrip-line structures on a printed circuit board (PCB). The substrate that we used is RO3010 from Rogers Corporation, with a dielectric constant of 10.2 and thickness of 25 mils. Normalized characteristic impedance of two transmission lines for −1 ns group delay is obtained from Table 1: that is, $Z_1 = 1.1503$, $Z_2 = 1.1052$. After that we fabricated the quarter-wave transformer according to this normalized characteristic impedance. Figure 3 shows the simulated results of three different NGD at 1 GHz for a two-stage quarter-wave transformer using the tabulated coefficients. In addition, it is worth mentioning that one can also design a three-stage transformer to generate prescribed NGD as shown in Figure 4 with the coefficients obtained from Table 1. For a given NGD, the bandwidth can be enhanced when more stages are cascaded.

In practical application, it is often desired to introduce the negative group delay between two ports. In fact, we can transfer the group delay from one port to another port by simply using a branch-line coupler. The schematic for the complete circuit is shown in Figure 5. The S parameters of the two-port network can be obtained by solving the branch-line coupler with odd and even mode analysis [12]. The port reduction technique [13] can be applied to get the reflection coefficient of two-port negative group delay circuit. Figure 5

shows all the voltages that we are required to find out the S matrix for the circuit. V_1^+ and V_1^- are the incident and reflected voltage at the input port 1. V_2^- is the reflected voltage at the output port 2. V_{A1}^+ represents incident and V_{A1}^- represents reflected voltage at junction A; similarly V_{B1}^+ represents incident and V_{B1}^{+-} represents the reflected voltage at junction B. We can calculate S_{21} by the following procedure:

$$S_{21} = \frac{V_2^-}{V_1^+}. \tag{9}$$

The reflected voltage from the output port 2, V_2^-, can be written in terms of reflected voltages V_{A1}^- and V_{B1}^- as

$$
\begin{aligned}
V_2^- &= \frac{e^{-j2\theta}}{\sqrt{2}} V_{A1}^- + \frac{e^{-j(2\theta - \pi/2)}}{\sqrt{2}} V_{B1}^- \\
&= \frac{e^{-j2\theta}}{\sqrt{2}} \Gamma_A V_{A1}^+ + \frac{e^{-j(2\theta - \pi/2)}}{\sqrt{2}} \Gamma_B V_{B1}^+ \\
&= \frac{e^{-j(2\theta + \pi/2)}}{2} \Gamma_A V_1^+ + \frac{e^{-j(2\theta + \pi/2)}}{2} \Gamma_B V_1^+ \\
&= \frac{e^{-j(2\theta + \pi/2)}}{2} V_1^+ (\Gamma_A + \Gamma_B).
\end{aligned}
\tag{10}
$$

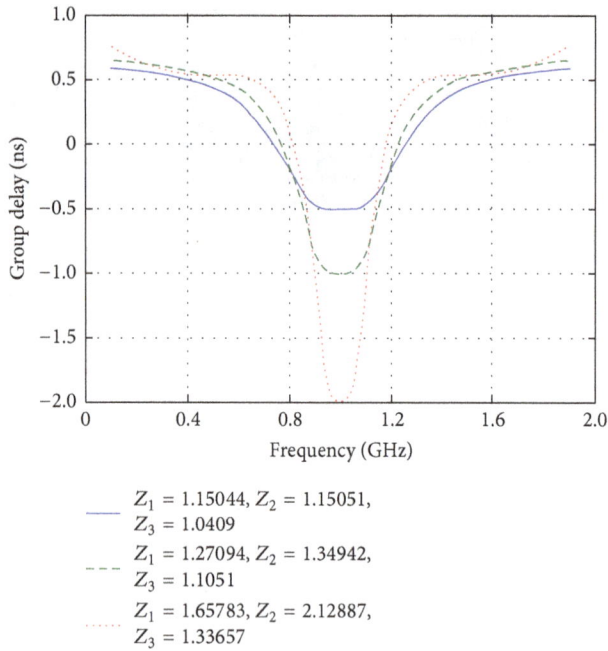

FIGURE 4: Simulated results for prescribed group delay of −0.5 ns, −1 ns, and −2 ns for a three-stage ($N = 3$) QWT.

Legend:
- $Z_1 = 1.15044$, $Z_2 = 1.15051$, $Z_3 = 1.0409$
- $Z_1 = 1.27094$, $Z_2 = 1.34942$, $Z_3 = 1.1051$
- $Z_1 = 1.65783$, $Z_2 = 2.12887$, $Z_3 = 1.33657$

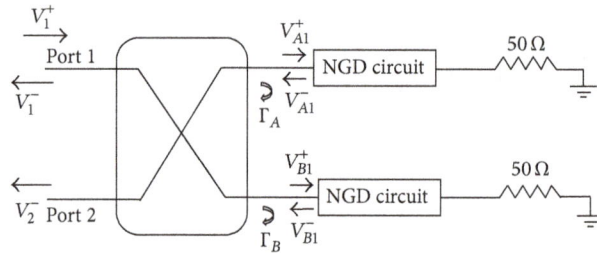

FIGURE 5: Schematic of two-stage quarter-wave transformers integrated with a branch-line coupler.

Therefore, $S_{21} = (e^{-j(2\theta+\pi/2)}/2)(\Gamma_A + \Gamma_B)$, and we know $\Gamma_A = \Gamma_B$. Let us assume $\Gamma_A = \Gamma_B = \Gamma$; this will give us

$$S_{21} = e^{-j(2\theta+\pi/2)}\Gamma. \qquad (11)$$

Similarly, S_{11} will be obtained as

$$V_1^- = \frac{1}{\sqrt{2}}V_{A1}^- + \frac{e^{-j\theta}}{\sqrt{2}}V_{B1}^- = \frac{1}{\sqrt{2}}\Gamma_A V_{A1}^+ + \frac{e^{-j\theta}}{\sqrt{2}}\Gamma_B V_{B1}^+$$
$$= \frac{1}{2}V_1^+\left(\Gamma_A + e^{-2j\theta}\Gamma_B\right). \qquad (12)$$

Thus, we can write

$$S_{11} = \frac{V_1^-}{V_1^+} = \frac{1}{2}\left(\Gamma_A + e^{-2j\theta}\Gamma_B\right). \qquad (13)$$

Since $\Gamma_A = \Gamma_B$, we can put $\Gamma_A = \Gamma_B = \Gamma$, which will give us $S_{11} = (\Gamma/2)(1+e^{-2j\theta})$. The reflection coefficient matrix for the

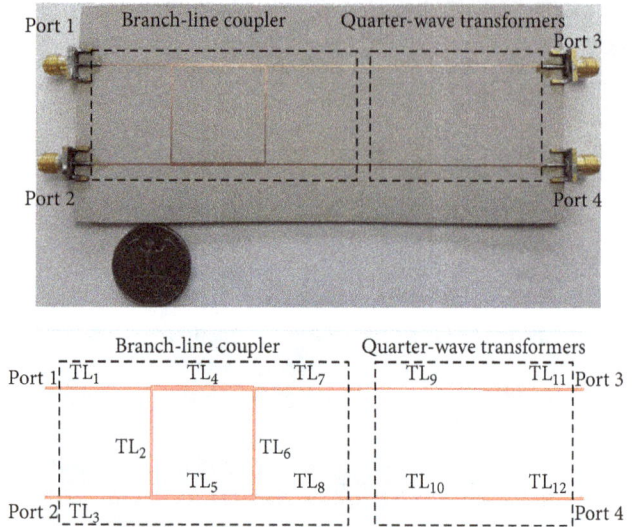

FIGURE 6: Fabricated prototype of NGD circuit using two-stage quarter-wave transformers integrated with a branch-line coupler.

two-port negative group delay circuit can thus be represented as

$$S = \begin{bmatrix} \frac{\Gamma}{2}\left(1 + e^{-2j\theta}\right) & e^{-j(2\theta+\pi/2)}\Gamma \\ e^{-j(2\theta+\pi/2)}\Gamma & \frac{\Gamma}{2}\left(1 + e^{-2j\theta}\right) \end{bmatrix}. \qquad (14)$$

From (7) and (14), the group delay at desired frequency becomes

$$-\frac{\partial \angle S_{21}}{\partial \theta}\bigg|_{\theta=\pi/2} = \tau_{gp} + 2. \qquad (15)$$

The constant term in (15) represents the extra group delay caused by the coupler, which also agrees with the result in [13]. The prototype of the design including a branch-line coupler along with two-stage quarter-wave transformers is shown in Figure 6. Transmission lines TL_1, TL_3, TL_7, and TL_8 are included in the prototype for interconnections, which will result in addition of group delay in the circuit. The effects can be nullified by fine-tuning the dimensions of the NGD circuit based on quarter-wave transformers. Table 3 indicates the length and width of the branch-line coupler and quarter-wave transformer as depicted in Figure 6.

Figure 7 depicts the comparison of group delays between the simulation and measurement in which we use the branch-line coupler. The simulated and measured results agree with each other very well. The resulting NGD of the entire circuit is around −1 ns at 1.1 GHz. Furthermore, the magnitude of corresponding S_{21} is plotted in Figure 8, which indicates around 23 dB signal attenuation during the NGD region.

4. Conclusion

In this paper, we present a systematic technique of generating prescribed NGD by simply using multistage quarter-wave

TABLE 3: Dimensions of branch-line coupler and quarter-wave transformers for group delay of −1 ns (substrates: Rogers RO3010 with dielectric constant of 10.2 and thickness of 25 mils).

| | Branch-line coupler | | Quarter-wave transformer | |
| | | | First stage | Second stage |
	TL_1, TL_2, TL_3, TL_6, TL_7, and TL_8	TL_4 and TL_5	TL_9 and TL_{10}	TL_{11} and TL_{12}
Width (mm)	0.575	1.111	0.417	0.560
Length (mm)	29.034	28.054	33.440	28.000

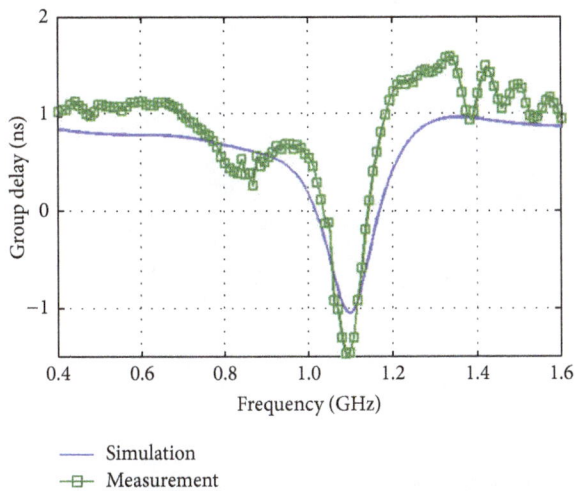

— Simulation
—□— Measurement

FIGURE 7: Simulation and measurement of group delay for two-stage quarter-wave transmission line with branch-line coupler NGD circuit.

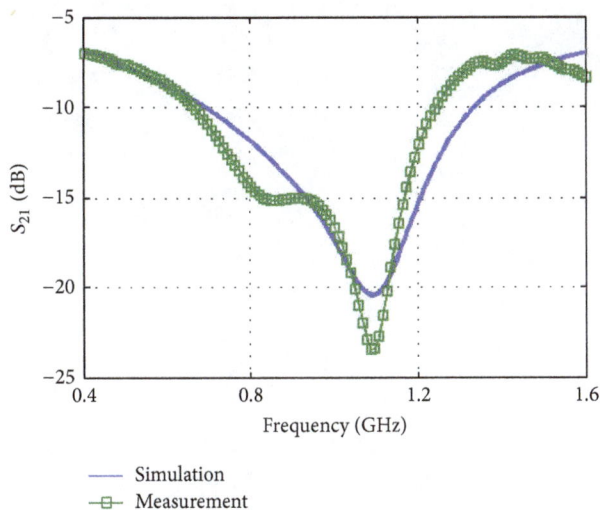

— Simulation
—□— Measurement

FIGURE 8: Comparison between magnitude of S_{21} in simulation and measurement for two-stage quarter-wave transmission line with branch-line coupler NGD circuit.

transformers to form microwave recursive filters. By properly choosing the impedance of transformer, we can realize the desired NGD. A table providing the associated values for desired NGD is given and utilized to synthesize the desired NGD. In addition, a branch-line coupler is used to transfer the group delay from one-port circuit to a two-port NGD circuit. The proposed method is promising to be further used in high frequency circuitries to synthesize any desired amount of NGD in order to compensate the undesired excessive group delay such as in feed-forward amplifiers and envelop tracking power amplifier, which can improve the system overall efficiency.

Conflict of Interests

The authors declare that there is no conflict of interests regarding the publication of this paper.

References

[1] L. Brillouin and A. Sommerfeld, *Wave Propagation and Group Velocity*, Academic Press, New York, NY, USA, 1960.

[2] C. G. B. Garrett and D. E. McCumber, "Propagation of a Gaussian light pulse through an anomalous dispersion medium," *Physical Review A*, vol. 1, no. 2, pp. 305–313, 1970.

[3] S. Chu and S. Wong, "Linear pulse propagation in an absorbing medium," *Physical Review Letters*, vol. 48, no. 11, pp. 738–741, 1982.

[4] L. J. Wang, A. Kuzmich, and A. Dogariu, "Gain-assisted superluminal light propagation," *Nature*, vol. 406, no. 6793, pp. 277–279, 2000.

[5] M. S. Bigelow, N. N. Lepeshkin, and R. W. Boyd, "Superluminal and slow light propagation in a room-temperature solid," *Science*, vol. 301, no. 5630, pp. 200–202, 2003.

[6] H. Choi, Y. Jeong, C. D. Kim, and J. S. Kenney, "Efficiency enhancement of feedforward amplifiers by employing a negative group-delay circuit," *IEEE Transactions on Microwave Theory and Techniques*, vol. 58, no. 5, pp. 1116–1125, 2010.

[7] S.-S. Oh and L. Shafai, "Compensated circuit with characteristics of lossless double negative materials and its application to array antennas," *IET Microwaves, Antennas and Propagation*, vol. 1, no. 1, pp. 29–38, 2007.

[8] H. Noto, K. Yamauchi, M. Nakayama, and Y. Isota, "Negative group delay circuit for feed-forward amplifier," in *Proceedings of the IEEE MTT-S International Microwave Symposium (IMS '07)*, pp. 1103–1106, June 2007.

[9] C.-T. M. Wu and T. Itoh, "Maximally flat negative group-delay circuit: a microwave transversal filter approach," *IEEE Transactions on Microwave Theory and Techniques*, vol. 62, no. 6, pp. 1330–1342, 2014.

[10] C. Rauscher, "Microwave active filters based on transversal and recursive principles," *IEEE Transactions on Microwave Theory and Techniques*, vol. 33, no. 12, pp. 1350–1360, 1985.

[11] G. Chaudhary and Y. Jeong, "Distributed transmission line negative group delay circuit with improved signal attenuation,"

IEEE Microwave and Wireless Components Letters, vol. 24, no. 1, pp. 20–22, 2014.

[12] D. M. Pozar, *Microwave Engineering*, John Wiley & Sons, 2009.

[13] S. Lucyszyn and I. D. Robertson, "Analog reflection topology building blocks for adaptive microwave signal processing applications," *IEEE Transactions on Microwave Theory and Techniques*, vol. 43, no. 3, pp. 601–611, 1995.

19

The Terahertz Controlled Duplex Isolator: Physical Grounds and Numerical Experiment

Konstantin Vytovtov,[1] Said Zouhdi,[2] Rostislav Dubrovka,[3] and Volodymyr Hnatushenko[1]

[1]*Department of Physics, Electronics and Computing Systems, Oles Honchar Dnipropetrovsk National University, Gagarina 72, Dnipropetrovsk 49010, Ukraine*
[2]*Laboratoire de Génie Electrique de Paris, SUPELEC, 91192 Gif-sur-Yvette, France*
[3]*Queen Mary University of London, UK*

Correspondence should be addressed to Volodymyr Hnatushenko; vvgnatush@gmail.com

Academic Editor: Dmitry Kholodnyak

Electromagnetic properties of an anisotropic stratified slab with an arbitrary orientation of the anisotropy axis under an oblique incidence of a plane harmonic wave are studied. The dependence of the eigenwave wavenumbers and the reflection coefficient on an anisotropy axis orientation and frequency is investigated. For the first time, the expression for the translation matrix is obtained in the compact analytical form. The controlled two-way dual-frequency (duplex) isolator based on the above described slab is presented for the first time. It is based on the properties of the anisotropic structure described here but not on the Faraday effect.

1. Introduction

The terahertz domain actively developed in the last decades [1–5]. Various devices have been created and numerous electromagnetic structures have been used in this frequency range. One of the most interesting classes of terahertz devices is the class of nonreciprocal devices [1–10]. Usually, the Faraday rotation effect itself is exploited naturally to achieve desired nonreciprocal performance [1]. The electrically tunable Faraday effect in a *HgTe* thin film has been reported in [2]. Faraday rotation has also been studied for pump pulse of terahertz radiation [3]. In [4], the heterostructures based on the piezoelectric and semiconductor layers have been proposed for radio frequency applications. Other examples of nonreciprocal properties of nonlinear devices in terahertz range have been presented in [5] where a Faraday isolator operating on the coupled microwave resonators circuit has been proposed. A theoretical study of the reflection of infrared radiation from antiferromagnets and *M*-type barium hexagonal ferrite using an attenuated total reflection geometry has been presented in [6]. Nonreciprocal devices using attenuated total reflection and thin film magnetic layered structures have been described in [7]. Thus, we

can clearly see that this topic is quite relevant in modern microwave science.

In this paper, we consider nonreciprocal properties of an anisotropic structure based on dependence of the reflection coefficient on an incident wave direction and an anisotropy axis orientation, but not on the Faraday effect. The physical grounds and numerical simulations of a controlled bidirectional dual-frequency isolator not based on the Faraday effect are presented. For this, the analytical investigation and numerical calculations of a stratified anisotropic structure are carried out. At first, the elements of the translation matrix are written in the compact analytical form (4) for the general case of an anisotropic medium for the first time. It is important that these elements are expressed in elementary algebraic functions. The obtained form is very useful in subsequent numerical experiments.

In this work, it also is found that the reflection coefficients are different for different incidence angles α and $-\alpha$. Thus, a structure shows nonreciprocal properties.

These properties allow us to build nonreciprocal elements in the terahertz domain. The physical principles of construction of a two-way double-frequency (duplex) isolator are described here. These principles are based on dependence

(a)

(b)

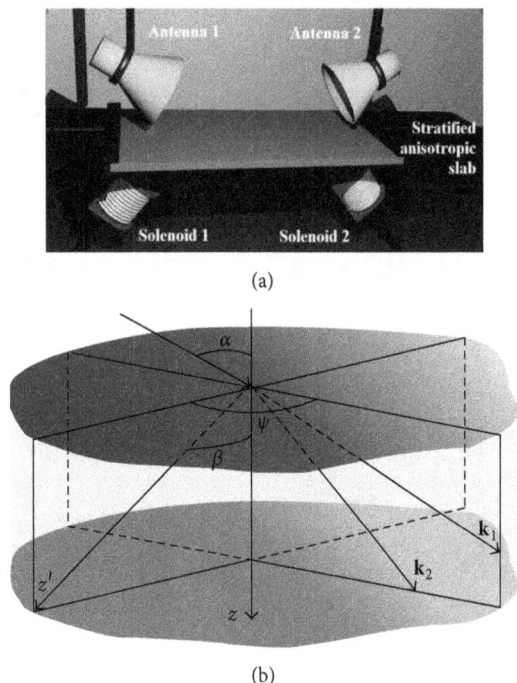

FIGURE 1: (a) Appearance of the controlled double-frequency bidirectional isolator; (b) geometry of an anisotropy layer.

of reflection coefficient on an incidence wave direction and an orientation of an anisotropy axis. It should be noted that the Faraday effect is also present in these structures and also defines nonreciprocal properties of anisotropic structures. Today, a wide variety of isolators is described in contemporary scientific literature [7–9]. A feature of the presented device is the fact that it is a two-way dual-frequency one. Switching an isolator (changing a direction and a frequency) is implemented by two solenoids. Its tuning is done mechanically by changing an orientation of external magnetic field. Note that here a very complicated geometry of the problem is used to create the isolator. But only this geometry allows us to obtain the pronounced minimums of reflectance for a given incidence angle, as well as a wide range of total reflection.

2. Description of the Isolator Scheme

A presentation of the offered duplex isolator is shown in Figure 1(a). This device contains two transceiver antennas, two solenoids generating a magnetizing field, and a stratified anisotropic slab (a nonreciprocal element).

An orientation of the axes within a homogeneous layer of a stratified anisotropic slab is presented in Figure 1(b). Here, z' is an anisotropy axis, z is a normal to a structure, $\mathbf{k}_1$ and $\mathbf{k}_2$ are wave vectors of forward refracted waves within an anisotropic layer, β is an inclination angle, ψ is an angle between an incidence plane and a plane containing an anisotropy axis, and α is an incidence angle. The choice of such a complicated geometry is due to the need to obtain an asymmetric (nonreciprocal) structure, the need to obtain wide range of total reflection, and the need to

control the structure. Of course nonreciprocal properties can be observed in the simplest case of orientation of an anisotropy axis along interfaces [10, 11], but in that case total reflection regions are not sufficiently wide and a structure has low selective properties in comparison with the given case. Additionally, here mechanical tuning of an isolator is assumed. And using such a complicate geometry gives us the additional freedom to control the device. Moreover, properties of a structure having such a geometry are poorly understood. Therefore, the results of these investigations have both practical and theoretical significance.

The layers of a stratified structure are described by a scalar permittivity and a dyadic permeability in the gyrotropic form:

$$\mu = \begin{vmatrix} \mu_{xx} & j\mu_{xy} & 0 \\ -j\mu_{xy} & \mu_{xx} & 0 \\ 0 & 0 & \mu_{zz} \end{vmatrix} \quad (1)$$

with

$$\mu_{xx} = 1 + \frac{\omega_M (\omega_0 + j/T)}{(\omega_0 + j/T)^2 - \omega^2},$$

$$\mu_{xy} = \frac{\omega_M \omega}{(\omega_0 + j/T)^2 - \omega^2}. \quad (2)$$

Here, $\omega_M = 4\pi\gamma M$, $\omega_0 = \gamma H_0$, γ is the gyromagnetic relation, $H_0 = 0.1T$ is an external magnetic field, and T is a relaxation time.

In our numerical investigation, we take into account losses and frequency dispersion of the ferrite material and the elements in (1) are written in the Lanfau-Lifshits form [12]. The dependence of the permeability elements on a frequency for the first and second layers of a period is given in Figure 2. Here, μ'_{xx} and μ'_{xy} are the real parts of the elements, and μ''_{xx} and μ''_{xy} are the imaginary parts of the elements. The resonance frequencies are 1.1 GHz and 1.8 GHz for the first and second layers accordingly.

3. Method of Calculation

The well-known translation matrix method and the reflection matrix method [13, 14] are used in this study of a stratified anisotropic slab. Both of these methods are accurate and analytical and these can be used in any frequency range for both lossless and lossy media without restrictions. According to the definition, a translation matrix relates tangential field components on both boundaries of a plane-parallel structure:

$$\begin{vmatrix} E_x(z) \\ E_y(z) \\ H_x(z) \\ H_y(z) \end{vmatrix} = \mathbf{M}(z) \begin{vmatrix} E_x(0) \\ E_y(0) \\ H_x(0) \\ H_y(0) \end{vmatrix}. \quad (3)$$

Here, $\mathbf{M}(z)$ is a 2×2 matrix for an isotropic medium or a 4×4 translation matrix for an anisotropic medium.

The method is described in the literature in detail for both isotropic [13] and anisotropic [14] media. The matrices have been found for a normal orientation of an anisotropy axis,

(a)

(b)

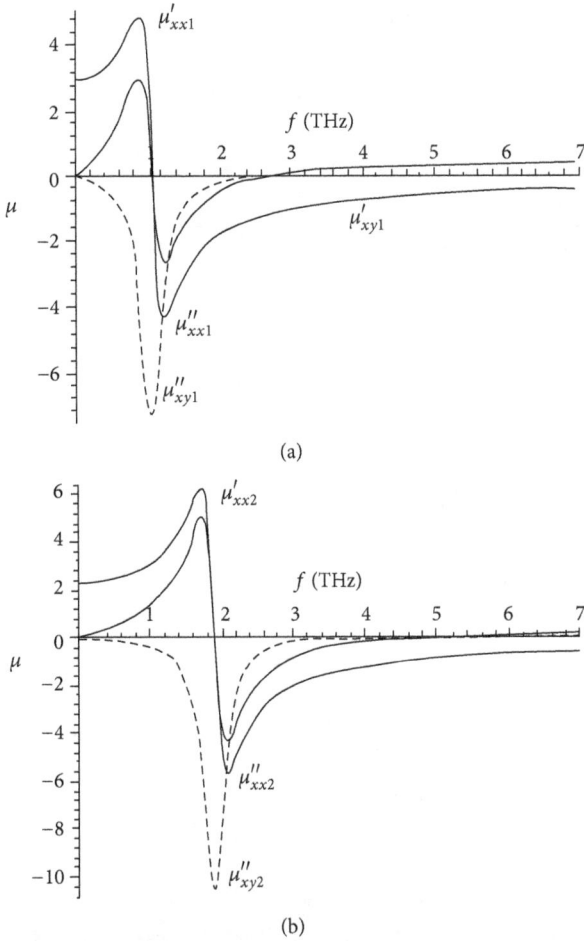

FIGURE 2: The dependence of permeability elements on frequency for the first (a) and second (b) layers of a period.

for a tangential orientation of an anisotropy axis, for arbitrary orientation of an anisotropy axis, and for a normal incidence of a plane harmonic wave. However, the expressions for the most general case of an anisotropic medium with an arbitrary orientation of the anisotropy axes under an oblique incidence of a wave have not been obtained in a compact analytical form. In this paper, the translation matrix elements of a homogeneous layer are presented for the first time as

$$M_{ml} = \sum_{i=1}^{4} (-1)^{m+i} \gamma_{mi} \frac{\det \mathbf{G}_{li}}{\det \mathbf{G}} \exp\left(-jk_i z\right). \tag{4}$$

Here, $\mathbf{G}_{li}$ is the minor of the element γ_{mi} of the 4×4-matrix $\mathbf{G}$. Matrix $\mathbf{G}$ is defined by the expression

$$
\begin{vmatrix} E_x \\ E_y \\ H_x \\ H_y \end{vmatrix} =
\begin{vmatrix} \gamma_{11} & \gamma_{12} & \gamma_{13} & \gamma_{14} \\ \gamma_{21} & \gamma_{22} & \gamma_{23} & \gamma_{24} \\ \gamma_{31} & \gamma_{32} & \gamma_{33} & \gamma_{34} \\ \gamma_{41} & \gamma_{42} & \gamma_{43} & \gamma_{44} \end{vmatrix}
\begin{bmatrix} A_1 \exp\left(-jk_1 z\right) \\ A_2 \exp\left(-jk_2 z\right) \\ A_3 \exp\left(-jk_3 z\right) \\ A_4 \exp\left(-jk_4 z\right) \end{bmatrix}
$$

$$
= \mathbf{G} \begin{bmatrix} A_1 \exp\left(-jk_1 z\right) \\ A_2 \exp\left(-jk_2 z\right) \\ A_3 \exp\left(-jk_3 z\right) \\ A_4 \exp\left(-jk_4 z\right) \end{bmatrix}, \tag{5}
$$

where A_i is a constant. Here also k_i is a wavenumber defined by well-known fourth-order dispersion relation [12–16]:

$$k^4 + a_3 k^3 + a_2 k^2 + a_1 k = 0. \tag{6}$$

The coefficients a_i in (6) must be obtained directly from Maxwell equations.

The expression of matrix $\mathbf{G}$ in (5) can be obtained directly from Maxwell equations. Actually, expression (5) is a general solution of Maxwell's equations for this case [12–14] (see Appendix). The achieved form of the translation matrix (4) essentially simplifies the solution of the programming problem and improves accuracy of numerical calculations. Moreover, no analytical expression of the translation matrix element has been presented in scientific literature for this complicate case.

The reflection matrix relates the tangential components of incident and reflected waves [15]:

$$
\begin{vmatrix} E_{x \, \text{refl}} \\ E_{y \, \text{refl}} \end{vmatrix} = \mathbf{R} \begin{vmatrix} E_{x \, \text{inc}} \\ E_{y \, \text{inc}} \end{vmatrix}. \tag{7}
$$

Here $E_{x \, \text{inc}}$ and $E_{y \, \text{inc}}$ are the components of an incident field, $E_{x \, \text{refl}}$ and $E_{y \, \text{refl}}$ are the components of a reflected field, and $\mathbf{R}$ is a 2×2 reflection matrix.

4. Nonreciprocal Properties of Anisotropic Structure

The main element of the isolator is a stratified anisotropic slab (Figure 1) possessing nonreciprocal properties. In this section, we consider the main physical features of a slab, which ensure its isolator properties. An orientation of the axes for the studied case is shown in Figure 1(b). Angle ψ is given in Figure 1(b). The presented complicated geometry of the problem gives us the interesting features of electromagnetic field in the structure. Such properties cannot appear in a case of more simple geometry. In particular, study of wavenumbers in a homogeneous anisotropic layer suggests that the wavenumbers can be real, imaginary, or complex even in a lossless medium. In the considered case, imaginary parts of wavenumbers indicate existence of a complex wave [16] but not in presence of losses in an anisotropic structure. Indeed a resulting structure of electromagnetic field in a layer is determined by a superposition of all four eigenwaves only, but not each eigenwave separately. In addition, these numbers are not the same for the incidence angles α and $-\alpha$ (Figure 3). The numerical calculations analytical investigations give us the following result (Figure 4): $k_1(\alpha) = -k_4(-\alpha)$, $k_2(\alpha) = -k_3(-\alpha)$, $k_3(\alpha) = -k_2(-\alpha)$, and $k_4(\alpha) = -k_1(-\alpha)$. Thus, we can see nonreciprocal properties of the medium in the considered case.

It also is clear (Figure 3) that two of the four wavenumbers can be real and two others can be complex for the certain parameters of the structure ($10° < \alpha < 16°$). It should also be noted that all four wave numbers have positive (negative) real part (Figure 3) in the certain range of the incidence angles ($\alpha > 16°$). Such unusual combinations of wavenumbers are possible only in the case of this complex geometry

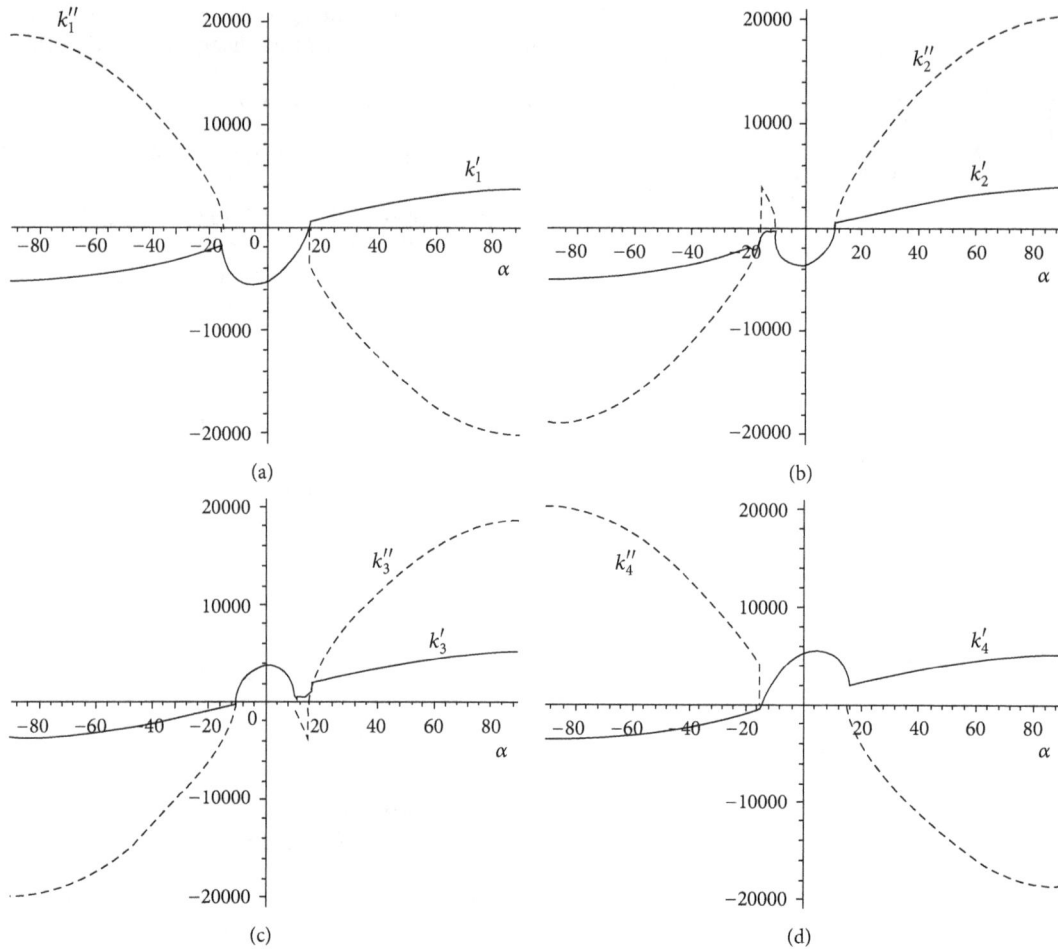

FIGURE 3: The dependence of the wavenumbers on an incidence angle ($\beta = 60°$, $\psi = 30°$, $f = 10$ THz, $\mu_{xx} = 4.9$, $\mu_{xy} = 2.9$, and $\mu_{zz} = 0.99$).

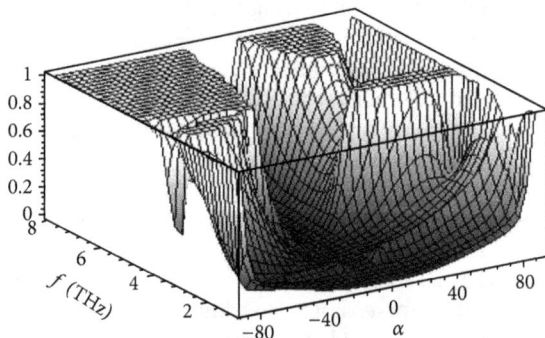

FIGURE 4: The dependence of the reflection on a frequency and an incidence angle ($\beta = 30°$, $\psi = 50°$), $d_1 = 0.78$ mm, and $d_1/d_2 = 1.3125$.

of the problem. Similarly, it also is verified that the field components are different for the incidence angles α and $-\alpha$.

Obviously, this kind of wavenumber affects reflection properties of a medium. So now we analyze the dependence of the reflection coefficient on a frequency and an anisotropy axis orientation. One of the possible variants of this dependence is presented in Figure 4. The study shows that there

is no regularity in the reflection coefficient dependence on a frequency and an orientation of the anisotropy axis. Here a very wide range of total reflection even for small incidence angles can be seen. Their existence is determined by multiple reflection phenomenon, existence of four eigenwaves with different wavenumbers and propagation angles in each layer, presence of complex waves, and the losses in the medium. The resonant character of the reflection coefficient is determined by the effects described above also. Note that there are no sharp edges of the reflection coefficient in the neighborhood of the unity. It also is shown (Figure 5) that the dependence of the reflection coefficient of a frequency is not symmetric with respect to $\alpha = 0$. This is due to the asymmetric dependence of wave numbers and the field components on the incidence angle.

5. Operation Principle of the Isolator

The operation principle of the device is based on an angle dependence of medium properties (in particular, the reflection coefficient) but not on the Faraday effect. Wherein for the certain structure parameters and an orientation of the anisotropy axis (an external magnetic field direction) it is possible to obtain the total reflection at an incidence angle

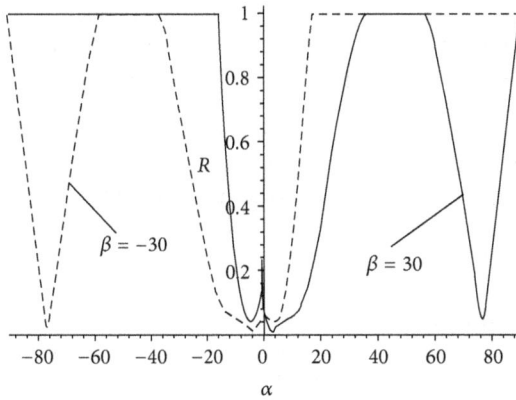

FIGURE 5: The dependence of the reflection coefficient on the incidence angle for $f = 6.35 \cdot 10^{12}$, $\psi = 50°$, $d_1 = 0.78$ mm, and $d_1/d_2 = 1.3125$; (1) $\beta = 30°$ solid line and (2) $\beta = -30°$ dash line.

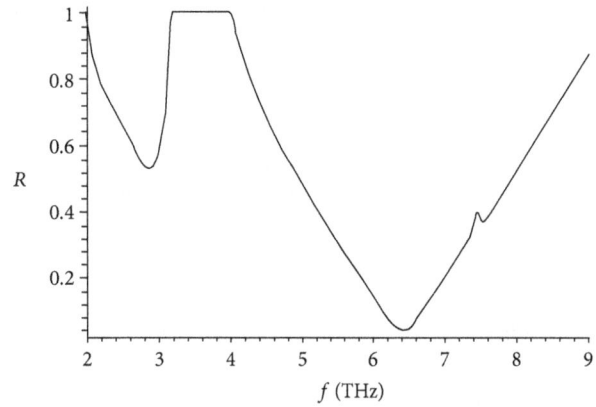

FIGURE 6: The dependence of reflection coefficient on a frequency at $\alpha = 76.4°$, $d_1 = 0.78$ mm, and $d_1/d_2 = 1.3125$.

α and the minimal reflection at an incidence angle $-\alpha$. Analogously, it is possible to choose an orientation of the external magnetic field for the given structure parameters such that it is possible to obtain the total reflection for $-\alpha$ and the minimal reflection for α. Thus, changing a direction of wave propagation (α or $-\alpha$) can be accomplished by changing a direction of an external magnetic field. Analogously it is possible to find the two frequencies that give such an effect.

Now let us describe, for clarity, the operation principle of a bidirectional double-frequency isolator shown in Figure 1. If a voltage is applied to solenoid 1, a signal transmits from antenna 1 to antenna 2 without attenuation at a frequency f_1 as the reflection coefficient is equal to unite, a signal cannot propagate from antenna 2 to antenna 1 at this frequency as the reflection coefficient is about zero, and a signal transmits from antenna 2 to antenna 1 without attenuation at a frequency f_2; however, inverse propagation is not possible at this frequency. When voltage is applied to solenoid 2, a signal transmits from antenna 2 to antenna 1 substantially without attenuation at a frequency f_1, and it transmits from antenna 1 to antenna 2 without attenuation at a frequency f_2. Circular aperture of the antennas is used because the Faraday effect (rotation of a polarization plane) is taken into account.

6. The Parameters of the Isolator

A very important characteristic of an isolator is its amplitude response. During the researches, numerous calculations have been carried out and the optimal structure for isolator is chosen. This structure includes the 12 double-layered periods. The first layer of the structure is FeF$_2$ and the second one is MnO. The antiferromagnetic resonance frequencies in the considered case are 1.1 THz and 1.8 THz for the first and second layers accordingly. The dependence of the element on a frequency is given in Figure 2. Here the operating frequency $f = 6.35$ THz and the angles $\beta = \pm30°$ and $\psi = 50°$ are also taken. In our problem, the operating frequency is essentially higher than the resonant one. The losses in FeF$_2$ and MnO can be neglected at this frequency (Figure 2). Here we consider an above resonance frequency and we have $\mu_{xx} = 1.415$ and μ_{xx}

$= -0.42$ for the first layer of the period at this frequency and $\mu_{xx} = 1.2$ and $\mu_{xx} = -0.63$ for the second layer. In the given work we consider a TE incident wave. The dependence of the reflection coefficient on an incidence angle for this structure (the amplitude characteristic) is presented in Figure 5. It is seen that the reflection coefficient is minimal ($R = 0.05$) at $\alpha = 76.4°$ and the one is equal to unit at $\alpha = -76.4°$ for $\beta = 30°$ (the solid line).

Thus, this structure has isolator properties and it passes a signal only in a forward direction and does not pass a signal in an opposite direction. The stopband for an opposite direction is 9.6° at the level 0.707, 15.8 at the level 0.5, and 27.7 at the level 0.1. Now, let us chose $\beta = -30°$ (the dash line). For this case, the minimum of the reflection coefficient ($R = 0.03$) is at $\alpha = -76.4°$.

The frequency dependence of the reflection coefficient is presented in Figure 6. This graph is presented for the incidence angle $\alpha = 76.4°$. It is obvious that this structure has bad frequency selective properties and low squareness coefficient. The bandwidth is 1.88 THz (29.38%) at the 0.707 level (a half power level), and the bandwidth is 3.13 THz (48.9%) at the 0.5 level of the maximum. The squareness is 1.66 at levels 0.5 and 0.707. It also is verified that changing orientation of magnetic field allows us to shift a reflection minimum in a small range. And changing value of this field does not give us any useful effect.

The results demonstrating the possibility of mechanical tuning of the isolator by changing an angle of an anisotropy axis inclination are shown in Figure 7. The minimum is shifted from $\alpha = 72°$ to $\alpha = 82°$ if an inclination angle is changed from $\beta = 25°$ to $\beta = 40°$. Simultaneously, the angle bandwidth is narrowed and it is equal to 28°, 10°, and 6° correspondingly. Decreasing the inclination angle less than $\beta = 25°$ and increasing it more than $\beta = 50°$ leads to disappearance of the angle selective properties of the structure.

Figure 8 shows the results of calculations, allowing us to improve the described isolator. It is seen in Figure 8 that the reflection minimum is observed at $\alpha = 76.4°$ when the frequency is 6.35 GHz and the reflection minimum

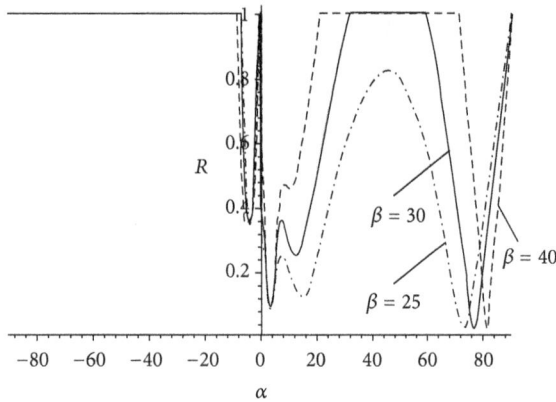

FIGURE 7: Dependence of the reflection coefficient on an incidence angle for difference inclination angles β, $d_1 = 0.78$ mm, $d_1/d_2 = 1.3125$, and $f = 6.35 \cdot 10^{12}$.

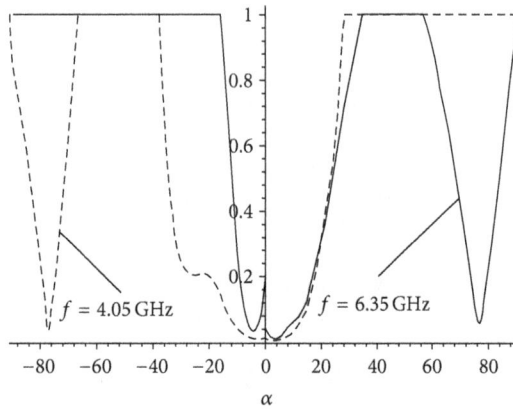

FIGURE 8: The dependence of the reflection coefficient on an incident angle for $d_1 = 0.78$ mm, $d_1/d_2 = 1.3125$, $\beta = 30°$; (1) $\psi = 50°$ and (2) $\psi = -50°$.

corresponds to $\alpha = -76.4°$ at the frequency 4.05 GHz. Thus, the structure can be used in duplex terahertz systems as a two-way isolator. Moreover, when the direction of the magnetic field is switched from $\psi = 50°$ to $\psi = -50°$ the above minima are reversed. Therefore, if voltage is applied to solenoid 1 then a signal transmits from antenna 1 to antenna 2 substantially without attenuation at frequency 6.35 GHz, and a signal transmits from antenna 2 to antenna 1 without attenuation at frequency 4.05 GHz. If voltage is applied to solenoid 2 then a signal transmits from antenna 2 to antenna 1 substantially without attenuation at frequency 6.35 GHz, and a signal transmits from antenna 1 to antenna 2 without attenuation at a frequency 4.05 GHz.

7. Conclusions

Electromagnetic properties of a stratified anisotropic slab with an arbitrary orientation of the anisotropy axis are studied analytically and numerically. The translation and reflection matrices methods are used. The analytical expression of the translation matrix is obtained in the convenient

compact form for the first time. The dependence of the eigenwave wavenumbers in an anisotropic slab and reflection coefficient on an anisotropy axis orientation and an incidence angle is investigated. It is shown that the structure shows nonreciprocal properties. It is important that these properties are not based on the Faraday effect. Although this effect is still inherently present in an anisotropic structure, the controlled two-way dual-frequency (duplex) isolator is presented here for the first time. Its operation principle is based on the properties of an anisotropic structure described here for the first time. The very complicated geometry of an anisotropic slab is used in this isolator. But this geometry (an arbitrary orientation of magnetic field) allows us to obtain the necessary properties of the isolator and to control a resonant frequency and an angle of a wave transmission mechanically.

Appendix

Finding a solution of Maxwell equations in the form $\exp(j(\omega t - k_x x - k_y y - k_z z))$, we can write

$$
\begin{vmatrix}
0 & -jk_z & jk_y & -j\omega\mu_{xx} & -j\omega\mu_{xy} & -j\omega\mu_{xz} \\
jk_z & 0 & -jk_x & -j\omega\mu_{yx} & -j\omega\mu_{yy} & -j\omega\mu_{xz} \\
-jk_y & jk_x & 0 & -j\omega\mu_{zx} & -j\omega\mu_{zy} & -j\omega\mu_{zz} \\
j\omega\varepsilon & 0 & 0 & 0 & -jk_z & jk_y \\
0 & j\omega\varepsilon & 0 & jk_z & 0 & -jk_x \\
0 & 0 & j\omega\varepsilon & -jk_y & jk_x & 0
\end{vmatrix}
\begin{vmatrix}
E_x \\ E_y \\ E_z \\ H_x \\ H_y \\ H_z
\end{vmatrix} \quad (A.1)
$$

$$= 0.$$

From (A.1), we have

$$
\begin{vmatrix}
0 & -jk_z & jk_y & -j\omega\mu_{xx} & -j\omega\mu_{xy} \\
jk_z & 0 & -jk_x & -j\omega\mu_{yx} & -j\omega\mu_{yy} \\
-jk_y & jk_x & 0 & -j\omega\mu_{zx} & -j\omega\mu_{zy} \\
j\omega\varepsilon & 0 & 0 & 0 & -jk_z \\
0 & j\omega\varepsilon & 0 & jk_z & 0
\end{vmatrix}
\begin{vmatrix}
E_x \\ E_y \\ E_z \\ H_x \\ H_y
\end{vmatrix}
$$

$$
= \begin{vmatrix}
j\omega\mu_{xz} \\
j\omega\mu_{yz} \\
j\omega\mu_{zz} \\
-jk_y \\
jk_x
\end{vmatrix} H_z. \quad (A.2)
$$

The solutions of (A.2) give us the expressions of the matrix elements in (5) in the form

$$\gamma_{mi} = \frac{\Delta_{mi}}{\Delta}, \quad (A.3)$$

where Δ is the determinant of the coefficient matrix in (A.3) and

$$\Delta_{1i} = \begin{vmatrix} j\omega\mu_{xz} & -jk_{zi} & jk_y & -j\omega\mu_{xx} & -j\omega\mu_{xy} \\ j\omega\mu_{yz} & 0 & -jk_x & -j\omega\mu_{yx} & -j\omega\mu_{yy} \\ j\omega\mu_{zz} & jk_x & 0 & -j\omega\mu_{zx} & -j\omega\mu_{zy} \\ -jk_y & 0 & 0 & 0 & -jk_{zi} \\ jk_x & j\omega\varepsilon & 0 & jk_{zi} & 0 \end{vmatrix},$$

$$\Delta_{2i} = \begin{vmatrix} 0 & j\omega\mu_{xz} & jk_y & -j\omega\mu_{xx} & -j\omega\mu_{xy} \\ jk_{zi} & j\omega\mu_{yz} & -jk_x & -j\omega\mu_{yx} & -j\omega\mu_{yy} \\ -jk_y & j\omega\mu_{zz} & 0 & -j\omega\mu_{zx} & -j\omega\mu_{zy} \\ j\omega\varepsilon & -jk_y & 0 & 0 & -jk_{zi} \\ 0 & jk_x & 0 & jk_{zi} & 0 \end{vmatrix},$$

$$\Delta_{3i} = \begin{vmatrix} 0 & -jk_{zi} & jk_y & j\omega\mu_{xz} & -j\omega\mu_{xy} \\ jk_{zi} & 0 & -jk_x & j\omega\mu_{yz} & -j\omega\mu_{yy} \\ -jk_y & jk_x & 0 & j\omega\mu_{zz} & -j\omega\mu_{zy} \\ j\omega\varepsilon & 0 & 0 & -jk_y & -jk_{zi} \\ 0 & j\omega\varepsilon & 0 & jk_x & 0 \end{vmatrix}, \quad \text{(A.4)}$$

$$\Delta_{4i} = \begin{vmatrix} 0 & -jk_{zi} & jk_y & -j\omega\mu_{xx} & j\omega\mu_{xz} \\ jk_{zi} & 0 & -jk_x & -j\omega\mu_{yx} & j\omega\mu_{yz} \\ -jk_y & jk_x & 0 & -j\omega\mu_{zx} & j\omega\mu_{zz} \\ j\omega\varepsilon & 0 & 0 & 0 & -jk_y \\ 0 & j\omega\varepsilon & 0 & jk_{zi} & jk_x \end{vmatrix}.$$

Conflict of Interests

The authors declare that there is no conflict of interests regarding the publication of this paper.

References

[1] R. J. Potton, "Reciprocity in optics," *Reports on Progress in Physics*, vol. 67, no. 5, pp. 717–754, 2004.

[2] A. Shuvaev, A. Pimenov, G. V. Astakhov et al., "Room temperature electrically tunable terahertz Faraday effect," *Applied Physics Letters*, vol. 102, no. 24, Article ID 241902, 2013.

[3] M. Shalaby, F. Vidal, M. Peccianti et al., "Terahertz macrospin dynamics in insulating ferrimagnets," *Physical Review B*, vol. 88, no. 14, Article ID 140301, 2013.

[4] M. Rotter, W. Ruile, and A. Wixforth, "Non-reciprocal SAW devices for RF applications," in *Proceedings of the IEEE Ultrasonics Symposium 2000*, vol. 1, pp. 35–38, San Juan, Puerto Rico, October 2000.

[5] J. Koch, A. A. Houck, K. L. Hur, and S. M. Girvin, "Time-reversal-symmetry breaking in circuit-QED-based photon lattices," *Physical Review A*, vol. 82, no. 4, Article ID 043811, 2010.

[6] N. R. Anderson and R. E. Camley, "Attenuated total reflection study of bulk and surface polaritons in antiferromagnets and hexagonal ferrites: propagation at arbitrary angles," *Journal of Applied Physics*, vol. 113, no. 1, Article ID 013904, 2013.

[7] T. J. Fal and R. E. Camley, "Non-reciprocal devices using attenuated total reflection and thin film magnetic layered structures," *Journal of Applied Physics*, vol. 110, no. 5, Article ID 053912, 2011.

[8] M. Shalaby, M. Peccianti, Y. Ozturk, and R. Morandotti, "A magnetic non-reciprocal isolator for broadband terahertz operation," *Nature Communications*, vol. 4, article 1558, 2013.

[9] S. Chen, F. Fan, X. Wang, P. Wu, H. Zhang, and S. Chang, "Terahertz isolator based on nonreciprocal magneto-metasurface," *Optics Express*, vol. 23, no. 2, pp. 1015–1024, 2015.

[10] M. R. F. Jensen, T. J. Parker, K. Abraha, and D. R. Tilley, "Experimental observation of magnetic surface polaritons in FeF$_2$ by attenuated total reflection," *Physical Review Letters*, vol. 75, no. 20, pp. 3756–3759, 1995.

[11] M. R. F. Jensen, S. A. Feiven, T. Dumelow et al., "Fourier transform spectroscopy of magnetic materials at terahertz frequencies," *International Journal of Terahertz Science and Technology*, vol. 2, no. 4, pp. 105–119, 2009.

[12] B. Lax and K. Button, *Microwave Ferrites and Ferrimagnetics*, McGraw-Hill Book, 1962.

[13] P. Yeh, A. Yariv, and C. S. Hong, "Electromagnetic propagation in periodic stratified media. I. General theory," *Journal of the Optical Society of America*, vol. 67, no. 4, pp. 423–436, 1977.

[14] D. W. Berreman, "Optics in stratified and anisotropic media: 4× 4 matrix formulation," *Journal of the Optical Society of America*, vol. 62, no. 4, pp. 502–510, 1972.

[15] S. Teitler and B. W. Henvis, "Refraction in stratified anisotropic media," *Journal of the Optical Society of America*, vol. 60, no. 6, pp. 830–834, 1970.

[16] F. Fedorov, *Optics of Anisotropic Media*, Editorial URSS, 2004.

20

Quad-Band Bowtie Antenna Design for Wireless Communication System Using an Accurate Equivalent Circuit Model

Mohammed Moulay,[1] Mehadji Abri,[1] and Hadjira Abri Badaoui[2]

[1]*Telecommunications Laboratory, University of Tlemcen, Tlemcen, Algeria*
[2]*STIC Laboratory, Faculty of Technology, University of Tlemcen, Tlemcen, Algeria*

Correspondence should be addressed to Mehadji Abri; abrim2002@yahoo.fr

Academic Editor: Giancarlo Bartolucci

A novel configuration of quad-band bowtie antenna suitable for wireless application is proposed based on accurate equivalent circuit model. The simple configuration and low profile nature of the proposed antenna lead to easy multifrequency operation. The proposed antenna is designed to satisfy specific bandwidth specifications for current communication systems including the Bluetooth (frequency range 2.4–2.485 GHz) and bands of the Unlicensed National Information Infrastructure (U-NII) low band (frequency range 5.15–5.35 GHz) and U-NII mid band (frequency range 5.47–5.725 GHz) and used for mobile WiMAX (frequency range 3.3–3.6 GHz). To validate the proposed equivalent circuit model, the simulation results are compared with those obtained by the moments method of Momentum software, the finite integration technique of CST Microwave studio, and the finite element method of HFSS software. An excellent agreement is achieved for all the designed antennas. The analysis of the simulated results confirms the successful design of quad-band bowtie antenna.

1. Introduction

The world of wireless telecommunications evolves rapidly. Wireless networking represents the future of computer and Internet connectivity worldwide. This technology enables two or more computers to communicate using standard network protocols. Broadband wireless technologies are increasingly gaining popularity by the successful global deployment of the Wireless Personal Area Networks WLAN in IEEE 802.11 a/b, Wireless Metropolitan Area Networks (WiMAX-IEEE 802.16a), and bands of the Unlicensed National Information Infrastructure (U-NII). Wireless networks provide all of the features and benefits of traditional local area network technologies such as Ethernet and Token Ring without the limitations of wires and cables.

Recently, WLANs have gained a strong popularity and great place in the local area network (LAN) market. Today, WLANs based on the IEEE 802.11 standard are considered a practical and interesting solution of network connection offering mobility, flexibility, and low cost of deployment and use. The IEEE 802.11a operates at three bands. The first band extends from 5.15 to 5.25 GHz, the second from 5.25 to 5.35 GHz, and the third from 5.725 to 5.825 GHz. Unlicensed National Information Infrastructure (U-NII) band is used in WLAN, Bluetooth, and Wi-Fi operations. The U-NII band can be divided into three subbands as U-NII low band (frequency range 5.15–5.35 GHz), U-NII mid band (frequency range 5.47–5.725 GHz), and U-NII high band (frequency range 5.725–5.875 GHz) [1–5].

Recently, wireless communication for wireless local area network (WLAN) and Worldwide Interoperability for Microwave Access (WiMAX) have experienced tremendous growth [6].

Antennas design for wireless communication systems has attracted a great interest during the last years [7, 8]. Printed antennas have a number of advantages, such as low profile, light weight, and low cost, which make them perfect to be used in wireless communication applications, and ease

(a)

(b)

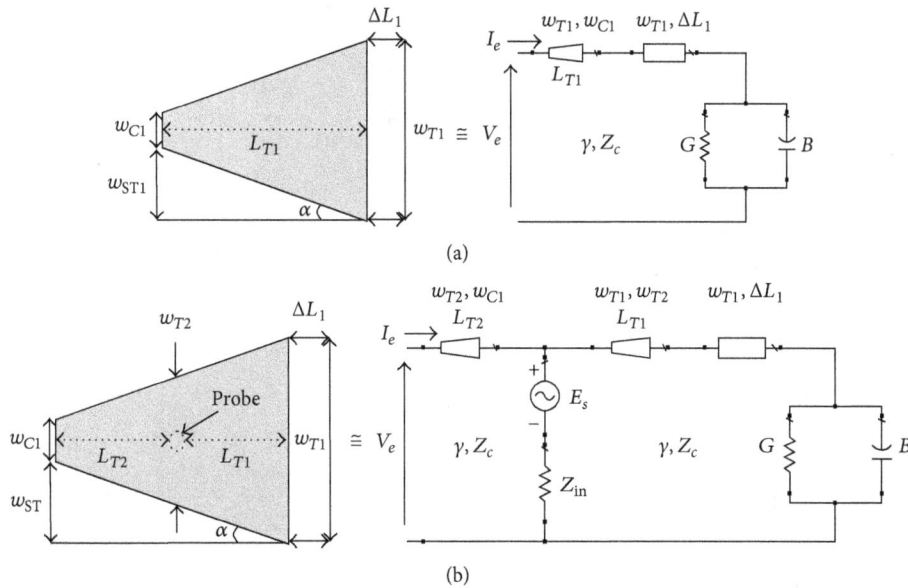

FIGURE 1: (a) Configuration of one taper bowtie antenna and the corresponding equivalent circuit model. (b) Configuration of bowtie antenna fed by coaxial probe and the corresponding equivalent circuit model.

of integration with microwave circuits [9–13]. A multiband antenna is attractive in many commercial applications as it is designed to have a single radiator with a capability to transmit and receive multiple frequencies. Nevertheless, a multiband antenna may not sufficiently cover the required operating bands. Therefore, an antenna which is able to operate with multiple independent frequency bands is required. This antenna should also provide ease in controlling the desirable resonance frequencies, impedance bandwidths, radiation patterns, and polarizations. These are obviously becoming the most important factors for the applications of antennas in both contemporary and future wireless communication systems.

Rigorous numerical methods such as integral equations solved by the method of moments are preferable to analyze such bowtie antenna structures. These methods offer better accuracy, but they require long and tedious calculations. As a result, these methods cannot be used in optimization. The equivalent circuit model is perfectly suited to this type of antennas.

In this paper, we propose a novel quad-band bowtie antenna covering the operating band of Bluetooth operating in the range frequency (2.40–2.484 GHz), two bands of the Unlicensed National Information Infrastructure (U-NII) low band (frequency range 5.15–5.35 GHz) and U-NII mid band (frequency range 5.47–5.725 GHz), and mobile Worldwide Interoperability for Microwave Access (WiMAX) operating in the range frequency (3.40–3.60 GHz). The proposed antenna was modeled and optimized using the proposed equivalent circuit model and the obtained results are compared with those given by moments method of Agilent software, the finite integration technique of the CST Microwave studio, and the finite element method of HFSS software. By adjusting the radiating elements dimensions and

adding other elements or tapers, broad bandwidth behavior suitable for Bluetooth, mobile WiMAX, and two bands of the Unlicensed National Information Infrastructure (U-NII) can be achieved. Details of the antenna design results are presented and discussed.

2. Monoband, Dual-Band, and Quad-Band Bowtie Antenna Equivalent Circuit

2.1. Monoband Bowtie Antenna Equivalent Circuit. In this section, the equivalent circuit applied to bowtie antenna is presented. This model represents the patch as a low-impedance microstrip line whose width determines the impedance and effective dielectric constant. A combination of parallel-plate radiation conductance and capacitive susceptance loads both radiating edges of the patch. This simple circuit is used directly from the transmission line model to compute the input impedance matching at any frequency providing the resonance. The geometry of one taper of the bowtie microstrip antenna and those fed by coaxial probe are shown in Figure 1.

L_{T1} represents the length, w_{T1} is the open end width, and w_{C1} is the width at the input of the structure. ΔL is the physical length of the radiating slot which forms a useful model for calculating the radiation field of the antenna. It is given by the following formula [14–18]:

$$\Delta L = h \frac{0.412 \left(\varepsilon_{\text{eff}} + 0.3\right) \left(w_{T1}/h + 0.264\right)}{\left(\varepsilon_{\text{eff}} - 0.258\right) \left(w_{T1}/h + 0.8\right)}. \tag{1}$$

ε_{eff} is the corresponding effective dielectric constant:

$$\varepsilon_{\text{eff}} = \frac{1 + \varepsilon + (\varepsilon - 1) / \sqrt{1 + 10 \left(h/w_{T1}\right)}}{2}. \tag{2}$$

(a)

(b)

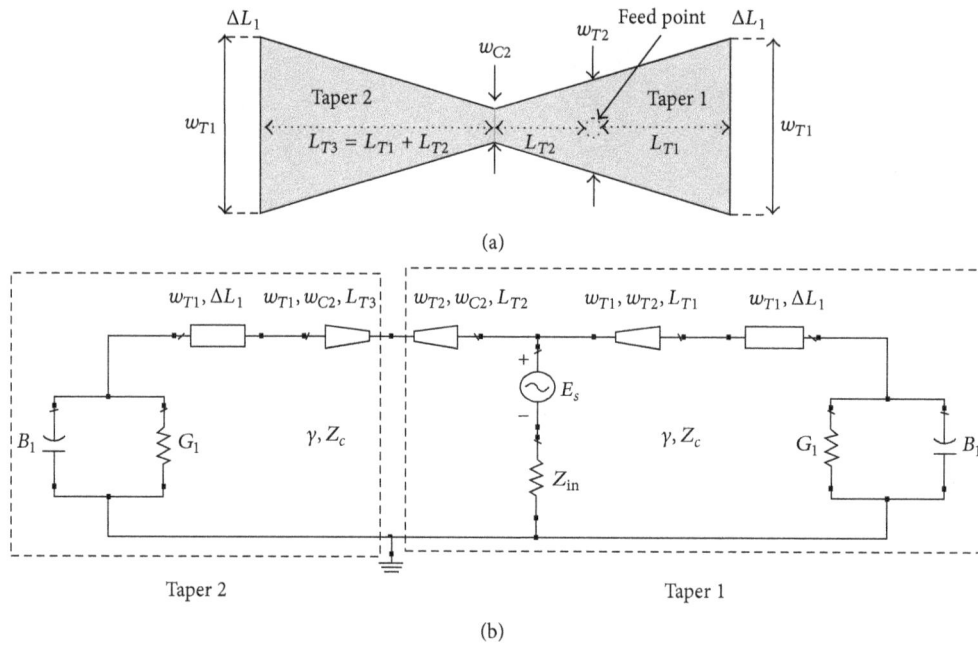

FIGURE 2: (a) Configuration of the proposed monoband bowtie antenna composed of two tapers with same dimension fed by coaxial probe. (b) The corresponding equivalent circuit.

γ is the constant propagation and Z_c is the characteristic impedance. It is given by the following rigorous formula:

$$Z_c = 120\pi \left(2\sqrt{\varepsilon_{\text{eff}}} \left(\frac{w_{T1}}{2h} + 0.082 \left(\frac{(\varepsilon_{\text{eff}} - 1)}{\varepsilon_{\text{eff}}^2} \right) \right. \right.$$
$$\left. \left. + \varepsilon_{\text{eff}} + 0.411 \right) \right)^{-1}. \tag{3}$$

B and G are capacitive and conductive components of the edge admittance Y. The susceptance B accounts for the fringing field associated with the radiating edge of the width w_{T1}, and G is the conductance contributed by the radiation field associated with each edge. Each radiating slot is represented by an equivalent parallel admittance (Y) [17]. The equivalent width of the slots of width w_{ST1} is calculated using the following formula:

$$w_{ST1} = \frac{(w_{T1} - w_{C1})}{2} = L_{T1} \tan \alpha. \tag{4}$$

The expressions of G and B are given by the relations below.

The angle α must not exceed $10°$ and w_{T1} must be less than 2 mm in optimization operation for the results accuracy [4]. Consider

$$G = \frac{60 (\pi h)^2}{Z_c^2 \lambda^2} \left[\frac{(\varepsilon_{\text{eff}} + 1)}{\varepsilon_{\text{eff}}} - \frac{(\varepsilon_{\text{eff}} - 1)^2}{2\varepsilon_{\text{eff}}\sqrt{\varepsilon_{\text{eff}}}} \cdot \log \left(\frac{\sqrt{\varepsilon_{\text{eff}}} + 1}{\sqrt{\varepsilon_{\text{eff}}} - 1} \right) \right], \tag{5}$$

where λ is the wavelength and h is height of the substrate.

V_e and I_e are, respectively, the voltage and current of the source.

When we associate the two tapers, a bowtie antenna is obtained with two radiating elements. Figure 2 shows the structure of the monoband bowtie antenna and its equivalent circuit fed by coaxial probe designed for single resonance, with double radiating elements with same sizes.

2.2. Dual-Band Bowtie Antenna Equivalent Circuit. The dual-band bowtie antenna fed by coaxial probe designed to operate in two resonant frequencies and its equivalent circuit are shown in Figure 3. The two tapers are of different dimensions in order to allow operation at different frequencies.

2.3. Quad-Band Bowtie Antenna Equivalent Circuit. In this section, an equivalent circuit model for quad-band bowtie antenna design is proposed. In order to allow the antenna to resonate at multiple frequencies, it is necessary to add several radiating elements. The quad-band bowtie antenna consists of four tapers with different sizes; each taper radiates at a specific resonance frequency. The configuration of the proposed quad-band bowtie antenna and its corresponding equivalent circuit model are depicted in Figure 4.

3. Simulation Results

To test the validity of the proposed equivalent circuit model for the bowtie antenna, several simulations were made and compared with those obtained by the moments method and the finite integration technique. The proposed optimized bowtie antennas are designed on an FR4 substrate with thickness 1.6 mm and relative permittivity 4.32 and a loss tangent of about 0.048 and 0.05 mm conductor thickness.

(a)

(b)

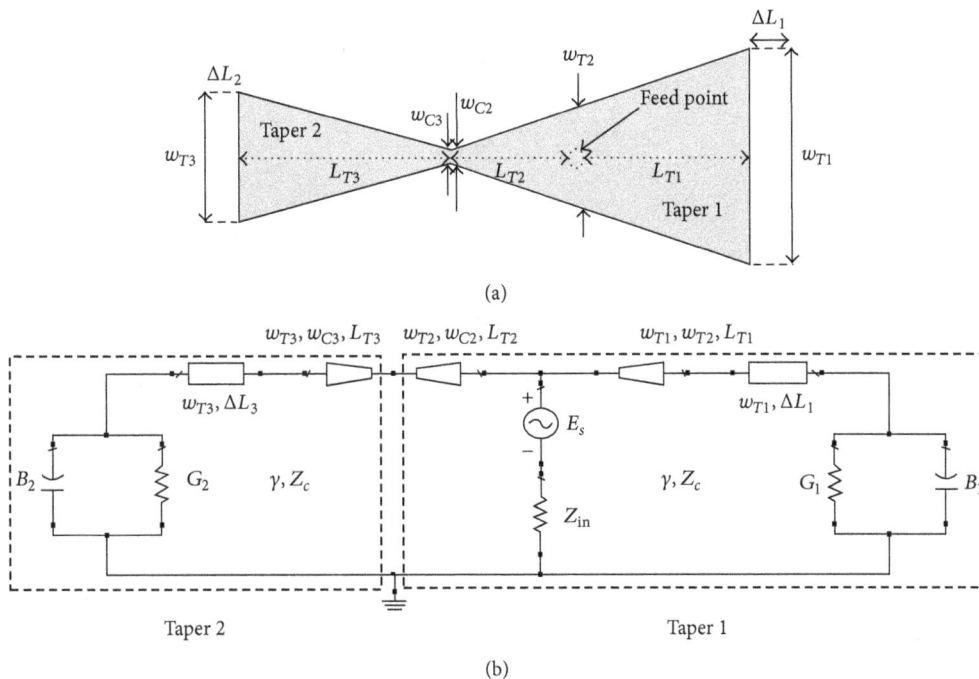

FIGURE 3: (a) Configuration of the proposed dual-band bowtie antenna with two tapers of different dimension fed by coaxial probe. (b) The corresponding equivalent circuit.

3.1. Single Band Bowtie Antenna for Bluetooth Applications. The reflection coefficient of monoband bowtie antenna is displayed from 2 to 3 GHz in Figure 5. The mask layout of bowtie antenna operating at centered frequency 2.44 GHz for Bluetooth applications is presented in the same figure.

Let us notice that the simulated reflection coefficient of the antenna is found below −20 dB at the desired frequency. An excellent agreement is observed between the proposed equivalent circuit model, the moments method, and the finite integration technique. The beam width is well covered with the three models. Notice that a peak of −33 dB at the resonant frequency is obtained by the transmission line model, of about −32.5 dB by, respectively, CST Microwave studio and HFSS software and −23 dB by the moments method of ADS software.

3.2. Dual-Band Bowtie Antenna. The antenna was simulated using the dual equivalent circuit and the results are compared with those obtained by the moments method and the finite integrated technique based on CST Microwave studio. The antenna was optimized to operate at 2.44 GHz and 5.6 GHz, respectively, for Bluetooth and U-NII mid bands. Figure 6 shows the simulated reflection coefficient versus frequency for the proposed dual-band bowtie antenna consisting of two tapers with different sizes.

Notice that a perfect concordance is achieved between the proposed equivalent circuit model, Momentum, CST, and HFSS software. It can be observed that the simulated two resonant frequencies 2.44 GHz and 5.6 GHz give a good impedance matching. The simulated reflection coefficients are below −10 dB in the bandwidth range from 2.4 GHz to

2.50 GHz with respect to the center frequency at 2.44 GHz, and also in the bandwidth range from 5.45 GHz to 5.725 GHz with respect to the center frequency at 5.69 GHz. These bands are practically wide bands and located at the Bluetooth and U-NII mid band.

3.3. Quad-Band Bowtie Antenna. In this section, numerical results of a quad-band printed bowtie antenna are presented. The geometry of the designed quad-band bowtie antenna consists of four tapers and is designed on the same substrate as the previous sections. The excellent performance of this bowtie antenna gives us the advantage that by modifying its structural parameters and by adding other resonant elements we can obtain several frequency bands for this antenna by introducing a proper optimization. A comparison of reflection coefficient obtained with the equivalent model, CST Microwave studio, Momentum, and HFSS is displayed in Figure 7 in the range frequency (2–6 GHz).

From the results shown in Figure 7, four resonances are observed and excellent coherence between the simulated results is observed. The equivalent circuit results are very similar to those simulated by Momentum, CST Microwave studio, and HFSS in all the range frequency. A very small shift between the simulations at the fourth band is recorded. The antenna can generate four operating bands to, respectively, cover the Bluetooth band, mobile WiMAX, and the Unlicensed National Information Infrastructure (U-NII) with two bands. The first resonance occurs at f_1 = 2.44 GHz, the second resonance occurs at f_2 = 3.5 GHz, the third resonance occurs at f_3 = 5.25 GHz, and the fourth resonance occurs at f_4 = 5.6 GHz. Let us notice the appearance of some pics at 3.1

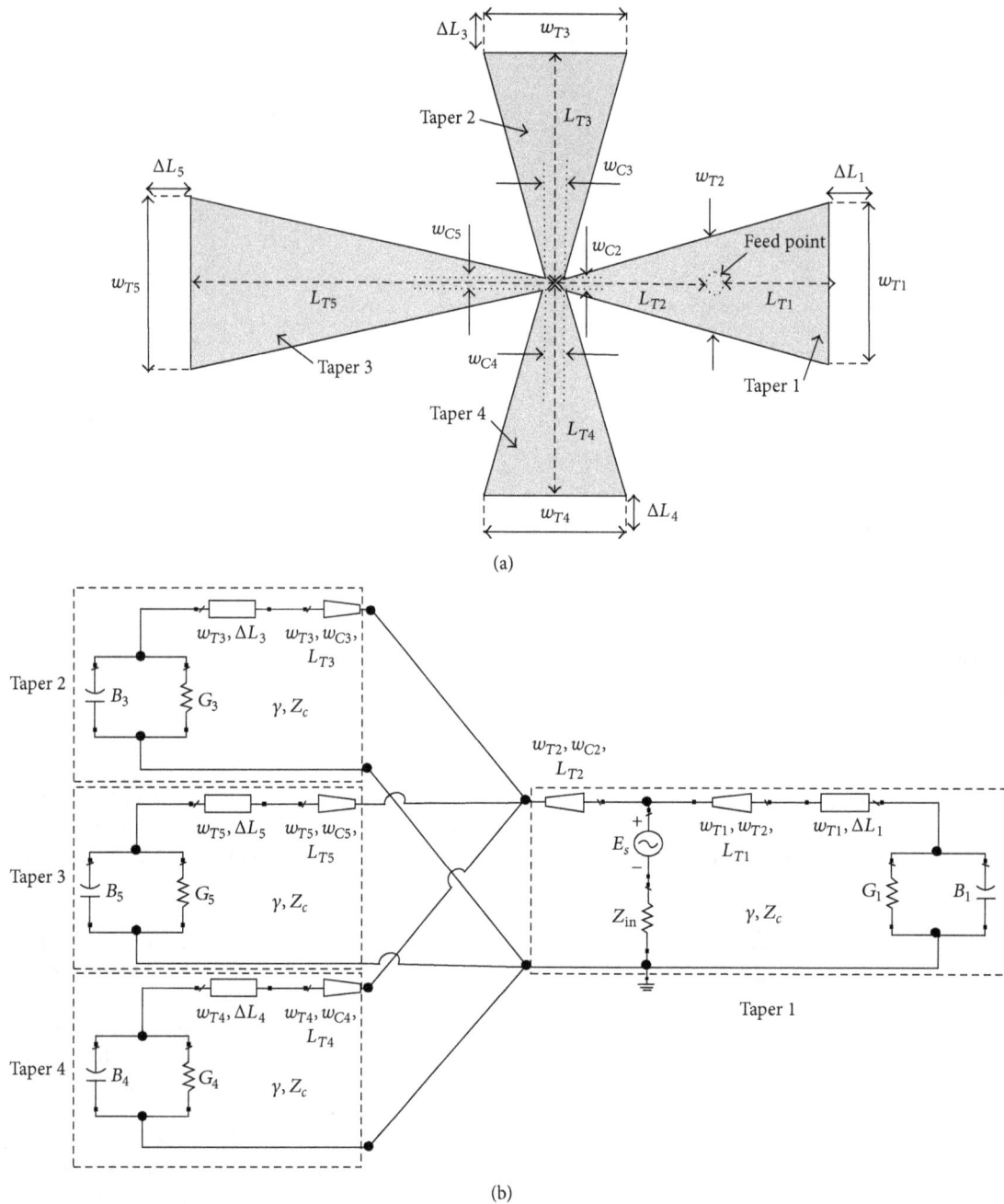

FIGURE 4: (a) Configuration of the proposed quad-band bowtie antenna. (b) Equivalent circuit model of the proposed quad-band bowtie antenna.

and 4.5 GHz which do not exceed −15 dB probably due to the mutual coupling between adjacent elements.

4. Conclusion

An accurate equivalent model is developed designated for microstrip bowtie antennas design fed via a coaxial probe. Various antenna configurations, single band, dual-band, and quad-band, have been optimized and designed in this paper. The validity of the equivalent circuit modeling is supported by comparing the values of the bowtie antenna reflection

coefficient with those obtained, respectively, with moments method computed by Momentum, the finite integration technique using CST software, and the finite element method of HFSS software. The simulated results show that the desired frequencies used in wireless communication systems, Bluetooth, mobile WiMAX, and two bands of the Unlicensed National Information Infrastructure (U-NII) are efficiently achieved and an excellent coherence between the equivalent circuit model and the moments method and finite integration technique is obtained. It can be concluded that the bowtie antenna is capable of generating multiple resonance modes,

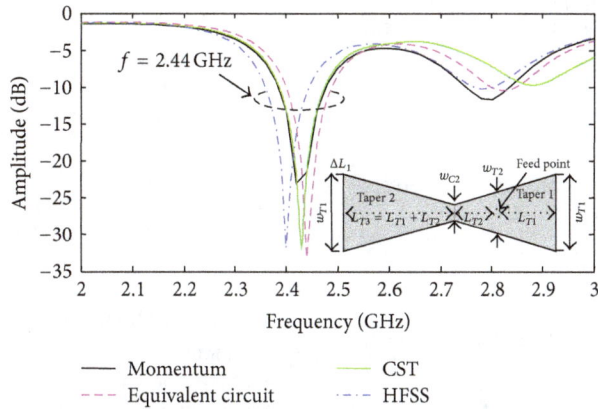

FIGURE 5: Computed return loss of the bowtie antenna designated for WLAN in IEEE 802.11 a/b applications. Geometry of the bowtie antenna under test is also presented. The antenna parameters are set as $w_{T1} = 24.21$ mm, $w_{C2} = 0.9$ mm, $L_{T3} = 66.10$ mm, $w_{T2} = 5.49$ mm, $L_{T1} = 53.09$ mm, and $L_{T2} = 13.01$ mm.

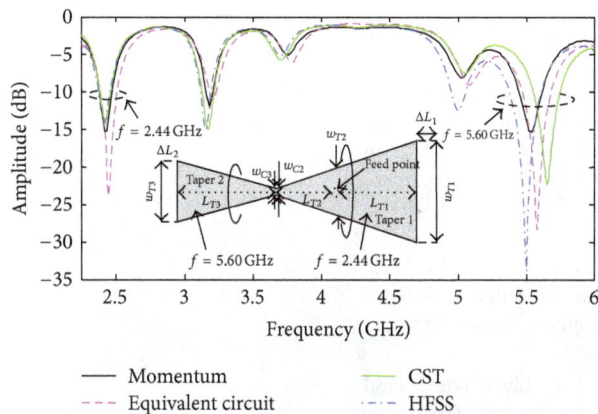

FIGURE 6: Simulated return loss for the dual-band bowtie antenna and the proposed dual-band bowtie antenna. The antenna parameters are set as $w_{T3} = 28.03$ mm, $w_{C3} = 0.85$ mm, $L_{T3} = 77.08$ mm, $w_{T1} = 17.41$ mm, $w_{T2} = 8.24$ mm, $w_{C2} = 1.79$ mm, $L_{T1} = 26$ mm, and $L_{T2} = 18.29$ mm.

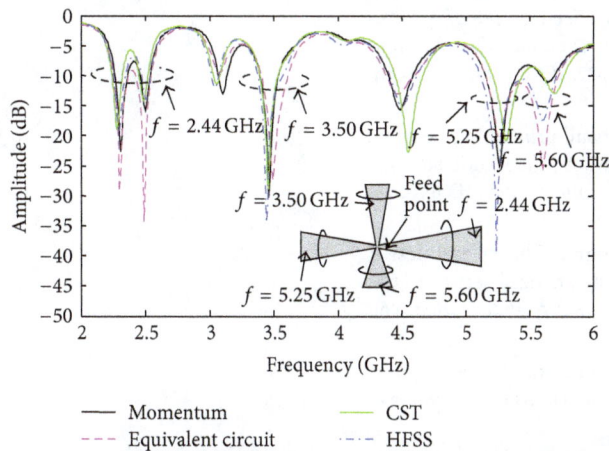

FIGURE 7: Comparison of the simulated return losses for the proposed quad-band bowtie antenna. The geometry of the proposed quad-band bowtie antenna is also displayed. The antenna parameters are set as $w_{T5} = 17.38$ mm, $w_{C5} = 1.62$ mm, $L_{T5} = 89.44$ mm, $w_{T3} = 21.67$ mm, $w_{C3} = 1.11$ mm, $L_{T3} = 57.87$ mm, $w_{T2} = 4$ mm, $w_{C2} = 1.78$ mm, $L_{T2} = 6.29$ mm, $w_{T1} = 28.26$ mm, $L_{T1} = 68.84$ mm, $w_{T4} = 20.97$ mm, $w_{C4} = 1.35$ mm, and $L_{T4} = 55.67$ mm.

by varying the dimensions and multiplying the number of tapers. The designed antennas can be employed for wireless applications.

Conflict of Interests

The authors declare that there is no conflict of interests regarding the publication of this paper.

References

[1] T.-H. Kim and D.-C. Park, "Compact dual-band antenna with double L-slits for WLAN operations," *IEEE Antennas and Wireless Propagation Letters*, vol. 4, no. 1, pp. 249–252, 2005.

[2] W. Diels, K. Vaesen, P. Wambacq et al., "Single-package integration of RF blocks for a 5 GHz WLAN application," *IEEE Transactions on Advanced Packaging*, vol. 24, no. 3, pp. 384–391, 2001.

[3] H. Labiod, H. Afifi, and C. Desantis, *Bluetooth Zigbee Wifi and Wimax*, Springer, 2007.

[4] C. A. Balanis, *Modern Antenna Handbook*, John Wiley & Sons, New York, NY, USA, 2008.

[5] M. A. Rahman, M. Hossain, I. S. Iqbal, and S. Sobhan, "Design and performance analysis of a dual-band microstrip patch antenna for mobile WiMAX, WLAN, Wi-Fi and bluetooth applications," in *Proceedings of the 3rd International Conference on Informatics, Electronics & Vision (ICIEV '14)*, pp. 1–6, Dhaka, Bangladesh, May 2014.

[6] S.-M. Zhang, F.-S. Zhang, W.-M. Li, W.-Z. Li, and H.-Y. Wu, "A multi-band monopole antenna with two different slots for WLAN and WiMAX Applications," *Progress in Electromagnetics Research Letters*, vol. 28, pp. 173–181, 2012.

[7] Y.-J. Wu, B.-H. Sun, J.-F. Li, and Q.-Z. Liu, "Triple-band omni-directional antenna for WLAN application," *Progress in Electromagnetics Research*, vol. 76, pp. 477–484, 2007.

[8] Y.-C. Lee and J.-S. Sun, "Compact printed slot antennas for wireless dual- and multi-band operations," *Progress in Electromagnetics Research*, vol. 88, pp. 289–305, 2008.

[9] I. E. Lager, M. Simeoni, and C. Coman, "Mutual coupling in non-uniform array antennas—an effective recipe," in *Proceedings of the 6th European Conference on Antennas and Propagation (EUCAP '12)*, pp. 1518–1522, IEEE, Prague, Czech, March 2012.

[10] G. A. Casula and P. Maxia, "A multiband printed log-periodic dipole array for wireless communications," *International Journal of Antennas and Propagation*, vol. 2014, Article ID 646394, 6 pages, 2014.

[11] F. A. Ghaffar, J. R. Bray, and A. Shamim, "Theory and design of a tunable antenna on a partially magnetized ferrite LTCC substrate," *IEEE Transactions on Antennas & Propagation*, vol. 62, no. 3, pp. 1238–1245, 2014.

[12] J. J. H. Wang and V. K. Tripp, "Design of multioctave spiral-mode microstrip antennas," *IEEE Transactions on Antennas & Propagation*, pp. 332–335, 1991.

[13] T. Chio and D. H. Schaubert, "Effects of slotline cavity on dual-polarized tapered slot antenna arrays," in *Proceedings of the Antennas and Propagation Society International Symposium*, pp. 130–133, IEEE, Orlando, Fla, USA, July 1999.

[14] H. A. Majid, M. K. Abd Rahim, M. R. Hamid, and M. F. Ismail, "Frequency reconfigurable microstrip patch-slot antenna with directional radiation pattern," *Progress in Electromagnetics Research*, vol. 144, pp. 319–328, 2014.

[15] M. R. Ahsan, M. T. Islam, M. Habib Ullah, H. Arshad, and M. F. Mansor, "Low-cost dielectric substrate for designing low profile multiband monopole microstrip antenna," *The Scientific World Journal*, vol. 2014, Article ID 183741, 10 pages, 2014.

[16] E. H. van Lil and A. R. van de Capelle, "Transmission line model for mutual coupling between microstrip antennas," *IEEE Transactions on Antennas and Propagation*, vol. 32, no. 8, pp. 816–821, 1984.

[17] C. A. Balanis, *Antenna Theory Analysis and Design*, John Wiley & Sons, 2nd edition, 1997.

[18] S. Didouh, M. Abri, and F. T. Bendimerad, "Corporate-feed multilayer bow-tie antenna array design using a simple transmission line model," *Modelling and Simulation in Engineering*, vol. 2012, Article ID 327901, 8 pages, 2012.

21

Comparative Assessment of GaN as a Microwave Source with Si and SiC for Mixed Mode Operation at Submillimetre Wave Band of Frequency

Pranati Panda, Satya Narayan Padhi, and Gana Nath Dash

Electron Devices Group, School of Physics, Sambalpur University, Jyoti Vihar, Burla, Sambalpur, Odisha 768019, India

Correspondence should be addressed to Gana Nath Dash; gndash@ieee.org

Academic Editor: Salvador Sales Maicas

The potentials of GaN, SiC, and Si for application as microwave sources in mixed tunnelling avalanche transit time mode operation at submillimetre wave (sub-mm wave) frequency around 0.35 terahertz (THz) are investigated using some computer simulation methods. Design criteria to choose width, doping concentration, and area are highlighted. From the results of our simulation we observed that the Si diode produces the least power output of 41 mW followed by the GaN diode with 760 mW and the SiC diode with 2.89 W. In addition, the GaN diode has more noise than the SiC diode (by 5 dB) as well as the Si diode (by 10 dB). The drastically different performance between the GaN and the SiC diode is attributed to the incorporation of disparate carrier velocity in GaN which were not being used by other authors. In spite of the low power and high noise of the GaN compared to the SiC diode, the presence of several peaks in the mean square noise voltage curves and the existence of several minima in the noise measure curves would open a new direction in the design of GaN low-noise ATT diodes capable of multifrequency tuning like a DAR diode.

1. Introduction

The potentials of GaN for avalanche transit time (ATT) devices have been explored by several authors [1–3]. But they are based on simulation results of symmetric diode structures where the hole saturation velocity is assumed to be the same as the electron saturation velocity. This assumption is however incorrect in view of reports [4, 5]. In report [4], Albrecht et al. have used Monte Carlo simulations of electron transport based upon an analytical representation of the lowest conduction bands of bulk, wurtzite phase GaN to develop a set of transport parameters for devices with electron conduction in GaN. On the other hand, in report [5], Oğuzman et al. have calculated the hole saturation velocity using an ensemble Monte Carlo simulator, including the full details of the band structure, and numerically determined phonon scattering rate based on empirical pseudopotential method. They found that the average hole energies are significantly lower than the corresponding electron energies believed to be due to the drastic difference in curvature between the uppermost valence bands and the lowest conduction band [5]. The

relatively flat valence band is responsible for hole heating, leading to low average hole energy and drastically low hole velocity compared to that for electrons. Thus there is a substantial difference in electron and hole velocities reported by the two groups. We for the first time used such disparate carrier velocities for the simulation of microwave properties of GaN MITATT (Mixed Tunnelling Avalanche Transit Time) diodes [6] and reported some interesting results from our preliminary study. The purpose of this paper is to substantiate our earlier work by extending the study and compare the results with those of the industry leader Si and the wide band gap rival SiC for operation as MITATT diodes in the same sub-mm wave band of frequency.

In avalanche transit time (ATT) diodes carrier velocities play an important role in generating the transit time phase delay which together with the avalanche phase delay produces the microwave negative resistance responsible for power production from the device. When the carrier velocities are equal, the electron and hole currents maintain the same phase leading the total current to preserve the required phase relationship with the voltage. This is the case with Si and

SiC avalanche transit time diodes where the carrier velocities are nearly equal. But such phase relationship between the *RF* voltage and *RF* current is disturbed when the electron and hole currents develop different amount of transit time phase delays from the diode active region due to disparate carrier velocities in materials like GaN. This has an adverse impact on the performance of the GaN ATT diode. We thus feel that comparing the microwave properties of the MITATT diodes based on the three materials will not only reveal their relative merits but also uncover the effect of disparate carrier velocity on the performance of the device. To start with we present the design methods for the diodes in the next section. A brief description of the simulation method is presented in Section 3 followed by results and discussion in Section 4. Finally we conclude our paper in Section 5.

2. Design Considerations

Four diode structures, three DDR (Double Drift Region) diodes based, respectively, on GaN, Si, and SiC and one SDR (single drift region) diode based on GaN, were designed for operation in sub-mm wave band at a frequency around 0.35 THz. The basic design parameters of the diode include the width, doping, and the area of cross section. The methods used to determine them are explained in the following subsections.

2.1. Width of the Active Region. For the determination of width, two criteria are generally followed. In one of them the thrust is to maximise the efficiency while the other aims at maximising the diode negative resistance. It has been seen that the efficiency is maximum when the IMPATT mode transit angle function $g_I(\theta) = (1 - \cos\theta)/\theta$ is the maximum [7]. A little amount of algebra will show that $g_I(\theta)$ will be maximum when (the appendix contains definition of symbols)

$$\theta = 2\pi f_d \tau = 0.74\pi, \qquad (1)$$

where $\tau = W_{n,p}/v_{sn,sp}$ is the transit time across the diode width, f_d is the design frequency, and $v_{sn,sp}$ is the saturation drift velocity of charge carriers. With this, (1) can be manipulated to get an expression for the diode width as

$$W_{n,p} = \frac{0.37 v_{sn,sp}}{f_d}. \qquad (2)$$

Now we come to the second criterion for width determination. We know that to maximise the negative resistance the phase delay between the RF voltage and RF current, θ, should be equal to π. Using this condition the expression for width becomes

$$W_{n,p} = \frac{0.5 v_{sn,sp}}{f_d}. \qquad (3)$$

Once the design frequency is decided, (2) or (3) can be used to determine the required width of the diode from the knowledge of experimental values of carrier velocities for the

semiconductor under consideration. Since we have chosen same frequency of operation, and since Si and SiC have equal carrier velocities, these two materials have symmetric diode structures. But, as a result of disparate carrier velocities the structures of GaN DDR diode have become asymmetric (Table 1).

2.2. Doping Concentrations. The doping profiles near the junctions can be made realistic on both donor and acceptor sides by using appropriate exponential functions. The expression used for the n-side is

$$N(x) = N_1 \left[1 - \exp\left\{ \frac{x - x_j}{s} \right\} \right], \qquad (4)$$

and that for the p side is

$$N(x) = N_2 \left[\exp\left\{ \frac{x_j - x}{s} \right\} - 1 \right]. \qquad (5)$$

Here x_j is the position of the junction and N_1 and N_2 represent the flat doping levels of the n- and p-sides, respectively. In order to match the doping profiles with practical structures, the constant "s" has been taken as 5 nm. The doping profiles at the interfaces of substrate and epitaxy correspond to the solution of Fick's equation and are given by complimentary error function profiles which can be closely approximated by using exponential function of the type [8]

$$N(x) = N_H \exp\left(-1.08\lambda - 0.78\lambda^2\right), \qquad (6)$$

for the n-side, and

$$N(x) = -N_H \exp\left(-1.08\lambda - 0.78\lambda^2\right), \qquad (7)$$

for the p-side, where $\lambda = x''/2\sqrt{Dt}$ and x'' is the distance from the surface. The doping level, N_H, is taken to be $10^{26}/m^3$ and $\sqrt{Dt}$ has been assumed to be $1\,\mu m$ for our analysis. N_1 and N_2 are adjusted through several computer runs so as to maximise the efficiency and minimise the avalanche zone width. Due to the reasons described in our earlier paper [6], the doping concentration of p-side is 5 times that of n-side in the designed GaN diodes considered in this paper.

2.3. Area of Cross Section. We have used the method indicated in [9] for the determination of area of cross section of each diode structure and the same is presented in this subsection. The transit time devices such as IMPATT and MITATT diodes exhibit a negative resistance property. The admittance per unit area of such a negative resistance device is a function of the frequency and can be written as

$$Y(\omega) = G(\omega) + jB(\omega), \qquad (8)$$

where $G(\omega)$ is the conductance and $B(\omega)$ the susceptance per unit area, respectively. The total diode admittance is then given by

$$Y_d(\omega) = G_d(\omega) + jB_d(\omega) = AY(\omega)$$
$$= AG(\omega) + jAB(\omega), \qquad (9)$$

TABLE 1: Design parameters of GaN DDR, GaN SDR, Si DDR, and SiC DDR diodes for operation in sub-mm wave band frequency around 0.35 THz.

Structures	Widths (nm)		Doping concentrations (10^{23} m^{-3})		Area (10^{-11} m^2)
	n-side	p-side	n-side	p-side	
GaN (DDR)	185	37	1.3	6.5	2.65
GaN (SDR)	185	—	1.3	—	0.68
Si (DDR)	117	117	3.5	3.5	2.86
SiC (DDR)	336	336	7.2	7.2	3.00

where A is the area of the device. Let us write the total diode impedance, which is the reciprocal of the diode admittance, as

$$Z_d(\omega) = \frac{1}{Y_d(\omega)} = R_d(\omega) + jX_d(\omega), \quad (10)$$

where the dynamic diode resistance can be evaluated to

$$R_d(\omega) = \frac{G(\omega)}{A\left[G^2(\omega) + B^2(\omega)\right]}, \quad (11)$$

and the dynamic diode reactance may be derived as

$$X_d(\omega) = \frac{-B(\omega)}{A\left[G^2(\omega) + B^2(\omega)\right]}. \quad (12)$$

For transit time devices at high operating frequencies it is found that $B(\omega) \gg G(\omega)$, for which (11) and (12) can be approximated, respectively, as

$$R_d(\omega) = \frac{G(\omega)}{A\left[B^2(\omega)\right]}, \quad (13)$$

$$X_d(\omega) = \frac{-1}{AB(\omega)}. \quad (14)$$

In addition to $R_d(\omega)$ the diode offers some series resistance R_S due to the finite conductivity of the semiconductor which is positive. The load offers a positive resistance R_L and a positive reactance X_L. The oscillation condition demands that $R_d(\omega) + R_S + R_L = 0$ and $X_d(\omega) + X_L = 0$. In other words $R_d(\omega)$ and $X_d(\omega)$ should both be negative which in turn implies that $G(\omega)$ is negative and $B(\omega)$ is positive. Therefore the operating frequency $\omega_p = 2\pi f_p$ is so chosen such that the device conductance has the maximum negative value while the device susceptance remains positive at that frequency; we call it the optimum frequency. For sustained oscillation therefore, we invoke (13) to obtain an expression for the diode area as

$$A = \frac{-G(\omega_p)}{\left[B^2(\omega_p)\right](R_S + R_L)}. \quad (15)$$

The area of each diode structure has been calculated using (15). The values of peak negative conductance $G(\omega_p)$ and positive susceptance $B(\omega_p)$ are obtained from small-signal simulation described in the next section. The value of $(R_S + R_L)$ is taken to be of the order of $10\,\Omega$. The areas of diode obtained from the calculation are input for simulation of noise program. The design parameters of all the four diode structures considered in this paper are listed in Table 1.

3. Simulation Methods

The behaviours of GaN, Si, and SiC diodes are studied by considering a one-dimensional model of diode having doping distribution of the form n$^+$npp$^+$. The microwave behaviours of the diodes are analysed for mixed mode operation. For the DC simulation we have followed the scheme described in [10]. It involves simultaneous solution of Poisson's equation, carrier continuity equation, and space charge equation. The outputs from the DC analysis are used as input for the small-signal analysis.

A small-signal method of analysis including tunnelling current to the conduction current and displacement current developed by Dash and Pati [10] is used for our study. In essence it solves two simultaneous second-order differential equations on the real and imaginary parts of the resistivity $(\rho_R(x,\omega), \rho_I(x,\omega))$ at any point x in the diode active layer, subject to the essential boundary conditions employing a double iterative computer simulation method which performs iterations over the initial values $\rho_R(0,\omega)$ and $\rho_I(0,\omega)$ since they are not known. When the iterations converge we get the final solutions $\rho_R(x,\omega)$ and $\rho_I(x,\omega)$ for a given frequency. Integrating these over the active layer of the diode one gets

$$R(\omega) = \int_0^W \rho_R(x,\omega)\, dx,$$
$$X(\omega) = \int_0^W \rho_I(x,\omega)\, dx, \quad (16)$$

from where the diode conductance and diode susceptance per unit area can be obtained as

$$G(\omega) = \frac{R(\omega)}{R^2(\omega) + X^2(\omega)},$$
$$B(\omega) = \frac{-X(\omega)}{R^2(\omega) + X^2(\omega)}. \quad (17)$$

The process is repeated for several frequencies within the frequency band for which the diode is designed to determine the optimum frequency and other small-signal diode characteristics.

MITATT mode noise simulation scheme developed by Dash et al. [11] is used to analyse the noise behaviour of the designed MITATT diodes. The computation starts by putting the noise source at the beginning of the generation region. The noise electric field corresponding to the location of the

TABLE 2: DC and microwave properties of GaN DDR, GaN SDR, Si DDR, and SiC DDR diodes at sub-mm wave frequency band.

Material and structure	E_0 (10^8 V/m)	V_B (V)	η (%)	f_p (THz)	$G_d(\omega_p)$ (10^{-3} S)	$R_d(\omega_p)$ (Ω)	$P_{RF}(\omega_p)$ (mW)	J_T/J_0 (%)
GaN DDR	3.16	50.2	20.7	0.32	−2.415	−20.92	760	2.24
GaN SDR	3.46	44.5	21.0	0.38	−0.386	−9.97	95	6.34
Si DDR	0.757	12.2	5.89	0.30	−2.23	−9.82	41	20.45
SiC DDR	5.04	182	17.2	0.36	−0.699	−13.63	2890	18.27

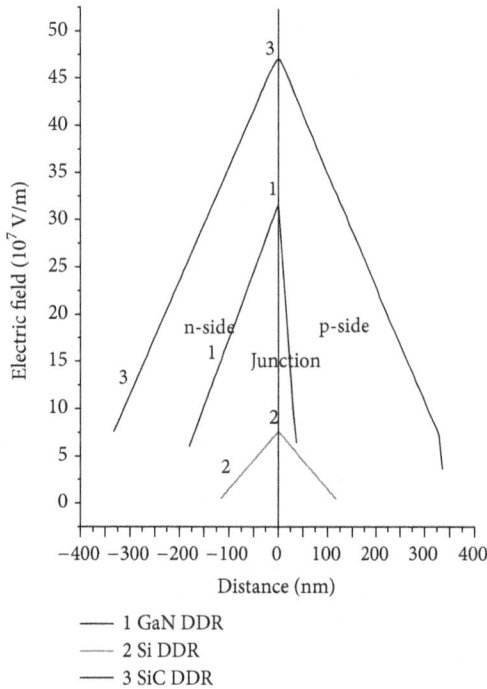

FIGURE 1: Electric field profiles of GaN, Si, and SiC flat doping profile DDR MITATT diodes considered in this paper.

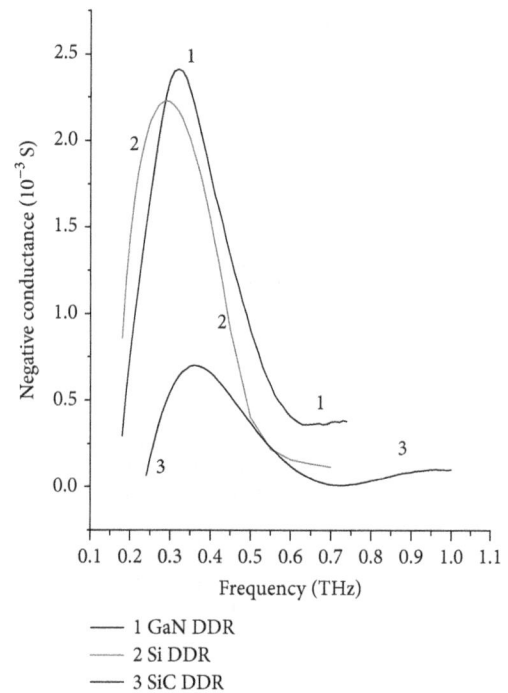

FIGURE 2: Negative conductance plots as a function of frequency for GaN, Si, and SiC DDR MITATT diodes referred to in Figure 1.

noise source is computed from which the terminal voltage and transfer impedances are determined. The noise source is then shifted to the next space step and the process is repeated until it covers the whole generation region. Then the mean square noise voltage and the "noise measure" (NM) are determined following the approach described in [11].

4. Results and Discussion

The DC and small-signal properties of the diodes are presented in Table 2. The design as well as the doping asymmetry of the GaN diode can be clearly seen from the table. The doping asymmetry results in not only asymmetry in the electric field profile as shown in Figure 1 but also a lower field maximum (E_0 in Table 2) for the GaN diode compared to the SiC diode. Concomitantly the SiC DDR diode has much higher breakdown voltage (V_B about 4 times) than that of the GaN diode. But, the breakdown voltage of Si DDR diode is about one-fourth of that of GaN diode. This can be understood from the fact that Si has a much lower

band gap compared to SiC and GaN resulting in higher energy and higher voltage for a breakdown of the latter diodes compared to the former. Further, it can be observed from Table 2 that the GaN diode is accompanied by the highest efficiency (η), the highest negative conductance [$-G_d(\omega_p)$], and the highest negative resistance [$-R_d(\omega_p)$] at the optimum frequency (f_p). These features are indicative of superior material performance of GaN. In spite of these facts, the power output [$P_{RF}(\omega_p)$] of GaN is much less than that of SiC. The reason for such degradation in power output in GaN diode can be attributed to the disparate carrier velocities leading to lower p-side width and lower voltage drop there. This in turn decreases the breakdown voltage and hence the input voltage of the GaN diode compared to the SiC diode. Thus, although the efficiency is high, the output power is only 790 mW in the former compared to a substantial 2890 mW in the latter.

The microwave negative conductance of the device [$-G_d(\omega)$] as a function of frequency is depicted in Figure 2.

TABLE 3: Noise behaviours of GaN DDR, GaN SDR, Si DDR, and SiC DDR diodes at sub-mm wave frequency band.

Material & structure	f_g (THz)	$\langle v^2 \rangle/df$ (V^{2}s)	f_l (THz)	NM at f_l (dB)	$\langle v^2 \rangle/df$ at f_p (V^{2}s)	NM at f_p (dB)
GaN DDR	0.11	7.00×10^{-13}	0.35	33.50	3.97×10^{-16}	34.40
GaN SDR	0.11	3.73×10^{-12}	0.67	31.59	4.38×10^{-16}	34.23
Si DDR	0.18	2.86×10^{-14}	0.62	20.68	2.45×10^{-17}	24.59
SiC DDR	0.14	4.86×10^{-14}	0.57	25.20	1.54×10^{-16}	29.70

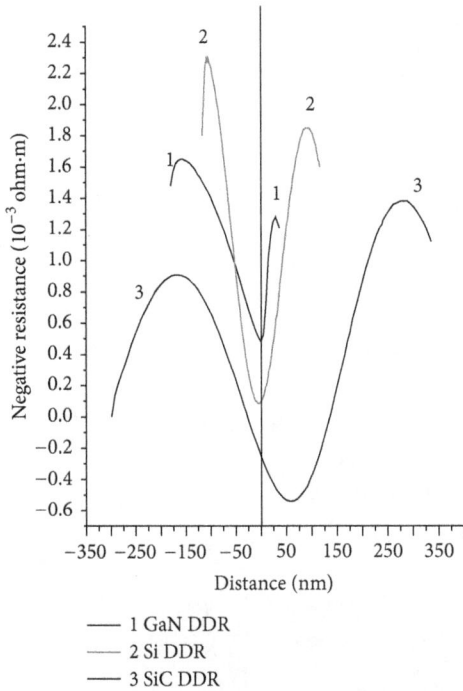

FIGURE 3: Negative resistivity profiles of the three DDR MITATT diodes at their respective optimum frequencies.

The GaN diode exhibits higher negative conductance compared to the Si and SiC diode over a wide range of frequencies around the design frequency of 0.35 THz. Again, the values of microwave negative resistivity as a function of distance x are computed to obtain the intensity of oscillation at each space point and to study the contribution of individual space step of the diode towards negative resistance. The microwave negative resistivity profiles $\rho_R(x, \omega_p)$ at the respective optimum frequencies are shown in Figure 3. From the figure it is observed that the $\rho_R(x, \omega_p)$ profile in each case possesses two maxima, one in each of the drift regions separated by a minimum near the diode junction. So, it is clear that the contribution to diode negative resistance mostly comes from the drift regions. The peaks in the profile of Si and SiC DDR diode are symmetrically placed with respect to the junction. But, in case of GaN DDR diode they are situated asymmetrically with the n-side peak at farther distance from the junction than the p-side peak. These features are due to equal carrier velocities in Si and SiC and inequality in carrier velocities of GaN. Further, in case of GaN and Si the magnitudes of n-side peaks are higher than that of p-side. This is because the ionisation rate of electron is higher than

that for hole both for Si and for GaN. But in case of SiC the ionisation rate for holes is more than that for electrons. So for SiC diode the magnitude of p-side peak is higher than that for n-side. While the SiC diode has positive resistance contribution from the avalanche zone, the GaN diode has negative resistance contribution from the entire depletion width. This can be understood in the following way.

In SiC diode, due to equal carrier velocities, electron and hole currents develop nearly the same phase delay of π from the transit time across the diode depletion width. When combined with the avalanche phase delay of $\pi/2$ the total current, therefore, develops a phase delay of $3\pi/2$ from the drift region giving rise to negative resistance contribution from the whole of drift regions. However, due to disparate carrier velocities in GaN, the electrons develop a phase delay of π from the transit time across the diode width whereas the holes develop the same phase delay from only 1/5th of the diode width. In other words the holes will develop a phase angle of 5π as they travel across the diode width. So the hole current does not depend on the avalanche phase delay for negative resistance; the transit time phase delay is sufficient for the purpose. The diode characteristic is a manifestation of the combined phase delay of the electronic and hole currents resulting in the observed behaviour.

The percentage of tunnelling current (J_T/J_0) recorded in Table 2 reveals an important piece of information. The high tunnelling current in the Si and SiC diodes shifts the optimum frequency to a very high value around 0.8 THz due to loss in avalanche phase delay associated with the tunnelling current. In order to draw a comparative assessment, the diodes must be operated in the same frequency band (around 0.35 THz). Therefore, the widths of the Si and SiC diodes have been modulated [12] to restore the frequency to a value near the GaN diode. This makes the width of the former diodes much higher than the latter (Table 1). The noise characteristics of the diodes are presented in Table 3. It can be seen from the table that, in spite of the lower electric field in the GaN diode, it generates an order of magnitude more noise than the Si and the SiC diode as evident from the peak mean square noise voltage. From the minimum noise measure consideration also GaN diode is observed to produce 10 dB and 13 dB more noise, respectively, than SiC and Si diodes. These features are understood to be due to lower tunnelling current in the GaN diode (Table 2). The mean square noise voltage per bandwidth is plotted in Figure 4. While Si and SiC MITATT diode show single peaks in the frequency range of 0.01 THz to 1 THz, the GaN MITATT diode shows several peaks of decreasing magnitude in the same frequency range. The latter fact is indicative of the existence of several negative

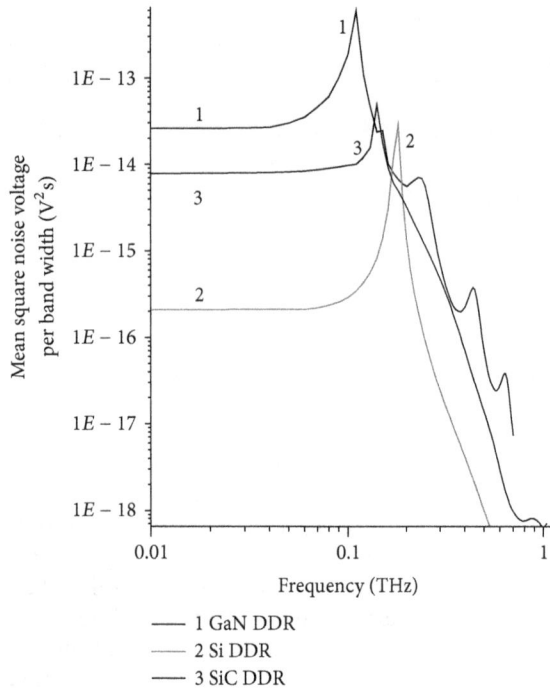

FIGURE 4: Mean square noise voltage per band width as a function of frequency for the three DDR MITATT diodes.

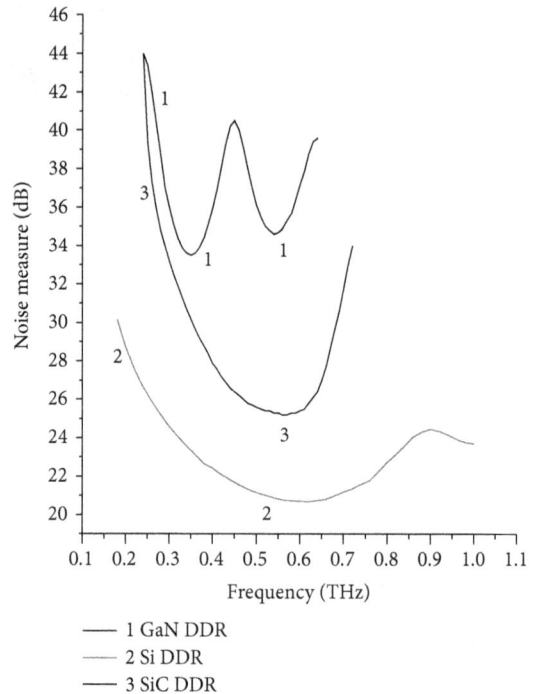

FIGURE 5: Noise measure characteristics of the three DDR MITATT diodes.

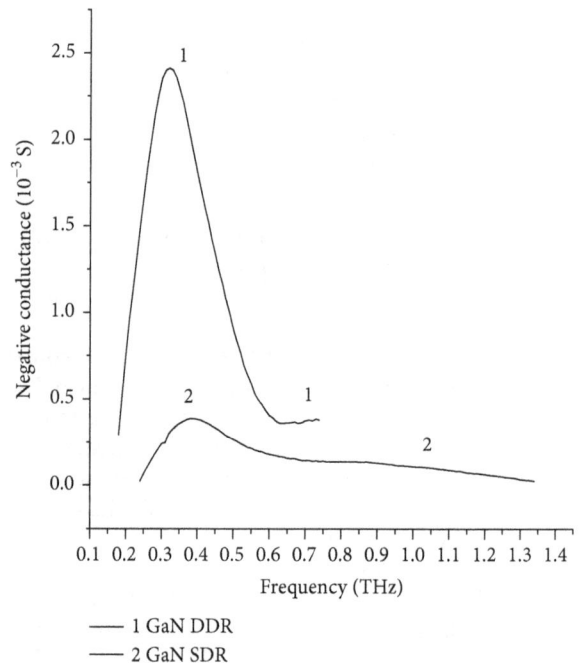

FIGURE 6: Negative conductance plots as a function of frequency for the GaN DDR and GaN SDR MITATT diodes considered in this paper.

conductance bands in the referred frequency range. This is due to disparate phase relationship between I_n and I_p with respect to the voltage. While I_n satisfies the required phase relation for negative conductance at around 0.35 THz I_p satisfies the same at several other frequencies giving rise to multiple negative conductance bands. This is a new feature similar to that observed in a double avalanche region (DAR) diode [13]. Thus, it is believed that the GaN diode is capable of multiple frequency tuning similar to that of a DAR diode. The disparate phase relationship of the electron and hole current has the additional effect of multiple noise measure minima shown in Figure 5. This is indicative of the fact that the GaN MITATT diode has the option of being operated at more than one frequency satisfying the condition of minimum noise measure, albeit at a much higher noise level compared to the Si and SiC MITATT diodes.

An SDR structure of GaN diode has also been designed and its performance is compared with that of GaN DDR diode. The peak electric field of the SDR diode is higher than that of the DDR diode (Table 2). But, due to lower width, the SDR diode has lower breakdown voltage compared to the DDR diode. This in turn results in considerably lower power output from the SDR diode compared to that of the DDR diode. The decrease in the power output is in conformity with decrease in the value of integrated value of negative conductance (Figure 6) and negative resistance (Figure 7) of GaN SDR as compared to that of GaN DDR diode.

It is further observed from Table 3 and Figure 8 that the peak mean square noise voltage in the GaN SDR diode is almost an order higher than that in GaN DDR diode. This is because of the higher electric field in case of the former

than the latter. It is worthwhile noting that the percentages of tunnelling current in GaN diodes are not so high. Therefore, the effect of tunnelling current in reducing noise level has not been observed in the mean square noise voltage of diodes with this material. Further when we consider the noise measure of the DDR and SDR structures we find that it is

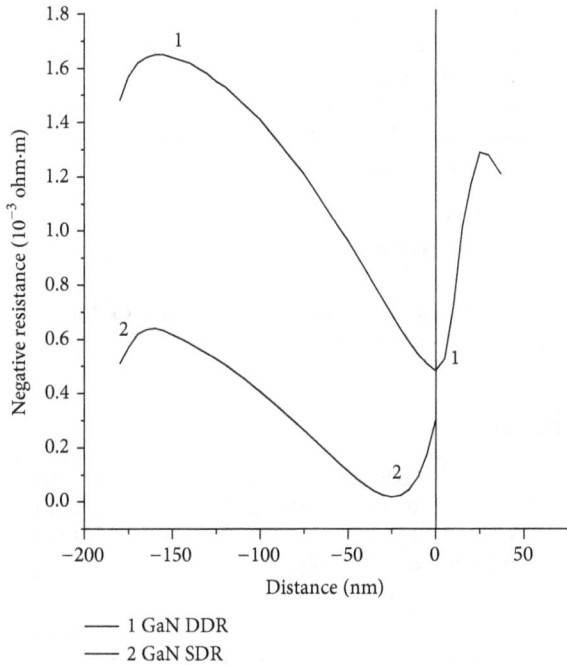

FIGURE 7: Negative resistivity profile of the GaN DDR and GaN SDR MITATT diodes.

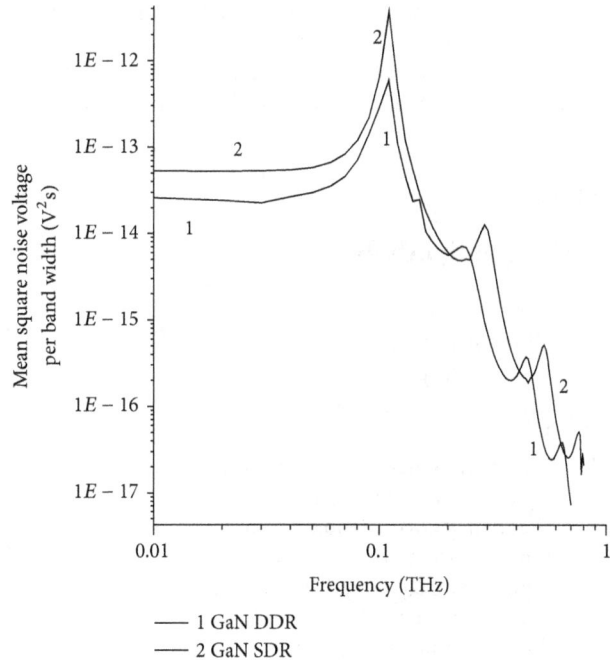

FIGURE 8: Mean square noise voltage per band width as a function of frequency for GaN DDR and GaN SDR MITATT diodes.

33.5 dB for the first one and 37.38 dB for the second at the operating frequency of 0.35 THz (Figure 9). Although the noise measure minimum of the SDR is 31.59 dB, which is lower than that of the DDR (33.5 dB), it is hardly of any use since it occurs at 0.67 THz where the RF properties of the diode are sufficiently degraded. Thus the effect of tunnelling current has no advantage in reducing the noise level in GaN diodes.

It is interesting to observe that the features like multiple mean square noise voltage peaks, multiple noise measure minima, and multiple negative conductance peaks observed in case of the GaN DDR diode are all present in case of the GaN SDR diode also. We consider it an additional confirmation that these features are due to the disparate carrier velocity in GaN.

5. Conclusion

The results obtained from mixed mode simulation of GaN DDR, GaN SDR, Si DDR, and SiC DDR diodes show the supremacy of GaN diodes over conventional Si diode at sub-mm wave frequency of operation in terms of both microwave power output and DC to microwave conversion efficiency. But GaN diodes are noisier than conventional Si diodes by around 13 dB with references to their minimum noise measures. Nonetheless, the microwave power output of the GaN diode is much less than that of the SiC diode. In addition, the former has 5 dB more noise than the latter at the optimum frequency. These two facts together are clear indication of the advantage of SiC over GaN for application as MITATT diode at sub-mm wave frequency. Such dismal

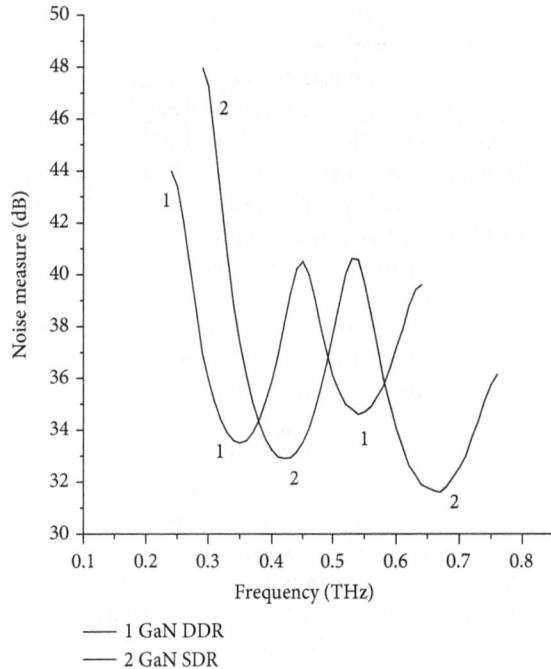

FIGURE 9: Noise measure characteristics of the GaN DDR and GaN SDR MITATT diodes.

performance of GaN is visibly attributable to the disparate carrier velocities. But as a silver lining in the dark cloud the disparate carrier velocities have unearthed a few interesting features of the GaN MITATT diodes. First, the presence of several peaks in the mean square noise voltage curves and the existence of several minima in the noise measure curves would open a new direction in the design of low-noise ATT diodes. Second, the presence of several peaks in the negative conductance plots will offer the GaN diodes with a multifrequency tuning facility like that in a DAR diode.

Appendix

Definitions of Symbols

A: Area of cross section (m^2)

$B(\omega)$:   Diode susceptance per unit area (Sm^{-2})

$B_d(\omega)$:   Diode susceptance (S)

E_0:   Peak electric field (Vm^{-1})

f_d: Design frequency (Hz)

f_g: Frequency corresponding to peak mean square noise voltage (Hz)

f_l: Frequency corresponding to minimum noise measure (Hz)

f_p: Optimum frequency, frequency at which ($-G$) attains peak (Hz)

$G(\omega)$:   Diode conductance per unit area at any frequency ω (Sm^{-2})

$G_d(\omega)$:   Diode conductance at any frequency ω (S)

J_T/J_0:   Ratio of tunnelling current to total current (%)

N_1:   Donor doping concentration (m^{-3})

N_2:   Acceptor doping concentration (m^{-3})

N_H: Doping level of substrate or epitaxy (m^{-3})

$N(x)$:   Impurity doping concentration at any point x, in the diode active layer (m^{-3})

$P_{RF}(\omega)$:   Power output at any frequency ω (W)

$R(\omega)$:   Integrated resistivity along the diode width at any frequency ω (Ωm^2)

$R_d(\omega)$:   Diode resistance at any frequency ω (Ω)

R_L: Load resistance (Ω)

R_S: Series resistance of the diode (Ω)

v_{sn}:   Saturated drift velocity of electron ($m\ s^{-1}$)

v_{sp}:   Saturated drift velocity of hole ($m\ s^{-1}$)

$\langle v^2 \rangle / df$:   Mean square noise voltage (V^2s)

V_B: Breakdown voltage of the diode (V)

W_n: Width of active layer on n-side (m)

W_p: Width of active layer on p-side (m)

x: Distance in the diode active layer (m)

x_j: Position of the junction (m)

$X(\omega)$:   Integrated reactivity along the diode width at any frequency ω (Ωm^2)

$X_d(\omega)$:   Diode reactance at any frequency ω (Ω)

X_L: Load reactance (Ω)

$Y(\omega)$:   Diode admittance per unit area at any frequency ω (Sm^{-2})

$Y_d(\omega)$:   Diode admittance at any frequency ω (S)

$Z_d(\omega)$:   Diode impedance at any frequency ω (Ω)

θ: Transit angle or transit time phase delay (rad)

τ: Transit time in the drift region of the diode (s)

η: Diode efficiency (%)

$\rho_R(x,\omega)$:   Real part of the diode resistivity at any space point x in the diode active layer and at frequency ω (Ωm)

$\rho_I(x,\omega)$:   Imaginary part of the diode resistivity at any space point x in the diode active layer and at frequency ω (Ωm).

Conflict of Interests

The authors declare that there is no conflict of interests regarding the publication of this paper.

References

[1] S. Banerjee, M. Mukherjee, and J. P. Banerjee, "Bias current optimisation of Wurtzite-GaN DDR IMPATT diode for high power operation at THz frequencies," *International Journal of Advanced Science and Technology*, vol. 16, pp. 11–19, 2010.

[2] B. Chakrabarti, D. Ghosh, and M. Mitra, "High frequency performance of GaN based IMPATT diodes," *International Journal of Engineering Science and Technology*, vol. 3, no. 8, pp. 6153–6159, 2011.

[3] A. Acharyya and J. P. Banerjee, "Prospects of IMPATT devices based on wide bandgap semiconductors as potential terahertz sources," *Applied Nanoscience*, vol. 4, no. 1, pp. 1–14, 2014.

[4] J. D. Albrecht, R. P. Wang, P. P. Ruden, M. Farahmand, and K. F. Brennan, "Electron transport characteristics of GaN for high temperature device modeling," *Journal of Applied Physics*, vol. 83, no. 9, pp. 4777–4781, 1998.

[5] I. H. Oğuzman, J. Kolník, K. F. Brennan, R. Wang, T.-N. Fang, and P. P. Ruden, "Hole transport properties of bulk zinc–blende and wurtzite phases of GaN based on an ensemble Monte Carlo calculation including a full zone band structure," *Journal of Applied Physics*, vol. 80, no. 8, pp. 4429–4436, 1996.

[6] G. N. Dash, P. Panda, and S. N. Padhi, "Effect of disparate carrier velocity in GaN on the terahertz characteristics of double drift region mixed tunnelling avalanche transit time diode," in *Proceedings of the IEEE International Conference on Electron Devices and Solid-State Circuits (EDSSC '15)*, pp. 800–803, Singapore, June 2015.

[7] B. Culshaw, R. A. Giblin, and P. A. Blakey, "Invited paper Avalanche diode oscillators III. Design and analysis: the future," *International Journal of Electronics*, vol. 40, no. 6, pp. 521–568, 1976.

[8] V. Stupelman and F. Filaretov, *Semiconductor Devices*, Mir Publishers, 1976.

[9] G. I. Haddad and R. J. Trew, "Microwave solid-state active devices," *IEEE Transactions on Microwave Theory and Techniques*, vol. 50, no. 3, pp. 760–779, 2002.

[10] G. N. Dash and S. P. Pati, "A generalized simulation method for MITATT-mode operation and studies on the influence of tunnel current on IMPATT properties," *Semiconductor Science and Technology*, vol. 7, no. 2, pp. 222–230, 1992.

[11] G. N. Dash, J. K. Mishra, and A. K. Panda, "Noise in mixed tunneling avalanche transit time (MITATT) diodes," *Solid-State Electronics*, vol. 39, no. 10, pp. 1473–1479, 1996.

[12] G. N. Dash, "A new design approach for MITATT and TUN-NETT mode devices," *Solid State Electronics*, vol. 38, no. 7, pp. 1381–1385, 1995.

[13] J. K. Mishra, G. N. Dash, and I. P. Mishra, "Simulation studies on the noise behaviour of double avalanche region diodes," *Semiconductor Science and Technology*, vol. 16, no. 11, pp. 895–901, 2001.

22

Multifrequency Oscillator-Type Active Printed Antenna Using Chaotic Colpitts Oscillator

Bibha Kumari and Nisha Gupta

Department of Electronics and Communication Engineering, Birla Institute of Technology, Mesra, Ranchi 835215, India

Correspondence should be addressed to Bibha Kumari; bibhakumari3@gmail.com

Academic Editor: Chien-Jen Wang

This paper presents a new concept to realize a multifrequency Oscillator-type active printed monopole antenna. The concept of period doubling route to chaos is exploited to generate the multiple frequencies. The chaotic Colpitts oscillator is integrated with the printed monopole antenna (PMA) on the same side of the substrate to realize an Oscillator-type active antenna where the PMA acts as a load and radiator to the chaotic oscillator. By changing the bias voltage of the oscillator, the antenna can be made to operate at single or multiple frequencies. To test the characteristics of the antenna at single and multiple frequencies of operation, two similar prototype models of printed monopole broadband antennas are developed. One of these antennas used at transmit side is fed by the chaotic Colpitts oscillator while the other is used as the receive antenna. It is observed that the antenna receives single or multiple frequencies simultaneously for particular values of the bias voltage of the oscillator at the transmit end.

1. Introduction

During the last decades, tremendous attention has been devoted to the design and applications of active antennas and chaotic circuits. The active antenna is an active microwave circuit in which the output or input port is free space instead of a conventional 50-Ω interface [1]. The integration of active devices with a radiating element forms an active antenna. The integration of active devices directly in the antenna structure is known as active integrated antenna (AIA). Depending on the functions of active devices used in the structure the active antennas are classified into three categories; amplifier type active antenna, Oscillator-type active antenna, and frequency conversion type active antenna [2]. The present work deals with the Oscillator-type active antenna employing a nonlinear device.

Most of the nonlinear devices exhibit chaotic modes of operation under certain parametric conditions. There are different routes to chaos. To achieve the multifrequency operation the concept of deterministic chaos exhibited by the chaotic Colpitts oscillator is used. Colpitts oscillator exhibits bifurcation phenomena for certain values of the circuit parameter and in this way it exhibits periodic, multiperiodic,

and chaotic motion [3]. The applications of chaotic circuits have been already reported in the past in chaotic communication, spread spectrum communication, low power communication, and so forth. In [4] a robust assessment for the Colpitts oscillator and the application of chaotic circuit in encrypted data transmission are presented. Over the last three decades different techniques have been used to analyze the behavior of an antenna with a nonlinear load. However, little work has been reported in the literature where such circuits are integrated with the antenna. Overfelt [5] has studied the effect of chaos on an electrically small dipole antenna operating at low frequency loaded with the nonlinear circuit known as a Chua's oscillator by assuming that the dipole could be modeled as a pure capacitance and an equivalent circuit of the combined antenna/load system is determined. However, the characteristic of the chaotic oscillators for multifrequency operation is unexploited. When this chaotic oscillator is integrated with the printed patch antenna on the same substrate, it can be viewed as an active integrated antenna [2]. An active antenna (Oscillator-type) consists of two terminal nonlinear devices such as Tunnel diode, Gunn diode, or three terminal devices such as BJT or FET loaded with printed patch antenna. Under certain parametric conditions these oscillators may exhibit chaotic phenomena due to presence of nonlinear elements in the circuit. The output of such oscillators exhibits sinusoidal oscillations initially and by changing a single parameter of the system the output period bifurcates. This process of bifurcation continues further if the parameter value is changed. As the parameter value changes, due to bifurcation, the period of the output signal becomes 2, 4, 8, 16, and so forth, and it continues till no more stable state is obtained and the system goes into the chaotic mode of operation. Chaotic oscillators have wide spectrum properties and this property can be used to design a multifrequency antenna. Multifrequency antennas are antennas which can operate at several frequencies at a particular instant of time. In this paper, for the first time the chaotic mode of operation of these nonlinear oscillators is exploited in generating multiple frequencies in the GHz range to be used in printed patch antennas for multifrequency operation. The designed Oscillator-type active antenna can be made to operate on single, dual, quadruple, octal frequencies, and so forth, or in the chaotic mode of operation by changing the bias voltage of the oscillator. When the oscillator operates in multiple frequency modes, it can be used in realizing multifrequency antennas; otherwise in the chaotic mode of operation it can be used in secure communication [4].

The paper is organized as follows. Section 2 presents the simulation and experimental study of the microwave chaotic Colpitts oscillator and its route to chaos. Section 3 presents the design of a printed rectangular monopole antenna along with the simulation and experimental results. Section 4 presents the design of an Oscillator-type active antenna by integrating the chaotic Colpitts oscillator with printed monopole antenna (PMA) to realize the antenna in multifrequency mode of operation. Conclusions are discussed in Section 5.

2. Microwave Colpitts Oscillator

The classical Colpitts oscillator is commonly used to generate sinusoidal signals. However with a certain set of circuit parameters one can generate chaotic waveform across the capacitors. Chaos in the Colpitts oscillator has been first reported by Kennedy [6] and, subsequently, various types of chaotic Colpitts oscillator circuits operating at both low and microwave frequencies have been realized [7–12]. The active device used in the oscillator is the BFG425W NPN bipolar transistor in CB configuration. The resonance loop has three energy storing elements: an inductor L and two capacitors C_1 and C_2. The parametric values are $C_1 = 3.3\,\text{pF}$, $C_2 = 3.3\,\text{pF}$, $R = 391\,\Omega$, $C_3 = 10\,\text{pF}$, and $L = 3.3\,\text{nH}$. Here C_3 is the coupling capacitor. The fundamental frequency of oscillation of the designed oscillator is 2.2 GHz. The fundamental frequency of oscillation of the circuit can be estimated by [8]

$$f_0 = \frac{1}{2\pi}\sqrt{\frac{C_1 + C_2}{C_1 C_2 L}}. \tag{1}$$

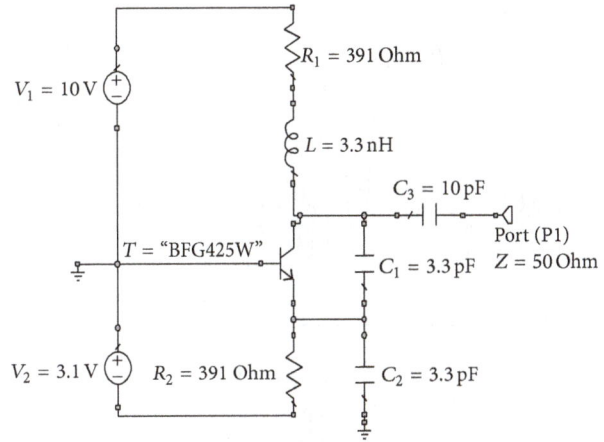

FIGURE 1: Schematic of microwave Colpitts oscillator.

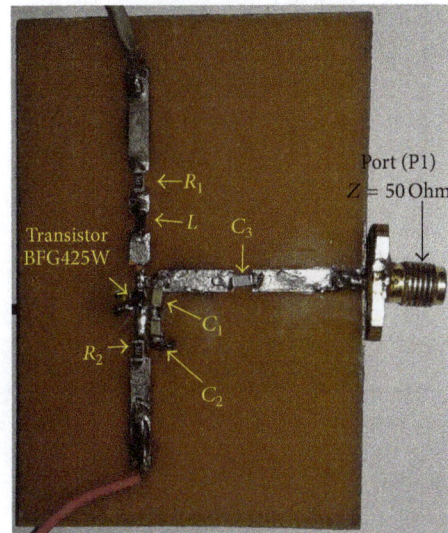

FIGURE 2: Microwave Colpitts oscillator fabricated on FR4 substrate.

The schematic of microwave Colpitts oscillator [12] is shown in Figure 1. The fabricated structure of microwave chaotic Colpitts oscillators on FR4 substrate is shown in Figure 2.

The chaotic Colpitts oscillator shown in Figure 2 exhibits different dynamical behavior for different values of emitter bias voltage V_2. In the circuit all parameters have been kept constant and the supply voltage V_2 is varied to observe the different dynamical behavior under periodic mode, period-doubled mode, and chaotic mode of operations. The simulation and experimental results for these modes of operation of the microwave chaotic Colpitts oscillator are shown in Figures 3 and 4.

2.1. Simulation Results. See Figure 3.

2.2. Experimental Results. The experimental results showing the behavior of microwave chaotic Colpitts oscillator are also presented for various modes of operations (see Figure 4).

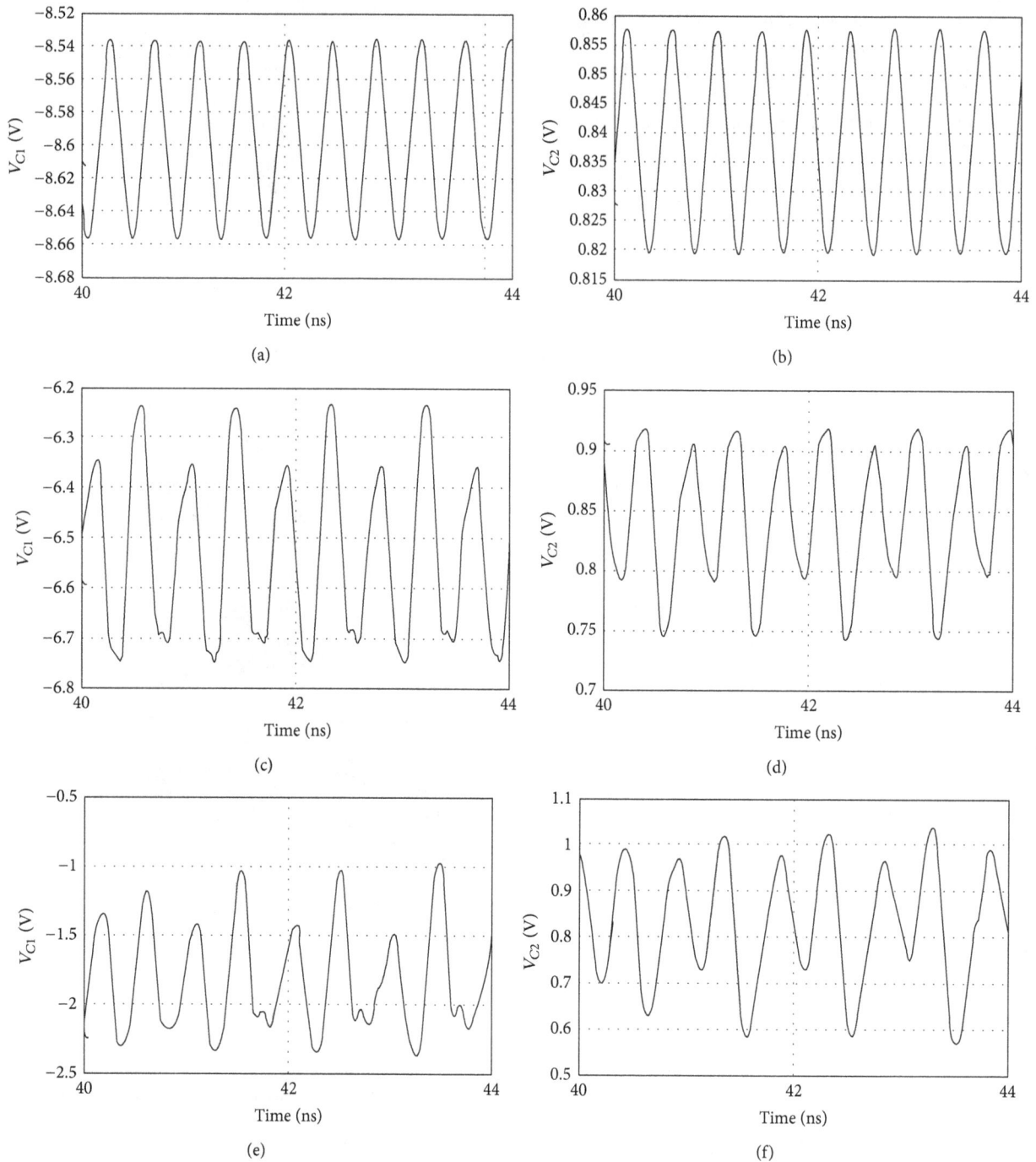

FIGURE 3: Time series plot of V_{C1} and V_{C2} with respect to emitter bias voltage V_2 of chaotic Colpitts oscillator: (a) and (b) periodic mode for $V_2 = 3.1$ V, (c) and (d) period-doubled mode for $V_2 = 5.2$ V, and (e) and (f) chaotic mode for $V_2 = 7$ V.

3. Printed Monopole Antenna

The selection of antenna is of prime importance in the active integrated approach. Planar type antennas such as the patch or slot are considered good for minimizing interconnects, as they are suitable for direct integration with microstrip or coplanar waveguide (CPW). The microwave Colpitts oscillator is designed at 2.2 GHz. This oscillator can generate several

frequencies following the route to chaos by varying the emitter bias voltage V_2. Further to transmit these frequencies a wide band antenna is required. Hence, a printed rectangular monopole antenna is selected and designed to achieve the broadband characteristics. The antenna parameters for the desired band of frequency are obtained as [13]. The patch dimensions are selected as follows: length of patch $(L) = 11.33$ mm, width of patch $(W) = 15.21$ mm, width of the feed

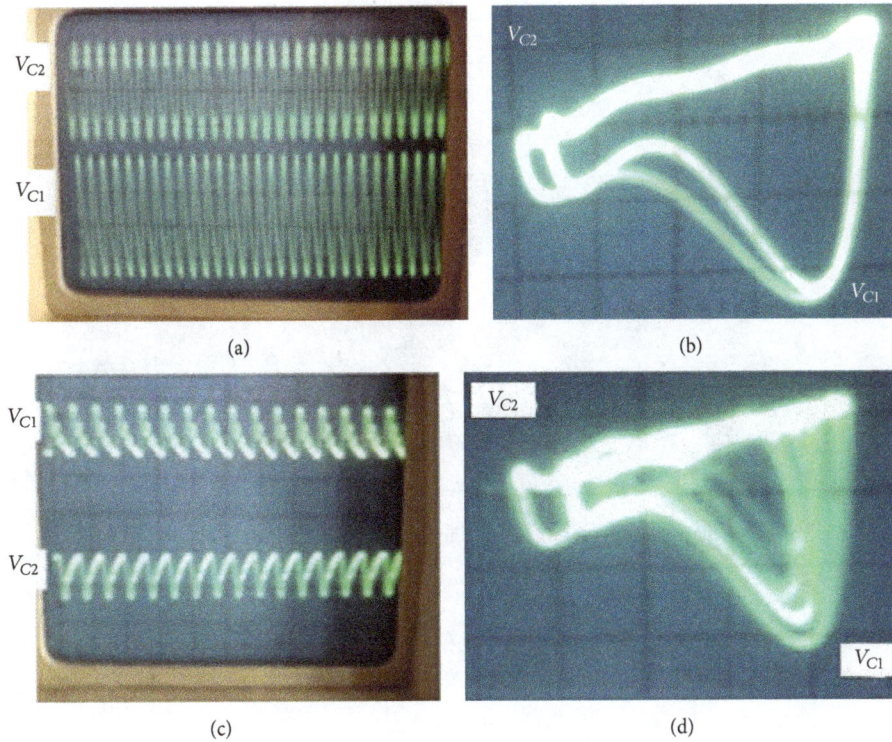

FIGURE 4: Oscilloscope display of time series plot of V_{C1} and V_{C2} and phase portrait of V_{C2} versus V_{C1}. (a) Periodic mode for $V_2 = 3.1$ V, (b) period-doubled mode, X-Y mode: two loops can be seen for $V_2 = 5.2$ V, and (c) and (d) chaotic mode for $V_2 = 7$ V.

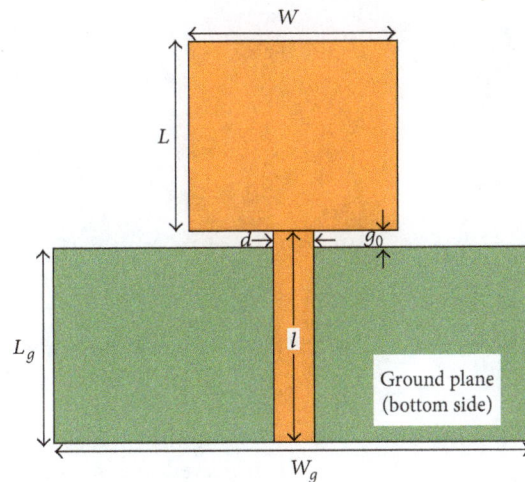

FIGURE 5: A rectangular monopole antenna structure.

line (d) = 3 mm, distance between patch and the ground plane (g_0) = 1 mm, length of the feed line (l) = 12.5 mm, length of the ground plane (L_g) = 11.5 mm, and width of the ground plane (W_g) = 35.21 mm. Figure 5 shows the geometry of printed rectangular monopole antenna structure. The antenna uses FR4 substrate of thickness 1.6 mm, relative dielectric constant as 4.4, and the loss tangent as 0.002. The designed printed rectangular monopole antenna is simulated using the IE3D electromagnetic simulator from Zeland, Inc., USA. The lumped equivalent circuit of the proposed printed rectangular monopole antenna is obtained as presented in [14, 15]. The lumped equivalent circuit model is simulated in the advanced design system (ADS) software. The prototype model and the lumped equivalent of the proposed printed rectangular monopole antenna are shown in Figures 6 and 7. A comparison of the reflection coefficient (S_{11}) parameter of simulation result obtained from IE3D EM simulator, experimental result, and simulation result of lumped equivalent of proposed antenna obtained in ADS is shown in Figure 8.

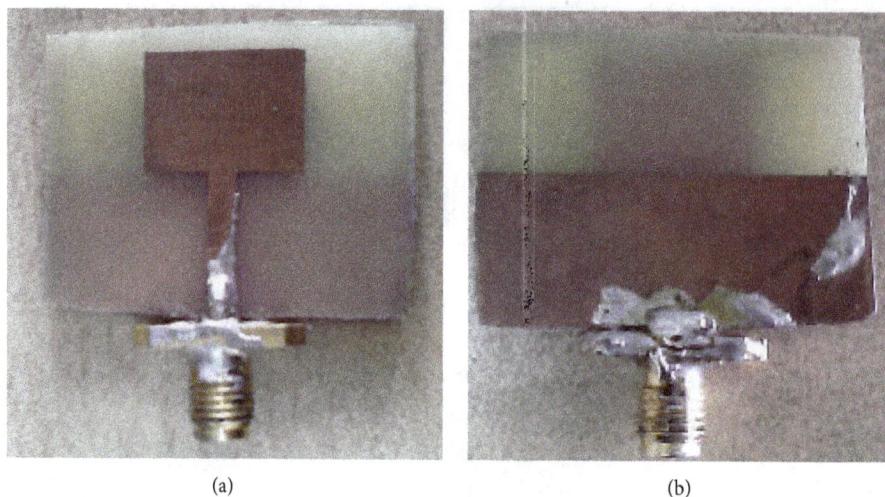

FIGURE 6: Prototype model of printed monopole antenna (a) top view and (b) bottom view.

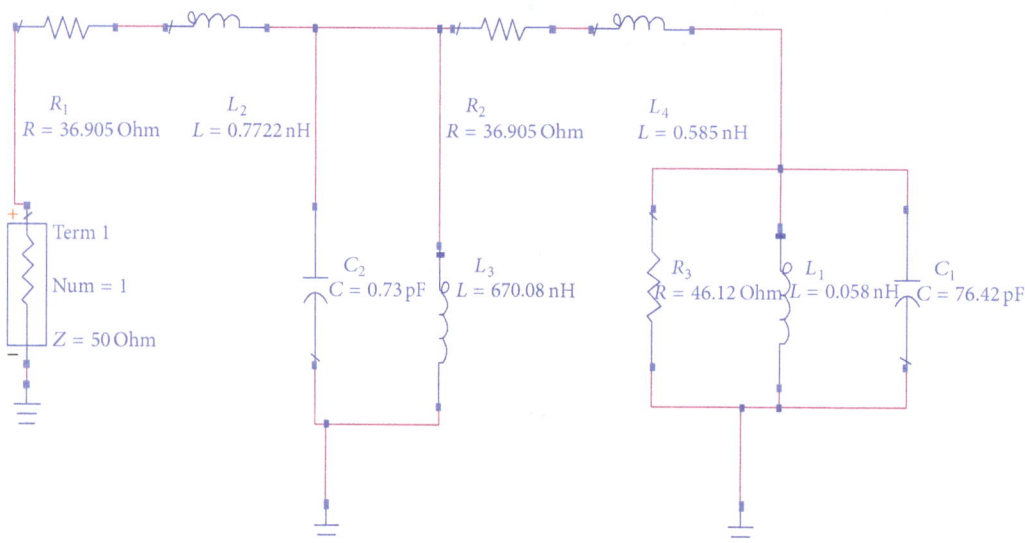

FIGURE 7: Lumped equivalent circuit model of the proposed printed monopole antenna.

4. Design of Oscillator-Type Active Antenna

Active devices can easily be integrated with printed antenna on a common substrate. Several different microwave oscillator circuits integrated with printed antenna have been reported in the literature where active nonlinear devices such as Gunn diode and IMPATT diode [16–26] have been used. Haskins et al. [26] have designed an active antenna using the Schottky diode as a tuning device and it is shown that the oscillation frequency can be varied by means of a bias voltage applied to the tuning device. In the present work, the concept of the oscillator loaded antenna is first demonstrated using the AWR Microwave Office simulation tool. Here a microstrip antenna (MSA) represented by a lumped equivalent circuit is loaded with a Colpitts oscillator and is shown in Figure 9.

From the simulation study carried out in this section, it is observed that for certain values of bias voltage, various types of periodic, intermittent, and chaotic state of operation occur across the lumped equivalent of MSA since the antenna is fed by a chaotic Colpitts oscillator. However, the narrow band characteristics of the MSA restrict the depiction of all the intermittent states of route to chaos, and only the few frequency components falling within the bandwidth of the MSA can be observed. Therefore to accommodate all the frequencies generated by the chaotic oscillator in chaotic mode of operation, a broadband printed monopole antenna [27] is preferred. Next, the chaotic Colpitts oscillator is integrated with the printed monopole antenna on the same side of the FR4 substrate. The prototype model of the Oscillator-type active printed antenna where the chaotic Colpitts oscillator and printed monopole

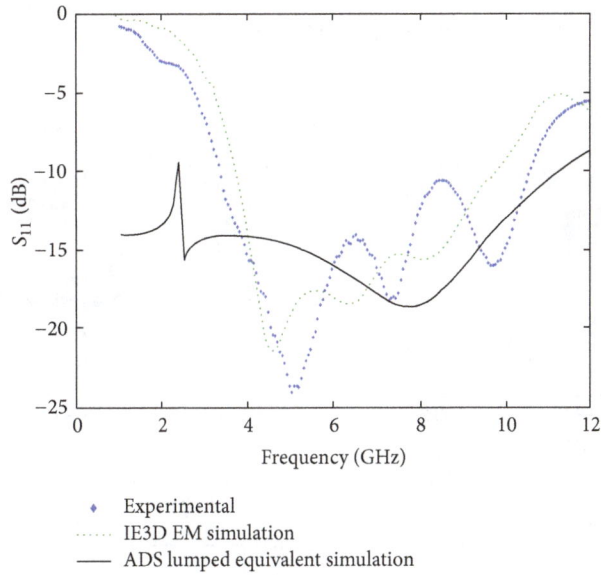

FIGURE 8: Comparison of results.

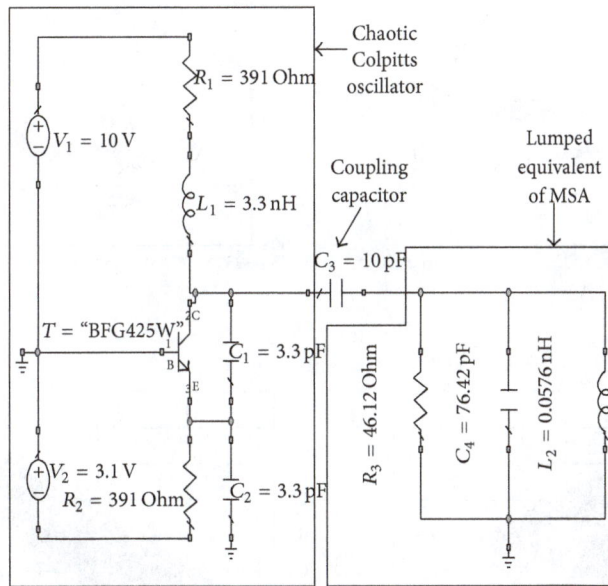

FIGURE 9: Schematic showing the integration of chaotic Colpitts oscillator and the lumped equivalent of MSA.

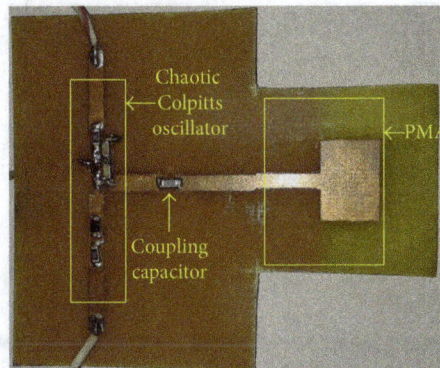

FIGURE 10: Prototype model of Oscillator-type active antenna fabricated on FR4 substrate.

(a)

(b)

(c)

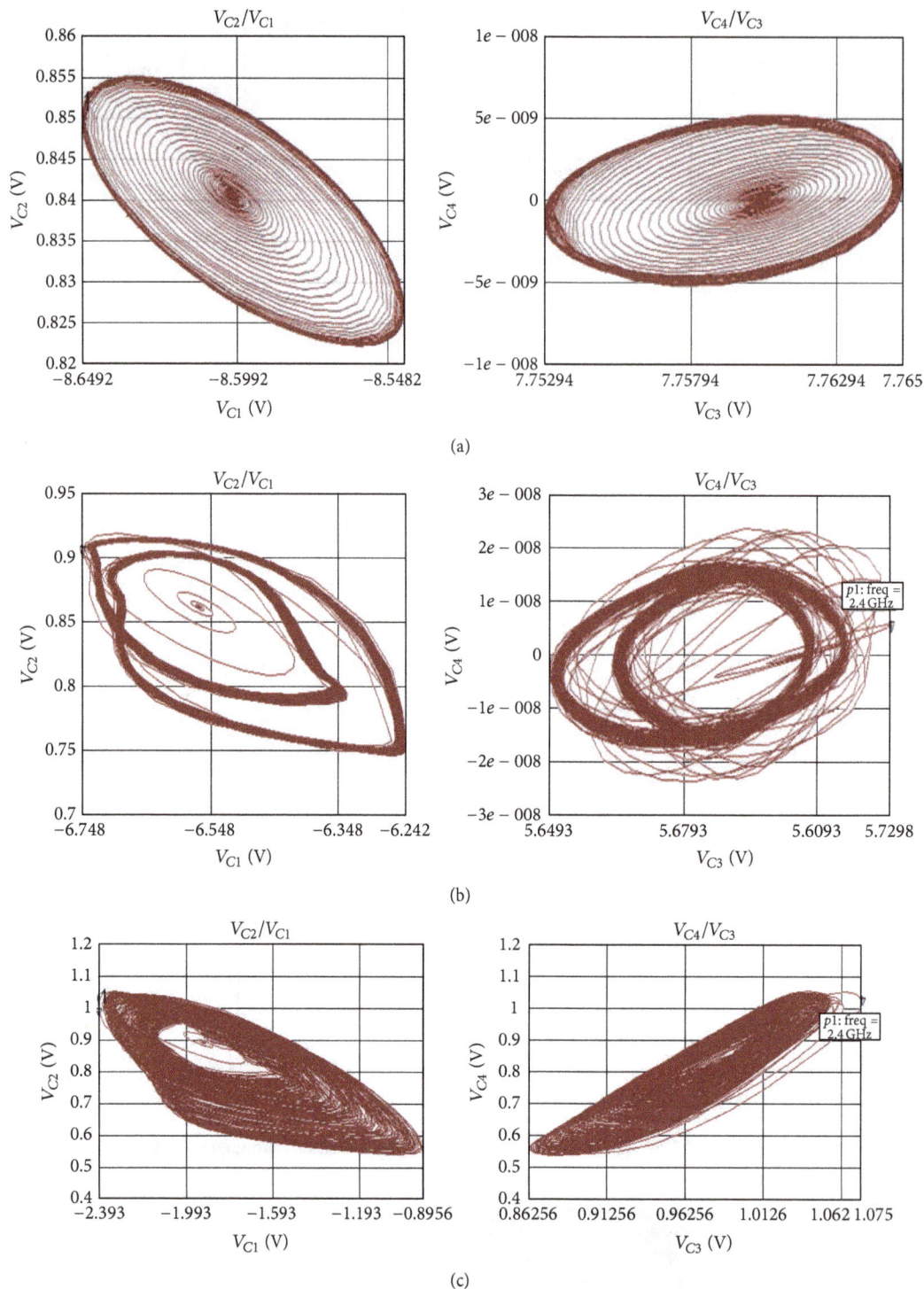

FIGURE 11: Phase portrait of V_{C2} versus V_{C1} and V_{C4} versus V_{C3} by changing the emitter biasing voltage V_2 of chaotic Colpitts oscillator (a) periodic mode for $V_2 = 3.1$ V, (b) intermittent state for $V_2 = 5.2$ v, and (c) chaotic mode for $V_2 = 7$ V.

antenna are integrated over a single substrate is shown in Figure 10.

4.1. Simulation Results. The phase portraits obtained in the simulation are shown in Figure 11. The phase portrait shows the graph of V_{C2} versus V_{C1} and V_{C4} versus V_{C3}. Here V_{C1},

V_{C2}, V_{C3}, and V_{C4} are the voltages across capacitor C_1, C_2, C_3, and C_4. Capacitors C_1 and C_2 are the frequency determining components of the Colpitts oscillator; C_3 is the coupling capacitor and the capacitor C_4 is one of the lumped equivalent components of MSA. The bifurcation parameter is the voltage source V_2. Figure 11(a) shows the periodic mode of operation

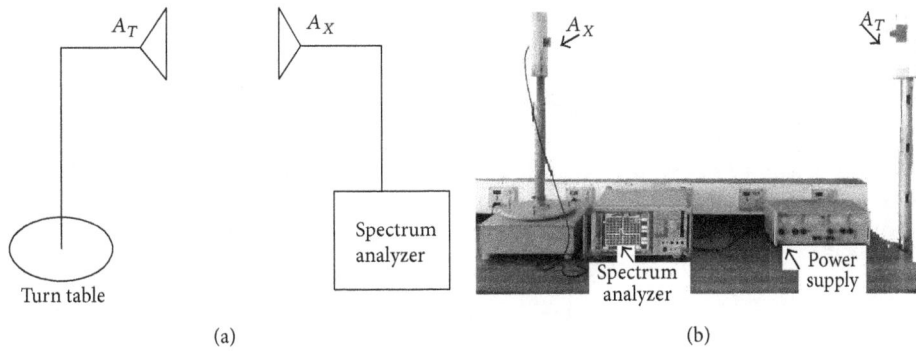

FIGURE 12: Experimental setup.

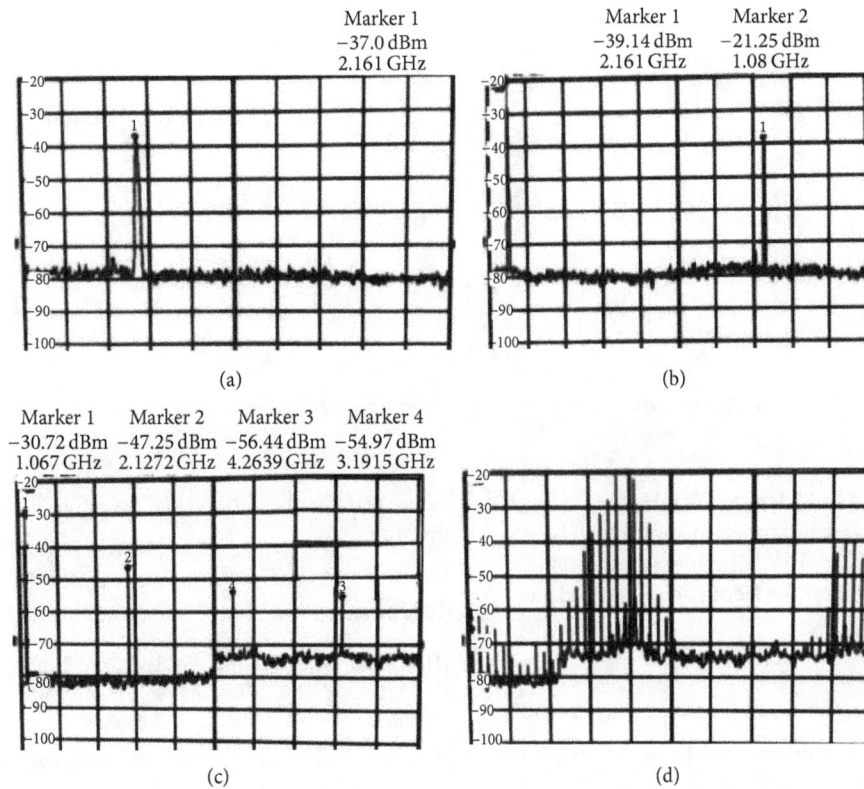

FIGURE 13: Spectrum analyzer display of Oscillator-type active antenna (a) periodic mode, (b) period-doubled mode, (c) period-four mode, and (d) chaotic mode.

for V_2 = 3.1 V as single loop can be seen. Figure 11(b) shows an intermittent state of operation for V_2 = 5.2 V as two loops falling within the BW of MSA can be seen. Figure 11(c) shows the chaotic mode of operation for V_2 ranges from 7 to 10 V, where a large number of frequency components falling within the BW of MSA are observed.

4.2. Experimental Results. By changing the bias voltage of the oscillator the antenna can be made to operate at single or multiple frequencies. Here the bifurcation parameter is the emitter bias voltage source V_2 of the chaotic Colpitts oscillator. Figure 12 shows the experimental setup for the

measurement of different frequencies. In Figure 12 A_T is the Oscillator-type active antenna which is the integration of chaotic Colpitts oscillator and PMA. A_X is the PMA only. To accommodate all the frequencies transmitted by this active antenna, a similar broadband PMA is used at the receiving side which is connected to the spectrum analyzer. The spectrum analyzer display is shown in Figure 13. Figure 13(a) shows the periodic mode of operation for V_2 = 3.1 as the single frequency component at 2.161 GHz can be seen. Figure 13(b) shows the period-doubled mode of operation for V_2 = 5.2 V as the two frequency components at 1.08 GHz and 2.161 GHz can be seen. Figure 13(c) shows the period-four mode of operation for V_2 = 8.9 V as the four

(a) E-Plane

(b) H-Plane

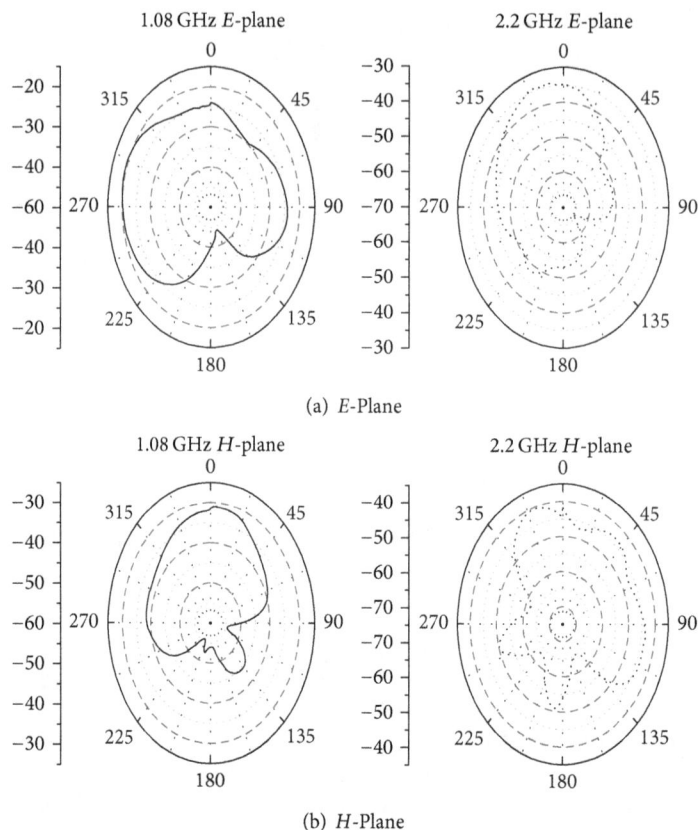

FIGURE 14: Experimental results: showing the measured radiation pattern.

frequency components at 1.067 GHz, 2.1272 GHz, 3.1915 GHz, and 4.2639 GHz can be seen. Figure 13(d) shows the chaotic mode of operation for $V_2 = 10$ V. The measured radiation pattern in E-plane and H-plane for frequencies 1.08 GHz and 2.2 GHz is shown in Figure 14.

5. Conclusion

The Oscillator-type active antenna has been designed by using the chaotic Colpitts oscillator. The multifrequency operation of the antenna is exploited based on the theory of route to chaos. Both simulated and experimental results confirm the validity of the proposed technique. It is evident that the antenna can be made to transmit multiple frequencies simultaneously by setting the bias voltage of the chaotic oscillator to a certain value. The proposed technique is found to be suitable for wireless devices.

Conflict of Interests

The authors declare that there is no conflict of interests regarding the publication of this paper.

Acknowledgment

The authors wish to acknowledge the financial support received from University Grant commission, New Delhi, India, for carrying out this research under SAP II programme.

References

[1] K. Chang, R. A. York, P. S. Hall, and T. Itoh, "Active integrated antennas," *IEEE Transactions on Microwave Theory and Techniques*, vol. 50, no. 3, pp. 937–944, 2002.

[2] I. Bahl and P. Bhartia, *Microstrip Antennas*, Artech House, 1980.

[3] D. F. Oscer and G. M. Maggio, "Bifurcation phenomena in the colpitts oscillator: robustness analysis," in *Proceedings of the IEEE International Symposium on Circuits and Systems (ISCAS '00)*, vol. 2, pp. 469–472, Geneva, Switzerland, 2000.

[4] G. Lombardo, G. Lullo, and R. Zangara, "Experiments on chaotic circuits and crypted data transmission," in *Proceedings of the 8th IEEE International Conference on Electronics, Circuits and Systems (ICECS '01)*, vol. 1, pp. 461–464, September 2001.

[5] P. L. Overfelt, "An electrically small dipole antenna loaded with Chua's oscillator: time-retarded chaos," in *Proceedings of the IEEE Antennas and Propagation Society International Symposium*, vol. 4, pp. 2872–2875, Orlando, Fla, USA, 1999.

[6] M. P. Kennedy, "Chaos in the Colpitts oscillator," *IEEE Transactions on Circuits and Systems I: Fundamental Theory and Applications*, vol. 41, no. 11, pp. 771–774, 1994.

[7] G. Mykolaitis, A. Tamaševičius, and S. Bumeliene, "Experimental demonstration of chaos from Colpitts oscillator in VHF and UHF ranges," *Electronics Letters*, vol. 40, no. 2, pp. 91–92, 2004.

[8] S. Zhigno, R. Lixin, and C. Kangsheng, "Simulation and experimental study of chaos generation in microwave band using colpitts circuits," *Journal of Electronics (China)*, vol. 23, no. 3, pp. 433–436, 2006.

[9] S. Wei and S. Donglin, "Microwave chaotic colpitts circuit design," in *Proceedings of the 8th International Symposium on Antennas, Propagation and EM Theory (ISAPE '08)*, pp. 1127–1130, Kunming, China, November 2008.

[10] B. E. Kyarginsky and N. A. Maxico, "Wideband microwave chaotic oscillators," in *Proceedings of the IEEE 1st International Conference on Circuits & Systems for Communications (ICCSC '02)*, pp. 296–299, 2002.

[11] A. Panas, B. Kyarginsky, and N. Maximov, "Single-transistor microwave chaotic oscillator," in *Proceedings of International Symposium on Nonlinear Theory & Its Applications*, vol. 2, pp. 445–448, 2000.

[12] B. Kumari and N. Gupta, "Effect of radiated electromagnetic interference on colpitts oscillators," *Proceedings of the 29th Annual Review of Progress in Applied Computational Electromagnetics*, March 2013.

[13] C. A. Balanis, *Antenna Theory: Analysis and Design*, John Wiley & Sons, New York, NY, USA, 1996.

[14] M. H. Badjian, C. K. Chakrabarty, and S. Devkumar, "Circuit modeling of an UWB patch antenna," in *Proceedings of the IEEE International RF and Microwave Conference (RFM '08)*, Kuala Lumpur, Malaysia, December 2008.

[15] S. Jangid and M. Kumar, "An equivalent circuit of UWB patch antenna with band notched characteristics," *International Journal of Engineering and Technology*, vol. 3, no. 2, 2013.

[16] N. Wang, S. E. Schwarz, and T. Hierl, "Monolithically integrated Gunn oscillator at 35 GHz," *Electronics Letters*, vol. 20, no. 14, pp. 603–604, 1984.

[17] Z. Ding, L. Fan, and K. Chang, "New type of active antenna for coupled Gunn oscillator driven spatial power combining arrays," *IEEE Microwave and Guided Wave Letters*, vol. 5, no. 8, pp. 264–266, 1995.

[18] R. A. York and R. C. Compton, "Dual-device active patch antenna with improved radiation characteristics," *Electronics Letters*, vol. 28, no. 11, pp. 1019–1021, 1992.

[19] W. W. Lam, Y. C. Ngan, and Y. Saito, "Millimeter-wave active patch antenna," in *Proceedings of the Antennas and Propagation Society International Symposium*, vol. 2, pp. 791–794, 1990.

[20] C.-C. Huang and T.-H. Chu, "Radiating and scattering analyses of a slot-coupled patch antenna loaded with a MESFET oscillator," *IEEE Transactions on Antennas and Propagation*, vol. 43, no. 3, pp. 291–298, 1995.

[21] J. Bartolic, D. Bonefacic, and Z. Sipus, "Modified rectangular patches for self-oscillating active-antenna applications," *IEEE Antennas and Propagation Magazine*, vol. 38, no. 4, pp. 13–21, 1996.

[22] G.-J. Chou and C.-K. C. Tzuang, "Oscillator-type active-integrated antenna: the leaky-mode approach," *IEEE Transactions on Microwave Theory and Techniques*, vol. 44, no. 12, pp. 2265–2272, 1996.

[23] A. Bhattacharya and P. Lahiri, "New design philosophy of active microstrip patch antenna," in *Proceedings of the Antennas and Propagation Society International Symposium*, vol. 3, pp. 1284–1287, 2000.

[24] P. S. Hall, "Analysis of radiation from active microstrip antennas," *Electronics Letters*, vol. 29, no. 1, pp. 127–129, 1993.

[25] F. Giuppi, A. Georgiadis, M. Bozzi, S. Via, A. Collado, and L. Perregrini, "Hybrid electromagnetic and non-linear modeling and design of SIW cavity-backed active antennas," *Applied Computational Electromagnetics Society Journal*, vol. 25, no. 8, pp. 682–689, 2010.

[26] P. M. Haskins, P. S. Hall, and J. S. Dahele, "Active patch antenna element with diode tuning," *Electronics Letters*, vol. 27, no. 20, pp. 1846–1848, 1991.

[27] R. S. Kshetrimayum, "Printed monopole antennas for multiband applications," *International Journal of Microwave and Optical Technology*, vol. 3, no. 4, 2008.

23

Turn Ratio, Substrates' Permittivity Characterization, and Analysis of Split Ring Resonator Based Antenna

Seyi S. Olokede,[1] Nurul A. Mohd-Razif,[2] and Nor M. Mahyuddin[2]

[1]*Department of Electrical & Electronic Engineering, Olabisi Onabanjo University, PMB 5026, Ifo, Ogun State, Nigeria*
[2]*School of Electrical & Electronic Engineering, Universiti Sains Malaysia, 14300 Nibong Tebal, Penang, Malaysia*

Correspondence should be addressed to Seyi S. Olokede; solokede@gmail.com

Academic Editor: Walter De Raedt

The turn ratio, coupling space between sections, and substrate permittivity effects on spilt ring resonator (SRR) are investigated. The analysis of the presented SRR with respect to the effects of substrate and number of gaps per ring to further characterize its peculiarities is experimented with miniaturized capability as our intent. Six different SRRs were designed with different turn ratios, and the sixth is rectangular microstrip patch centre-inserted. Different numbers and gap sizes are cut on the SRRs while the gap spacing between the conductors of the SRR was varied to determine their effects taking cognizance of the effects of different substrates. The designs were investigated numerically using 3D finite integration technique commercial EM solver, and the resulting designs were prototyped and subsequently measured. Findings indicate that the reflection coefficient of the MSRR with centre-inserted patch antenna is better compared to MSRR without the patch antenna irrespective of the laminate substrate board, and so is its gain.

1. Introduction

The explosive growth in the demand for wireless communication and information transfer using mobile phone, satellites, and personal communications devices has created the need for advanced antenna design technology. Microstrip planar antennas are smart solutions for compact and cost effective wireless communication systems, in particular due to its significant attractive features of low profile, light weight, easy fabrication, low volume, low profile, ease of integration with printed circuit boards, low power handling capability of printed circuits, and conformability to mounting hosts [1]. Thus, these types of antennas are flexible due to their conformability. As such, smaller antennas are feasible based on the electrodynamics of the patch antenna, though they are limited by their electrical length, in particular the conventional patch antenna.

SRR and its derivatives have been one of the burning issues on antenna miniaturization at microwave and millimeter wave systems in recent times. One very notable attribute that makes SRRs stands out is the increasing need for the optimum usage of space in modern microwave circuits. High Q, low cost, and low radiation loss are extra additives that makes them common technology for filter designs [2–6], antenna design [7–11], and recently applicable in metamaterials [12, 13].

Theoretically, SRR structure exhibits the most common negative permeability characteristics. When the alternating magnetic field is applied perpendicular to the SRR plane, the electromagnetic feature of the SRR can be excited by a time-varying magnetic field with a nonnegligible component applied parallel to the ring axis. This will form a resonant circular current in the ring or rings and hence dictate the resonant frequency. It therefore behaves like a magnetic field driven by inductor and capacitor (*LC*) resonant tanks that are externally driven by the magnetic field [14].

Hence in a rectangular multiple SRR (MSRR), the conductor trace with opposite sides of the SRR may be treated as coupled line sections. The total inductance is the sum of self-inductance of the sections and the mutual inductance between the turns on assumption that the magnitude and phase of the current across the sections are constant.

On the other hand, SRR has maximum electric field density near the slit and, interestingly, the nature and the intensity of the fringing fields between the gaps, and also the number of rings determines the nature and the strength of the coupling. Ironically, the SRR has the maximum electric charge densities at the gap edges with opposite signs, whereas the current flow is optimum along the ring ribs opposite to the rib.

Therefore the effect of the magnetic field excitation at the perpendicular plane of the SRR with respect to its electromagnetic feature vis-à-vis, its resonance frequency, and the spacing between the parallel coupled sections and the effect of electric field intensity near the slits are examined in this work. Much more, the nature and the extent of the fringing field with respect to different dielectric substrate are investigated.

2. Materials and Methods

The proposed SRR structure consists of N number of concentric split ring resonators to obtain magnetic resonances at distinct frequencies. The value of each frequency can be adjusted by changing the design parameters such as the metal width (w) and gap size (g) for each ring as well as ring to interrings spacing (d). As a result, increasing the number of rings (N) will increase the resonance frequency. Self-inductance also increases when the side length of the metal ring increases with a corresponding decrease in the LC resonance frequency of the resonator, as corroborated by Turkmen et al. [15].

The frequency response of the SRR is obtained using (1), where L_T represent the total equivalent inductance while C_{eq} represent the equivalent capacitance. The L_T value is obtained by using (2a), whereas C_{eq} is obtained using (2b) [16]:

$$f_0 = \frac{1}{2\pi \left(L_T C_{eq} \right)^{0.5}}, \tag{1}$$

$$L_T = 2 \times 10^{-4} \ell \left(2.303 \log_{10} \frac{4\ell}{c} - \gamma \right), \tag{2a}$$

$$C_{eq} = \frac{1}{2} \left(C_0 + C_g \right), \tag{2b}$$

where $\ell = 8a_{ext} - g$, $C_0 = (4a_{avg} - g)C_{pul}$, and, $a_{avg} = a_{ext} - w - d/2$. C_0 is the series capacitance between two adjacent rings, C_{pul} is the capacitance per unit length of the ring, γ constant (= 2.853) for wire loop of square geometry, and g is the slit gap, whereas d is the gap between the rings. The equation to calculate C_{pul} is stated in (3) while (4) defines the gap capacitance (C_g) between the slits, whereas ε_r is the relative dielectric permittivity of the substrate, Z_0 is the characteristic impedance (set at 50 Ω), and c is the speed of light

$$C_{pul} = \frac{(\varepsilon_r)}{cZ_0}, \tag{3}$$

$$C_g = \frac{\varepsilon_0 w}{g}. \tag{4}$$

The dimensions of the SRR were determined using the above equations with g = 3 mm, d = 0.3 mm, and w = 0.5 mm.

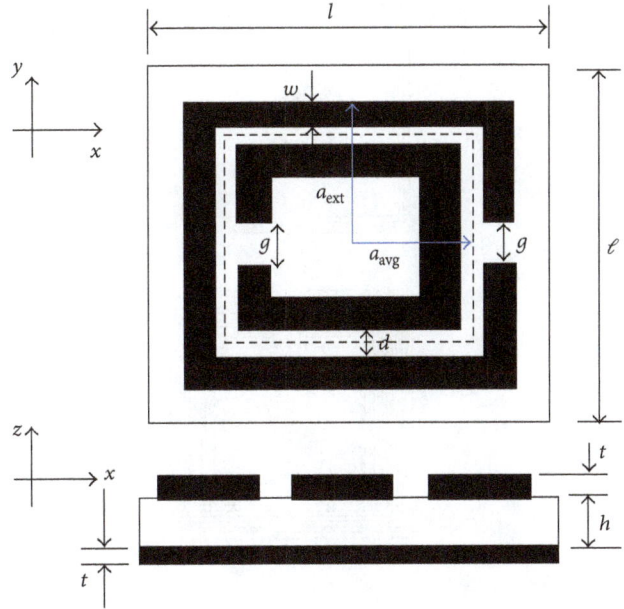

FIGURE 1: Topology of the proposed SRR.

Figure 1 depicts the geometry of the SRR, Figure 2 is the proposed designs simulated using 3M EM microwave solver, and Figure 3 is the fabricated designs of the propose design. The optimized prototypes of the proposed were photoetched on (1) Roger duroid laminate microwave substrate of permittivity (ε_r) of 3.38, height of 0.813 mm, and metallization of 35 μm; (2) duroid permittivity (ε_r) of 6.45, height of 2.5 mm, and metallization of 35 μm; (3) FR4 epoxy substrate of permittivity (ε_r) of 2.4, height of 1.6 mm, and metallization of 35 μm; and, finally, (4) FR4 of permittivity (ε_r) of 4.4, height of 5 mm, and metallization of 35 μm. The resulting designs were tested using the HP 8720D (50 MHz–20 GHz) network analyzer in order to measure the reflection coefficients, the antenna impedance, and the voltage standing wave ratio (VSWR). For this measurement process, HP 8720D (50 MHz–20 GHz) network analyzer was used to measure the reflection coefficient, resonant frequency, bandwidth, and input impedance. For bandwidth measurement, −10 dB point was used to determine the range of frequency. Prior to measurement, a kit is required to calibrate the network analyzer at the frequency required for one port only. The recorded data from the network analyzer measurement was plotted and compared with simulated results. The connection configuration of the network analyzer is shown in Figure 4(a).

The radiation pattern characteristics of these fabricated antennas represent graphically their radiation properties as a function of space coordinates. These properties are the power density, radiation intensity, field strength, directive phase, and polarization.

The measurement of the radiation patterns was done using HP 83620B (10 MHz to 20 GHz) signal generator, transmitting antenna, a rotating machine, and Agilent 8565E (9 KHz–50 GHz) spectrum analyzer, whereas each antenna under measurement (AUT) was attached to the rotating machine where the AUT acts as a receiving antenna.

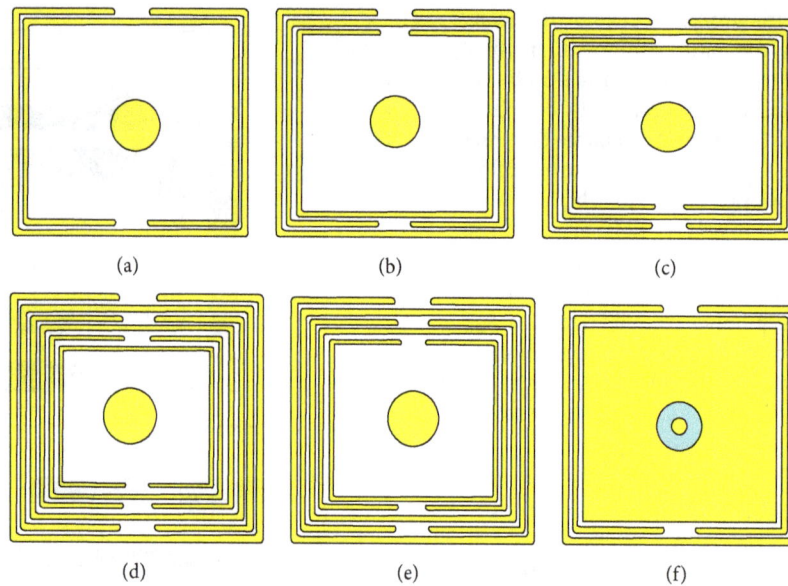

FIGURE 2: The proposed SRR designs (a) $N = 2$, (b) $N = 3$, (c) $N = 4$, (d) $N = 5$, (e) $N = 6$, and (f) $N = 2$ patch inserted.

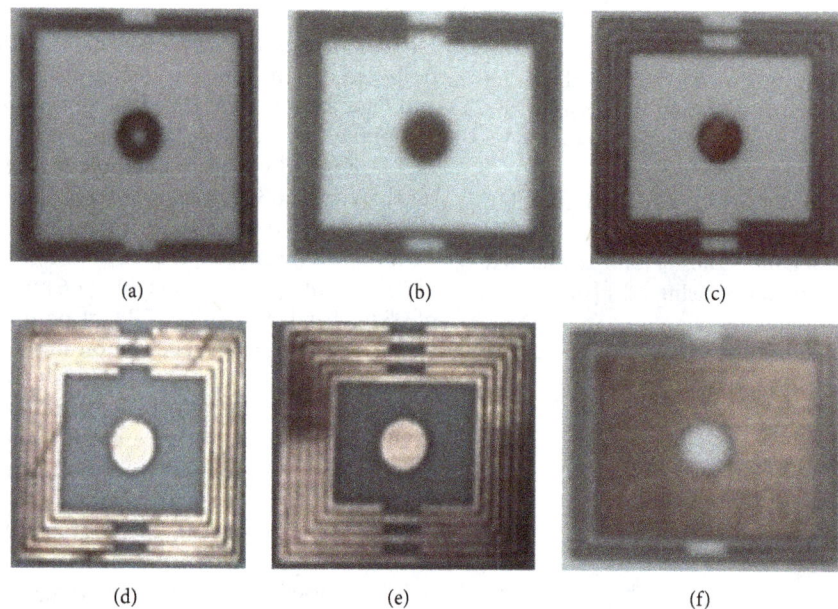

FIGURE 3: The fabricated SRR designs (a) $N = 2$, (b) $N = 3$, (c) $N = 4$, (d) $N = 5$, (e) $N = 6$, and (f) $N = 2$ centre patch inserted.

The transmitting antenna used is a standard dipole antenna. Both receiving and transmitting antenna were placed such that both of them were aligned with each other. The normalized input power data of each 10° was subsequently used to plot radiation pattern on the polar graph for each of the designed antennas. The measured radiation patterns were finally compared with the radiation patterns from the 3D EM simulated results. The measurements were done two times for each antenna design. First, the copolarization in the E plane was measured. Both the receiving and transmitting antennas were placed in vertical positions where the receiving antenna was rotated 360° while the measured data was

recorded. The next step was to measure the copolarization in the H-copolarization. Both receiving and transmitting antennas were placed in horizontal positions. Then the receiving antenna was rotated 360° while the measured data was recorded. Then, normalized input power data for each 10° was used to plot radiation pattern on the polar graph for each of designed antennas. The testing arrangement for gain measurement is the same as in Figure 4(b). According to IEEE Standard Test Procedures for Antennas, ANSI/IEEE Std 149-1979, the most commonly employed method for antenna gain measurement is gain-transfer (gain-comparison) method. This technique utilizes a gain standard antenna to determine

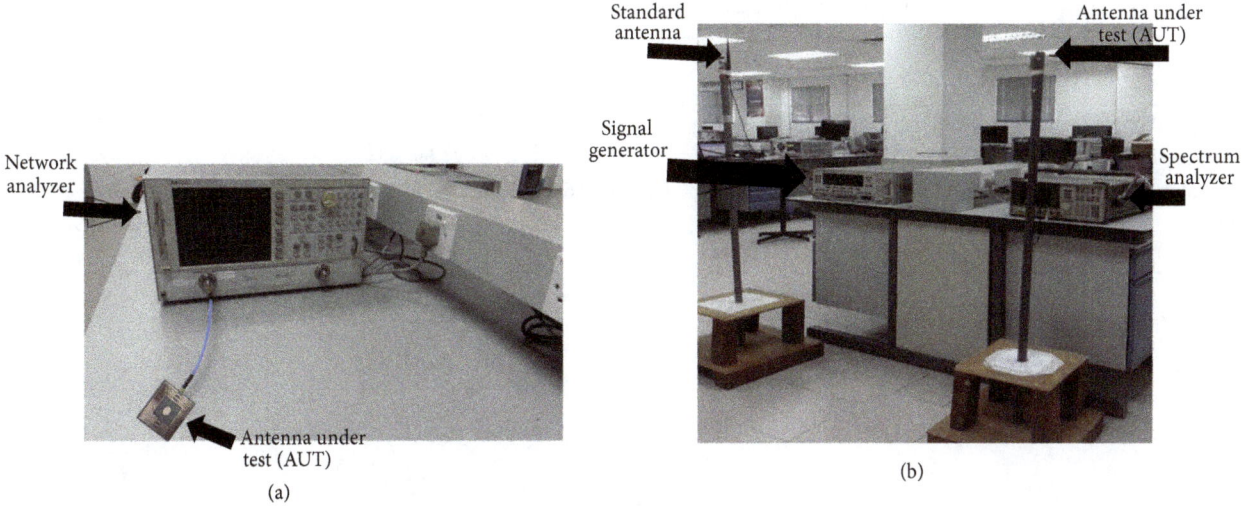

FIGURE 4: Equipment setup measurement. (a) Reflection coeff. (b) Radiation pattern and impedance matching.

the absolute gains. The procedure requires two sets of measurements. In one set, using the test antenna as receiving antenna, the received power (P_T) was recorded. In the other set, the test antenna was replaced by the standard gain antenna and the received power (P_S) was recorded. In both sets, the geometrical arrangement was maintained intact, and the input power was maintained the same. This method is deployable both in the anechoic chamber and in open space. Equation (5) was used to calculate the gain of the AUT as follows:

$$(G_T)_{dB} = (G_S)_{dB} + 10 \log_{10} \left(\frac{P_T}{P_S} \right), \quad (5)$$

where G_T is the gain of tested antenna, G_S is the gain of standard antenna, P_T is the power received by tested antenna, and P_S is the power received by the standard antenna. Equation (5) can be written as

$$G_T \text{ (dB)} = P_T \text{ (dBm)} - P_S \text{ (dBm)} + G_S \text{ (dB)}. \quad (6)$$

If the test antenna is circularly or elliptically polarized, gain measurements using the gain-transfer method can be accomplished by at least two different methods. One way would be to design a standard gain antenna that possesses circular or elliptical polarization. This approach would be attractive in mass productions of power-gain measurements of circularly or elliptically polarized antennas (ANSI/IEEE Std 149-1979, 1979). The other way would be to use a standard linearly polarized antenna. Thus, the gain of the AUT has to be measured in two orthogonal orientations. In these types of measurements, the first method would require a standard known gain helical antenna. However, this is outside the scope of our work seeing that the propose designs are linearly polarized. The effects of number of turns (N), gap size (g), and inter-ring spacing (d) and, finally, effect of different substrates permittivity (ε_r) were examined both experimentally and through numerical method. These effects were first parametrically investigated using 3D EM commercial solver and, secondly, were investigated analytically. The findings are presented in Section 3.

3. The Results and Discussions

Figure 5 depicts the simulated and measured reflection coefficient of the multiple SRR (MSRR) with respect to different substrates. Figure 5(a) is the simulated frequency pattern, whereas measured frequency response is shown in Figure 5(b). Table 1 summarizes the performance profile of the proposed based on Roger duroid 4003C substrate. Figures 5 and 6 are similarly based on the same substrate. The resonance occurred at an average frequency of 12.4 GHz as against 12 GHz target frequency with an average differential of 3.33% notwithstanding the value of the turn ratio (N). The reflection coefficients are also reasonable and agreed to a large extent.

The input impedances are largely resistive and marginally reactive. Therefore, instead of storing energy, the electromagnetic energy is properly radiated as supported by the reasonable gain exhibited by the designs across the numbers of turn ratios (N). The standing wave ratios (VSWR) are also commensurate and appropriate. It is observed that resonance occurred at 12 GHz for $N = 2$ in both simulated and measured patterns.

In particular, it is observed that, as $N > 2$, the frequency responses shift to the right hand, indicating higher frequency though with a marginal value of 0.3 GHz. By antenna theory, the only feasible reason for this is if a smaller size dimension is obtained. On the contrary, the aperture size becomes larger as N increases. In Figure 5(b), the shift was left hand side by an amount equal to 0.5 GHz. Seeing that these observations contradict themselves when comparing both simulated and measured frequency patterns, it is definitely certain that there is another factor that could be responsible other than aperture effect. It was observed that as inter-ring spacing d decreases, the coupling becomes stronger and the reflection coefficient improves significantly. In effect, this also shifts the frequency. In addition, the fringing fields' effect is experienced at the gap ends of the gap edges, which inadvertently increases the effective length of the SRR conductor trace, thus shifting the frequency as a result. Much more, the conductor trace forms

TABLE 1: The performance results of the propose.

Parameters	Output results					
	$N = 2$	$N = 3$	$N = 4$	$N = 5$	$N = 6$	Average
Centre freq., GHz	12.38	12.34	12.39	12.43	12.44	12.396
Reflection coeff., S_{11} (dB)	−23.53	−24.16	−25.1	−24.59	−25.68	−24.612
Input impedance, Ω	$46.68 + j0.4$	$49.4 + j0.1$	$44.94 + j0.6$	$50.23 + j0.4$	$50.67 + j0.3$	$48.384 + j0.36$
Gain (dBi)	6.55	6.58	6.59	6.59	6.93	6.698
VSWR	1.14 : 1	1.13 : 1	1.12 : 1	1.12 : 1	1.11 : 1	1.124 : 1

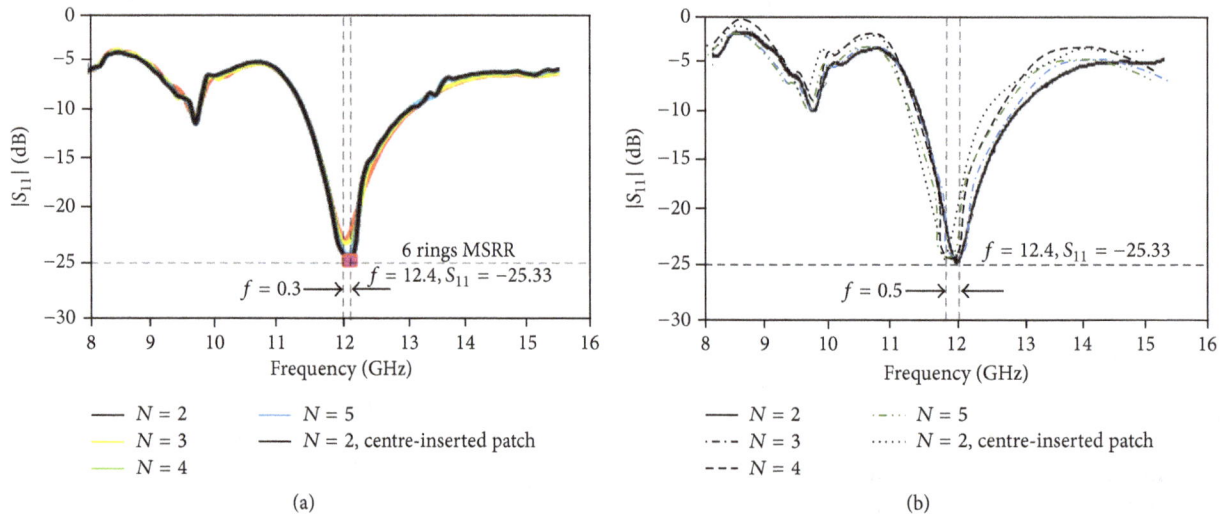

FIGURE 5: The proposed SRR designs: (a) simulated and (b) measured.

lumped inductance with magnetic field effect and as such the inductive contribution. The gaps between the rings also form the capacitive effects and thus their respective capacitive contributions. This value increases as the turn ratio increases. In effect, these properties affect the performance of these designs significantly.

In Figure 6, the simulated and measured radiation patterns across the impedance bandwidth of 11.5–12.4 GHz are presented in both xz- and yz-planes. In all, the sole maxima occurred at the boresight, and the radiated energy in all is generally directional, though marginal values of back lobes in few of the patterns were noticed. It is noted that the gain of the antennas increases with increases in the number of rings (N). There are not any established findings where variance of gap size (g) and inter-ring spacing (d) influences the gain in any way. Ironically, the effects of gap size (g) on resonance frequency with respect to different substrate are demonstrated in Figure 7, whereas the effects of inter-ring spacing (d) on resonance frequency with respect to different substrate are depicted in Figure 8.

In both figures, the numerical results using 3D EM computer simulation, theoretical results, and experimental results using the stated equations are plotted in order to compare the level of agreement. In Figure 7, resonance responses vary marginally with varying gap sizes and are most noticeable for substrate with substrate dielectric constant of 3.38 notwithstanding the resonant frequency of consideration. Similar observation was noticed in Figure 8 but much significant for

TABLE 2: The performance results of the propose.

Number of gaps/rings	Varied gap size (mm)	Turn ratio N	Simulated S_{11}
1	0.25	2	−23.58
1	0.50	3	−24.16
1	0.75	4	−24.59
1	1.00	5	−25.68
1	1.25	6	−24.61

substrate with substrate dielectric constant of 6.42 as inter-ring spacing increases. Similar observation was noticed for substrate with substrate dielectric constant of 2.40.

Table 2 also summarizes the effects of the number of gap sizes on the reflection coefficients. Finding indicates that neither the numbers of turn ratios (N) nor the number of gap sizes (g) per turn influences the reflection coefficients significantly.

4. Conclusions

The effects of turn ratio (N), gap size (g), inter-ring spacing (d), and, finally, the dielectric constant of the substrate (ε_r) were examined both analytically and numerically. These values were varied and the resulting designs were fabricated on four types of substrates, namely, Roger duroid 4003C, 6010,

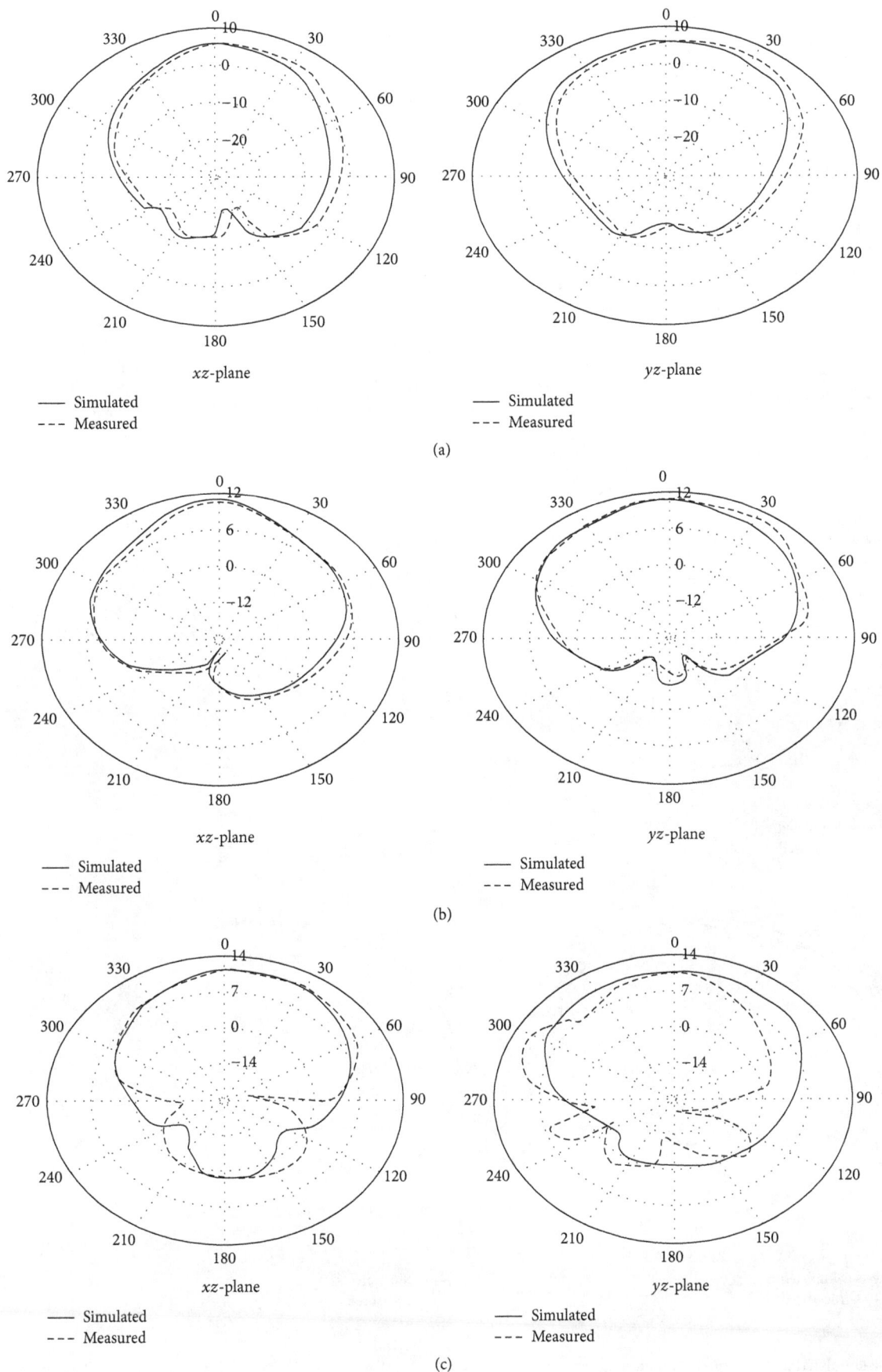

xz-plane

—— Simulated
--- Measured

yz-plane

—— Simulated
--- Measured

(a)

xz-plane

—— Simulated
--- Measured

yz-plane

—— Simulated
--- Measured

(b)

xz-plane

—— Simulated
--- Measured

yz-plane

—— Simulated
--- Measured

(c)

FIGURE 6: Continued.

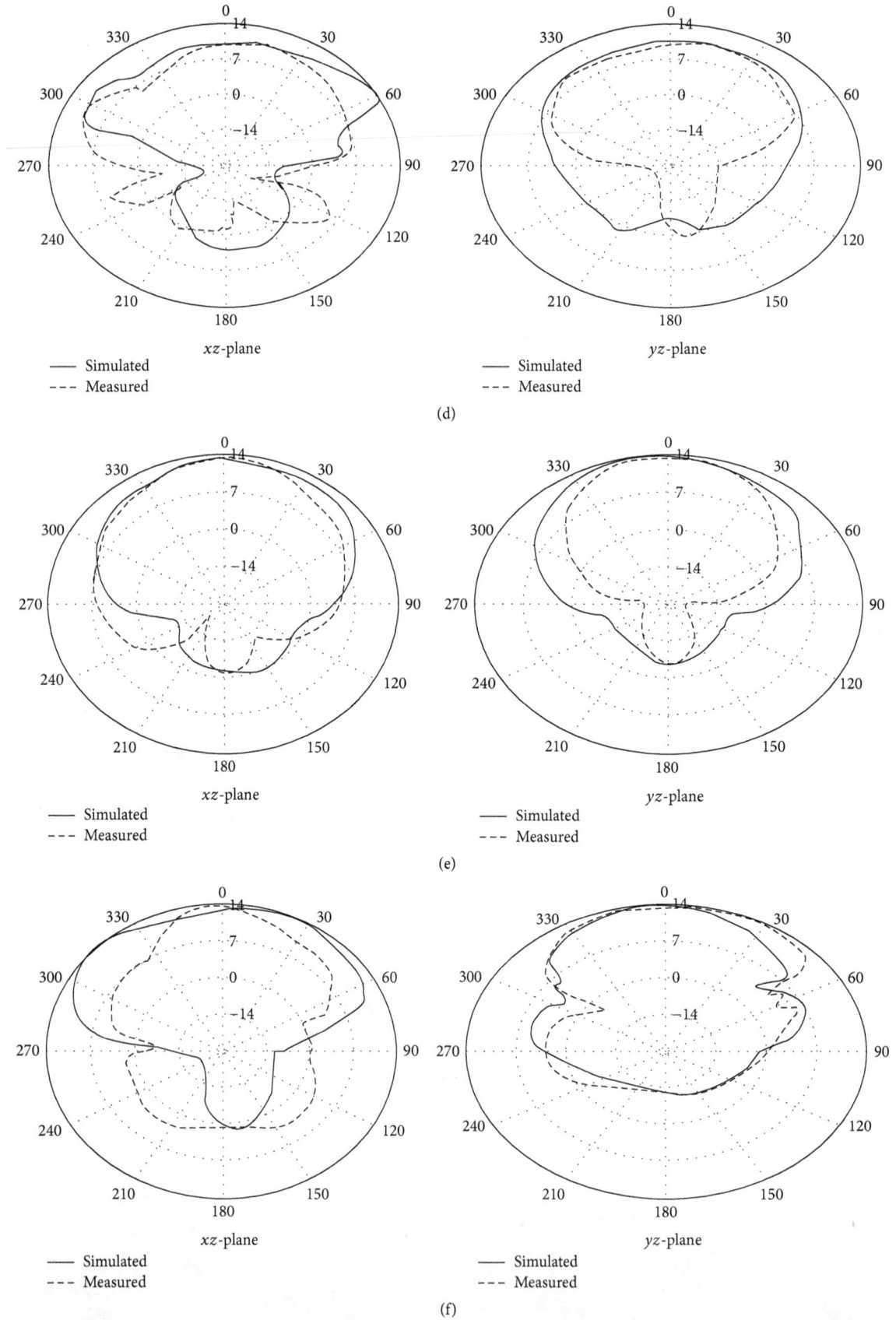

FIGURE 6: The simulated and measured radiation pattern of the proposed SRR. (a) $N = 2$, (b) $N = 3$, (c) $N = 4$, (d) $N = 5$, (e) $N = 6$, and (f) $N = 2$ patch inserted.

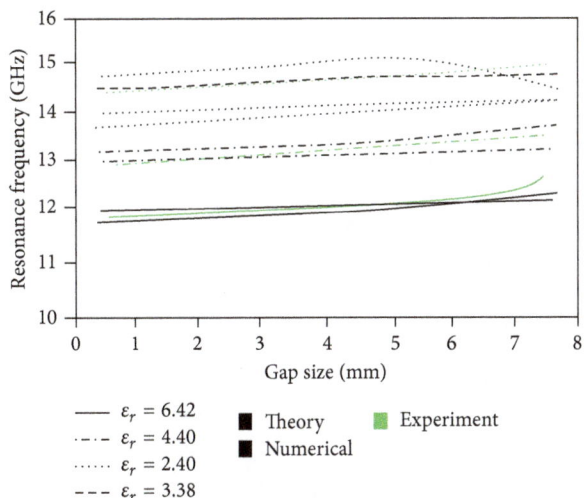

FIGURE 7: The effect of gap size (g) on the resonance frequency.

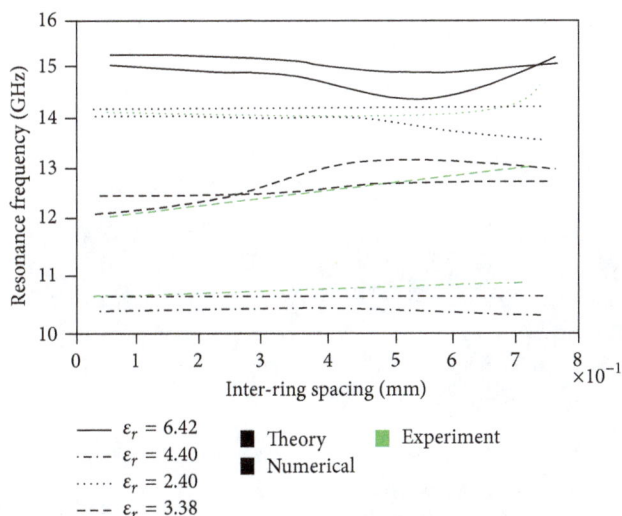

FIGURE 8: The effect of interring spacing (d) on the resonance frequency.

and Epoxy FR4 with dielectric constants of 2.4 and 6.42. Their respective effects on the designed antennas were investigated using numerical solution (via computer simulation) and theoretical formulas. Findings indicate that (1) the reflection coefficients are affected by neither the number of turn ratios, the gap size, nor the number of gap sizes per turn; (2) the gain improves as the turn ratio increases, whereas the gain of the centre-inserted patch proves to be better; and (3) the resonance frequency shifts away from the frequency of consideration as the gap sizes vary.

Conflict of Interests

The authors declare that there is no conflict of interests regarding the publication of this paper.

References

[1] Z. Yu and W. Qi, "A multi-band small antenna based on deformed split ring resonators of left-handed meta-materials," in *Proceedings of the IEEE International Symposium on Microwave, Antenna, Propagation and EMC Technologies for Wireless Communications (MAPE '07)*, pp. 531–534, IEEE, Hangzhou, China, August 2007.

[2] F. Falcone, T. Lopetegi, J. D. Baena, R. Marqués, F. Martín, and M. Sorolla, "Effective negative-ε stop-band microstrip lines based on complementary split ring resonators," *IEEE Microwave and Wireless Components Letters*, vol. 14, no. 6, pp. 280–282, 2004.

[3] A. Vélez, F. Aznar, J. Bonache, M. C. Velázquez-Ahumada, J. Martel, and F. Martín, "Open Complementary Split Ring Resonators (OCSRRs) and their application to wideband cpw band pass filters," *IEEE Microwave and Wireless Components Letters*, vol. 19, no. 4, pp. 197–199, 2009.

[4] J. Bonache, I. Gil, J. García-García, and F. Martín, "Novel microstrip band pass filters based on complementary split-ring resonators," *IEEE Transactions on Microwave Theory and Techniques*, vol. 54, no. 1, pp. 265–271, 2006.

[5] J. Martel, J. Bonache, R. Marqués, F. Martín, and F. Medina, "Design of wide-band semi-lumped bandpass filters using open split ring resonators," *IEEE Microwave and Wireless Components Letters*, vol. 17, no. 1, pp. 28–30, 2007.

[6] R. H. Geschke, B. Jokanovic, and P. Meyer, "Filter parameter extraction for triple-band composite split-ring resonators and filters," *IEEE Transactions on Microwave Theory and Techniques*, vol. 59, no. 6, pp. 1500–1508, 2011.

[7] H. Zhang, Y.-Q. Li, X. Chen, Y.-Q. Fu, and N.-C. Yuan, "Design of circular polarisation microstrip patch antennas with complementary split ring resonator," *IET Microwaves, Antennas & Propagation*, vol. 3, no. 8, pp. 1186–1190, 2009.

[8] H. Zhang, Y.-Q. Li, X. Chen, Y.-Q. Fu, and N.-C. Yuan, "Design of circular/dual-frequency linear polarization antennas based on the anisotropic complementary split ring resonator," *IEEE Transactions on Antennas and Propagation*, vol. 57, no. 10, pp. 3352–3355, 2009.

[9] Y. Dong, H. Toyao, and T. Itoh, "Design and characterization of miniaturized patch antennas loaded with complementary split-ring resonators," *IEEE Transactions on Antennas and Propagation*, vol. 60, no. 2, pp. 772–785, 2012.

[10] B. D. Braaten and M. A. Aziz, "Using meander open complementary split ring resonator (MOCSRR) particles to design a compact UHF RFID tag antenna," *IEEE Antennas and Wireless Propagation Letters*, vol. 9, pp. 1037–1040, 2010.

[11] B. D. Braaten, "A novel compact UHF RFID tag antenna designed with series connected Open Complementary Split Ring Resonator (OCSRR) particles," *IEEE Transactions on Antennas and Propagation*, vol. 58, no. 11, pp. 3728–3733, 2010.

[12] M. Gil, J. Bonache, J. Selga, J. García-García, and F. Martín, "Broadband resonant-type metamaterial transmission lines," *IEEE Microwave and Wireless Components Letters*, vol. 17, no. 2, pp. 97–99, 2007.

[13] M. Durán-Sindreu, A. Vélez, F. Aznar, G. Sisó, J. Bonache, and F. Martín, "Applications of open split ring resonators and open complementary split ring resonators to the synthesis of artificial transmission lines and microwave passive components," *IEEE Transactions on Microwave Theory and Techniques*, vol. 57, no. 12, pp. 3395–3403, 2009.

[14] F. Karshenas, A. R. Mallahzadeh, and J. Rashed-Mohassel, "Size reduction and harmonic suppression of parallel coupled-line bandpass filters using defected ground structure," in *Proceedings of the 13th International Symposium on Antenna Technology and Applied Electromagnetics and the Canadian Radio Science Meeting (ANTEM/URSI '09)*, pp. 1–6, IEEE, Toronto, Canada, February 2009.

[15] O. Turkmen, E. Ekmekci, and G. T. Sayan, "A new multi-ring SRR type metamaterial design with multiple magnetic resonances," in *Proceedings of the Progress in Electromagnetics Research Symposium (PIERS '11)*, pp. 315–319, Marrakesh, Morocco, March 2011.

[16] C. Saha, J. Y. Siddiqui, and Y. M. M. Antar, "Theoretical investigation of the square split ring resonator," in *Proceedings of the IEEE WIE National Symposium on Emerging Technologies (WieNSET '07)*, Kolkata, India, June 2007.

24

High Power Combline Filter for Deep Space Applications

A. V. G. Subramanyam,[1] D. Siva Reddy,[1] V. K. Hariharan,[1] V. V. Srinivasan,[1] and Ajay Chakrabarty[2]

[1] ISRO Satellite Centre (ISAC), Bangalore 560017, India
[2] Department of Electronics and Electrical Communications Engineering, Indian Institute of Technology (IIT), Kharagpur 721302, India

Correspondence should be addressed to A. V. G. Subramanyam; avgsub@isac.gov.in

Academic Editor: Mirco Raffetto

An S-band, compact, high power filter, for use in the Mars Orbiter Mission (MOM) of Indian Space Research Organization (ISRO), has been designed and tested for multipaction. The telemetry, tracking, and commanding (TT&C) transponder of MOM is required to handle continuous RF power of 200 W in the telemetry path besides simultaneously maintaining an isolation of greater than 145 dBc to its sensitive telecommand path. This is accomplished with the help of a complex diplexer, requiring high power, high rejection transmit path filter, and a low power receive path filter. To reduce the complexity in the multipaction-free design and testing, the transmit path filter of the diplexer is split into a low rejection filter integral to the diplexer and an external high rejection filter. This paper highlights the design and space qualification phases of this high rejection filter. Multipaction test results with 6 dB margin are also presented. Major concerns of this filter design are isolation, insertion loss, and multipaction. Mission performance of the on-board filter is normal.

1. Introduction

Telemetry, tracking, and commanding (TT&C) transponder in a spacecraft meant for deep space mission requires a high power (100 s of watts) transmitter, a high sensitive (~135 dBm) receiver, and respective high gain antenna systems. Since both the transmitter and receiver require similar antenna systems, it will be highly taxing in terms of on-board weight and volume, if we use two antennas independently. It is economical to use a common antenna system for both uplink and downlink, with the help of a diplexer [1]. Diplexer is a passive component that connects the common antenna feed, simultaneously, to both transmitter and receiver with proper isolation. It consists of a high power transmit filter, a receive filter, and a combining network as shown in Figure 1.

Multipactor [2] is an electron resonance phenomenon which occurs at radio frequencies in high power components like filters and resonators and transmission lines operating in vacuum. It represents a possible payload failure mechanism for communications satellites since it can destroy microwave components or transmission lines, or it can significantly

raise noise levels [3]. Multipactor effect has been known for many years; it still presents a critical problem and more constraints in satellite communication system applications in terms of transmit power, number of carriers, and wider bandwidth. Multipactor occurs whenever the electrons are sufficiently energized by the RF waves; they are driven back into a surface and secondary electrons are produced. The effective secondary electron yield depends on the impact energy and angle of incidence, the surface properties, and the direction of the RF field at the time of impact [4]. Under certain conditions, the phase of the secondary electrons remains locked with the RF field driving the impacts, so that secondary electron emissions tend to be in phase with the applied RF field resulting in multiplication of the electrons. This finally results in reducing the power output of the component and increasing its return loss.

Main aim of this paper is to prevent multipaction breakdown in the transmitter path and also realize the hardware in compact size and lesser weight. This paper highlights the intricacies involved in the design, realization, and space qualification of the high power transmit filter, in a short time,

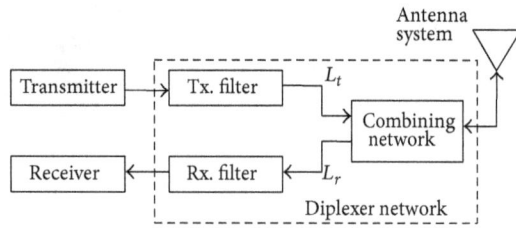

FIGURE 1: TT&C system configuration of a spacecraft.

TABLE 1: Required specification and simulated results of the ridged guide filter-F1.

Parameter	Requirement
Carrier frequencies (MHz)	2217 2231
1 dB bandwidth (MHz)	100
Rejection (dB) [@2041 & 2054 MHz]	145 min.
Insertion loss (dB)	0.7 max.
Return loss (dB)	15 min.
Power handling requirement (W) CW	225 (53.52 dBm) min.
I/O interface	Quarter height WR340

for a deep space mission requirement. Simulated and measured performances of the filter are presented. Multipaction analysis and test results are also discussed.

2. High Power Filter Design, Multipaction Analysis, and Fabrication

2.1. Design and Optimization. The primary purpose of the high power filter is to suppress the receiver band frequency spurious coming from the TWTA amplifier and pass the transmit band RF signal with minimum loss. It is required to provide an isolation of at least 120 dBc to the receive signal and thereby it will aid in simplifying the diplexer design by sharing the rejection requirement. Otherwise the diplexer alone has to provide this higher isolation (>145 dBc) for the receive signal, which increases its design, realization, and testing complexity.

The filter design also aims at achieving the required performance in a compact and low weight system. Accordingly suitable design-cum-fabrication techniques and material reduction were also implemented.

Compact implementation of filtering structures for space applications [5] is based on evanescent mode waveguides, whose typical configuration using symmetrical metal ridges is as shown in Figure 2. These filters, originally proposed in [6, 7] and refined in [8], are essentially composed of a hollow, below cut-off waveguide housing, which transmits the energy between standard waveguide access ports through shunt capacitive elements (ridges). The below cut-off waveguide sections placed between consecutive ridges can be modeled as impedance inverters and shunt inductances that, when combined with the cited capacitances, provide the required filter resonances. Evanescent mode filters are a good choice for the input and output stages of satellite payloads, since they can provide moderate bandwidth responses with excellent out-of-band performance and sharp selectivity. In addition these filters are very competitive in terms of mass and volume due to their below cut-off waveguide housing [9–11].

Initially, when the MOM system was evolving, independent polarizations for uplink and downlink were planned and there was no necessity for a diplexer. To meet these requirements, specifications for the high power filter were derived as in Table 1, for which a 12-pole double-side ridged waveguide filter (F1) was designed. Based on [12], a rigorous field theory description of the ridge waveguide is utilized to formulate the modal scattering matrix of the waveguide-to-ridge-waveguide discontinuity, which is the basic building block in the design. Also with the help of Fritz Arndt's Wasp-Net Software [5], the ridged waveguide filter integrated with the quarter height waveguide to double-side ridged waveguide transitions was designed and overall optimization was carried out. Figure 2 shows the structure of the filter-F1 and its simulated response.

Simulated response on HFSS was meeting all the required electrical specifications (rejection of 154 dBc), including power handling, and has the cross section as 35 mm × 35 mm with the overall length as 622 mm. It is easy to fabricate this filter into two symmetric halves and assemblies. But the sensitivity analysis shows that this structure is vulnerable to fabrication tolerances and after fabrication the scope for tuning is very remote. For this structure, small perturbations in the penetration depths of the upper and lower metal insert seriously affect the ridge gap dimension, which is identified to be very sensitive parameter. Additionally, slight misalignments can easily spoil the required symmetry of this filter design. Hence this filter-F1 was not fabricated.

To overcome such drawbacks, a potential alternative consisting of an asymmetrical configuration, with metal inserts placed only in one wall of the waveguide housing [13], a combline based structure, filter-F2, is planned as shown in Figure 4. Combline resonator is a hybrid structure where a coaxial transmission line is formed by a partial height post in series with a gap capacitor placed in a rectangular or circular cavity. Combline cavities use a below cut-off waveguide structure and hence are small compared to their counter parts in normal propagating waveguide structures.

Unlike the traditional comb-line filter, which has no metallic obstacle between resonators, this filter employs rectangular cavities with coupling slots. This further reduces resonator lengths and thereby the overall filter length. This structure yields a somewhat higher unloaded quality factor because more of the field is enclosed. The coupling between resonators is controlled via the width of the slot, which in turn control the bandwidth of the filter.

Relatively this structure has better symmetry in the skirt response and the required rejection could be met with eleven resonators [14, 15]. The cavity size/dimension is decided by the centre frequency and spurious suppression response of the filter. To realize the given filter specifications in an iris-coupled combline structure, the transmission and reflection

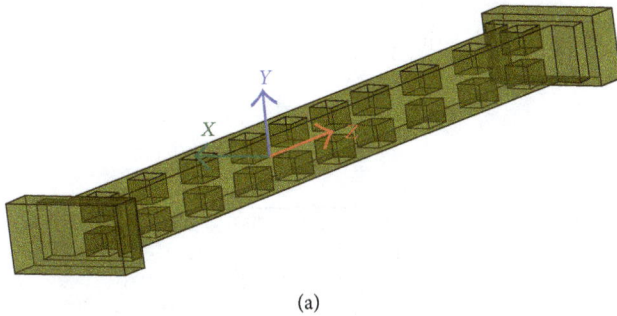

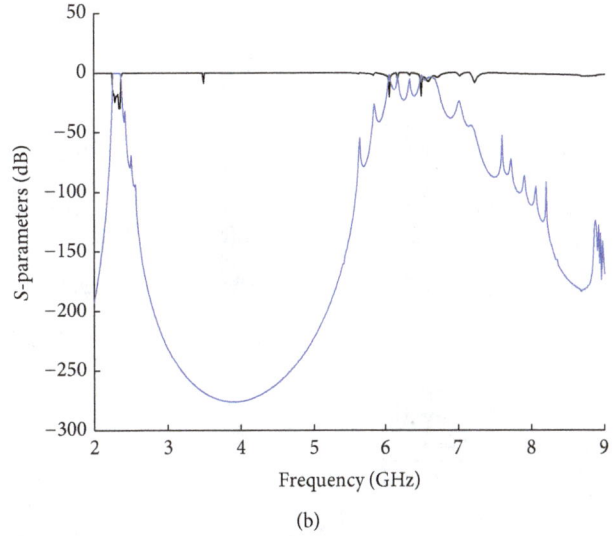

FIGURE 2: Double-side ridged waveguide filter-F1 and its simulated response.

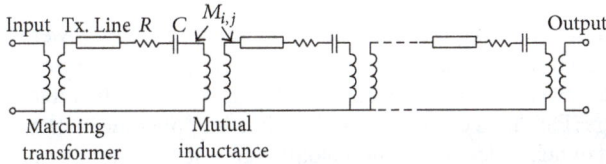

FIGURE 3: Equivalent circuit of the iris-coupled combline filter.

parameters of Chebyschev transfer function and the coupling coefficient matrix $K_{i,j}$ from the low-pass prototype elements [14, 16] are determined. Using these values, the parameters of the equivalent circuit shown in Figure 3 can be calculated [1].

This type of iris-coupled combline filters is also presented recently in [1]. The basic structure of the present filter is the same as that in [1], that is, iris-coupled combline filters, usually meant for low power applications but intended for high power satellite applications. Hence the equivalent circuits are the same except for the input/output interface (here quarter height waveguide interface instead of coaxial interface). Similarly, multipactor analysis is also based on the resonator voltages that are obtained from 3D simulators like HFSS.

In this circuit, the resonators are represented by transmission lines, short-circuited at one end, in series with a lumped capacitor. The iris-coupling between the resonators is inductive and is modeled by a mutual inductance; the input/output coupling to the filter is also inductive and is modeled by an ideal transformer. This transformer matches the quarter height guide of WR340 to the combline guide of 35 mm by 32 mm. The insertion loss of the filter is represented by a resistor in series with the lumped capacitor.

The transmission-line (Tx. Line) parameters are characteristic impedance Z_0 and electrical length θ at the center frequency f_c. At the resonant frequency, the series capacitance C will be

$$C = \frac{1}{2\pi f_c Z_0 \tan\theta}. \tag{1}$$

Reactance slope parameter X for this resonator is given by [1]

$$X = \frac{Z_0}{2}\left[\theta(1 + \tan\theta)^2 + \tan\theta\right]. \tag{2}$$

In this equation, the electrical length θ is expressed in radians. The mutual inductances $M_{i,j}$ are related to coupling coefficients $K_{i,j}$, the reactance slope parameter X, and the frequency f_c by

$$M_{i,j} = 2\pi f_c X K_{i,j}. \tag{3}$$

The simulated S-parameter response of this 11-pole filter-F2 is as shown in Figure 4. The structure is optimized using library elements based Wasp-Net software [5] and verified in HFSS software [17] before fabrication.

The power handling of combline filter is limited by the gap capacitance formed between the resonator open end and ground plane where the electric field has its maximum intensity [1]. This gap is typically a small fraction of the wavelength, making it susceptible to corona or multipaction. Due to this reason combline filters are usually not used for high power applications. However [1] shows one way of designing (dielectric filling), whereas this paper shows the other simple and easy way (increasing critical gap) at the design stage, by which the power handling capability can be increased still maintaining the combline performance.

Overall length of the present filter-F2, structure got reduced to almost 450 mm and the cavity cross-sectional dimension, with lots of trials, has been finalized to be 35 mm × 32 mm for the optimum performance. Due to impedance mismatch, these optimum dimensions cannot be directly interfaced with the preceding and succeeding sections that carry high RF powers. Waveguide routing on both sides of the filter-F2 are of standard WR 340 with reduced height (86.36 × 10.8 mm^2), to save on weight and volume. To match the filter impedance to the rest of the routing, suitable impedance transformers were designed and attached on either side. The simulated filter-F2 transmission loss and return loss response shown in Figure 4 includes that of the transformer also.

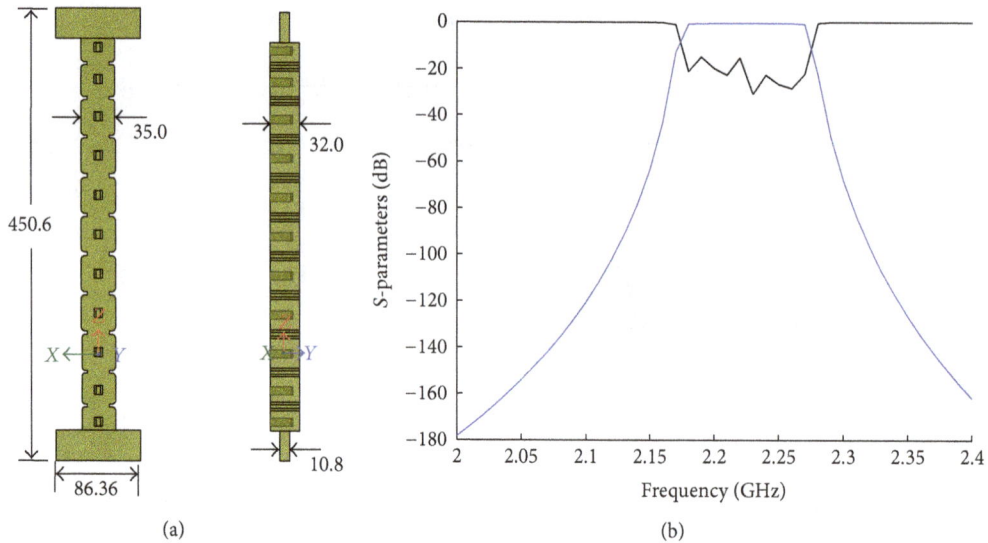

FIGURE 4: The structure and response of 11-pole iris-coupled combline filter-F2.

In an attempt to increase the power handling capability of the filter-F2, the field pattern was studied using HFSS. It is found that the peak values of field are concentrated near the sharp edges of the posts towards the grounding planes. These sharp edges were rounded off with 1 mm radius and field pattern is observed again. It is noted that rounding of these edges improved the power handling of the filter by 57%. Hence this technique was adopted in all later versions of the filter design.

At the time of loading this filter-F2 for fabrication, TT&C frequencies of the mission got changed. Also to exercise the option of launch support from external agencies (outside India) a single polarization scheme for both uplink and downlink was adopted, which needs a diplexer to separate transmit and receive signals with proper isolation. Therefore a complex diplexer has to be designed to provide the required rejection of >145 dBc between the high power transmit and high sensitive receive channels.

The design complexity of the diplexer is shared by designing a simple diplexer with about 40 dBc rejection and a high power filter in tandem for the rest of the rejection requirements. With this requirement a new set of specifications for the high power filter (F3) were derived as enumerated in Table 4. Meanwhile the 11-pole filter-F2 that got fabricated was taken up as qualification model (QM) for all stringent tests including high power tests, at initial TT&C frequencies.

For the new specifications of Table 4, a filter design with 10-pole Chebyschev polynomial and 0.01 dB ripple in the pass-band was chosen. These filter-F3 characteristics have been realized using a similar combline structure with iris coupling and impedance transformers (see (1)–(3)), without tuning elements. Size of the filter-F3 is the same as the previous one and the overall length is reduced to just 374 mm, as in Figure 5. Its simulated response is shown in Figure 6.

2.2. Multipactor Analysis. Estimation of multipactor threshold for this combline structure-F3 is based on the procedure

given in [1]. The voltages of each resonator of the high power filter-F3 were calculated using HFSS model and 200 W of rms input power. This is achieved from the 3D model of the filter using HFSS "field calculator" and by drawing "polyline" at the point of field calculation; the absolute value of peak resonator can be evaluated for the respective resonators. These are reported in Table 2 at centre frequency for both filters F2 and F3, which also shows the input voltage V_{in} corresponding to P_{in} = 200 W rms input power and the voltage magnification factors (VMFs). V_{in} is calculated from

$$V_{in} = \left(2 \times 50 \times P_{in}\right)^{1/2} = 141.4 \, \text{V}. \tag{4}$$

It can be noted that maximum peak voltage is built up in the 6th resonator for F3.

After determining the voltages of the combline resonators, multipaction safety margin can be calculated for different gaps using multipactor threshold voltage V_M given by [18]

$$V_M = 63 * f * d, \tag{5}$$

where frequency f is in GHz and gap d is in millimeters.

Filters are silver plated inside and hence, in the above equation, the slope of 63 V/(GHz·mm) for silver has been used. Safety margin for the multipaction can be calculated from

$$\text{Margin} = 20 \times \log\left(\frac{V_M}{V_r}\right). \tag{6}$$

Table 3 shows the computed multipaction margins for both filters F2 and F3. The analysis is based on the parallel plate model and does not take into account the quality of the plating process in multipaction breakdown. However it can be used to determine the indicative multipaction thresholds and not the exact values. The critical gap area of the combline filter, between the resonator post open end and top cover,

TABLE 2: Calculation of voltage magnification factor (VMF) for filters F2 and F3.

Resonator Number	Peak input voltage, V_{in}	Peak resonator voltage, V_r		VMF	
		F2	F3	F2	F3
1	141.4	973	911	6.88	6.44
2	141.4	1169	1317	8.27	9.31
3	141.4	1480	1331	10.47	9.41
4	141.4	1274	1412	9.01	9.98
5	141.4	1509	1326	**10.67**	9.38
6	**141.4**	1286	1420	9.09	**10.04**
7	141.4	1479	1293	10.46	9.14
8	141.4	1265	1399	8.95	9.89
9	141.4	1417	1184	10.02	8.37
10	141.4	1150	939	8.13	6.64
11	141.4	913	—	6.46	—

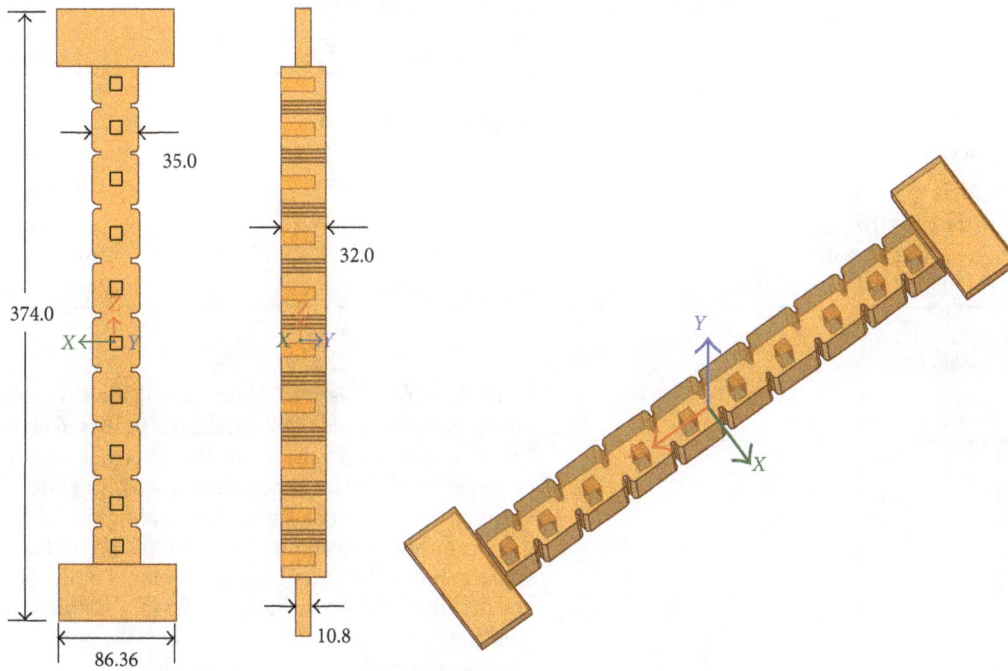

FIGURE 5: Different views of the final filter-F3 configuration with internal dimensions.

is far from infinite parallel plate. In fact the aspect ratio of the gap (height/length) is >1. Therefore it is expected that multipactor threshold will be higher than the values calculated, due to significant fringing fields in the region [1].

From Table 3, though filter F2 consists of tuning screws for RF optimization purposes, here in the computations, these are not taken into account. It is observed that the worst case margin for F2 is −3.58 dB, which is poorer by more than 1 dB. As tuning elements were introduced in the case of F2, the critical gap between the post and top wall further reduces, resulting in multipactor margins worse than those computed in Table 3. In spite of the above negative margins, the filter-F2 has demonstrated more than 3 dB margin (measured); further, to achieve the required ECSS

6 dB margin, drastic physical changes such as increasing the gaps/size are not warranted due to the restricted physical constraints (component size).

In contrast, as the multipaction margins for F3 are not worrisome (worst case margin is −2.1 dB only), it necessitates multipaction testing for higher power levels. Hence even with little negative margins (highly approximated) for the multipactor, we have proceeded for the fabrication of this new filter F3 and could achieve the required result.

2.3. Fabrication. Filter is fabricated by milling a block of space grade Aluminum alloy 6061. The adopted mechanical design as shown in the assembly drawing of Figure 7 gives better mechanical stability and environmental protection.

TABLE 3: Estimation of multipaction margin for filters F2 and F3.

Res. Number	Gap (mm)		V_M		Multipactor margin (dB)	
	F2	F3	F2	F3	F2	F3
1	7.26	7.91	1017.7	1110.7	0.39	1.72
2	6.93	7.72	971.4	1083.9	−1.61	−1.69
3	7.12	7.91	998.0	1110.6	−3.42	−1.57
4	7.13	7.94	999.4	1115.2	−2.11	−2.05
5	7.13	7.94	999.4	1115.2	**−3.58**	−1.50
6	7.13	7.94	999.4	1115.2	−2.19	**−2.10**
7	7.13	7.94	999.4	1115.2	−3.40	−1.29
8	7.13	7.91	999.4	1110.6	−2.05	−2.00
9	7.12	7.72	998.0	1083.9	−3.04	−0.77
10	6.93	7.91	971.4	1110.7	−1.47	1.46
11	7.26	—	1017.7	—	0.94	—

TABLE 4: Specifications and measured parameters of filter F3.

Parameters	Specification	Measured
Frequency f_c (MHz)	2295.5	2295.5
1 dB band-width	100 MHz ± 10 MHz	106 MHz
Insertion loss at f_c ± 10 MHz	<0.7 dB	0.40 dB
Return loss at f_c ± 10 MHz	>15 dB	23.5 dB
Rejection at Rx band 2114 ± 10 MHz	>120 dBc	141.0 dBc
Power handling requirement @2298.48 MHz	185 W	740 W (6 dB margin)
I/P & O/P interface	WR-340 quarter height	WR-340 quarter height

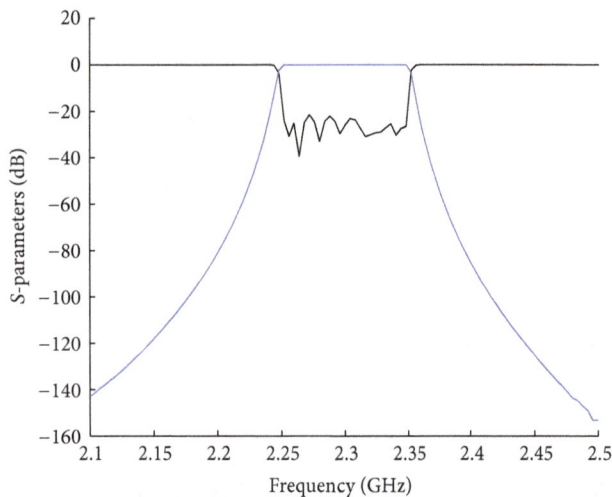

FIGURE 6: Simulated S-parameter response of the final filter configuration.

The input and output are interfaced with WR-340 quarter height waveguide. The internal surface of the high power filter has been silver-plated to improve the insertion loss. The photograph of the assembled high power filter is shown in Figure 8. With 200 watts as the input power to this filter, it can dissipate 17.6 watts (0.4 dB max. measured insertion loss) in terms of heat. This input along with its physical dimensions, metallic properties, and the mounting location on the carrier plate were subjected to thermal analysis. Based on the thermal analysis the outer surface of the filter has only been black painted, as it does not generate any considerable amount of heat energy that necessitates any cooling mechanism.

Vent hole requirement was addressed and implemented based on the ESA calculator estimation. For the given volume of the filter eight vent holes of 1.6 mm diameter are drilled on the input and output interface flanges at appropriate places where there is less possibility for radiation. For the given volume of the structure with eight vent holes it takes roughly 500 sec to completely vent to the level of 10^{-5} torr. The filter design also aims at achieving the required performance in a compact and low weight system. Accordingly suitable material reduction and fabrication techniques were also implemented.

3. Measurements and Environmental Tests

3.1. PNA Measurements. Figure 9 shows the measured pass band S-parameters of the filter-F3 using Agilent's Precision Network Analyzer (PNA). Similarly Figure 10 shows the typical rejection response of the filter at receive frequency, which reads to be better than 123 dBc (120 dBc spec.) on PNA. The measured results are listed in Table 4 against the system specifications. All the parameters are comfortably meeting the specifications. Higher bandwidth is considered to achieve lower insertion loss and better margin for multipaction.

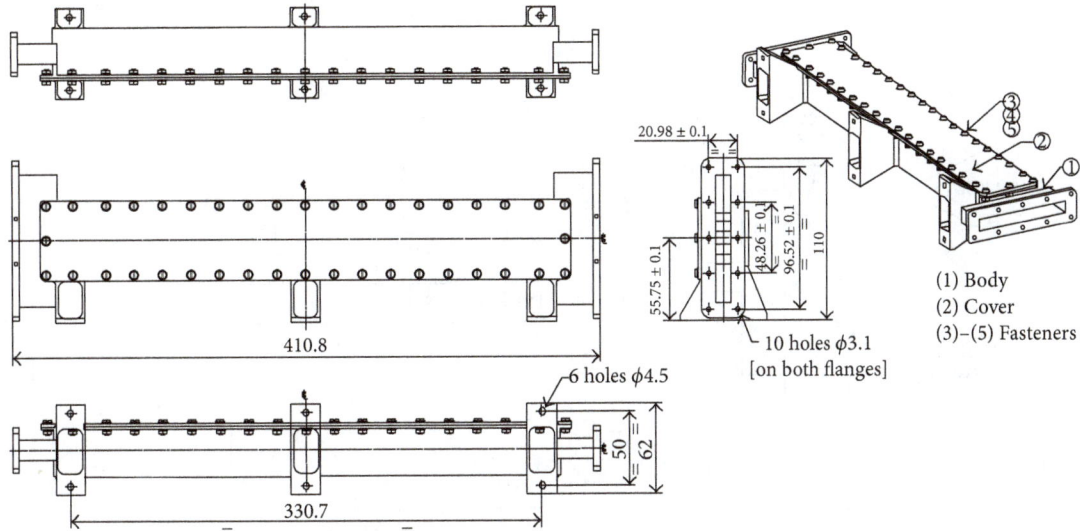

FIGURE 7: Assembly drawing of high power filter.

(1) Body
(2) Cover
(3)–(5) Fasteners

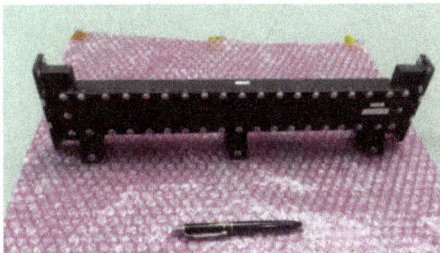

FIGURE 8: Photograph of the high power filter.

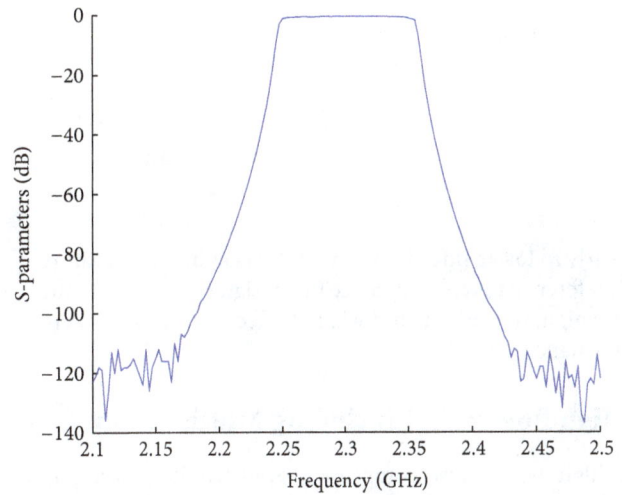

FIGURE 10: Measured out-of-band rejection response of the filter-F2.

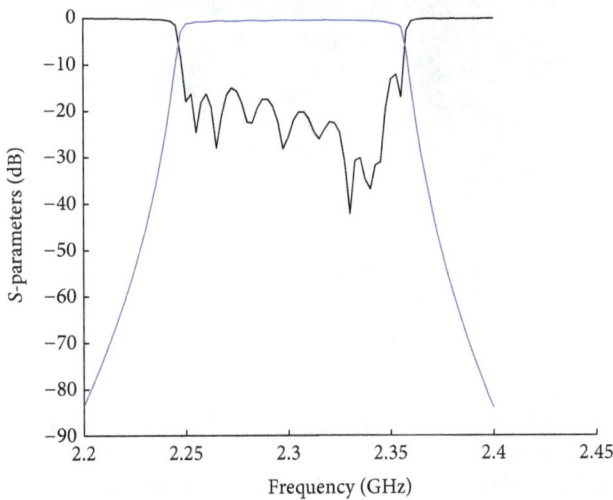

FIGURE 9: Measured pass band response of the filter-F2.

3.2. Rejection Measurements. In fact the rejection of the filter is about −140 dBc as per simulation. Due to the dynamic range limitation of the PNA, measurement was not feasible to find the actual rejection offered by the filter at receive frequency. To perform this measurement, a high power source and high dynamic range spectrum analyzer were used. Measurement was carried out at +47.5 dBm and filter rejection at receive frequency is found to be 141.0 dBc.

3.3. RF Characterization with Different Environmental Tests. RF characterization of the filter involves the following tests starting with functional test; they are initial bench test, postpassive thermal cycle test (10 cycles of 1 hour dwell at −15° to +80°C), postvibration test (sine, random vibrations, and mechanical shock at stipulated higher "*g*" values), and high power test followed by final bench test. In addition, RF characterization of the filter was carried out at the extreme hot (+80°C) and cold (−15°C) temperatures, to estimate the shift in filter response due to temperature. During this test, the frequency response has been shifted to higher side almost by 2.0 MHz in cold soak with reference to ambient and

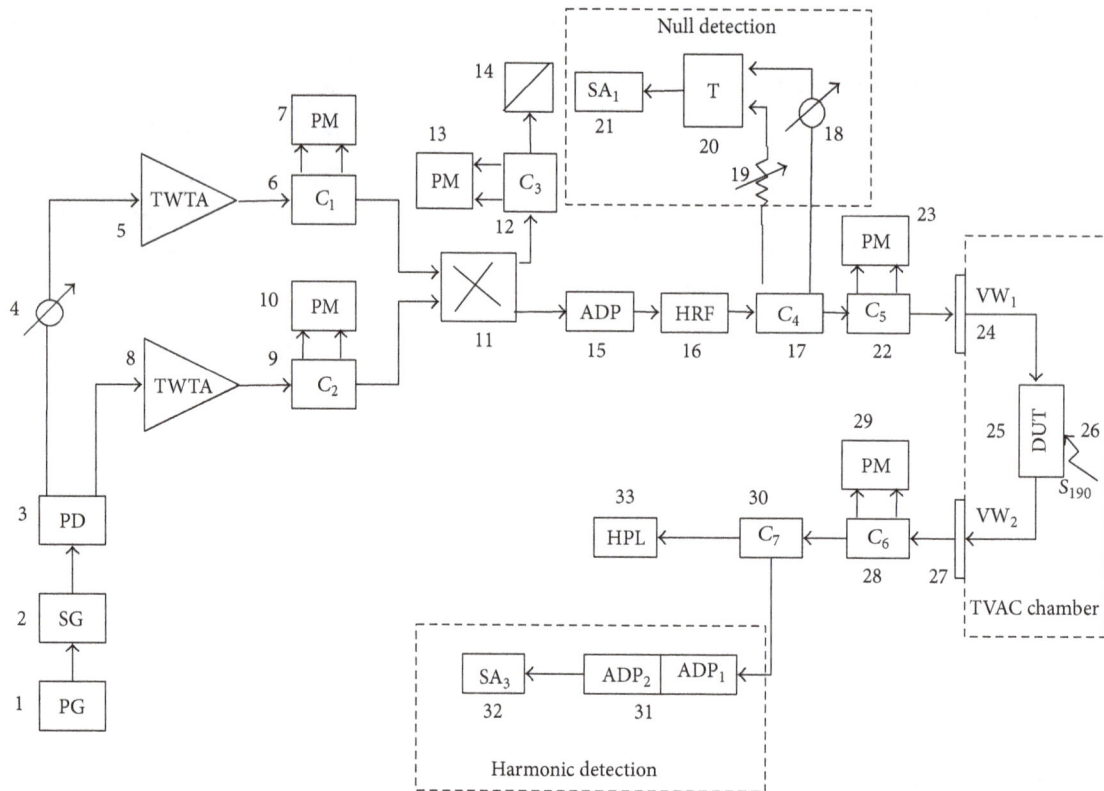

FIGURE 11: Multipaction and power handling test setup.

4. High Power Test Including Multipaction

The high power test comprises power-handling test and multipaction test at hard vacuum. The block diagram in Figure 11 shows the high power test setup. The serial numbers in the diagram indicate the instrumentation used. In the block diagram, pulse generator (PG) sends pulses to signal generator for modulating the RF carrier. Through power divider (PD) and phase shifter equiphase signals are fed to TWT amplifiers. Outputs of amplifiers are combined in a hybrid/combiner (11) and fed to the DUT that is placed in the thermovacuum chamber (TVAC) through filters. Null detector (ND) is connected in this path. The variable attenuator and the phase shifter in the reflection path were adjusted to achieve a return loss null on Spectrum Analyzer (SA$_1$). The DUT output from the TVAC is connected to a high power load after being tapped for the third harmonic detection (THD) on a spectrum analyzer (SA$_3$). Forward and reflected signal flow is monitored at various points in the chain using dual directional couplers (C), power meters (PM), and spectrum analyzers (SA).

For electron seeding, a radioactive source is placed next to the filter and directed to the vent holes in the close proximity to the high field region. Almost the same setup

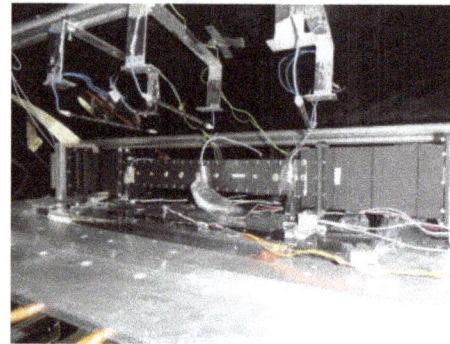

FIGURE 12: Filter-F2 inside TVAC chamber with associated components for the multipaction test.

is used for both power handling and multipaction tests of the filter, except that the power handling test is done with CW RF power (from a single TWTA) whereas multipaction test is done with pulsed power. The setup consists of two mechanisms for detecting the onset of multipaction, namely, null detection and third harmonic detection. Simultaneous observation of abnormality in both the detectors confirms the occurrence of multipaction.

To validate the test setup, an S-band gap-sample is used and multipaction is simultaneously detected on both the third harmonic and null detectors as shown in Figure 13. Figure 12 shows the photograph of the filter-F2 mounted

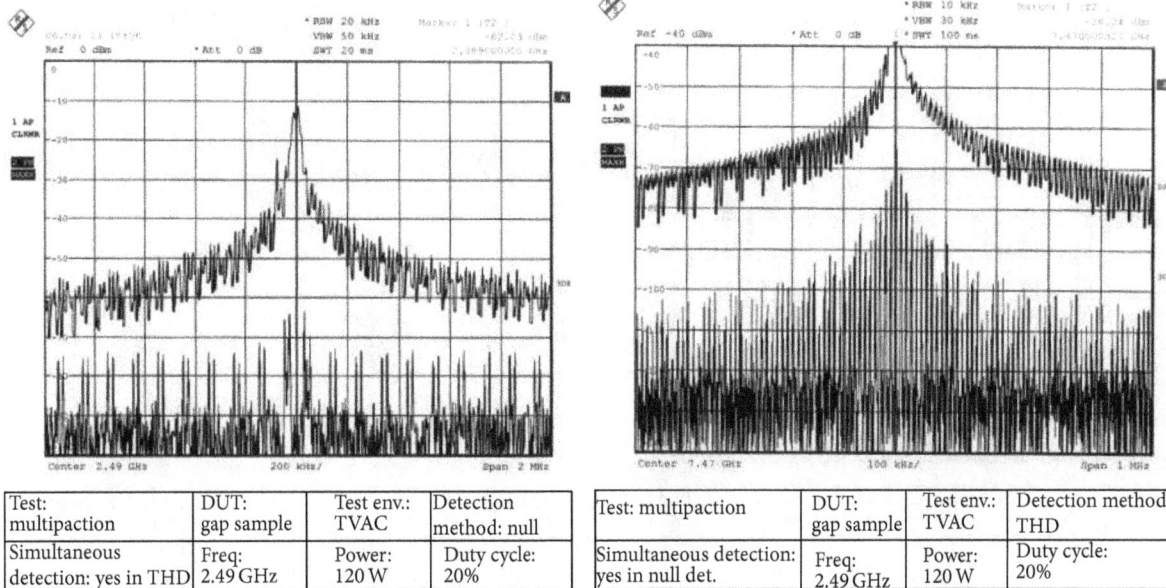

Test: multipaction	DUT: gap sample	Test env.: TVAC	Detection method: null
Simultaneous detection: yes in THD	Freq: 2.49 GHz	Power: 120 W	Duty cycle: 20%

Test: multipaction	DUT: gap sample	Test env.: TVAC	Detection method: THD
Simultaneous detection: yes in null det.	Freq: 2.49 GHz	Power: 120 W	Duty cycle: 20%

FIGURE 13: Gap-sample test for validating multipaction test setup.

inside thermovacuum (TVAC) chamber along with IR lamps, thermal sensors, waveguide transitions, and so forth.

A vacuum bake-out at +85°C for 12 hr was performed prior to multipaction test to prevent out-gassing. Initially the high power test has been carried out on the filter-F2 which has been designed at old TT&C frequencies (2217 & 2231 MHz). The package has successfully completed the power handling test with nominal power (250 W) for 12 hours. But it has failed during multipaction test at RF input power level of 420 W. The corresponding null detector and third harmonic plots are shown in Figure 14. Subsequently one more Qualification Model Filter-F3 has been planned and developed exactly at MOM frequencies (identical to flight model).

Reasons for the failure of the filter-F2 were thoroughly studied. (a) Critical dimensions and gaps were measured and ensured that the new design had better margins, (b) tuning elements might have introduced critical gaps with dissimilar metal junctions and hence were avoided in new design, and (c) extra precaution in maintaining cleanliness and handling was adopted in addition to regular procedures. Final power reaching at the input of the filter-F3 is estimated to be 185 W after considering the actual TWTA output and en route on-board losses. Subsequent to basic characterization and vacuum-baking, the new filter-F3 is subjected to power-handling test at nominal power (200 W) for 6 hrs duration at extreme hot temperature (+80°C). The multipaction test has been carried out, at the same elevated temperature, initially at 50 W of RF power and then slowly peak power is increased to 100, 200, 370, 600, and then 740 W.

The intermediate test data after a few minutes dwell, for lower power levels, and 1-hr dwell period for 370 W (3 dB margin) and 740 W (6 dB margin), are recorded using printer connected to spectrum analyzers. Temperature variation at three points on the DUT is continuously monitored during the test and found to be within limits. No abnormalities in the detectors are observed, confirming nonoccurrence of multipaction event. The new filter-F3 has successfully completed the multipaction test and paved the way for realizing the flight model. The corresponding high power test plots of the detectors are shown in Figure 15. On the same lines main and redundant flight model filters were realized and flown successfully with the spacecraft.

5. Conclusion

Filter design is aimed at meeting the primary specifications for multipaction, insertion loss, rejection, mass, and size. Accordingly the number of sections, cavity cross section, and other parameters were optimized to meet the primary specifications. This resulted in a pass band bandwidth of about 100 MHz. This also caters to cover both the main and redundant frequencies in transmit/receive bands, dispersion due to temperature, and fabrication tolerances. This higher bandwidth helped (1) to reduce the insertion loss and (2) to extend the multipaction threshold margin to higher powers. The disadvantage of the increased bandwidth lies in the increased number of sections for the same rejection and thereby physical length. But the disadvantage of having lesser bandwidth is that it requires loose coupling between adjacent resonators which demands more spacing between them and thereby increases the overall filter length. In light of these points, a trade-off bandwidth of 100 MHz for the designs is chosen.

To take care of high power problems, especially after the failure of the first component, several additional precautions were implemented including the regular steps, most of which are summarized as increasing the critical gaps, avoiding tuning screws, avoiding dissimilar metal joints and uniform

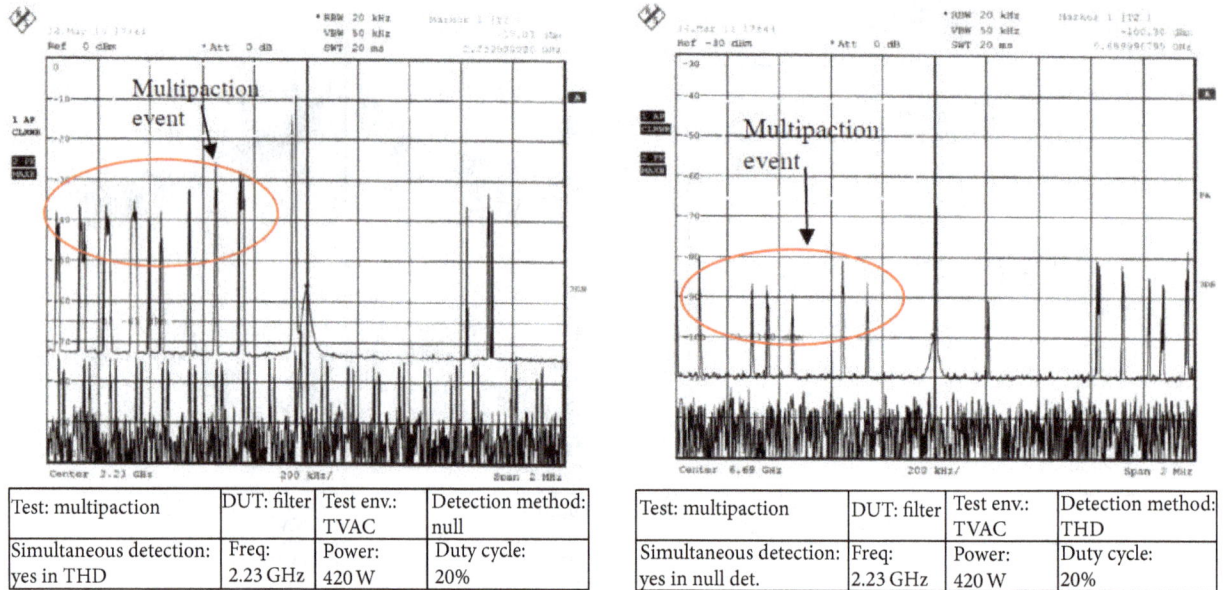

| Test: multipaction | DUT: filter | Test env.: TVAC | Detection method: null |
| Simultaneous detection: yes in THD | Freq: 2.23 GHz | Power: 420 W | Duty cycle: 20% |

| Test: multipaction | DUT: filter | Test env.: TVAC | Detection method: THD |
| Simultaneous detection: yes in null det. | Freq: 2.23 GHz | Power: 420 W | Duty cycle: 20% |

FIGURE 14: Multipaction test on high power filter-F2.

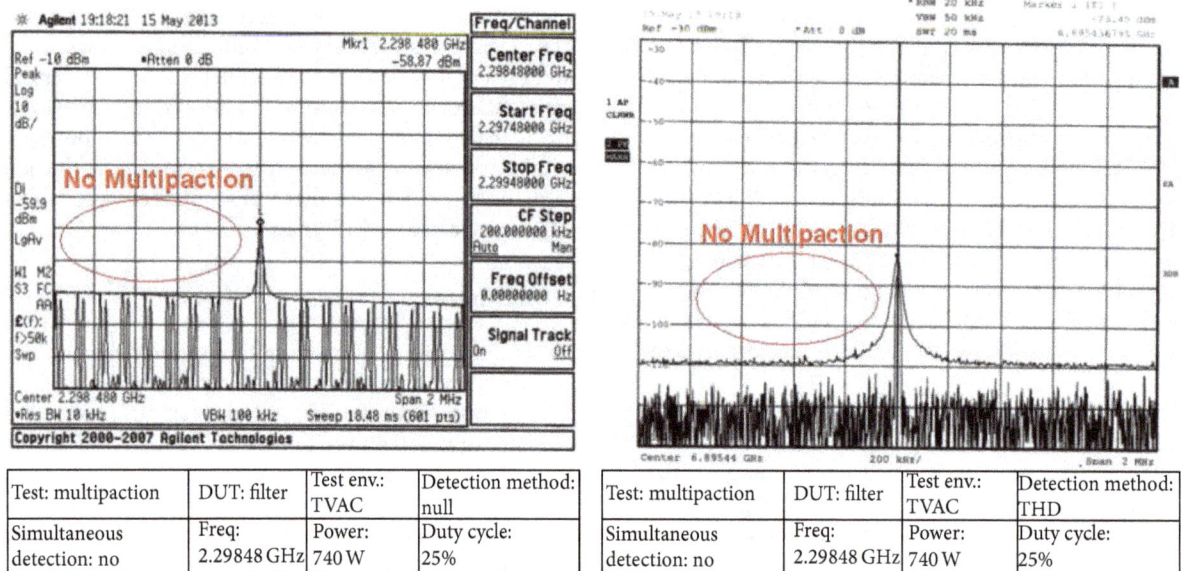

| Test: multipaction | DUT: filter | Test env.: TVAC | Detection method: null |
| Simultaneous detection: no | Freq: 2.29848 GHz | Power: 740 W | Duty cycle: 25% |

| Test: multipaction | DUT: filter | Test env.: TVAC | Detection method: THD |
| Simultaneous detection: no | Freq: 2.29848 GHz | Power: 740 W | Duty cycle: 25% |

FIGURE 15: Successful multipaction test @ 740 W on the final filter-F3.

silver-plating the inner surface, proper matching at the input/output, rounding off the sharp edges, providing vent holes, using black paint on the outer surface for better thermal control, and last but not least maintaining cleanliness and proper handling of the filter throughout the developmental stages. After successful tests the high power filter is integrated with the spacecraft and flown. Presently, its on-board mission performance, in deep space en-route Mars orbit, is normal and is providing seamless communication link to ground stations on the Earth.

Conflict of Interests

The authors declare that there is no conflict of interests regarding the publication of this paper.

Acknowledgments

Authors thank Mr. K. Krishnamoorthy and Mr. K. M. Subramanyam, Mechanical section, ISAC, for their meticulous planning and fabrication of the component in short time and Dr. V. Vamsi Krishna, Passive Systems Section, ISAC, for his support in measurements and helpful discussions. Authors also thank Mr. H. K. Arora and his team of Space Application Centre, Ahmedabad, for their timely cooperation and support in Multipaction testing.

References

[1] K. Shamsaifar, T. Rodriguez, and J. Haas, "High-power combline diplexer for space," *IEEE Transactions on Microwave Theory and Techniques*, vol. 61, no. 5, pp. 1850–1860, 2013.

[2] J. R. M. Vaughan, "Multipactor," *IEEE Transactions on Electron Devices*, vol. 35, no. 7, pp. 1172–1180, 1988.

[3] N. Rozario, H. F. Lenzing, K. F. Reardon, M. S. Zarro, and C. G. Baran, "Investigation of Telstar 4 spacecraft Ku-band and C-band antenna components for multipactor breakdown," *IEEE Transactions on Microwave Theory and Techniques*, vol. 42, no. 4, pp. 558–564, 1994.

[4] S. Riyopoulos, D. Chemin, and D. Dialetis, "Effect of random secondary delay times and emission velocities in electron multipactors," *IEEE Transactions on Electron Devices*, vol. 44, no. 3, pp. 489–497, 1997.

[5] F. Arndt, *WASP-NET Software*, MiG GmbH & Co.KG, Bremen, Germany, 1996–2011.

[6] G. F. Craven, "Waveguide Bandpass Filters using evanescent modes," *Electronics Letters*, vol. 2, no. 7, pp. 251–252, 1966.

[7] G. F. Craven and C. K. Mok, "The design of evanescent mode waveguide bandpass filters for a prescribed insertion loss characteristic," *IEEE Transactions on Microwave Theory and Techniques*, vol. 19, no. 3, pp. 295–308, 1971.

[8] R. V. Snyder, "New application of evanescent mode waveguide to filter design," *IEEE Transactions on Microwave Theory and Techniques*, vol. 25, no. 12, pp. 1013–1020, 1977.

[9] R. V. Snyder, "Broadband waveguide filters with wide stopbands using a stepped-wall evanescent mode approach," in *Proceedings of the IEEE MTT-S International Microwave Symposium Digest*, pp. 151–153, Boston, Mass, USA, 1983.

[10] C. K. Mok, "Design of evanescent-mode waveguide diplexers," *IEEE Transactions on Microwave Theory and Techniques*, vol. 21, no. 1, pp. 43–48, 1973.

[11] R. Levy, H. W. Yaos, and K. A. Zaki, "Transitional combline/evanescent mode microwave filters," in *Proceedings of the IEEE MTT-S International Microwave Symposium Digest*, pp. 461–464, April 1996.

[12] J. Bornemann and F. Arndt, "Transverse resonance, standing wave, and resonator formulations of the ridge waveguide eigen value problem and its application to the design of E-plane finned waveguide filters," *IEEE Transactions on Microwave Theory and Techniques*, vol. 38, no. 8, pp. 1104–1113, 1990.

[13] V. E. Boria and B. Gimeno, "Waveguide filters for satellites," *IEEE Microwave Magazine*, vol. 8, no. 5, pp. 60–70, 2007.

[14] G. Matthaei, L. Young, and E. M. T. Jones, *Microwave Filters, Impedance-Matching Networks, and Coupling Structures*, Artech House, Dedham, Mass, USA, 1980.

[15] R. Levy, R. V. Snyder, and G. Matthaei, "Design of microwave filters," *IEEE Transactions on Microwave Theory and Techniques*, vol. 50, no. 3, pp. 783–793, 2002.

[16] G. Craven and R. Skedd, *Evanescent Mode Microwave Components*, Artech House, 1989.

[17] *High Frequency Structure Simulator-HFSS*, M/s Ansoft, Atlanta, Ga, USA, 2002.

[18] ESA/ESTEC Multipactor Calculator, ver. 1.6, April 2007.

Permissions

All chapters in this book were first published in IJMST, by Hindawi Publishing Corporation; hereby published with permission under the Creative Commons Attribution License or equivalent. Every chapter published in this book has been scrutinized by our experts. Their significance has been extensively debated. The topics covered herein carry significant findings which will fuel the growth of the discipline. They may even be implemented as practical applications or may be referred to as a beginning point for another development.

The contributors of this book come from diverse backgrounds, making this book a truly international effort. This book will bring forth new frontiers with its revolutionizing research information and detailed analysis of the nascent developments around the world.

We would like to thank all the contributing authors for lending their expertise to make the book truly unique. They have played a crucial role in the development of this book. Without their invaluable contributions this book wouldn't have been possible. They have made vital efforts to compile up to date information on the varied aspects of this subject to make this book a valuable addition to the collection of many professionals and students.

This book was conceptualized with the vision of imparting up-to-date information and advanced data in this field. To ensure the same, a matchless editorial board was set up. Every individual on the board went through rigorous rounds of assessment to prove their worth. After which they invested a large part of their time researching and compiling the most relevant data for our readers.

The editorial board has been involved in producing this book since its inception. They have spent rigorous hours researching and exploring the diverse topics which have resulted in the successful publishing of this book. They have passed on their knowledge of decades through this book. To expedite this challenging task, the publisher supported the team at every step. A small team of assistant editors was also appointed to further simplify the editing procedure and attain best results for the readers.

Apart from the editorial board, the designing team has also invested a significant amount of their time in understanding the subject and creating the most relevant covers. They scrutinized every image to scout for the most suitable representation of the subject and create an appropriate cover for the book.

The publishing team has been an ardent support to the editorial, designing and production team. Their endless efforts to recruit the best for this project, has resulted in the accomplishment of this book. They are a veteran in the field of academics and their pool of knowledge is as vast as their experience in printing. Their expertise and guidance has proved useful at every step. Their uncompromising quality standards have made this book an exceptional effort. Their encouragement from time to time has been an inspiration for everyone.

The publisher and the editorial board hope that this book will prove to be a valuable piece of knowledge for researchers, students, practitioners and scholars across the globe.

List of Contributors

Ahmed Boutejdar
Microwave Engineering Department, University of Magdeburg, 39106 Magdeburg, Germany

Ahmed A. Ibrahim
Electronic and Communication Engineering Department, Minia University, Minia 61519, Egypt

Edmund P. Burte
Micro and Sensor Department, University of Magdeburg, 39106 Magdeburg, Germany

Mohammad Asif Zaman and Md. AbdulMatin
Department of Electrical and Electronic Engineering, Bangladesh University of Engineering and Technology, Dhaka 1000, Bangladesh

J. A. Ansari, Sapna Verma and Ashish Singh
Department of Electronics and Communication, University of Allahabad, Allahabad, Uttar Pradesh 211002, India

Zhang Yun-feng and Zhou Zhong-shan
Collaborative Innovation Center on Forecast and Evaluation of Meteorological Disasters, Nanjing University of Information Science & Technology, Nanjing 210044, China
Key Laboratory for Aerosol-Cloud-Precipitation of China Meteorological Administration, Nanjing University of Information Science,
No. 219, Ningliu Road, Nanjing 210044, China

Su Zhi-guo, Wang Rong-zhu and Chen Ze-huang
Key Laboratory for Aerosol-Cloud-Precipitation of China Meteorological Administration, Nanjing University of Information Science, No. 219, Ningliu Road, Nanjing 210044, China

Ramkumar Uikey, Ramanand Sagar Sangam and Rakhesh Singh Kshetrimayum
Department of Electronics & Electrical Engineering, IIT Guwahati, Assam 781039, India

Kakumanu Prasadu
Ford Motor Pvt. Ltd., Dr. MGR Road, Perungundi, Chennai 600096, India

Swarup Das, Debasis Mitra, and Sekhar Ranjan Bhadra Chaudhuri
Department of Electronics & Telecommunication Engineering, Indian Institute of Engineering Science and Technology, Shibpur, Howrah 711 103, India

Jahnavi Kachhia, Amit Patel, Alpesh Vala, Romil Patel and Keyur Mahant
Department of Electronics and Communication Engineering, Charotar University of Science & Technology, Changa, Anand, Gujarat, India

Pravin Ratilal Prajapati
A. D. Patel Institute of Technology, Department of Electronics and Communication Engineering, Gujarat 388121, India

Guido Biffi Gentili
Department of Information Engineering, University of Florence, Via di S. Marta 3, 50139 Florence, Italy

Cosimo Ignesti
Biomedical Srl, Via G.B. Lulli 43, 50144 Florence, Italy

Vasco Tesi
WaveComm S.r.l., Loc. Belvedere, Ingresso 2, 53034 Colle Val d'Elsa, Siena, Italy

I. M. Fabbri
Department of Physics, University of Milan, Via Celoria 16, 20133 Milan, Italy

Maryam Shafiee
Department of Electrical Engineering, Arizona State University, Tempe, AZ 85287, USA

Mohammad Amin Chaychi Zadeh and Homayoon Oraizi
Department of Electrical Engineering, Iran University of Science and Technology, Narmak, Tehran 16846 13114, Iran

Ram M. Narayanan, Sonny Smith and Kyle A. Gallagher
The Pennsylvania State University, University Park, PA 16802, USA

Steffen Scherr, Serdal Ayhan, Grzegorz Adamiuk, Philipp Pahl and Thomas Zwick
Institut für Hochfrequenztechnik und Elektronik, Karlsruhe Institute of Technology (KIT), Kaiserstraße 12, 76131 Karlsruhe, Germany

Yu-Hsin Kuo and Jean-Fu Kiang
Department of Electrical Engineering and the Graduate Institute of Communication Engineering, National Taiwan University, Taipei 106, Taiwan

Mohammad Ashraf Ali and Chung-TseMichael Wu
Department of Electrical and Computer Engineering, Wayne State University, 5050 Anthony Wayne Drive, Detroit, MI 48202, USA

Mohammed Moulay and Mehadji Abri
Telecommunications Laboratory, University of Tlemcen, Tlemcen, Algeria

Hadjira Abri Badaoui
STIC Laboratory, Faculty of Technology, University of Tlemcen, Tlemcen, Algeria

Pranati Panda, Satya Narayan Padhi and Gana Nath Dash
Electron Devices Group, School of Physics, Sambalpur University, Jyoti Vihar, Burla, Sambalpur, Odisha 768019, India

Bibha Kumari and Nisha Gupta
Department of Electronics and Communication Engineering, Birla Institute of Technology, Mesra, Ranchi 835215, India

Seyi S. Olokede
Department of Electrical & Electronic Engineering, Olabisi Onabanjo University, PMB 5026, Ifo, Ogun State, Nigeria

Nurul A.Mohd-Razif and Nor M. Mahyuddin
School of Electrical & Electronic Engineering, Universiti Sains Malaysia, 14300 Nibong Tebal, Penang, Malaysia

A. V. G. Subramanyam, D. Siva Reddy, V. K. Hariharan and V. V. Srinivasan
ISRO Satellite Centre (ISAC), Bangalore 560017, India

Ajay Chakrabarty
Department of Electronics and Electrical Communications Engineering, Indian Institute of Technology (IIT), Kharagpur 721302, India

www.ingramcontent.com/pod-product-compliance
Lightning Source LLC
Chambersburg PA
CBHW080514200326
41458CB00012B/4201